ARID AND SEMI-ARID GEOMORPHOLOGY

Based on four decades of research by Professor Andrew S. Goudie, this volume provides a state-of-the-art synthesis of our understanding of desert geomorphology. It presents a truly international perspective, with examples from all over the world. Extensively referenced and illustrated, it covers such topics as the importance of past climatic changes, the variability of different desert environments, rock breakdown, wind erosion and dust storm generation, sand dunes, fluvial and slope forms and processes, the role of the applied geomorphologist in desert development and conservation and the Earth as an analogue for other planetary bodies. This book is destined to become the classic volume on arid and semi-arid geomorphology for advanced students and researchers in physical geography, geomorphology, Earth science, sedimentology, environmental science and archaeology.

ANDREW S. GOUDIE is director of the China Centre at the University of Oxford and an honorary Fellow of Hertford College and St Cross College, University of Oxford. Goudie was professor of geography and a Fellow of Hertford College from 1984 to 2003. A distinguished physical geographer, he is the recipient of a Royal Medal from the Royal Geographical Society in 1991, the DSc from Oxford in 2002, the Prize of the Royal Belgian Academy in 2002, the Geological Society of America's Farouk El-Baz Prize for desert research in 2007 and the David Linton Award of the British Society for Geomorphology in 2009. From 2005 to 2009, he was president of the International Association of Geomorphologists and has also been president of the Geographical Association, president of Section E of the British Association and chairman of the British Geomorphological Research Group. He is also a former delegate of Oxford University Press, former pro–vice chancellor and was Master of St Cross College, Oxford (2003–2011). He has authored, co-authored or edited more than thirty books on physical geography, including *The SAGE Handbook of Geomorphology* (2011), *Landscapes and Geomorphology: A Very Short Introduction* (2012), *The Oxford Companion to Global Change* (2009), *Encyclopedia of Geomorphology* (2004) and *The Nature of the Environment* (2001).

ARID AND SEMI-ARID GEOMORPHOLOGY

ANDREW S. GOUDIE
University of Oxford

CAMBRIDGE
UNIVERSITY PRESS

CAMBRIDGE UNIVERSITY PRESS
Cambridge, New York, Melbourne, Madrid, Cape Town,
Singapore, São Paulo, Delhi, Mexico City

Cambridge University Press
32 Avenue of the Americas, New York, NY 10013-2473, USA

www.cambridge.org
Information on this title: www.cambridge.org/9781107005549

First published 2013

Printed in the United States of America

A catalog record for this publication is available from the British Library.

Library of Congress Cataloging in Publication Data

Goudie, Andrew.
Arid and semi-arid geomorphology / Andrew Goudie, University of Oxford.
pages cm
Includes bibliographical references and index.
ISBN 978-1-107-00554-9 (hardback)
1. Geomorphology. 2. Deserts. 3. Arid regions. I. Title.
GB611.G674 2013
551.410915′4–dc23 2012047937

ISBN 978-1-107-00554-9 Hardback

Contents

Preface

In previous books I have written about particular phenomena (e.g. dust storms) or processes (e.g. salt weathering) that occur in deserts. I have also undertaken a regional survey of the great warm deserts of the world, in which I explored what made particular deserts distinctive in terms of their landscapes and evolution. The purpose of this new volume is to produce a systematic synthesis on the landforms and land-forming processes that occur in the world's deserts. It has involved the digestion of a huge amount of literature, as is demonstrated by the size of the reference list, but it also includes material derived from my own travels and research over a period of years. I have not attempted to cover the literature on high-latitude deserts.

Acknowledgements

Over the years I have been fortunate to be able to benefit from the assistance of a wide range of desert scientists who have published with me, talked with me, sent me their publications, been my students or traveled with me in the field. Among those I would like to mention are:

Adrian Parker, Alayne Perrott, Alexander Shaw, Alice Goudie, Amy Carter, Amy Goudie, Andrew Warren, Andrew Watson, A.S. Alsharhan, Ashok Singhvi, Asma Al-Farraj, Ben Hickey, Bob Allison, Bridget Allchin, Cameron Petrie, Charles Koch, Chris Pearson, Christian Velder, Cordula Robinson, Dan Yaalon, David Jones, David Price-Williams, David Thomas, Denys Brunsden, Derek Kennett, Detlef Busche, Dick Grove, Ed McKee, Elaine Wright, Farouk El-Baz, Frank Eckardt, Geneviève Coudé-Gaussen, Gordon Wells, Graham Evans, Hang Gao, Heather Viles, Helen Rendell, Ian Evans, Ian Livingstone, J. Al-Awadhi, Jean Poesen, Jennifer Lalley, Joe Henschel, Joe Prospero, John Doornkamp, John Whitney, John Wilkinson, Jon Holmes, Joseph Ballard, Julie Rich, Ken Pye, Kevin White, Klaus Heine, K.T.M. Hegde, Lai ZhongPing, Mahmoud Ashour, Marilyn Robertson-Rintoul, Mark Taylor, Martin Sands, Martin Todd, Mary Bourke, Mary Seely, Mike Summerfield, Monique Mainguet, Nabil Embabi, Neil Roberts, Nick Drake, Nick Lancaster, Nick Middleton, Nick Rosser, Nigel Winser, Olly Atkinson, Owais A. El-Rashidi, Paolo Paron, Pete Magee, Peter Bush, Peter Fookes, Piotr Migoń, Raymond Allchin, Richard Bailey, Richard Washington, Rita Gardner, Rob Sarre, Ron Cooke, Ron Peel, Sebastian Englestaedter, Simon Armitage, S.N. Rajaguru, S. Samieh, Stephen Stokes, Steve Fryberger, Surendra Singh, Susan Conway, Tex Reeves, Warren Wood, Yang Xiaoping and Zhou Leping.

I am grateful to the following for permission to use their figures: Figure 2.19, Professor Frank Eckardt of the University of Cape Town; Figure 3.3, John Wiley and Sons; Figure 3.4, Elsevier; Figure 3.7, Elsevier; Figure 4.3, Nature Publishing Group; Figure 4.15, Elsevier; Figure 5.8, Elsevier.

1
Introduction

The world's arid lands are some of the most beautiful landscapes on Earth, and their paucity of vegetation often makes it possible to see landforms and structures with great clarity (Figure 1.1). In addition, they have many characteristics both in terms of landforms and geomorphological processes that render them different from other major environments (see, for example, Laity, 2008; Parsons and Abrahams, 2009; Thomas, 2011a). In this section we first consider how ideas about dryland geomorphology have evolved, before moving on to a consideration of what desert environments are like now and how they have varied over time.

1.1 The History of Ideas

The study of the geomorphology of arid lands has a long history and has been truly international in scope. Its history has been characterised by rapid shifts in the importance attributed to such processes as wind erosion and physical weathering. It has also been characterised by differences in approaches adopted by different national groups, by a tendency to concentrate on the bizarre and by a propensity to develop a complex multinational terminology (Cooke et al., 1993). It also needs to be appreciated that drylands cover around one-third of the Earth's land surface and have varied settings and environmental histories (Goudie, 2002), so different approaches and emphases have developed in different countries. Some classic papers that reflect this have been collected together in Goudie (2004).

In many desert regions, the landforms and processes are not necessarily so diagnostic of aridity or so different from landforms and processes encountered in more 'normal' humid environments. Most arid areas have experienced a great range of climatic changes that have caused them to both expand and shrink. Nonetheless, the roles of wind and salt are plainly very significant in certain drylands, and the limited vegetation cover is a critical control of the operation of fluvial and slope processes.

Figure 1.1 The Barstow Syncline in the Mojave Desert of California and an erosional unconformity in the upper part of the section are revealed with great clarity as a result of the limited vegetation cover. (ASG)

The impact of these changes on their landscapes and those of their neighbours has often been profound.

Detailed studies of desert geomorphology began with exploration and colonial expansion in the second half of the nineteenth century. John Strong Newberry, one of the greatest explorers of the Colorado Plateau in the 1850s, recognized these classic desert landscapes as having been 'formerly much better watered than they are today' (1861, p. 47). Such work reached a climax when Gilbert and Russell examined the desiccated lake basins that were such a feature of the western United States (see Orme, 2008, for a full discussion). Geologists of the U.S. Geological Survey made other highly important investigations in the American Southwest. Especially influential was the work of John Wesley Powell and Clarence Dutton on the landforms of the Colorado Plateau. American scientists also contributed greatly to the development of knowledge on desert aeolian processes (Udden, 1894; Free, 1911). Also remarkable was the work of W.P. Blake on stone pavements, desert varnish, old lake basins, calcretes (caliche) and wind grooving of rock surfaces (e.g. Blake 1855, 1904). Moreover, it was in the American West that W.J. McGee (1897) drew attention to the role of sheetfloods on

Figure 1.2 Johannes Walther was a German pioneer of desert geomorphology in the early years of the twentieth century. (Source: http://www.geolsoc.org.uk/gsl/info/collections/archives/page5150.html) (accessed March 11, 2011).

pediment surfaces. Also notable were Gilbert's studies in the Colorado Plateau on rates of denudation in arid regions (Gilbert, 1876) (see Section 5.4).

The French acquisition of its North African territories – Algeria, Tunisia and Morocco – led to some major expeditions into the great sand seas of the Sahara and some fundamental studies of dune forms (Goudie, 1999c). As early as 1864, Henri Duveyrier classified the main types of dune, related their orientation to that of the winds, estimated their area and argued that much of the sand was the result of intense rock disintegration (Duveyrier, 1864). French scientists were very active in the western Sahara and accumulated a great deal of vital information on the full range of desert landforms (see Gautier 1908 and Chudeau, 1909).

The Germans for their part took over South West Africa (now Namibia), and individual German geomorphologists worked in the Middle East (e.g. J. Walther) and in the Kalahari (e.g. Passarge, 1904). They made major studies of weathering phenomena, wind erosion and the development of inselberg landscapes. Walther (Figure 1.2) studied the deserts of North Africa, Sinai, the United States and Australia. His *Das Gesetz der Wüstenbildung in Gegenwart und Vorzeit* (1900) was the first full-scale book devoted to desert geomorphological processes, and he championed the role of such mechanisms as thermal fatigue weathering, salt weathering and deflation.

In central Asia, the Swedish explorer Sven Hedin (1903) discovered wind erosional landforms – yardangs – and the American geographer E. Huntington found dramatic evidence for changes of climate in historical times (Huntington, 1907). Other influential work included that of Sir Aurel Stein on wind erosion (Stein, 1912) and Berkey and Morris (1927) on pans and weathering in Mongolia.

With respect to the British, the most sustained work on desert geomorphology in this pioneer era was undertaken by surveyors and geologists in the Western (Libyan) Desert of Egypt, of whom Ball, Beadnell, Hume and King were the leading figures, particularly in the study of dunes and weathering (Goudie, 2008b).

In Australia, Jutson was initially a major exponent of the role of wind in moulding desert surfaces (see, for example, Jutson, 1917), contributing to the development of salt lakes and leading to the wearing back of scarps. He recognized that wind operated in tandem with salt weathering. Later, however, Jutson (1934) recognized the role of fluvial processes in moulding the planation surfaces of the arid landscapes of Western Australia (Brock and Twidale, 2011).

In the first six decades of the twentieth century, geomorphology as a whole was often dominated by either those with an interest in long-term landscape evolution or those who were concerned with the development of ideas of climatic geomorphology. Towering figures such as W.M. Davis (1905, 1938) and W. Penck (1953) developed models of landscape development that were relevant to arid regions, and debates raged about the relative importance of wind and water erosion in creating desert plains (see, for example, Bryan, 1923). The maverick American C. Keyes (1912) was an especially vigorous and repetitive exponent of the power of wind erosion (Goudie, 2012).

In continental Europe, the French and German schools of climatic geomorphology sought to establish the broad links between climate and morphogenetic regions, and notable figures included Birot, Dresch, Tricart, Cailleux and de Martonne in France, and Passarge, Mortensen and Büdel in Germany. A body of important French work is reviewed by Tricart and Cailleux (1969).

W.M. Davis is perhaps best known for his evolutionary model of landform development – the cycle of erosion. This was originally developed in an essentially humid temperate environment, but Davis recognized that it needed to be modified in other types of environment where processes were different. Building on the work of Passarge, a German geomorphologist who had worked in the Kalahari (Passarge, 1904), Davis saw wind action as a factor in the cycle's operation under arid conditions, especially in its later stages. He also, however, recognized the role of water, especially in the earlier stages of the arid cycle. Thus he did not adopt the extreme views of Keyes, but neither did he reject the role of aeolian denudation as Penck was to do (1953 translation, p. 327). The views of Davis (1905) can be appreciated by considering these statements which relate to the progressive evolution of an arid area:

In the early stage of the arid cycle the relief is slowly diminished by the removal of waste from the highlands, and its deposition on the lower gentler slopes and on the basin beds of all the separate centripetal drainage systems.... Streams, floods, and lakes are the chief agencies in giving form to the aggraded basin floors, as well as to the dissected basin margins in the early stages of the cycle; but the winds are also of importance. (pp. 383–84)

He then goes on to describe the mature stage:

The obliteration of the uplands, the development of graded piedmont slopes, and the aggradation of the chief basins will be more and more extensive. (p. 387)

As the processes thus far described continue... the initial relief will be extinguished... the plains will be interrupted only where parts of the initial highlands and masses of unusually resistant rocks here and there survive as isolated residual mountains. (p. 388)

Finally, in old age:

During the advance of drainage integration the exportation of wind-borne waste is continued.... The tendency of wind-action to form hollows wherever the rocks weather to a dusty texture would be favoured by the general decrease of the surface slopes, and by the decrease of rainfall and of stream-action resulting from the general wearing down of the highlands.... [R]ock masses that most effectively resist dry weathering will remain as monadnocks – Inselberge, as Bornhardt and Passarge call them in South Africa. (pp. 390 and 392)

By the 1940s, however, the role of wind erosion in moulding desert landscapes was becoming the subject of doubt, and Cotton (1942, p. 3), for example, remarked that 'few if any major relief forms owe their origin or shape to wind scour and that the sculpture by wind of features even of minor detail in the landscape is rare and exceptional'.

In the 1930s, the High Plains of America, stretching up from Texas to the Dakotas, had a run of years that were torridly dry and hot. They coincided with a phase of agricultural intensification and extension that was facilitated by the widespread introduction of the tractor, the combine and the truck after the First World War. This created conditions for the Dust Bowl, which saw 'black blizzards' of topsoil being stripped off agricultural lands recently subjected to 'the busting of the sod'. This disastrous decade led to the establishment of the Soil Conservation Service under the directorship of H.H. 'Big Hugh' Bennett. In 1935, he addressed a Senate committee in Washington, D.C., about the need for a soil conservation act. As he was speaking, the sky darkened with the passage of a dust storm originating from the Great Plains to the west, and so the act was recommended (Brink, 1951). This marked the start of intensive work on the nature and dynamics of wind erosion of soil.

Of particular note in this respect was the work conducted by W.S. Chepil and his collaborators at the U.S. Department of Agriculture's Wind Erosion Research Center at Kansas State University, established in 1947. They were concerned with establishing the fundamentals of soil movement by wind, the properties of soils which influenced their susceptibility to wind erosion, the sedimentary characteristics of dust storms and the effects of various land-cover treatments (mulches, field size, maintenance of crop residues, type of ploughing, etc.). They also developed technology for advancing aeolian research, including dust samplers and portable wind tunnels. This type of work was further developed by Dale Gillette and colleagues (Marticorena, 2008).

In addition, after the Second World War a number of countries established research stations in their own desert regions. These permitted long-term monitoring and provided bases for sustained investigations. Such stations included those at Jodphur in India, Bardai in the central Sahara, Gobabeb in the Namib, Sidi Boqer in the Negev, Fowlers Gap in New South Wales (Australia), the Jornada Experimental Range in New Mexico, the Walnut Gulch experimental watershed in Arizona, the Zzyzx station of California State University, the Desert Institute of Turkmenistan, the Taklamakan Desert Research Station in north-west China and the Lanzhou Institute of Desert Research. Indeed, a feature of the last three decades has been the impressive growth of high-quality research by Chinese colleagues.

In Australia, the Commonwealth Scientific and Research Organization (CSIRO) undertook land-resource surveys of the interior drylands, and these surveys had a major geomorphological component. In the UK, groups of geomorphologists were employed as consultants to advise on building developments in the Middle East that resulted from the oil boom of the 1960s onwards, developing studies of, for example, flood hazards, slope instability and salt weathering of foundations (Cooke et al., 1982).

One of the most striking developments in recent decades, however, has been the growth of process studies. In some respects this mirrors developments in geomorphology as a whole, but in other respects desert geomorphology was ahead of the rest of the discipline, largely because of the fundamental studies of sediment movement by wind undertaken in the field as well as the wind tunnel initiated by R.A. Bagnold in the 1930s (Bagnold, 1935, 1936, 1937, 1941). Wind tunnel research has generated a great deal of fertile research, some of it on the scale of individual grain transport. Nickling and McKenna-Neuman (1999) provide good reviews of this type of work. Recent years have also seen many detailed studies of weathering processes, including salt weathering (see reviews in Goudie and Viles, 1997), sediment movement on slopes and in channels and dust transport and deposition (see the review by McTainsh, 1999). The data-logger revolution has facilitated process studies in the field by enabling monitoring of wind conditions, temperature and humidity cycles as well as sediment movement (Livingstone et al., 2007).

In the 1960s and 1970s, some remarkable work was done in the Negev Desert (see Section 5.15). Scientists from the Hebrew University in Jerusalem applied to arid environments the developments in quantitative process geomorphology that were being made at that time. Through intensive monitoring of conditions on slopes and in channels, they began to give a clear indication of how runoff and erosion occurred in dryland basins. A leading figure was A. Schick, who developed studies of the effects of floods and who set up experimental catchments that provided some decades of data (see, for example, Schick and Lekach, 1993). Also important was the work done by Evenari et al. (1982) on the hydrological conditions that had permitted runoff farming in arid areas by Nabatean farmers.

Yet another major influence on studies of desert geomorphology in recent years has been the development of remote sensing. Air photography was essential for providing information on large tracts of terrain that were inaccessible on the ground, and it was especially useful in ascertaining dune patterns (see Goudie, 1999c, pp. 7–8 for a review). Satellite-borne remote sensing became increasingly important in the 1970s for, inter alia, mapping dunes at a regional and subcontinental scale, for tracing dust events and for investigating fluctuations in areas inundated by lakes. Livingstone et al. (2010) provide an illustration of the wealth of databases on landform morphology that can now be assembled based on remote-sensing and digital-elevation models.

Beginning in the 1970s, geomorphologists sought analogues for Martian features on Earth, and this gave a considerable boost to studies of a range of arid zone processes and phenomena, including sapping phenomena (Laity and Malin, 1985), salt weathering (Malin, 1974) and aeolian forms (see Section 6.17). The U.S. space programme enabled the geomorphology of Mars to be investigated in detail for the first time. The images revealed volcanoes, lava plains, immense canyons, cratered areas, evidence of surface water and a whole range of wind-formed features. The aeolian phenomena of Mars were indeed both diverse and impressive (Wells and Zimbelman, 1997). Dust events range in size from dust devils to dust storms that may obscure the entire planet. Extending from many topographic highpoints, especially crater rims, are light (depositional) and dark (erosional) wind streaks. Yardangs are also plentiful, especially on the equatorial plains. The dunes on Mars, many of which occur in a large dunefield that encircles the northern polar cap, are largely barchanoid and transverse forms. Aeolian features are known from other parts of the solar system, including Titan, and have been reviewed by Greeley and Iversen (1985) and Craddock (2011).

Since the Second World War, the ability of desert geomorphologists to date phenomena has expanded markedly. Such techniques as radiocarbon and uranium series dating have been applied to desert sediments as they have been to other environments, but luminescence dating has proved to be particularly useful for dating dune sands and other aeolian materials. Optical dating (Figure 1.3) is now used routinely

Figure 1.3 The collection of samples from aeolian deposits (such as this dune sand in the United Arab Emirates) for optical dating has revolutionized our knowledge of the timing and rates of dune deposition. (ASG)

to provide dates that enable phases of dune accumulation to be established as well as the rates at which dunes accumulate (Singhvi and Porat, 2008). On longer timescales, cosmogenic nuclides have been used extensively to date surfaces, to estimate rates of erosion, to date lake shorelines and to estimate sand residence times in dunefields (see, for example, Wells et al., 1995; Fujioka et al., 2005; Nishiizumi et al., 2005; Kober et al., 2007; Hall et al., 2008; Vermeesch et al., 2010; Kurth et al., 2011; L.A. Owen et al., 2011) (see also Section 5.21).

Finally, environmental change has become an increasingly important field of research. Geomorphologists have become very involved with the study of anthropogenic degradation of desert surfaces (desertification). They have also become increasingly interested in the evidence for and causes of natural changes in climate, a quest that has been facilitated by increasing availability of high-resolution dating techniques (e.g. optical dating) and by studies of long-term sediment sequences in ocean cores and lake basins. There is also a fascination regarding how deserts may be impacted by possible future global warming (see Section 6.13).

1.2 Climatic Conditions: Aridity

Drylands, which cover about a third of Earth's land surface, occur in every continent (Goudie, 2002). Predominantly, because precipitation is low, there is a severe shortage of moisture. In some deserts, aridity also results from high temperatures, which means that evaporation rates are great (Mainguet, 1999). Good reviews of desert meteorology and climates are provided by Warner (2004) and Nicholson (2011).

Aridity can be defined by the water balance concept. This is the relationship that exists between the inputs of water in precipitation (P), the losses arising from evaporation and transpiration (evapotranspiration) (E_t) and any changes that occur in storage (soil moisture, groundwater, etc.). In arid regions there is an overall deficit in the annual water balance, and the size of that deficit determines the degree of aridity. The actual amount of evapotranspiration (AE_t) that occurs varies according to whether there is available water to evaporate, so the concept of potential evapotranspiration (PE_t) has been devised. This is a measure of the evapotranspiration that could occur from a standardized surface never short of water. The volume of PE_t varies in response to four climatic factors: radiation, humidity, temperature and wind. Thornthwaite (1948) developed a general aridity index based on PE_t: When $P = PE_t$ throughout the year, the index is 0. When $P = 0$ throughout the year, the index is –100. When P greatly exceeds PE_t throughout the year, the index is +100. Areas with values below –40 are classified as arid, those between –20 and –40 as semi-arid and those between 0 and –20 as subhumid (Meigs, 1953). The arid category can be subdivided into arid and extreme arid, with the latter being defined as the condition in any locality where at least twelve consecutive months without any rainfall have been recorded, and in which there is not a regular seasonal rainfall rhythm. Extremely arid areas, such as the Atacama, Namib, inner Arabia, the central and eastern Sahara and the Taklamakan, cover about 4 per cent of Earth's land surface, arid about 15 per cent and semi-arid about 14.6 per cent.

In addition, deserts can be classified on the basis of their proximity to the oceans or their continentality. Coastal deserts, such as the Namib or the Atacama, have very different temperatures and humidities from those of continental interiors. They have modest daily and seasonal temperature ranges and are subject to fogs. They also have very low rainfalls. In addition to the coastal and inland deserts of middle and low latitudes, there are also the cold polar deserts. The precipitation of the Arctic regions can be as low as 100 mm per year, and at Vostok in Antarctica it can be less than 50 mm.

1.3 Causes of Aridity

Most deserts are dry because they occur where there is subsiding air, relative atmospheric stability and divergent air flows at low altitudes associated with the presence of great subtropical high-pressure cells around latitude 30° (Nicholson, 2011). Such areas are only infrequently subjected to precipitation-bearing disturbances and depressions – either from the Intertropical Convergence Zone (ITCZ) or from the belt of mid-latitude depressions associated with the circumpolar westerlies. The trade winds that blow across these zones cause evaporation, and because of the trade wind inversion they are areas of subsidence and stability.

These global tendencies are often reinforced by more local factors. Of these, continentality can be dominant and plays a part in the location and character of the deserts of areas such as central Asia. The rain shadow produced by mountains ranges can create arid areas in their lee, as in Patagonia, where the Andes have an influence. Other deserts are associated with cold currents offshore (e.g. Namib and the Atacama). Winds that blow onshore tend to do so across cold currents (e.g. the Benguela and the Peruvian) and so are stable because they are cooled from beneath; they also have a relatively low moisture-bearing capacity. They reinforce the stability produced by the dominance of subsiding air. Aridity may also be reinforced by the high reflectivity (albedo) of desert surfaces themselves. This may cause net loss of radiative heat, create a horizontal atmospheric temperature gradient along the desert margin and induce circulation systems that either induce or reinforce subsidence. Finally, atmospheric dust palls may have a positive feedback effect, being both a consequence and an accentuator of aridity (Xu et al., 2011).

1.4 Desert Rainfall

The main characteristics of deserts are caused by the very low levels of rainfall, and these are a particular feature of some coastal deserts. For example, mean annual totals at Callao in Peru are only 30 mm, at Swakopmund in Namibia only 15 mm and at Port Etienne in Mauritania only 35 mm. Years may go by in such areas of extreme aridity during which no rain falls at all. The most intense aridity occurs in northern Chile, which receives less than 10 mm of rainfall per annum. Indeed, the climate station at Quillagua (mean annual rainfall 0.05 mm) can lay claim to be the driest place on Earth (Middleton, 2001). Very low precipitation amounts are also found in the centre of some great deserts. In Egypt there are stations where the mean annual precipitation only amounts to 0.5 mm, and parts of the Tarim Basin in China have only 17 mm of precipitation per annum (Dong et al., 2011).

Another very important characteristic of desert rainfall is its highly variable temporal character. This inter-annual variability (*V*) can be expressed as a simple index:

$$V\ (\%) = \frac{\text{the mean deviation from the average}}{\text{the average}} \times 100$$

Whereas European humid temperate stations may have a variability of less than 20 per cent, variability in the Sahara ranges from 80 to 150 per cent. There are some differences in the variability of rainfall for any given mean annual rainfall between deserts, with a general, although not universal, tendency for low latitude, summer rainfall deserts to be significantly more variable than higher latitude, winter rainfall ones (Van Etten, 2009). Four regions – the Thar, Namib-Kalahari, Somali and northern Australian deserts – are significantly more variable than all others. In the case of the Lake Eyre Basin in Australia, the coefficient of variation of the rainfall is 60 per cent greater than that found for stations located in arid regions in the rest of the world (McMahon et al., 2008a).

A graphic illustration of the inter-annual variability of rainfall in some dryland areas is provided by the record for Los Angeles, California. For the period from 1877–78 to 2006–07, annual rainfall during the hydrological year ranged from as high as 970 mm (in 1883–34) to only 110 mm (in 2001–02).

The considerable variability in rainfall amounts and intensities is reflected in changes in vegetation cover, runoff and sediment yields. Polyakov et al. (2010), for example, found that in semi-arid Arizona, during a thirty-four-year period, annual sediment yields varied between 0.85 t ha^{-1} and 6.69 t ha^{-1}, while Nichols (2006), also working in Arizona, found that in one catchment during a forty-seven-year period, sediment yield ranged from a low of 1.2 m^3 ha^{-1} to a high of 5.32 m^3 ha^{-1}.

The high temporal variability of desert rainfall means that from time to time, although mean rainfall levels are so low, there can still be individual storms of surprising size, and flash floods can play a very important role in sediment mobilization (Vanmaercke et al., 2010). Indeed, maximum falls in twenty-four hours may approach or exceed the long-term annual precipitation values. For example, at Chicama in Peru, where the mean annual precipitation over previous years had been a paltry 4 mm, in 1925, 394 mm fell in one storm. Similarly, at El Djem in Tunisia (mean annual precipitation 275 mm), 319 mm fell in three days in September 1969, causing severe flooding and creating great geomorphological changes. In June 1965, Plum Creek in Colorado, which has a mean annual rainfall of c 400 mm, received 360 mm of rainfall in just four hours (Osterkamp and Costa, 1987). In July 1981, Bassi in the Thar Desert of India received 560 mm in twenty-four hours, 93 per cent of its mean annual rainfall (Dhar et al., 1982). In August 2006, Jaisalmer (average annual rainfall 210 mm) received two separate daily falls of 130 and 140 mm each (Rao et al., 2011). The Pakistan floods of July–August 2010 were associated with large rainfall events in the Indus catchment. For instance, 274 mm of rain fell on July 29 in Peshawar, where the mean annual rainfall is 274 mm. In October 2004, Sedom in the Negev, which has a mean annual rainfall of 46 mm, received 74 mm in about two hours (Greenbaum et al., 2010). In April 2006, the desert town on Luderitz in southern Namibia received 102 mm, about six times its average annual rainfall (Muller et al., 2008), while further north, at Gobabeb on March 11, 2011, 49.5 mm of rain occurred in just two hours, more than double the annual mean. In June 2007, Cyclone Gonu struck eastern Oman, and dumped up to 610 mm in one storm in an area where the mean annual rainfall is around 70 mm (Abdalla and Al-Abri, 2010). In the Atacama Desert, storms in the summer of 2001 produced more than 400 mm of rainfall in areas where the mean annual rainfall was around 150 mm (Houston, 2006). Severe El Niños, like that of 1997–78, can have a remarkable effect on rainfall amounts. This was shown with particular clarity in the context of Peru (Bendix et al., 2000), where normally dry locations suffered huge storms. At Paita (mean annual rainfall 15 mm) there were 1,845 mm of rainfall, while at Chulucanas (mean annual rainfall 310 mm) there were 3,803 mm. Major floods resulted (Magilligan and Goldstein, 2001).

Not all desert rainfall occurs as storms of such ferocity, however. Indeed, contrary to a common perception, most of it falls in storms of low intensity. This is clear when one considers the rainfall statistics for the Jordanian Desert and Death Valley in California (Figure 1.4). Both these areas have very low rainfall in terms of mean annual levels (102 and 67 mm, respectively), yet on average rain falls on twenty-six and seventeen days respectively, so that the mean rainfall event tends to be only 3–4 mm – which is much the same as for London. Table 1.1 shows the average rainfall per rainy day for a range of deserts, with figures ranging from around 4 mm per rainy day for temperate deserts to more than 9 mm per rainy day for tropical deserts. Nicholson (2011, p. 191–2) reports that the contribution of daily rainfall events less than 5 mm to mean annual rainfall ranges from 76 per cent in the Chihuahuan Desert to 85 per cent in the Sonoran and to 95 per cent in the Mojave.

Precipitation in arid zones, in addition to showing temporal variability, also shows considerable spatial variability. For this reason it is often described as being ‘spotty’. The spottiness is especially common in areas where localized convectional cells occur (Sharon, 1972).

In coastal deserts, with cold currents offshore, the moisture provided by fogs may augment that produced by rain (see Section 2.5). In the coastal fringes of Namibia,

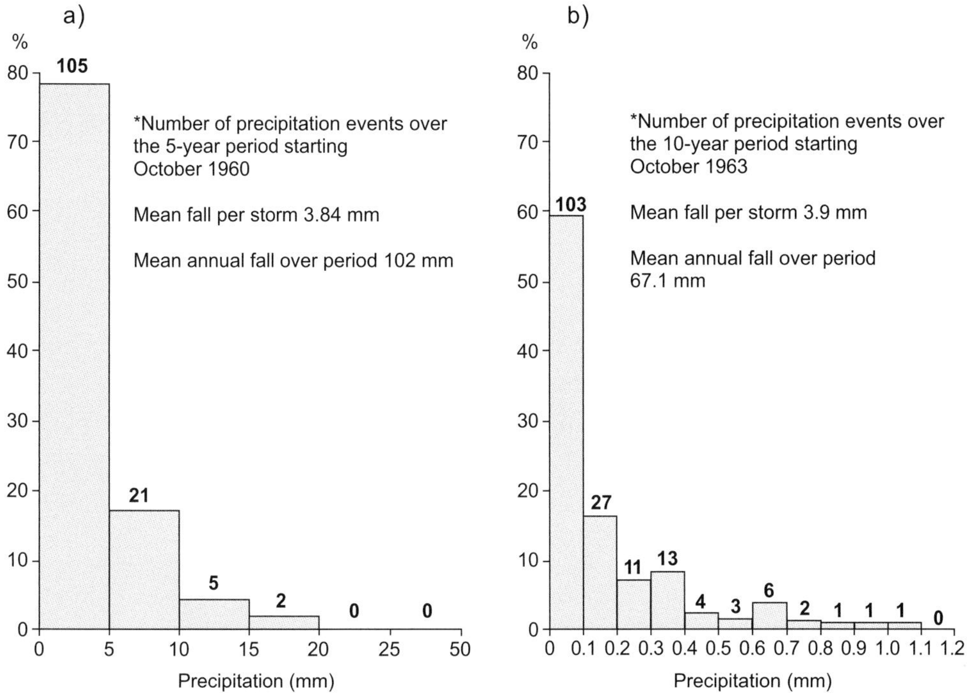

Figure 1.4 The amount of rain falling in each rainy day for two desert locations: (a) H4, Jordan: (b) Death Valley, California, USA. (From Goudie, 1984a, fig. 5.2)

the mean annual fog precipitation (35–45 mm) may exceed that from rain, and the fog may occur on up to 200 days in the year and extend more than 100 km inland. In Peru, the fogs and low cloud provide enough moisture to support a growth of vegetation in what would otherwise be a hyperarid habitat.

Table 1.1 *Rainfall per rainy day in arid areas*

Area	Number of stations in arid zones	Range of mean annual rainfall (mm) (30-year average)	Range of number of rainy days >0.1 mm per year	Average rainfall per rainy day over 30 years (mm)
Former USSR	12	92–273	42–125	3.56
China	6	84–396	33–78	4.51
Argentina	11	51–542	6–155	5.41
North Africa	18	1–286	1–57	3.82
West Africa	20	17–689	2–67	9.75
Kalahari	10	147–592	19–68	9.55
Total	77	1–689	1–155	6.19

Table 1.2 *Average mean monthly temperatures (°C) of coastal deserts*

Month	Tarfaya (Morocco)	Arica (Chile)	Walvis Bay (Namibia)
J	16	23	17
F	16	23	18
M	17	22	17
A	18	20	16
M	18	18	16
J	19	17	15
J	20	16	14
A	20	16	13
S	20	17	13
O	20	18	13
N	19	20	15
D	16	21	16
Range	**4**	**7**	**5**

1.5 Desert Temperatures

According to their type, deserts have a wide range of temperature conditions. Interior deserts can be subjected to extremes of temperature, both seasonally and diurnally, that are not equalled in any other climatic region, while coastal deserts tend to have relatively low seasonal and diurnal ranges. Here, the climate is modified and moderated by the presence of cold currents and upwelling. Average monthly temperature ranges over the year are low (Table 1.2), being about 4°C. Daily ranges in such stations are also low, often around 11°C, and only about half what one would expect in the Sahara. The mean annual temperature values are also generally moderate (c 19°C in the Atacama and 17°C in the Namib). By contrast, great extremes of temperature can occur in interior deserts, with maximum shade temperatures exceeding 50°C. Temperatures in excess of 37°C may occur for many days on end in the summer, but because of the clear skies there may be a marked reduction of temperature at night, and daily ranges of 17–22°C are normal. In the winter in high-altitude interior deserts, frost can occur frequently.

Whereas some deserts show great daily and seasonal ranges in air temperatures, ground surface temperatures show even greater ranges, which may have many implications both for rock weathering and for plant and animal life. Satellite-borne sensors now enable us to have a global picture of maximum land surface temperatures, and desert areas such as western North America, the Sahara, Egypt, India, the Middle East, the Gobi and much of Australia regularly exceed 60°C. Tracts of the Iranian deserts regularly exceed 70°C in the summer months, and this regularly has the largest contiguous area of surface temperatures above 65°C on Earth. Other notable temperature hotspots include the interior of Queensland and the Turpan Basin in China (Mildrexler et al., 2011).

1.6 The Antiquity of Deserts

Desert areas have changed through time, so present conditions of the type just outlined may not necessarily be those that have moulded desert surfaces. As Williams et al. (1998, p. 174) wrote:

> Many desert landforms are exceedingly old. The vast desert plains of the central and western Sahara have been exposed to subaerial denudation for well over 500 million years, . . . as have the Precambrian shield deserts of the Yilgarn Block and the Pilbara in Western Australia. It is misleading to consider such well-known desert monoliths as Ayers Rock (Uluru) in central Australia, or the granite inselbergs of the Sahara, as diagnostic of aridity, for they owe their present morphology to prolonged and repeated phases of weathering and erosion under a succession of former climates, few of which are particularly arid.

Indeed, the German climatic geomorphologist Büdel (1982) believed that the great landscapes of Australia and the Sahara were largely moulded by moist, seasonal tropical conditions that had characterised the Tertiary. He felt that Pleistocene aridity had been too short lived to achieve much landscape change. It is therefore necessary to consider their palaeoclimatic history.

Although formerly many deserts were regarded as a result of Holocene (post-glacial) progressive desiccation, it is now clear that our present deserts are old (Goudie, 2002). The climatic development of the Namib and the Atacama coastal deserts was closely related to post-Jurassic plate tectonics and sea-floor spreading in that the degree of aridity must have been largely controlled by the opening up of the seaways of the Southern Ocean, the location of Antarctica with respect to the South Pole and the development of the cold Benguela and Peruvian currents offshore. Arid conditions have existed in the Namib for some tens of millions of years, as is indicated by the Tertiary Tsondab Sandstone – a lithified mass of dune sand that underlies the current sand sea (Ward, 1988).

The Atacama's aridity may have started in the Eocene, becoming more profound in the middle to late Miocene (Alpers and Brimhall, 1998; Rech, Quaid and Betancourt, 2010). There were perhaps two crucial factors responsible for its initiation: the uplift of the Andes during the Oligocene and early Miocene and the development around 15–13 Ma of the cold offshore Peruvian current as a result of ice buildup in Antarctica. The former produced a rain-shadow effect (Placzek et al., 2009; Rech et al., 2010) and helped to stabilize the south-eastern Pacific anticyclone, while the latter provided the cold waters that are necessary for hyperaridity to develop (Alonso et al., 1999). Of these two factors, the development of a cold offshore current is given greater significance by Garreaud et al. (2010). One line of evidence for this early initiation of the Atacama, in comparison with many of the world's deserts, is the existence of gypsum crusts preserved beneath an ignimbrite deposit that has been dated to c 9.5 million years old (Hartley and May, 1998). The ready solubility of gypsum implies the existence of aridity ever since that time. The long-term stability/persistence of an arid climate is suggested by the fact that cosmogenic nuclide studies show some of the oldest exposure ages found anywhere on Earth, ranging between 9 and 37 million years (Kober et al., 2007; Placzek et al., 2009). Studies of lake basins also indicate drying in the late Miocene (Saez et al., 1999 Alonso et al., 1999; Diaz et al., 1999; Gaupp et al., 1999; May et al., 1999). However, there is some controversy relating to this issue on sedimentological grounds. Hartley and Chong (2002) have argued that the development of hyperaridity was a late Pliocene phenomenon, associated, as in other deserts, with global climate cooling. They recognized, however, that a semi-arid climate persisted from 8 to 3 Ma, punctuated by a phase of increased aridity at around 6 Ma. Amundson et al. (2012) also believed that although the area had been dry since the Miocene, full hyperaridity set in only in the late Pliocene to early Holocene, causing stream incision and erosion by water to be reduced to insignificance. In Argentina, sustained aridity appears to have set in by 5 Ma (Bywater-Reyes et al., 2010).

The timing of desert initiation in North America is still a matter of considerable uncertainty (see Wilson and Pitts, 2010, for a review), although it is possible that some may have existed as far back as 15 Ma. Uplift of the Sierra Nevada was an important factor in producing a rain-shadow effect (Smith, 2009).

In the Sahara, sediment cores from the Atlantic contain dust-derived silt indicating that a well-developed arid area, producing dust storms, existed in North Africa in the early Miocene, around 20 Ma (Diester-Haass and Schrader, 1979). Desert deposits are also found in the Mio-Pliocene strata of the Chad Basin (Schuster et al., 2009). It is possible that uplift of the Tibetan Plateau at this time (see the next paragraph) played a role in this by creating a strong counterclockwise spiral of winds that drove hot, dry air out of the interior of Asia across Arabia and northern Africa (Ruddiman, 2001, p. 388).

In China, Miocene uplift and a resulting transformation of the monsoonal circulation is one of the mechanisms that caused aridification (Zhang and Sun, 2011; Miao et al., 2012). The uplift of mountains and plateaux in Tibet and North America may have caused a more general change in precipitation in the late Miocene, as is made evident by the great expansion of C4 grasses in many parts of the world (Pagani et al., 1999) and the microbial lipid evidence for alkalinity and drought in the Tibetan Plateau region (Xie et al., 2012). The aeolian red clays and loess of China may have started to form around 7.2–8.5 Ma (Qiang et al., 2001), and while the Tarim Basin also has aeolian dune sediments dating to c 8 Ma (Zheng et al., 2010), it is now recognized that some aeolian deposits in the far north-west of China may even date back to c 24–25 Ma (i.e. the late Oligocene) (Sun et al. 2010; Qiang et al., 2011). Dupont-Nivet et al. (2007) have suggested that some aridification of the Tibetan Plateau was associated with widespread global cooling that took place around 34 Ma at the Eocene-Oligocene transition. Another contributing factor may have been the late Eocene retreat of the Paratethys Sea from the Tarim Basin (Bosboom et al., 2011; Zhuang et al., 2011).

In India and Australia, latitudinal shifts caused by sea-floor spreading and continental drift led to moist conditions during much of the Tertiary. For example, plant fossils show that rainforest covered much of central Australia until 25–30 Ma (Fujioka and Chappell, 2010). However, India and Australia entered latitudes where conditions were more arid in the late Tertiary. Isotopic studies in the Siwalik foothills of Pakistan illustrate increasing aridity in the late Miocene, where C3/C4 analyses show a change from a C3 (mainly forested) setting to a C4 (mainly grassland) setting at about 7 Ma (Quade et al., 1989). The upward and outward growth of the Tibetan Plateau may also have contributed to decreasing monsoon rainfall over north-western India since c 10 Ma (Molnar and Rajagopalan, 2012).

1.7 Increasing Aridity

In many regions aridity intensified in the late Pliocene and Pleistocene. It became a prominent feature of the Sahara in the late Cenozoic, partly because of ocean cooling and partly because the buildup of ice caps created a steeper temperature gradient between the equator and the poles. This led to an increase in trade-wind velocities and in their ability to mobilize dust. DeMenocal (1995) recognized that there was an acceleration in dust loadings in ocean cores off the Sahara and Arabia after 2.8 Ma and attributed this to decreased sea surface temperatures associated with the initiation of extensive Northern Hemisphere glaciation. Likewise, loess deposition became more vigorous in China after around 2.5 Ma (Ding et al., 1992), and between 3.6 and 2.6 Ma the loess became coarser and more widespread (Lu et al. 2010). Sediments from the central North Pacific indicate that dust deposition became more important in the late Tertiary, accelerating greatly between 7 and 3 Ma (Leinen and Heath, 1981), but it was around 2.5 Ma ago that there occurred the most dramatic increase in dust sedimentation. In Australia, some desert features started to develop 2–4 Ma, with major dune development occurring at around 1 Ma (Fujioka et al., 2009) and the desiccation of the palaeo-mega-lake Bungunnia after c 1.5 Ma (McLaren and Wallace, 2010; McLaren et al. 2012). Krebs et al. (2011) suggest that late Pliocene aridification of Australia may have been caused by tectonically induced changes in ocean circulation through Indonesia. In the Atacama, fluvial activity became greatly reduced in the late Pliocene as hyperaridity became pronounced (Amundson et al., 2012).

1.8 Quaternary Fluctuations

All deserts show the impact of Quaternary climatic changes (Anderson et al., 2007), and we now know that there were huge shifts in the strength of the tropical monsoon

Table 1.3 *Evidence for palaeoenvironmental reconstruction in drylands*

Evidence	Inference
1. Geomorphological	
Fossil dune systems	Past aridity
Breaching of dunes by rivers	Increased humidity
Discordant dune trends	Changed wind direction
Lake shorelines	Balance of hydrological inputs and outputs
Old drainage lines	Integrated hydrological network
Fluvial aggradation and siltation	Desiccation
Colluvial deposition	Reduced vegetation cover and stream flushing
Karstic (e.g. cave) phenomena	Increased hydrological activity
Frost screes	Palaeotemperature
2. Sedimentological	
Lake floor sediments	Degree of salinity, etc.
Swamp ore and rhizoconcretion deposits	Presence of swampy conditions
Lee dune (lunette) stratigraphy	Hydrological status of lake basin
Spring deposits and tufas	Groundwater activity
Duricrusts and palaeosols	Chemical weathering under humid conditions
Dust and river sediments in ocean cores	Amount of aeolian and fluvial transport
Loess profiles and palaeosols	Aridity and stability
Cave sediments	Hydrological activity
3. Biological and miscellaneous	
Macro-plant remains, including charcoal (in packrat or hyrax middens, bat guano, etc.)	Vegetation cover
Pollen and phytolith analysis of sediments	Vegetation cover
Faunal remains	Biomes
Disjunct faunas	Biomes
Isotopic composition of groundwater and speleothems	Palaeotemperatures and recharge rates
Distribution of archaeological sites	Availability of water
Drought and famine record	Aridity
Dendrochronology	Moisture conditions
Chloride concentrations in dunes	Recharge
Isotopic composition of calcretes, etc.	Presence of C3 or C4 plants
Microbial lipids	Alkalinity and drought

systems (Lézine, 2009; Preusser, 2009). There are many techniques for establishing such past environmental changes (Table 1.3).

Sand deserts expanded from time to time, covering areas that are now heavily vegetated. As a consequence, stabilized sand seas occur in areas where rainfall levels are currently in excess of 500–800 mm. That some dunes are fossil rather than active is indicated by features such as deep weathering and intense iron-oxide staining, clay and humus development, silica or carbonate accumulation, stabilization by vegetation, gullying by fluvial action and degradation to angles considerably below that of the

Table 1.4 *Indicators that dunes are relict*

Geomorphological
Degraded slopes, gully development, colluvial mantles, flooding by lakes, submergence in alluvium
Pedological
Calcification, clay and silt accumulation, soil horizon development
Biological
Mantled in vegetation
Archaeological
Covering of human artefacts of known age

angle of repose of sand (Table 1.4). Sometimes archaeological evidence has been employed to show that sand deposition is no longer progressing at any appreciable rate, whereas elsewhere dunes have been found to be flooded by lakes, to have had lake shorelines etched on their flanks and to have had clays deposited in interdune depressions.

If one compares the extent of old dunefields, using the types of evidence outlined in Table 1.4 with the extent of currently active ones, it is possible to appreciate the substantial changes in vegetation and rainfall that have taken place in many parts of the tropics. This is all the more striking when one remembers that decreased temperatures during glacials would have caused a reduction in evapotranspiration rates and thus led to increased vegetation cover. This would, if anything, have tended to promote some dune immobilization. Dune movement might, however, have been accentuated by higher glacial trade-wind velocities (Ruddiman, 1997), which could in some circumstances have led to dune building and sand transport (see Section 3.3) without any great reduction in rainfall (Wasson, 1984). It is also possible that in some cases increased fluvial sediment transport could have led to a much greater sand supply to river channels so that dune accretion may have been greater even if the climate was moister than it is today (Ellwein et al., 2011).

Stabilized sand seas occur on the south side of the Sahara between Senegal and Sudan (Grove and Warren, 1968) (Figure 1.5), while in southern Africa the Mega Kalahari (Grove, 1969) extended as far north as the Congo Basin (Figure 1.6). Relict dunes also occur in South America, including parts of Amazonia, the Llanos and the São Francisco valley in the north (De Oliveira et al., 1999; Latrubesse and Nelson, 2001; Filho et al., 2002) and the Pampas in the south (Tripaldi and Forman, 2007). The High Plains of America have extensive areas of stabilized dunes, the most notable examples of which are the Nebraska Sandhills. In north-west India the dunes of the Mega Thar can be traced from Rajasthan southwards into Gujarat and eastwards towards Delhi (Allchin et al., 1978), while in Australia large linear dunes can be

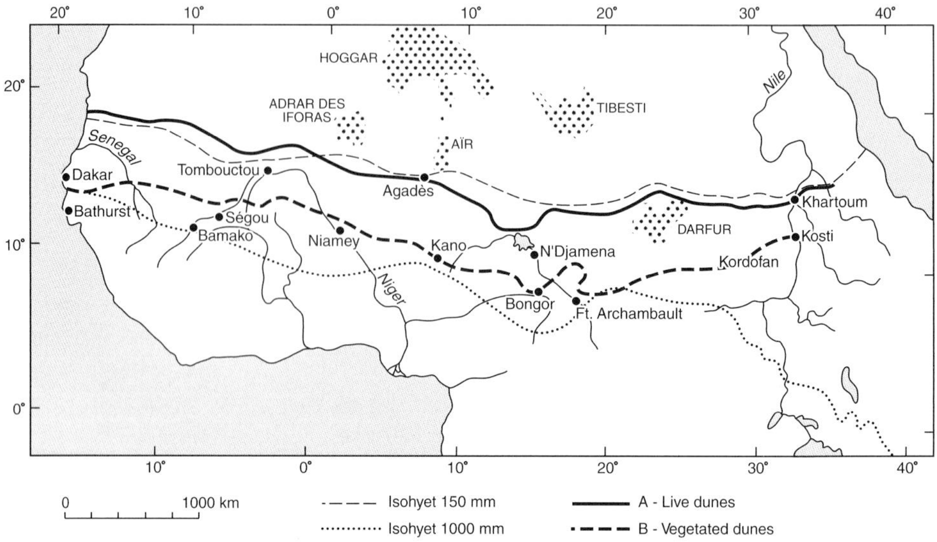

Figure 1.5 Limits of the live and vegetated dunes on the south side of the Sahara. (From Goudie, 2002, fig. 4.11)

found in the Kimberleys and elsewhere in the tropical north (Goudie et al., 1993) (Figure 1.7).

The development of optically stimulated luminescence (OSL) dating means that there are now many dates for periods of dune accumulation (e.g. Fitzsimmons et al.,

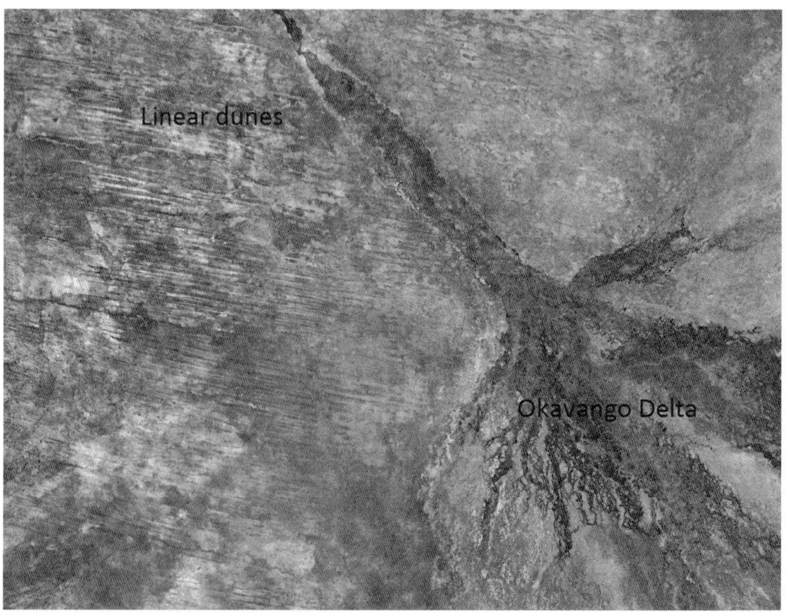

Figure 1.6 A Landsat image of ancient linear dunes, now forested, in the vicinity of the Okavango Delta in north-west Botswana. (Courtesy of NASA)

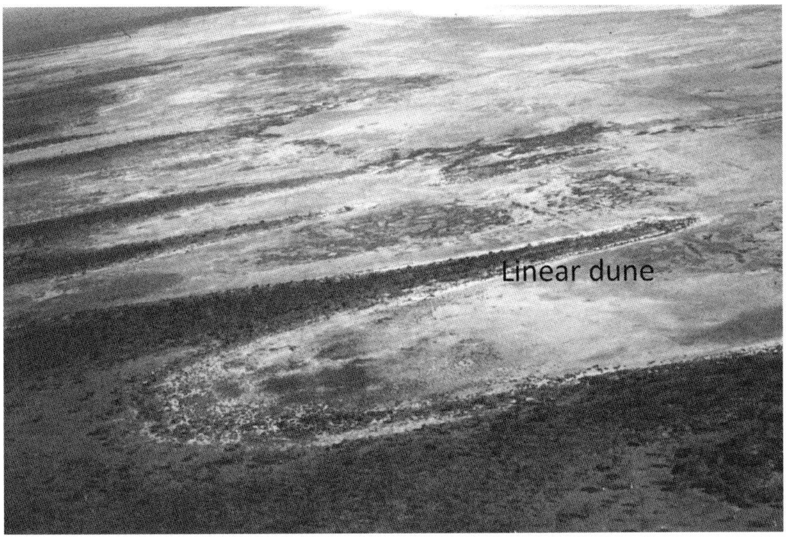

Figure 1.7 In the late Pleistocene, tropical aridity led to the expansion of sand deserts. These vegetated linear dunes, now engulfed by Holocene coastal alluviation, are located in King Sound near Derby, Western Australia. (ASG)

2012; Stone and Thomas, in press). These dates can, under favourable conditions, extend back for some hundreds of thousands of years. It is important to point out, however, that the dates are generally for periods of sand accumulation and stabilization rather than for periods of actual dune movement (Thomas, 2011b). This work has shown that many areas have had repeated phases of dune accumulation, some in the Holocene, as in the High Plains of the United States (Halfen et al., 2010; Hanson et al., 2010; Rich and Stokes, 2011; Halfen et al., 2012), or the Pampas of Argentina (Tripaldi and Forman, 2007), and some in the late Pleistocene. For example, in Australia, Fitzsimmons et al. (2007) identified four periods of dune activity at 73–66, 35–32, 22–18 and 14–10 ka, while Lomax et al. (2011) found dunes that were at least 380 ka old and with phases of substantial dune deposition at 72–63 and 38–18 ka. In the Arabian Peninsula, Preusser (2009) identified periods of dune accumulation at c 150, 110, 65 and 20 ka years ago, and Atkinson et al. (2011, 2012) identified periods at 106, 51, 22–20, 16–10 and 7–2 ka. In the Negev, there appear to have been three main periods of dune accretion: (1) 18–11.5 ka, (2) 2–0.8 ka and (3) modern. In the Thar, Singhvi et al. (2010) found no less than 12 cycles of dune accretion over the last 200 ka. In southern Peru, Londoño et al. (2012) identified four major aeolian depositional episodes at c 55–45 ka, 38–27 ka, 22–16 ka and 12 ka. In many areas, dune activity appears to have been considerable in the period between the Late Glacial Maximum (LGM) and the early Holocene wet phase, and a cluster of studies have demonstrated this in the context of, inter alia, the Mongolian and Chinese deserts (Li et al., 2002; Hülle et al., 2010), the Rub 'Al Khali of Arabia (Goudie et al., 2000), the Negev (Roskin et al., 2011), the Thar of India (Singhvi and Kar, 2004; Singhvi et al., 2010; Saini and Mujtaba, 2012), and the Witpan area of the Kalahari (Telfer and Thomas, 2007).

There are numerous problems in interpreting OSL dates on dunes, including the fact that erosional phases may cause the record to be incomplete. Bioturbation can also be a problem (Bateman et al., 2007). Moreover, the record obtained from dunes in close proximity to each other may not always be the same, and as more and more dates are obtained from an area, some of the supposed intervals of non-accumulation of sand may turn out to be unreal. Equally, a more intensive vertical sampling strategy may have a similar outcome (Stone and Thomas, 2008). It is also necessary to appreciate that aeolian accumulation may not always relate solely to aridity but also to windiness and sand availability (Roskin et al., 2011), and that OSL dates of dunes may not necessarily relate to periods of dune activity but to phases of dune stabilization (Lu et al., 2011). Rivers may be an important source of dune sand, so their state is important (Londoño et al., 2012. Broad, seasonal, sandy channels that dry out may provide ample sand for dune construction, whereas perennially wet or deeply incised channels may provide little sand (Wright et al., 2011).

The flow of rivers showed great variability during the Quaternary. For example, cores extracted from the Mediterranean offshore from the Nile record the strength of the monsoon over East Africa, and indicate greatly increased Nile flow between 18 and 8 ka years ago, and 98–72 ka years ago (Revel et al., 2010). Indeed, sapropel (organic-rich mud) layers in the eastern Mediterranean give a good picture of Nile flow variability in the late Pleistocene. They were created by considerably enhanced flow, which caused collapse of deep water ventilation and/or the elevated supply of nutrients, which then fuelled enhanced productivity (Scrivner et al., 2004). Sapropel development appears to have coincided systematically with Northern Hemisphere insolation maxima related to the orbital cycle of precession, which intensified the African monsoon (Tuenter et al., 2003). Ethiopian rivers, fed by strong monsoon rains in summer, caused a strong seasonal flood in the Nile, and eleven sapropel events have taken place during the last 465,000 years. The most recent period of sapropel deposition (Sapropel 1) started at c 9000 BP and ended after c 6000 years BP and was, like earlier sapropels, caused by a massive input of Nile freshwater (Freydier et al., 2001)

1.9 Lakes of the Quaternary

Deserts have many types of lake basin. These have a variety of names, including playa, pan (see Section 3.6), kavir, salar, daya (see Section 2.10), and so forth. They have a multitude or origins, including deflation, solution of limestone (as in the case of dayas) and evaporites, meteorite impact, aeolian blocking of drainage and tectonic subsidence. A great former lake in south-eastern Australia, Lake Bungunnia,

Table 1.5 *Areal extent (km^2) of some pluvial mega-lakes*

Basin	Area
Caspian/Aral, central Asia	1,100,000
Mega-Chad, Sahara	350,000–400,000
Mkagdikgadi, Botswana	120,000
Mega-Fazzan, Sahara	76,250
Lop Nor, China	55,000
Bonneville, USA	51,640
Bungunnia, Australia	>50,000
Jilantai-Hetao, China	34,000
Lahontan, USA	30,000–35,000
Eyre, Australia	25,260
West Nubia, Sudan	7,000
Mega-Frome, Australia	6,500
Lakes El Fresnal and Santa Maria, Mexico	5,650
Suguta, Kenya	2,150

for example, developed because of tectonic influences on the Murray River Basin (McLaren et al., 2011).

Most notably, there are large areas of internal drainage and many basins without outlets. This was pointed out very clearly by de Martonne (1927), who produced a diagram which showed the clear latitudinal correlation between an index of aridity and the distribution of closed drainage basins. Details of the hydrography and geomorphology of such basins are given by Currey and Sack (2009a and b) and by Shaw and Thomas (1997).

Pluvial lakes are bodies of water that accumulated in such basins because of previous greater moisture availability resulting from changes in temperature and/or precipitation. Their study developed in the second half of the nineteenth century (Flint, 1971). Closed-basin lakes have various advantages for environmental reconstruction – there are many of them, their deposits are amenable to dating and they are sensitive recorders of the local moisture balance (Quade and Broecker, 2009). As Table 1.5 shows, some pluvial lakes attained enormous dimensions. Equally, some lake basins contracted markedly during drought phases, as happened with Lake Malawi, which lost more than 95 per cent of its volume during some of the drier portions of the late Pleistocene (Scholz et al., 2010).

In the United States, the Great Basin contained eighty pluvial lakes during the Pleistocene, occupying an area at least eleven times greater than that of today. Lake Bonneville (Figure 1.8) (Benson et al., 2011) was roughly as big as present-day Lake Michigan and was about 370 m deep and covered 51,640 km^2. It reached a peak at about 18.5 kyr. Lake Lahontan was more complex in form, covered 23,000 km^2,

Figure 1.8 The shorelines of Lake Bonneville, a massive pluvial lake in Utah, USA. (ASG)

and reached a depth of about 280 m. It was nearly as extensive as present-day Lake Erie. River courses became integrated and lakes overflowed from one sub-basin to another (Reheis, 1999; Tchakerian and Lancaster, 2001). In Searles Lake, groundwater bubbling up under its floor produced great tufa pinnacles (Figure 1.9), and its history over the last 3.2 million years has been established by Smith (2009), who identified the importance of 413 kyr precipitation cycles related to the Earth's orbital fluctuations during that period. Smith and Street-Perrott (1983) demonstrated that many basins had particularly high stands during the period that spanned the LGM, between about 25,000 and 10,000 years ago. The high lake levels during the LGM, many of which are indicated by stranded lake features (Figure 1.10), may have resulted from a combination of factors, including lower temperatures and evaporation rates as well increased precipitation levels. Pacific storms associated with the southerly branch of the polar jet stream were deflected southwards compared to today. The lake basins also contain high shorelines that date back to earlier in the Pleistocene (Kurth et al., 2011). For example, a large pluvial lake developed in the Chihuahuan Desert in northern Mexico in the early Holocene (Castiglia and Fawcett, 2006).

Pluvial lakes also developed in the Atacama and Altiplano of western South America (Lavenu et al., 1984). Impressive algal accumulations at high levels occur above the present saline crusts of depressions such as Uyuni (Rouchy et al., 1996). However, a great deal of confusion exists about climatic trends in this region, not least with respect to the situation at the LGM and in the mid-Holocene (Placzek et al., 2001). Nonetheless, estimates have been made of the degree of precipitation change that the

Figure 1.9 Tufa pinnacles formed by groundwater discharge on the floor of Searles Lake, California, USA. (ASG)

high lake stands imply. Pluvial Laguna Lejíca, which was 15–25 m higher than today at 13.5 to 11.3 kyr MP and covered an area of 9–11 km^2 compared to its present extent of 2 km^2, had an annual rainfall of 400–500 mm, whereas today it has only around 200 mm. Pluvial Lake Tauca had an annual rainfall of 600 mm compared with 200–400 mm today.

In the Sahara, there are many pluvial lakes both in the Chotts of Tunisia and Algeria, in Mali (Petit-Maire et al., 1999), in Libya and in the south (e.g. Mega-Chad). Many lakes grew during the African Humid Period, which started at c 13.5 kyr BP in the eastern Sahara and at c 10.5 kyr in the western Sahara, and lakes and wetlands reached their maximum development between 7.5 and 9.5 kyr BP (Lézine et al., 2011). In the Libyan (Western) Desert of Egypt and the Sudan, playa sediments indicate that they once contained substantial bodies of water, which attracted Neolithic settlers. The mid-Holocene wet phase has been called the Neolithic pluvial. A large lake – the West Nubian Palaeolake – formed in the far north-west of Sudan (Hoelzmann et al., 2001). It was especially extensive between 9500 and 4000 years BP, and may have covered as much as 7,000 km^2. If it was that big, then a large amount of precipitation would have been needed to maintain it – possibly as much as 900 mm compared to the less than the 15 mm it receives today. In the Fazzan Basin of Libya, another

Figure 1.10 Bedded lake silts overlying a thin layer of calcareous oncoids on an old shoreline of Lake Manix, Mojave Desert, California, USA. (ASG)

large lake, with an area of 76,250 km^2, formed (Armitage et al., 2007). Mega-Chad (Schuster et al., 2009) reached a peak sometime before 7,000 years ago, when it was more than 173 m deep and had an area of at least 350,000–400,000 km^2 – bigger than the current Caspian Sea, the biggest lake on Earth today (Drake and Bristow, 2006). At that time, large spits developed on its shores (Bouchette et al. 2010). The Sahara may have largely disappeared in the early to mid-Holocene during what has been called the Greening of Africa and the African Humid Period, and rock paintings depict all the big African fauna, including elephant, rhinoceros, hippopotamus and crocodile.

Other lakes developed in the East African Rift Valleys (Figure 1.11), and the shorelines of Lake Chew Bahir in southern Ethiopia are marked by expansive spreads of algal stromatolites. Lake Turkana was also high in the early Holocene and overflowed into the White Nile system (Garcin et al., 2012). To the south, in one of the driest parts of the Rift Valley, a large lake occupied the Suguta basin, attaining a depth of 300 m between 16.5 and 8.5 ka BP, after which it disappeared (Garcin et al., 2009).

In the Kalahari of southern Africa, Lake Palaeo-Makgadikgadi was more than 50 m deep and covered 120,000 km^2, vastly greater than the present area of Lake

Figure 1.11 Algal stromatolites formed on a high shoreline of Lake Chew Bahir, southern Ethiopia. (ASG)

Victoria (68,800 km^2). This makes it second in size in Africa only to Lake Chad at its Quaternary maximum. OSL dates for high stands of Lake Ngami, Lake Makgadikgadi and the Mababe Depression, going back 300,000 years, are provided by Burrough et al. (2009). At times of low stand, barchan dunes accumulated on its floor (Burrough et al., 2012). The Etosha Basin was also filled by large bodies of water in the Pleistocene and at various stages in the Holocene (Brook et al., 2007).

In the Middle East, expanded lakes occurred in the currently arid Rub 'Al-Khali (Parker et al., 2004; Preusser, 2009; Lézine et al., 2010; Rosenberg et al., 2011; Preston et al., 2012) and in Anatolia (Roberts, 1983). The Dead Sea was also greatly expanded during the last glacial period and encompassed the rift valley from the Sea of Galilee to south of the present lake. This large lake, called Lake Lisan, was at its largest between 27 and 15 kyr, and then shrank dramatically, attaining its lowest late Pleistocene stand at about 14 kyr (Stein et al., 2010). The shorelines of Lake Lisan, which occurred at altitudes at up to c –150 m above world sea level (compared to c −422 m today), are well marked by a remarkable series of treads and risers and also by the presence of stromatolites and other deposits (Abu Ghazleh and Kempe, 2009).

In central Asia, the Aral-Caspian system also expanded. During the late Pleistocene (Late Valdai), the level of the lake rose to around 0 m (present global sea level) compared to −27 m today, and it inundated a huge area, particularly to its north. In the early Valdai it was even more extensive, rising to about +50 m above sea level, linking up to the Aral Sea, extending some 1,300 km up the Volga River from its present mouth and covering an area greater than 1.1 million km^2 (compared to 400,000 km^2 today). It may have overflowed into the Black Sea. In general, such lake expansions were probably associated with warming and large-scale influxes of meltwater (Mamedov, 1997), but they were also a feature of glacial phases when there was a decrease in evaporation, and blocking of groundwater by permafrost occurred. Conversely, regressions occurred during interglacials, and in the early Holocene, the Caspian's level dropped to −50 to −60 m.

In north-west India, numerous lake basins contained freshwater bodies in the past, especially in the early and mid-Holocene (see, for example, Achyuthan et al., 2007). Large pluvial lakes also occurred in China and Tibet, where levels were, in general, high from 40,000 to 25,000 BP (Li and Zhu, 2001; Yang and Scuderi, 2010). The Jilantai-Hetao mega-lake covered some 34,000 km^2 along the Yellow River (Fan et al., 2010), and Lop Nor was even bigger (Dong et al., 2011). Mega-lakes were especially well developed at about 30 kyr (Yang, Scuderi, Paillou, Liu, Li and Ren, 2011).

Similarly, the interior basins of Australia, including Lake Eyre, expanded and contracted, with high stands tending to occur in interglacials (Harrison and Dodson, 1993). Lake Eyre was large during the period from 130–80 kyr, attaining its peak at Marine Isotope Stage 5e (MIS 5e) (Webb, 2009) but was largely dry during the LGM. Another major palaeolake – Lake Mega-Frome – also developed in MIS 5 as a result of the coalescence of Lakes Frome, Blanche, Callabonna and Gregory (Cohen et al., 2011).

As these regional examples demonstrate, pluvial lakes were widespread (even in hyperarid areas) and reached enormous dimensions. Pluvials were not in phase in all regions and in both hemispheres (Spaulding, 1991), and a recent analysis of conditions during the Heinrich Stadial 1 at 16–19 kyr BP has demonstrated considerable variability in conditions in different parts of the tropics. In very general terms, however, dry conditions during the LGM at 19–23 kyr and during the Heinrich Stadial 1, together with humid conditions during part of the early to mid-Holocene, appear to have been characteristic of tropical deserts, though not of the south-west United States (Street and Grove, 1979; Lézine, 2009; Quade and Broecker, 2009; Lézine et al., 2010; Stager et al., 2011). Many areas were also moist during the Eemian interglacial, including southern Arabia (Fleitmann et al. 2011).

1.10 The Geography and Frequency of Change

In a mid-latitude location like the south-west region of the United States (Smith and Street-Perrott, 1983), there was greatly increased effective moisture at the time of the LGM. This was caused by decreased evaporation and also by an intensified zonal circulation and the equatorward displacement of mid-latitude westerlies and associated rain-bearing depressions, particularly in winter. Conversely, there may also have been mega-droughts during interglacials (Fawcett et al., 2011), in contrast to the situation in Arabia mentioned in Section 1.9. In the early Holocene (c 11,000 to 8,000 years ago), mid-latitude central Asia was drier than it is now, whereas the south and east Asian monsoonal areas were moister than today (Jin et al., 2012).

The tropics were much less influenced by the displaced full glacial westerlies than mid-latitudes. They experienced relatively dry conditions at that time but subsequently experienced a major pluvial in the early to mid-Holocene (Grove and Goudie, 1971). Under warmer Holocene conditions, monsoonal circulation was intensified, and in the Northern Hemisphere the ITCZ would have shifted north, bringing rainfall into areas such as West Africa, Sudan, Ethiopia, Arabia and the Thar. At around 9000 years BP, Milankovitch orbital precessional forcing led to Northern Hemisphere summers with almost 8 per cent more insolation than today (Kutzbach and Street-Perrott, 1985). Higher insolation caused greater heating of the land, stronger convection, more inflow of moist air and a higher summer monsoonal rainfall. In contrast, weaker insolation maxima around 35,000 and 60,000 years ago would have created weaker monsoons (Ruddiman, 2001).

Changes in insolation receipts at 9000 BP can help to explain the Northern Hemisphere low-latitude pluvial, but they have less direct relevance to the Southern Hemisphere (Tyson and Preston-Whyte, 2000). Also important in determining the spatial and temporal patternings of precipitation change are sea surface temperature conditions associated with the buildup and disintegration of the great ice sheets (Shi et al., 2000). In addition, changes in snow and ice cover over Asia, including Tibet and the Himalayas, could have had a major effect on the monsoon (Zonneveld et al., 1997).

In addition to being severe, Quaternary climatic changes in deserts were frequent. Dunes were repeatedly reactivated and stabilized – both during the Pleistocene and the Holocene – lakes rose and fell over short spans of time, and pulses of dust were deposited in the oceans. The frequency of change is indicated by high-resolution studies of ocean and lake cores and because of the increasing availability of multiple optical luminescence dates for aeolian sediments. The multiple glaciations and deglaciations of high latitudes, and the multiple fluctuations within them, all indicate the instability of Quaternary climates. They were all associated with major changes in the oceans, pressure systems and wind belts, which in turn had an impact on deserts. Lake level fluctuations in arid basins may indicate short-lived fluctuations that correlate with Heinrich events (Benson et al., 1998) and Dansgaard-Oeschger cycles (Lin et al., 1998), while the Younger Dryas also seems to have its counterparts in arid regions (e.g. Zhou et al., 2001). The mechanisms causing changes in atmospheric circulation would have been both numerous and complex, and there will have been lagged responses (e.g. slow decay of ice masses, gradual falls of groundwater, etc.). It is also apparent that there were differing hemispheric and regional responses to change (DeMenocal and Rind, 1993). For example, Arabia and northeast Africa may have been especially sensitive to changes in North Atlantic sea surface temperatures, while monsoonal Asia would have been especially affected by snow and ice conditions in the Himalayas and Tibetan Plateau. Lake level fluctuations may have lagged climate changes because of the lag effects of groundwater recharge (Lézine et al., 2011). Large lakes may also have acted as local moisture sources and so have stimulated precipitation, leading to a positive climatic feedback (Krinner et al., 2012).

1.11 Some Holocene Events

The Holocene experienced abrupt and relatively brief climatic episodes. Mid-Holocene aridity in the U.S. High Plains caused population declines (Louderback et al., 2010). Some of these dry phases, which may have occurred during the warm phases of the Atlantic Multidecadal Oscillation (Oglesby et al., 2012) caused dune mobilization (Arbogast, 1996a; Halfen et al., 2010, Rich and Stokes, 2011). For instance, large linear dunes developed in Nebraska c 800–1,000 years ago, during the Medieval Warm Period (Sridhar et al., 2006), and in Kansas a similar picture has been attained (Halfen et al., 2012). In Arizona, four periods of dune accumulation occurred in the Holocene (Wright et al., 2011). There were also alternations of humid and intense arid phases in tropical Africa, and there was an abrupt climate event at 8,200 years ago (the 8.2 ka event) that interrupted an otherwise humid period (Brooks, 2006; Fleitman et al., 2003, 2007). In southeastern Arabia, this event is recorded in the Hoti Cave speleothem record, where a positive shift in the $\delta^{18}O$ of calcite records a precipitation minimum (Neff et al., 2001; Fleitmann et al., 2007), and the Awafi Lake sediment record, where a large influx of aeolian material entered the lake system (Parker et al., 2006a, 2006b, 2006c). The 8.2 ka event may represent the final collapse of the Laurentide Ice Sheet into the North Atlantic and a reduction in salinity and sea surface temperatures (Alley et al., 1997; Bond et al., 1997). The impact on the climate system was global in magnitude and extent and has been observed in North Africa (Gasse and Van Campo, 1994; DeMenocal et al., 2000; Kuper and Kröpelin, 2006), the Near East (Bar-Matthews et al., 1997) and the Arabian Sea (Gupta et al., 2003).

Around 4,200–4,100 cal yr BP, evidence from a number of regions of the globe for another major shift in climate has been recognized. Dramatic drying out of Saharan lakes occurred at c 4.5 kyr BP (Lézine et al., 2011). A pronounced dry event is recorded from Red Sea sediments at around this time (Arz et al., 2006) and also from a core in the Gulf of Oman (Cullen et al., 2000), where mineralogical and geochemical analyses of the sediments revealed a large increase in windblown dust derived from Mesopotamia. A further abrupt climatic event took place in the Middle East around 3,500 to 2,500 years ago, causing devastation of some cultures and settlements (Kaniewski et al.,

2008). In Arizona, fan aggradation occurred from 3,200 to 2,300 years ago when there was an intensification of the El Niño-Southern Oscillation (ENSO) circulation (see Section 1.12). In the Mojave Desert, fans and lakes responded to the climate changes of the Medieval Warm Period and the Little Ice Age (Miller et al., 2010); widespread mobilization of dunes occurred in the High Plains during the former (Cook et al., 2011). In north-western China, the Little Ice Age seems to have been a time of relatively high precipitation associated with more frequent mid-latitude cyclone activity (Chen et al., 2010).

1.12 Short-Term Climate Fluctuations: ENSO and Other Phenomena

ENSO is the primary mode of climatic variability in the two-to-seven year time band. El Niño is an extensive warming of the upper ocean in the equatorial eastern Pacific lasting up to a year or even more, while the negative or cooling phase is called La Niña. In the twentieth century, there were around twenty-five warm events of differing strengths, with that of 1997–98 being especially strong (Changnon, 2000), although ENSO was relatively quiescent from the 1920s to the 1940s (Kleeman and Power, 2000). ENSO may, according to detailed tree-ring analyses in the south-western United States, exhibit a quasiregular cycle of fifty to ninety years (Li et al., 2011), and century-scale precipitation changes have been identified by tree-ring studies in the South American Altiplano (Morales et al., 2012).

The Holocene history of El Niño has been controversial (Wells and Noller, 1999), but Grosjean et al. (1997) have discovered more than thirty debris flow events caused by heavy rainfall between 6.2 and 3.10 kyr BP in the northern Atacama. The stratigraphy of debris flows has also been examined by Rodbell et al. (1999), who have been able to reconstruct their activity over the last 15 kyr. Between 15 and 7 kyr BP, the periodicity of deposition was equal to or greater than fifteen years and then progressively increased to a periodicity of 2 to 8.5 years. The modern periodicity of El Niño may have been established about 5 kyr BP, possibly in response to orbitally driven changes in solar radiation (Clement et al., 2000; Liu et al., 2000; Quigley et al., 2010).

Lake levels vary greatly in response to changes in rainfall into their catchments and outputs from changing amounts of evapotranspiration (Nicholson, 1998). For example, El Niño warming in 1997 led to an increase in rainfall over East Africa that caused lake levels to rise (Birkett et al., 1999). The abrupt change in the level of the Caspian (2.5 m between 1978 and 1995) has also been attributed to ENSO phenomena (Arpe et al., 2000). Similarly, the 3.7 m rise in the level of the Great Salt Lake (Utah, United States) between 1982 and 1986 was at least partly related to the record rainfall and snowfall in its catchment during the 1982–83 El Niño (Arnow and Stephens, 1990). The enormous changes that occur in the volume of Lake Eyre result from ENSO-related changes in inflow, with the greatest flooding occurring during

La Niña phases (Kotwicki and Allan, 1998) such as those of 1949–52, 1974 and 2010–11.

ENSO events produce streamflow anomalies (Arnell, 2002; Gayo et al., 2012). In the western United States there is a tendency for the Southwest to be wet and the Northeast to be dry during the El Niño warm phases (Negative Southern Oscillation Index), and vice versa for La Niña (Cayan et al., 1999). In central Australia, the flow regime of Cooper Creek has been found to be associated with ENSO, with a forty-eight-year hydrograph record showing floods clustered in La Niña episodes (Puckridge et al., 2000).

Links have also been established between the North Atlantic Oscillation (NAO) and streamflow. Because the NAO governs the path of Atlantic-derived mid-latitude storm tracks and precipitation across the eastern Mediterranean into the Middle East, it has a marked impact on the flow of the Tigris and Euphrates (Cullen et al., 2000) and on climatic conditions in the arid regions of China (Chen et al., 2010).

In the Atacama, El Niño events have caused serious flooding (Goldstein and Magilligan, 2011), and Vargas et al. (2000) have related twentieth-century debris flows in Chile to El Niño events; that of 1991 was highly destructive in Antofagasta (Sepúlveda et al., 2006). Similarly, Trauth et al. (2000) have suggested that there is an ENSO control of landslides in the eastern Argentine cordillera. High rainfall events together with increased pore-water pressures have caused accelerated undercutting of valley side slopes along deeply incised, narrow valleys.

ENSO can be associated with intensified drought and so influences the activity of dust storms and dunes, particularly in areas which are at a threshold for dust entrainment or dune activation. Such areas are those where in wet years there is just enough vegetation to stabilize ground surfaces. In the United States, dust emissions in the period 1983–84 were greatly reduced following the heavy rainfall of the 1982 El Niño (Lancaster, 1997). Likewise, Forman et al. (2001) have found that Holocene phases of dune activity in the Great Plains were associated with La Niña occurrence and weakened cyclone development over central North America. The Atlantic Multidecadal Oscillation, however, has also had a major impact on drought conditions (Nigam et al., 2011).

The extent of the Sahara has shown considerable inter-annual variability as determined by remote sensing observations since 1980. The greatest annual north–south latitudinal movement of the southern Sahara boundary was 110 km between 1984 and 1985. This resulted in a decrease in the desert's 724,000 km^2 area (Tucker et al., 1991). About 75 per cent of the inter-annual variation in its area can be accounted for by the combined effects of the NAO and ENSO (Oba et al., 2001). The drought that began in the mid-1960s also led to reductions in the flow of rivers such as the Senegal and Niger, and to an increase in dust storm activity (Middleton, 1985). The extent of Lake Chad's water surface has also fluctuated dramatically (Nicholson, 1996), falling from c 25,000 km^2 in 1960 to just a tenth of that figure in the mid-1980s (Birkett, 2000).

Figure 1.12 The patchy nature of desert vegetation illustrated by Joshua Tree National Park, California, USA. The main shrub here is *Larrea tridentata* (the creosote bush). (ASG)

1.13 Vegetation Cover and Animal Activity

Degree of aridity is a major control on vegetation cover. In deserts, the vegetation cover is generally low; a closed cover is seldom encountered (Figure 1.12). Deserts have a low biomass, often 100 times lower than that of an equivalent area of temperate forest. Water is the vital influence on plant growth, of course, and is responsible for this low biomass level. Although the relationship between annual precipitation and above-ground peak biomass may vary from region to region, there is nonetheless a clear tendency for it to increase linearly with precipitation (Bullard, 1997). Water not only controls the volume of plant matter produced, however – it also controls the distribution of plants within an area of desert; some areas – because of their soil texture, topographic position or distance from rivers or groundwater – have virtually no water available to plants, whereas others do. The nature and extent of vegetation cover may within a small area vary with aspect, with slopes facing the equator being more xeric than those facing the poles (see Section 5.20). Vegetation cover is a fundamental control of many geomorphological processes in deserts, including runoff and water erosion (Abrahams et al., 1995; Durán Zuazo et al., 2004). Rogers and Schumm (1991), for example, found that sediment yield increases rapidly from 43 to

15 per cent as vegetative cover decreases; however, with less than 15 per cent vegetative cover, the rate of increase of sediment yield diminishes markedly. Snelder and Bryan (1995) established that there was a critical threshold of 55 per cent vegetation cover, below which erosion rates rapidly increased. A very important issue here is the relative role of grasslands and shrublands in affecting the hydrological response and erosion of land surfaces on desert margins (Belski, 1996; Ravi et al., 2009). Vegetation cover also controls the degree of dune activity and dust emissions, and severe droughts can enable 'stable' dunes to become active (Hesse and Simpson, 2006). The balance between grasslands and shrublands may also be a factor in controlling dust emissions (Floyd and Gill, 2011).

Animals play a significant role in some desert areas (Whitford and Kay, 1999). Burrowing, trampling (Eldridge, 1998), wallowing, digging and geophagy can have significant impacts on the landscape and on soil properties (Kinlaw, 1999). Termites and ants can affect soil properties, infiltration capacities and runoff (Cammeraat et al., 2002). Indeed, termites are prominent – if not visible – consumers of biomass in most deserts (Schaefer and Whitford, 1981), and have a clear influence on soil infiltration capacities, which tend to be higher where they are present (Elkins et al., 1986). They may translocate a great deal of surface material by creating their soil sheets (Bagine, 1984). Mammals such as gophers, prairie dogs, wombats, porcupines, echidnas, mole rats, kangaroo rats and gerbils play a role in soil churning (Yair, 1995; Neave and Abrahams, 2001; Schooley and Wiens, 2001; Eldridge et al., 2011) and create certain types of patterned ground. Fine material brought to the surface may be removed by deflation, while the presence of subterranean galleries may increase water infiltration into the soil. Domestic stock can effect vegetation cover, biological crust wetability, soil compaction and infiltration capacity (du Toit et al., 2009).

1.14 The Importance of Plate Tectonic Setting

Tectonic history is a fundamental control of much of the macro-scale relief and climatic evolution of deserts. In North America (see Section 7.12) the style of tectonism explains the widespread development of closed depressions, high plateaux and extensive alluvial fans, especially in the Basin and Range Province. In South America, the Andes' role is crucial because of their climatic impact and their morphological effects (Lamb et al., 1997). Tectonic uplift and the eastward migration of the Andean volcanic arc, associated with the subduction of the oceanic Nazca plate beneath the South American continental plate, have created great contrasts in altitude, abundant volcanic activity, severe slope instability and the many closed depressions of the Altiplano (see Section 7.13). In northern Africa, the Atlas Mountains, the central Saharan highlands (e.g. Tibesti and Hoggar), the uplift of the Red Sea Hills and the evolution of the Nile have all been affected by plate tectonic processes and are major controls and features of the area's topography (Williams, 1994; Lamotte et al., 2009).

The evolution and geomorphology of the Namib (see Section 7.5) owe much to the opening of the South Atlantic by sea-floor spreading in the late Jurassic and early Cretaceous as well as the presence of the great hotspot associated with the Walvis Ridge (Goudie and Eckardt, 1999). In addition, the Great Escarpment and the basin form of the Kalahari (see Section 7.6) are associated with passive-margin evolution. In the Middle East (see Section 7.7), features such as the Red Sea rift, the Dead Sea transform fault, the Zagros Mountains and the ophiolitic ranges of Oman result from being at a crossroads of major plate boundaries (Guba and Glennie, 1998). In Asia, the uplift of the Himalayas and the Tibetan Plateau radically modified climatic conditions (see Section 1.6) (Raymo and Ruddiman, 1992) and accounts for the development of towering mountain ranges in close proximity to enormous closed basins (see Section 7.10). By contrast, Australia (see Section 7.14) is a low, dry, ancient continent with a long history of comparative orogenic stability over large areas, so that many of the present landscape features are inherited from a great variety of climates that may go back to the Jurassic or earlier (Twidale, 2000).

The tectonic settings of contemporary drylands have been discussed by Rendell (1997), who identified five types: (1) cratons (shield and platform areas), (2) active continental margins associated with Cenozoic orogenic belts, (3) older, Phanerozoic orogenic belts, (4) inter-orogenic basin and range and inter-cratonic rift zones, and (5) passive continental margins (Table 1.6).

1.15 Two Main Types of Desert Topography

Desert topography may be conveniently classified into two main types: shield and platform deserts and basin and range deserts. The former occur in areas of relative tectonic stability and are principally associated with the vast, stable plains of Africa (including much of the Sahara and southern Africa), large tracts of the Arabian Peninsula and parts of central Asia, central India and Australia (Cooke et al., 1993, p. 18). These areas have great planation surfaces that are often cut across basement igneous rocks (as in Western Australia and southern Africa), by huge shallow basins (e.g. Lake Eyre in Australia and Lake Chad in Africa) or by extensive surfaces on horizontally bedded and gently warped sediments (as in the Colorado Plateau, central Asia or the Nubian Sandstone areas of North Africa). The plains are occasionally surmounted by isolated hills or mountains associated with more resistant rocks or the eroded rumps of former mountains. In places, such broad plains may be entrenched with canyons, providing a distinctive mixture of plateaux and valley-side escarpments, as in the Colorado Plateau and northern Namibia.

Basin and range deserts, however, are dominated by alternating mountains and systems of enclosed drainage. They are most common in areas of tectonic activity, such as the south-western United States, the dry coastal deserts of Chile and Peru, the majority of Iran, Afghanistan (Table 1.7[c]) and Pakistan and parts of central Asia.

Table 1.6 *Tectonic settings of arid zones*

Contemporary tectonic setting	Examples	Comments
1. Cratons	Kalahari Great Karoo Australian Deserts Saudi Arabia	Relative stability since the late Tertiary
2. Active continental margins and Cenozoic orogenic belts	Atacama (Peru-Chile) Sahara (Atlas Mountains) Sinai-Negev Arabia-Zagros Karakoram valleys Baluchistan	Compressional setting, thrust and transcurrent faulting
3. Older orogenic belts	Sahara China Thar (Aravallis)	Some reactivation of existing fault zones
4. Inter-orogenic, inter-cratonic	East Africa (Afar, Ethiopia) Mojave Great Basin (W. USA) Sonora Desert Chihuahua Desert Monte Desert	Extensional tectonic setting, 'pull-apart' basins
5. Passive continental margins	Namib Patagonian Desert	

Source: Modified from Rendell (1997, table 2.1).

Table 1.7(a) *Percentages of different landform types in deserts*

Landform type	South-western USA	Sahara	Libyan desert	Arabia
Desert mountains	38.1	43	39	47
Playas (base-level plains)	1.1	1	1	1
Desert flats	20.5	10	18	16
Bedrock fields (including hamadas)	0.7	10	6	1
Regions bordering through flowing rivers	1.2	1	3	1
Dry washes	3.6	1	1	1
Fans and bajadas	31.4	1	1	4
Dunes	0.6	28	22	26
Badlands and subdued badlands	2.6	2	8	1
Volcanic cones and fields	0.2	3	1	2
Total	100	100	100	100

Source: Clements et al. (1957) in Cooke et al. (1993).

(b) *Extent of desert landform types in Australia*

	% of arid zone
Mountain and piedmont desert	17.5
Riverine desert	4.0
Shield desert	22.5
Desert clay plains	13.0
Stony desert	12.0
Sand desert	31.0
Total	100.0

Source: Mabbutt (1971).

(c) *Extent of landform types in Afghanistan (%)*

Sand sheets	1
Playas	1
Broad river valleys	1
Plateaux	4
Alluvial plains	7
Pediments	9
Sand seas/dunes	9
Alluvial fans	10
Badlands	13
Mountain highlands	45

Source: Bacon et al. (2008).

(d) *Extent of landform types in the Monte Desert, Argentina (%)*

High and low mountains	14.9
Dominant volcanic landforms	7.4
Peneplains, hilly country and depressions	9.6
Alluvial-aeolian plains	12.1
Intermontane (bolson)	8.4
Alluvial plains and fans	14.0
Aeolian plains	6.1
Active dunes	0.5
Miscellaneous others	27.0

Source: Abraham et al. (2009).

Both shield and basin and range deserts include mountains and plains, although their relative proportions differ. In the latter the mountain to plain ratio may approach one, whereas in shield deserts it may approach infinity. Typically, the mountains are characterised by steep-sided valleys, and they are locally terminated by relatively steep

Table 1.8 *Population of selected dryland countries (millions)*

Country	1950	2000
Egypt	21.80	67.90
Yemen	4.36	18.35
Saudi Arabia	3.20	20.35
Burkina Faso	3.96	11.54
Peru	3.52	11.35
Chad	2.66	7.89
Jordan	0.47	4.91
Libya	1.09	5.29
Oman	0.46	2.54
Kuwait	0.15	1.91
UAE	0.07	2.61
Namibia	0.51	1.76
Botswana	0.39	1.54
Djibouti	0.06	0.63
Qatar	0.03	0.57
Western Sahara	0.01	0.25
Algeria	8.75	30.29
Syria	3.50	16.19
Iran	16.91	70.33
Pakistan	39.66	141.26
Total	111.56	417.76

Source: UN data processed by author.

slopes at the mountain/piedmont junction (the piedmont angle). Typically, too, the piedmont may be subdivided into plains cut in bedrock (pediments) (see Section 5.3) and plains formed of alluvium. The latter may include alluvial fans (see Section 5.4) and closed basins (playas). In addition, aeolian deposits in the form of sand seas (ergs) occur on the plains, almost always in topographic lows.

Within regions of these two major types of desert, some areas are dominated by erosion and others by deposition, water action, wind action, salt accumulation or

Table 1.9 *Percentage change in population of selected dryland states in the United States between 1990 and 2000*

Arizona	40.0
Nevada	66.3
New Mexico	20.1
Utah	29.6

Source: UN data processed by author.

Figure 1.13 Urbanisation is an increasingly important aspect of the human occupation of drylands. This is Kuwait City. (ASG)

salt removal. The nature and location of the processes which operate are strongly influenced by the following major topographic settings (Mabbutt, 1969):

Desert uplands, where geological controls of relief are important, bedrock is exposed, and relief is high

Desert piedmonts, which are zones of transition separated from the uplands by a break of gradient but which none the less receive runoff and sediments from the uplands, and which have both depositional (e.g. alluvial fans) and erosional forms (e.g. pediments)

Stony deserts, which consist of stony plains and structural plateaus, and may have a cover of stone pavement [see Section 2.28]

Desert rivers and floodplains (features of *desert lowlands*)

Desert lake basins, which are terminal sumps to which the disorganised drainage progresses, and which are often salty, but may have held large freshwater bodies during pluvial [see Section 1.9]

Sand deserts, which tend to be beyond the limits of active fluvial activity but often derive their materials by wind action removing erodible material from floodplains or lake basins; they are characterised by dunes [see Chapter 4].

The proportions which particular landform assemblages occupy in different deserts are listed in Tables 1.7(a), (b), (c) and (d).

1.16 The Human Dimension

An increasingly important factor in the geomorphology of arid lands is the human one. The growth in human population, its concentration in ever-larger urban centres

Table 1.10 *Estimated population of major dryland urban areas, 2010 (millions)*

Cairo	17.29
Los Angeles	14.78
Karachi	13.09
Tehran	8.17
Lima	8.00
Lahore	7.11
Baghdad	5.85
Khartoum	5.18
Riyadh	4.74
Alexandria	4.30
Jeddah	3.18
Jaipur	3.05

Source: UN data processed by author.

Table 1.11 *Population of selected dryland cities in 1950 and 2010 (millions)*

City and country	1950	2010
Cairo, Egypt	2.41	12.66
Ouagadougou, Burkina Faso	0.03	2.55
Ndjamena, Chad	0.04	1.58
Lanzhou, China	0.32	2.10
Alexandria, Egypt	1.04	5.53
Jodhpur, India	0.18	1.22
Tehran, Iran	1.04	8.71
Esfahan, Iran	0.18	3.92
Alma Ata, Kazakhstan	0.32	1.29
Kuwait City, Kuwait	0.09	1.51
Bamako, Mali	0.06	2.13
Karachi, Pakistan	1.03	16.61
Lima, Peru	0.97	8.84
Riyadh, Saudi Arabia	0.11	4.59
Jeddah, Saudi Arabia	0.12	2.75
Damascus, Syria	0.37	3.10
Las Vegas, USA	0.04	1.12
Phoenix, USA	0.22	2.86
Los Angeles, USA	4.05	13.86
Dubai, UAE	0.02	1.70
Abu Dhabi, UAE	0.01	1.09
Baghdad, Iraq	0.58	5.44
Windhoek, Namibia	0.02	0.23
Total	13.37	105.38

Source: UN data processed by author.

and its increasing human impact are of great geomorphological and environmental significance. In terms of population growth, while the global population increased by 2.38 times between 1950 and 2000, that of the dryland countries (shown in Table 1.8) was considerably greater, averaging a growth of 3.74 times. Within some individual countries, dryland states grew more than the average. In the United States, for example, the states of Arizona, Nevada, New Mexico and Utah showed a remarkable growth in population during the 1990s (Table 1.9). Drylands have also seen rapid rates of urbanisation (Figure 1.13). Indeed, as Table 1.10 shows, there are now some enormous urban agglomerations. The growth of cities has also been substantial; as Table 1.11 suggests, the average size of major dryland cities expanded 7.9 times between 1950 and 2000. The footprint of such cities is enormous. They require water, fuel, building materials, food, space for construction and room for recreation, and all these needs have geomorphological impacts.

2

Rock Weathering and Desert Surfaces

In this chapter we will consider the processes that cause rocks to become weathered, some of the landforms and materials that are produced by such weathering and some of the other surface types that are characteristic of desert environments.

2.1 Rock Weathering Processes

Desert weathering is pervasive, selective and frequently superficial (Cooke et al., 1993). These characteristics result from extreme variations of microclimate and from the fact that, because the proportion of exposed rock is higher than in most other climatic zones, rock properties impose a strong differentiation on the effectiveness of weathering. Flaking, spalling, splitting, pitting, granular disintegration and bizarre forms abound. Weathering also prepares bedrock for the subsequent operation of erosion, thereby providing the wherewithal for landform evolution (Viles and Goudie, 2007). In some locations weathering rates can be very high, as in the salty and foggy coastal Namib, but elsewhere, superbly preserved archaeological inscriptions of the type found in the Nubian Sandstone at Abu Simbel in Egypt suggest that relatively little weathering has occurred over a period of around 3,000 years (Figure 2.1). Good introductory reviews of desert weathering processes and forms are provided by Smith (2009) and Viles (2011).

2.2 Insolation, Thermal Fatigue and Dirt Cracking

Traditionally, many desert weathering phenomena (Figure 2.2) have been attributed to physical weathering caused by severe temperature changes. Insolation weathering (also known as thermoclasty or thermal stress fatigue) is the rupturing of rocks and minerals primarily as a result of diurnal temperature cycles but also as a result of shorter thermal shocks. Areas that are heated expand relative to the cooler portions of the rock, and minerals expand and contract to degrees that depend on their different

Figure 2.1 A wonderfully fresh inscription, probably around 3,300 years old, carved in Nubian Sandstone at Abu Simbel, southern Egypt. It shows a captured African soldier (left) with earring and distinctive hair style. One partial reason for its preservation is that it may have been covered in dune sand for some of its history. (ASG)

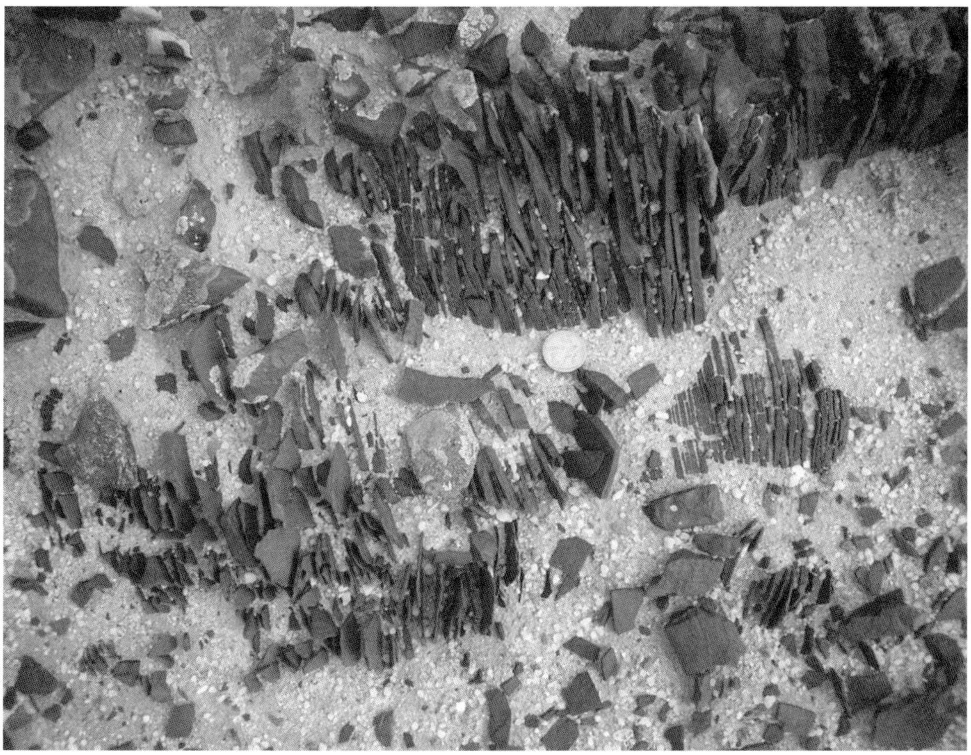

Figure 2.2 Rock outcrops, such as this dolerite in the central Namib, can be intricately split, a phenomenon that has often been attributed to insolation weathering. Coin for scale. (ASG)

Table 2.1 *Average linear coefficients of thermal expansion* ($\times 10^{-6}/^{\circ}C$)

Minerals	
Quartz	11.5–12.0
Orthoclase, microline	6.5–7.5
Pyroxenes, amphiboles	6.5–7.5
Olivine	6.0–9.0
Albite	5.0–6.0
Calcite	4.5–5.0
Oligoclase, andesine	3.0–4.0
Labradorite, bytownite	3.0–4.0
Anorthite	2.5–3.0
Rocks	
Quartzite, silica shale, chert	11.0–12.5
Sandstones	10.5–12.0
Argillaceous shales	9.5–11.0
Dolomite	7.0–10.0
Granites and gneisses	6.5–8.5
Marbles	4.0–7.0
Dense, crystalline, porous limestone	3.5–6.0

Source: N.G. Zoldners in Lane (2006).

coefficients of expansion, details of which for a range of rock types can be found in Weiss et al. (2004) and in Table 2.1. It can be seen that quartz and various types of quartz-rich rocks (e.g. sandstones) have especially high coefficients and so may be particularly susceptible to insolation effects. As expansion and contraction takes place, stresses may exceed the tensile strength of the rock and so cause splitting, spalling and granular disintegration. In deserts, under clear skies, rock surfaces can experience extreme diurnal temperature ranges (in excess of 50°C) and high daytime values (greater than 70°C) (Table 2.2). The highest ground surface temperature reported is 87.8°C for Death Valley, California (Roof and Callagan, 2003).

There have been reports of rocks ‘exploding’ like pistol shots as they cool rapidly after sunset. David Livingstone (1857, pp. 149–50) experienced this in what is now Botswana and attributed these bangs to the splitting of boulders as they cooled in the evening. Such a process is facilitated by the difference in temperature response between the outer and inner portions of a rock mass, with the surface experiencing more severe temperature regimes than the interior. This might promote exfoliation (onion skin weathering) and spheroidal weathering. Another important concern is the differential response of different minerals to heating and cooling. Rocks with many different sorts of minerals with different coefficients of expansion and contraction, with crystals of different sizes and with different colours and albedos (which determine their response to solar heating) may develop stresses that cause granular disintegration (Gómez-Heras et al., 2006). The presence of light-transmissive (translucent) minerals

Table 2.2 *Maximum ground and air surface temperature (°C) in deserts under clear skies*

Location	References	Month	Surface type	Air	Ground
Yuma (Arizona)	George (1986)	June	Rock	39	56
Tibesti	Jäckel and Dronica (1976)	March	Rock (sandstone)	30	57
			Sand	28	59
Tibesti	Peel (1974)	August	Varnished sandstone	47	79.3
			Pale sandstone	47	78.8
			Basalt	47	78.5
Sudan	Cloudsley-Thomson and Chadwick (1964)	September	Sand	41	82
			Rock	40	72.5
Karakorams	Whalley et al. (1984)		Basalt	17.4	41
		July/August			
			Basalt	14.3	33.5
			Varnish	26.5	46.3
			Sandstone	41.0	54.0
Tunisia	Heatwole and Muir (1979)	June	Soil	35	52
		July	Soil	35	55
		August	Soil	36	54
Tucson (Arizona)	Sinclair (1922)	June/ September	Adobe (silty mud)	49.7	71.5
Namib	Desert Ecological Research Unit (pers. comm.)	Mean	Soil	28.5	53.6
Egypt	Happold (1984)	August	Sand	31.1	51.1
Oman	Royal Geographical Society (1986)	February	Sand	26	42
Namib	Holm and Edney (1973)	January	Sand	33	59
California	Chappell and Bartholomew (1981)	June	Soil	52.5	72.5
Namib	Darborn (2000)	March	Granite	27.9	36.4
			Granite	27.5	36.3
			Granite	28.9	47.1
California			Granite	27.3	44.3
Namibia	Eppes and Griffing (2010)	June	Marble	31.6	60
Atacama	Sumner et al. (2004)	–	Granite gneiss	42.9	64.4
	McKay et al. (2009)	–	Stone pavement	33	45

may also play a role (Hall, 2011) by affecting heat transmission. Where no light can penetrate the rock, the heat exchange from incoming radiation is at the rock surface, and this can lead to steep thermal gradients within the rock. Where light can penetrate, however, radiation-to-heat transformation is not just at the rock surface but occurs within the rock, causing the thermal gradient to be less steep.

That heating and cooling of rock could cause its disintegration was evident to early investigators. In a pioneering text, Merrill (1897, p. 180) wrote:

> As temperatures rise, each and every constituent expands and crowds with almost resistless force against its neighbour; as temperatures fall, a corresponding contraction takes place. Since in but a few regions are surface temperatures constant for any great period of time, it will be readily perceived that almost the world over there must be continuous movement within the superficial portions of a rock.

The most important work undertaken by British investigators on the weathering effects of temperature variations was done in Egypt (see Hume, 1925, chapter 2). German workers were especially sure of the role of insolation in deserts, and this was particularly true of Walther. His *Das Gesetz der Wüstenbildung in Gegenwart und Vorzeit* was probably the most important textbook on desert geomorphology to appear in any language before the First World War. It has recently been translated into English (Walther 1997). As he wrote (p. 96):

> Innumerable small or larger fissures are created in the desert under the influence of sun rays.... If a block is heated up to 60–80° C during the day and cools down slowly at night, its outer zone will expand and contract in constant change, and it will depend on the nature of the rock whether this differential expansion causes the splitting off of only a few mm or 10 cm thick sheets.

Walther was especially struck by the disintegrated state of granites in Sinai (pp. 99–100):

> The differently colored constituents heat up at different rates under the influence of insolation and after sunset they contract again at a different speed. Thereby the rock bond gradually becomes looser and the same granite that is considered a symbol of firmness in a humid climate disintegrates easily into its elements in the desert.

By the 1920s, there was a wide degree of agreement between American, British and German geomorphologists that insolation weathering was a potent force of rock disintegration. Since then, however, doubt has been cast on the effectiveness of this mechanism (Schattner, 1961). Most notably, Blackwelder's (1925) experimental simulations failed to produce rock disintegration when heating and cooling cycles were applied in the absence of moisture. Equally, Griggs (1936) used photomicrography to look at samples that had been subjected to simulated insolation weathering and found no change. He averred (p. 796) that 'the effect of temperature changes over a thousand years is not sufficient to cause any exfoliation or disintegration of granite'. His simulations, however, produced breakdown when water was present in the equivalent of 'two and a half years'. Second, supposedly characteristic forms, such as spheroidal weathering, have been found under debris where temperature fluctuations are modest (Barton, 1916) and under tropical wet conditions, while exfoliation, a possible product of insolation, is much rarer in deserts than is often proposed (Blackwelder, 1925). Third, rock disintegration can be achieved by many other processes which do not involve chemical alteration of rock minerals, including unloading (pressure release), salt crystallization and hydration or moisture expansion. Fourth, the study of places where weathering appears severe often reveals a local environment with high moisture availability (Barton, 1916). Fifth, moisture (rain, fog, dew, atmospheric humidity and groundwater seepage) is more prevalent than has often been assumed and may play a potent role in weathering (see Section 2.5).

By the 1960s, many geomorphologists denied that insolation was effective. More recently, however, the process has received some rehabilitation, partly because of new laboratory simulations (Aires-Barros, 1977; Vargas et al., 2012), re-evaluations of the limitations of earlier simulations (Rice, 1976) and field observations (Ollier, 1963). Important in this rehabilitation has been the observation in a range of different deserts that rock cracks show a preferential orientation that correlates with aspect-related differences in solar energy receipt (McFadden et al., 2005; Moores et al., 2008). Moreover, engineering and ceramic studies have shown that a threshold value for thermal shock approximates to a rate of temperature change of 2°C min^{-1}, and data-logger studies show that such rates can occur in deserts (Hall and André, 2001; Molaro and McKay, 2010). In addition, fracture patterns observed on rock in cold, dry environments show very similar forms to those produced in thermal shock experiments (Hall, 1999). The ground monitoring of acoustic emissions from strained rocks may also help to indicate the fracturing of rocks in response to extreme temperature cycling (Swami, 2011). It is also important to remember that rocks exposed at desert surfaces will have experienced a stress history that will have weakened them and made them susceptible to insolation attack. Sources of pre-stressing include subjection to salt and frost, pressure release on exposure at the land surface and various types of tectonic stress (Warke, 2007; Smith, 2009, pp. 74–75; Viles et al., 2010).

Because of the temperature response of calcite crystals, marble seems to be especially prone to thermal degradation (Royer-Carfagni, 1999). On heating, calcite expands much more in the direction of its optical axis than perpendicular to it. Indeed, calcite may, when heated, expand along one axis and contract along another. This leads to particular stresses between contiguous grains (Logan, 2004). Experimental simulations combined with scanning electron microscopy (SEM) and Grindosonic studies have demonstrated that marbles decline in strength when heated and cooled (Goudie and Viles, 2000). Thermal stresses in marble may contribute to the formation of corestones and grus (Eppes and Griffing, 2010) and increase the porosity and number of fractures so that chemical weathering processes may become more active (Luque et al., 2011).

Recently, data loggers have become available for monitoring ground surface temperatures over long periods at regular and frequent intervals, and this has added greatly to our ability to understand weathering processes. Viles (2005) used them in a 100 km transect from the coastal plain inland across the central Namib over a three-year period. She was able to compare air and rock surface temperatures every three hours, and found at Ganab, for example, that there were at least fifty days in most years having maximum rock surface temperatures greater than 50°C and occasional days when rock surface temperatures fell below 0°C. She also did a short study in which data were collected every minute over a twenty-four hour cycle at Gobabeb on two rock blocks (marble and granite). Neither block experienced changes in temperature of 2°C or greater per minute, which is often regarded as the threshold rate for the production of thermal shock damage (Hall and André, 2001). Such thermal shock might, however, occur if rocks are cooled by sudden precipitation events (Viles, 2005; Sumner et al., 2007). This could help to explain the measurable decomposition (desquamation) that occurred on a granite boulder at Gobabeb when it was sprayed with water (Besler, 1979).

In coming decades, it is likely that ground surface temperature cycles will be monitored at high spatial and temporal resolutions all over the world through the use of various satellite-borne sensors (see, for example, Pinker et al., 2007; Trigo et al., 2008; Wan, 2008). This will enable great strides in our understanding of temperature-related weathering processes.

A physical weathering mechanism that is difficult to categorise in that it may involve thermal expansion, wetting and drying, calcrete accumulation and salt weathering is a process that Ollier (1965) termed ‘dirt cracking’. He argued that if a rock contained some cracks, dirt (e.g. windblown silt) would accumulate in them. He suggested that if differential thermal expansion between the core and outside of the rock mass caused the cracks to alternately open and close, then dirt particles would penetrate more deeply into the crevices when the crack opened. The particles then would prevent the crack from closing to its former position, putting the rock under strain. In addition, the dirt particles themselves would be affected by expansion and contraction, possibly in response to changes in moisture. Such would be the case for any expansive clays or salts. Crevices might also suffer strain because of the buildup of calcrete (see also Rothrock, 1925; Young, 1964). Dirt cracking probably does not initiate fractures but rather magnifies existing ones. A recent study of dirt cracking (Dorn, 2011) largely supports Ollier’s proposals.

2.3 Fire

Fire produces high and rapid temperature changes in rock and so leads to cracking and spalling. Bush fires are frequent in many semi-arid environments, particularly where resinous shrubs and trees are involved. Large tracts of Australia's dry interior, for example, are burnt with a reasonably high level of frequency, and fire scars are readily identified on satellite images (Strong et al., 2010). Between 1998 and 2004, almost 27 per cent of the region was burnt at least once (Turner et al., 2008), and Greenville et al. (2009) suggested that the mean minimum return interval in part of the Simpson Desert was twenty-six years. Fire return intervals in the Californian chaparral range between c ten and thirty years (Brunelle et al., 2010). Unfortunately, we have very sparse data on long-term fire frequencies and fire temperatures for most drylands, and current frequencies may be unrepresentative because of factors such as changing human impacts, including the proliferation of introduced annual weeds (Glenn and Finley, 2010). It appears, however, that because of a lack of vegetation cover, fires are unimportant in truly arid ecosystems, unless a run of wet years causes the explosive growth of grass. Equally, they are rare in very humid regions (van der Werf et al., 2008). The temperatures attained in grass or shrub fires may be less than those obtained in great forest fires, and this may restrict their importance as weathering agents. In the semi-arid mallee shrublands of Australia, soil surface fire temperatures were only c 60–120°C (Bradstock et al., 1992). However, values of between 538 and 1100°C have been recorded in chaparral fires (Allison and Goudie, 1994, table 3.1). Blackwelder (1926) attributed rock spalling in forested areas of the south-western United States to fire, something confirmed by Dorn's (2003) work in Arizona. Ollier and Ash (1983) and Spath (1977) demonstrated in Australia that fires cracked granodiorite outcrops. Laboratory experiments have also pointed to the power of fire to cause rock breakdown (Goudie et al., 1992; Allison and Goudie, 1994). Fires may also prepare rocks for subsequent attack by other weathering processes, such as salt (McCabe et al., 2007).

2.4 Frost Weathering

Frost action is present today in high-altitude deserts where temperatures regularly fall below freezing. On the southern Colorado Plateau, for instance, there are 100–140 days a year when the freeze-thaw boundary is passed (Laity, 2008, p. 125). Great screes on valley sides in the Karakoram Mountains of Pakistan may result from rock falls produced by frost attack of bare slopes. It is necessary, however, for freezing temperatures to occur at times when moisture is present, so the process is most likely to be effective in areas with a winter rainfall regime. There is also a high probability of more effective frost action during cooler, moister periods of the Quaternary. Tricart and Cailleux (1969) believed there was a fundamental difference between hot deserts and those with winter frosts. They believed that landform evolution was basically very slow in the former and that many landforms were inherited from past climatic conditions. In the latter they believed that both now, and in an even more intensive manner during glacial periods, frost weathering was a predominant cause of rock breakdown and slope evolution.

There is some experimental evidence to support the idea that the presence of salts in solution – a common feature of desert weathering environments – may either inhibit (if salt supply is limited) or enhance (if salt supply is plentiful) frost damage (McGreevy, 1982), depending on whether the minimum rock temperature falls below the appropriate eutectic point (Jerwood et al., 1987).

2.5 Wetting and Drying Weathering

There is empirical evidence that wetting and drying (hygric weathering) of rocks such as shales (Goudie et al., 1970; Fahey, 1986) and marbles (Koch and Siegesmund, 2004) can cause them to disintegrate. Ravina and Zaslavsky (1974) argued that cracking could be produced by pressures arising from the very high electrical gradients in the water of the electrical double layers at rock/water interfaces. The pressures are said to vary with relative humidity and diurnal temperature ranges, both of which are often high in deserts.

Also of importance, however, is the disintegration of susceptible rocks or loosely consolidated material following either the introduction of water or exposure to the atmosphere. This can cause shrinkage and swelling (Weiss et al., 2004) and is called slaking. Clays (especially smectites) and mudrocks are prone to this form of failure, especially in the presence of saline waters. Tests are available for determining the durability of slaking-prone materials (Czerewko and Cripps, 2001). Materials with high Exchangeable Sodium Percentages (ESP), including some colluvia, may be susceptible, and slaking is an important process on many badland surfaces. It is possible that wetting and drying, salt dissolution and re-precipitation can work together, particularly where mudstones have an original salt content associated with their environment of deposition (Cantón et al., 2001). Wetting and drying weathering has applied implications. The opening of tombs in the Valley of Kings at Thebes in Egypt has exposed them to floods and to humidity changes that have caused destruction by repeated swelling and shrinkage of the Esna Shales (Wüst and McLane, 2000).

In spite of their aridity, desert surfaces may be wetted by rainfall, fog dew or fluctuating groundwater on more occasions that one might intuitively think. A potentially very important source of moisture for weathering process – particularly in coastal deserts – is fog, as in Morocco (Lekouch et al., 2010). Fogs are also frequent in the Namib (Figure 2.3), although their significance is reduced inland (Table 2.3). A zone of high fog frequency ($>$50 fog days per year) 'hugs the coast over almost the entire length of the Namib' (Olivier, 1995, p. 132), although the highest frequency

Table 2.3 *Fog data for the central Namib Desert*

Station	Mean annual fog precipitation (mm)	Mean annual days of precipitating fog	Mean quantity of precipitation per foggy day (mm)
Floden Moor	65.13	55.50	1.17
Ganab	2.67	2.76	0.97
Gobabeb	30.79	37.23	0.83
Narabeb	35.91	38.45	0.93
Rooibank	80.19	75.64	1.06
Swakopmund	33.94	64.68	0.52
Swartbank	183.62	86.71	2.12
Vogelfederberg	183.48	77.37	2.37
Zebra Pan	15.11	16.00	0.94

occurs near Walvis Bay, with nearly 140 fog days per year. Inland from Walvis Bay and Swakopmund, the number of fog days is reduced to c 40 within the first 40 km and to c 10 at a distance of 100 km inland. The amount of fog precipitation exceeds rainfall at the coast and mean annual fog precipitation (Lancaster et al., 1984, table 5) amounts to c 34 mm at Swakopmund, rising to c 183 mm at some of the inland hills and then declining to around 3–15 mm at a greater distance inland (e.g. at Ganab

Figure 2.3 In coastal deserts, such as the central Namib, frequently precipitating fogs contribute to weathering. (ASG)

and Zebra Pan). Shanyengana et al. (2002) provide data on daily fog precipitation amounts for three sites along the Kuiseb Valley, again demonstrating the considerable quantities of fog that may be deposited on exposed surfaces: 3,308 ml/m^2 at Klipneus, 2,390 at Swartbank and 508 at Gobabeb. Moisture may be intercepted preferentially by lichen-covered surfaces (Maphangwa et al., 2012). Not all fog days precipitate fog moisture, but data on fog precipitating days per year is given by Lancaster et al. (1984), with 64.7 for Swakopmund, 86.7 for Swartbank and 2.8 for Ganab.

In the Atacama of Chile, some ridge sites in close proximity to the Pacific have as many as 189 fog days in the year (Cereceda and Schemenauer, 1991), although the quantity of fog water that is deposited falls off sharply as one moves inland (Cáceres et al., 2007), with a flux of about 8.5 l m^{-2} per day at the coast, and 1.1 l m^{-2} at 12 km inland (Cereceda et al., 2002). Fog clouds are important up to elevations of 1,000 m above sea level (Cereceda et al., 2008). Westbeld et al. (2009) estimated that in the coastal Atacama, annual fog water deposition amounted to about 25 l m^{-2}, which is an appreciable quantity.

In some desert areas, appreciable amounts of dewfall may occur on a surprisingly large number of nights in the year (Agam and Berliner, 2006) and may therefore provide conditions for salts to hydrate and for hygroscopic salts to take up moisture (Davila et al., 2008). In south-west Morocco, 178 dew events were recorded in a year and produced almost 19 mm of moisture (Lekouch et al., 2010). Henschel and Seely (2008, p. 364) stated that dew in the Namib 'occurs very frequently during most of the year'. They report that in 2001, dew occurred during fifty-three nights at Gobabeb. Ashbel's (1949) pioneering work in Palestine demonstrated that some stations received dew on more than 200 occasions in the year (Table 2.4) and that amounts of dewfall could exceed 100 mm per year. Recent observations of dews in the Negev Desert have been made by Zangvil (1996) based on observations at Sede Boker. The total amount of moisture deposited each year averages about 17 mm, and the number of dew hours per night averages between around 2.5 and 5 hours, depending on season. The number of dew events ranges between about eight and twenty-six days per month, with minimum numbers in April and maximum numbers in September. The amount of dew deposited depends on a number of local factors, including aspect, slope angle and altitude (Kidron, 1999, 2005), so that dewfall amounts are highly variable even across small areas (Kidron et al., 2000). Dew also feeds lichens (Termina and Kidron, 2011) and so may contribute to biological weathering.

In addition to fog, rain and dew, another important source of moisture can be groundwater discharge. In Namibia this is particularly true of the coastal zone to the north of Swakopmund, where there are large areas of moist ground that are readily identifiable on satellite imagery. Groundwater discharge is also evident at some localities further inland, however (Day and Seely, 1988; Brain and Koste, 1993).

Table 2.4 *Number of dew nights for selected stations in Palestine*

Station	Lat N	Total for year
Gevulote	31°13′	218
Beth Eshel	31°15′	135
Dorot	31°31′	182
Negba	31°40′	134
Sarafand	31°57′	200
Kefar Masaryk	32°54′	255
Nahalal	32°41′	242
Carmel	32°48′	176
Kefar Etsyon	31°39′	131
Jerusalem	31°47′	113
Tirat Tsevi	32°25′	162
Dan	33°14′	240
Amman	31°57′	70

Source: Ashbel (1949, table 1).

If one considers the combined occurrence of rain, fog precipitation and dewfall, it is apparent that in the central Namib moisture is deposited on rock surfaces on about 40 per cent of the days of the year (Henschel and Seely, 2008, p. 364). The Namib also shows large diurnal swings in relative humidity, which are important in terms of salt crystallization and hydration cycles. Monitoring by Viles (2005) indicates that at Vogelfederberg, an inselberg in the central Namib, there may be more than days in the month when relative humidity values cross various crucial salt crystallization thresholds.

2.6 Salt Weathering

It has been known since antiquity that salt attacks rock (Figure 2.4) and building materials (Goudie and Viles, 1997), but major developments in the study of salt weathering only began at the end of the nineteenth century (see Evans, 1970 for a review). Mortensen (1933) developed an equation that enabled the calculation of salt hydration pressures, and an influential contribution to the theory of salt crystallization was made by Correns (1949). A renewed spasm of interest arose beginning in the 1950s.

It has become clear that salt weathering comprises a range of mechanisms, some chemical and some mechanical (Table 2.5). Experimental studies have demonstrated that, of the mechanical processes, crystal growth from solution in rock pores and cracks is the most important (Evans, 1970; Goudie, 1974; Sayward, 1984). This is promoted by a decrease in solubility as temperature falls, by evaporation of solutions or by

Table 2.5 *Mechanisms of salt attack of rocks and building materials*

Physical changes
Crystallization
Hydration
Thermal expansion
Electrical slaking and double layer effects associated with hygroscopicity
Chemical changes
Silica mobilization under alkaline conditions
Etching of calcite under acid conditions
Changes to concrete mineralogy (e.g. ettringite formation)
Corrosion of incorporated iron and steel
Moisture-related chemical weathering associated with hygroscopicity
Gypsum/silicate replacement

mixing of different salts in solution (the common iron effect). Data on crystallization pressures are given in Winkler and Singer (1972).

For example, some salts rapidly decrease in solubility as temperatures fall (Ruiz-Agudo et al., 2007). This is particularly true of sodium sulphate, sodium carbonate, magnesium sulphate and sodium nitrate, although not to the same degree as calcium sulphate and sodium chloride. Thus nocturnal cooling could cause salt crystallization

Figure 2.4 Raised beach gravels near Iquique in the Atacama of Chile are cemented by salt, which has also caused splitting to occur. Beer can for scale. (ASG)

Table 2.6 *Hydration volume increase for selected common salts*

Salt	Molecular weight	Hydrate	Formula weight of hydrate	Density	Density of hydrate	Volume change (%)
Na_2CO_3	106.00	$Na_2CO_3 \cdot 10H_2O$	286.16	2.53	1.44	374.7
Na_2SO_4	142.00	$Na_2SO_4 \cdot 10H_2O$	322.20	2.68	1.46	315.0
$CaCl_2$	110.99	$CaCl_2 \cdot 2H_2O$	147.03	2.15	0.84	241.1
$MgSO_4$	120.37	$MgSO_4 \cdot 7H_2O$	246.48	2.66	1.68	223.2
$CaSO_4$	136.14	$CaSO_4 \cdot H_2O$	172.17	2.61	2.32	42.3

Source: Modified after Goudie (1977, table VI).

to occur. Such crystallization of a salt solution on a temperature fall affects a much larger volume of salt per unit time than crystallization induced by evaporation, which is a more gradual process (Kwaad, 1970). Nevertheless, evaporation does help to create saturated solutions from which crystallization can occur, and when this happens highly soluble salts will produce large volumes of crystals.

Disruptive stresses may also be exerted by anhydrous salts, dehydrated in high desert temperatures, which become hydrated from time to time. As a change of phase takes place to the hydrated form, water is absorbed. This increases the volumes of the salt and thus develops pressure against pore walls. The volume increases for some common salts are given in Table 2.6, where it can be seen that the change may be appreciable, with sodium carbonate and sodium sulphate both undergoing a volume change in excess of 300 per cent as they hydrate. These exert a force on the walls of the containing fissure or pore when they are wetted (Winkler and Wilhelm, 1970). Field evidence of the effects of hydration is limited (although see Smith and McAlister, 1986). Stresses created by the hydration of entrapped salts can be repeated many times seasonally or perhaps diurnally, depending on environmental conditions and the salts involved. For some salts, a change of phase occurs at the temperatures encountered widely in nature; sodium sulphate's transition temperature is 32.4°C for a pure solution and falls to 17.9°C in a NaCl saturated environment. Moreover, for some salts the transition may be rapid. At 39°C the transition from thenardite to mirabilite may take no longer than twenty minutes (Mortensen, 1933).

The number of occasions on which surface temperatures cycle across the temperature thresholds associated with the change of phase is probably substantial. If one assumes that an air temperature of c 17°C translates into a rock surface temperature of c 32°C (the transition temperature for sodium carbonate and sodium sulphate), then that value is crossed daily between five and nine months of the year depending on the station selected. In other words, in a typical desert there may usually be 150 to 270 days in the year in which rock temperatures favour salt hydration.

Table 2.7 *Coefficients of volume expansion between 20°C and 100°C*

Salts	
Halite	0.963
$NaNO_3$	1.076
Gypsum	0.58
Rock Minerals	
Calcite	0.105
Microline	0.128
Olivine	0.21
Orthoclase	0.049
Plagioclase	0.14
Quartz	0.36

Source: Skinner (1966).

Many common desert salts have high coefficients of volumetric expansion that are greater than those of common rocks such as granite (Cooke and Smalley, 1968) (Table 2.7). Until recently, the salt expansion process seemed only to be a theoretical probability, but empirical observations of cliff weathering under different conditions of insolation (Johannessen et al., 1982) pointed strongly towards its effectiveness.

In addition to these three main categories of mechanical effects, salt causes chemical weathering (Schiavon et al., 1995), and experimental studies using a range of salts and SEM showed that salts could cause chemical etching of quartz grains after only short periods of exposure (Magee et al., 1988).

Some saline lakes (although by no means all) are very alkaline and have elevated pH levels (Table 2.8). This is significant because silica mobility increases at pH values greater than 9, due to the ionisation of H_4SiO_4. At the other extreme, highly acidic saline lakes are known, as in parts of Australia, and can have pH values that fall as low as 1.7 (Bowen and Benison, 2009). The presence of sodium chloride may also affect the solubility and reaction velocity of quartz solutions. The growth of crystals may cause pressure solution of silicate grains in rocks, for silica solubility increases as pressure is applied to it. The attraction of moisture into the pores of rocks or concrete by hygroscopic salts (e.g. sodium chloride) can also accelerate the operation of chemical weathering. Salts have an adverse impact on engineering structures made of concrete and iron reinforcements through corrosion of iron and sulphate reactions with cements to produce expansive minerals such as ettringite and thaumasite (see Section 6.5).

Desert climates promote surface and near-surface evaporation and saline concentration of groundwater as well as capillary migration of water above the water table. The capillary fringe may intercept the surface to create a zone of efflorescence and crusting and cause salt weathering in features that rise out of the surface – the so-called wick

Table 2.8 *pH values for selected alkaline lake waters*

Lake	pH
Pyramid, Nevada	9.4
Hayq, Ethiopia	9.0
Van Gulu, Turkey	9.9
Corangamite, SE Australia	8.2–9.1
Lagoons, Tasmania	9.1–9.5
Kivu, Africa	9.5–9.7
Turkana, Kenya	10.9–11.0
Wadi Natrun, Egypt	8.5–11.0
Shala, Ethiopia	7.4–7.7
Great Salt Lake, Utah	7.4–9.6
Harney, Oregon	9.8
Abert, Oregon	9.8
Mono, California	10.0
Eastern Mongolia	10.0
Beloye, Russia	8.3–9.2
Alchichica, Mexico	9.0

effect (Goudie, 1986). Salty and foggy deserts may be especially prone to salt attack, as in the Namib (Goudie et al., 1997) and Atacama (Goudie et al., 2002). In areas such as the Arabian Gulf, sabkhas (intertidal flats) are loci of extensive salt accumulation (Fookes et al., 1985) and are among the most humid and wettest locations in deserts in terms of runoff, seawater and groundwater. Thus, they are significant regions of salt weathering.

The response of rocks varies with their properties. Experiments (e.g. Goudie et al., 1970, Smith et al., 1987; Angeli et al., 2008) suggest that of these properties, porosity, and particularly the proportion of micropores smaller than 5μm (Yu and Oguchi, 2009), are important. Experiments have shown that igneous and metamorphic rocks are more resistant than sandstones or limestones (Goudie, 1974). The combination of properties that make some rocks resistant to salt attack is complex, however (Goudie, 1999a). It is also interesting to note that the salt resistance of rocks is not the same as for frost (Goudie, 1999b), and that ranking the effects of different salts can give different results for different rock types (Yu and Oguchi, 2009).

Salts may be derived from various sources (Table 2.9) and subjected to a wide range of influences (Ewing et al., 2006): saline aerosols; dust (Shiga et al., 2011) or volcanic gases; dried-out lakes and lagoons; exposed fossil salt deposits; groundwater, which in turn derives its salt either from rainwater solutions or stratified salt deposits; subsurface penetration of saline seawater into coastal flats; and chemical weathering of rocks. The types and mixtures of salts present in any location (Goudie and Cooke, 1984) are variable. Calcium carbonate (especially in semi-arid areas), calcium sulphate (mainly

Table 2.9 *Some influences on salt lake chemistry*

Inputs
- Seawater
- Dust
- Rainfall
- Volcanic emissions
- Solutions derived from weathering of bedrock and ancient evaporites
- Submarine gas emissions (e.g. H_2S)

Changes within basin
- Cationic exchanges with clays
- Organic influences (e.g. algae, bacteria)
- Temperature changes
- Humidity changes
- Chemical reactions
- Tectonic changes
- Vertical leaching processes
  - (i) Affected by climate
  - (ii) Topographic position
- Deflational removal
- Evaporative concentration
- Capillary migration

in arid regions) and sodium chloride (chiefly in coastal locations) are common, but many other types occur.

The weathering effectiveness of the salts varies with their properties, such as solubility. Experiments (Table 2.10) (Goudie, 1993b; Yu and Oguchi, 2009) suggest that

Table 2.10 *Experimental ranking of the power of different salts (most effective salt at top of column)*

Pedro (1957)	Kwaad (1970)	Goudie et al. (1970)	Goudie (1974)	Goudie (1974)	Goudie (1986)	Goudie (1993b) (Wadi Digla Cycle)	Goudie (1993b) (Negev Cycle)	Yu & Oguchi (2009)
$NaNO_3$	Na_2SO_4	Na_2SO_4	Na_2SO_4	Na_2SO_4	Na_2CO_3	$NaNO_3$	Na_2SO_4	Na_2SO_4
Na_2SO_4	Na_2CO_3	$MgSO_4$	$MgSO_4$	Na_2CO_3	$MgSO_4$	Na_2CO_3	Na_2CO_3	$MgSO_4$
$Mg(NO_3)$	$MgSO_4$	$CaCl_2$	$CaCl_2$	$NaNO_3$	Na_2SO_4		$NaNO_3$	Na_2CO_3
K_2SO_4	NaCl	Na_2CO_3	Na_2CO_3	$CaCl_2$	NaCl		$MgSO_4$	NaCl
KNO_3	$CaSO_4$	NaCl	$NaNO_3$	$MgSO_4$	$NaNO_3$		NaCl	$Al_2(SO_4)_3$
Na_2CO_3				NaCl	$CaSO_4$			
K_2CO_4								
$MgSO_4$								
$CaSO_4$								
$Ca(SO_3)_2$								

by far the most effective (in isolation) are sodium sulphate, sodium carbonate and magnesium sulphate (see also Balboni et al., 2011). Equally, the mixture of salts appears to be important in determining weathering effectiveness (Robinson and Williams, 2000). The principal reason for the varying effectiveness of salts lies in their crystallization and hydration properties. Sodium sulphate, for example, is a particularly effective agent: hydration of thenardite (Na_2SO_4) to mirabilite ($Na_2SO_4.10H_2O$) involves volume changes; its solubility is relatively sensitive to temperature change, being more than five times greater at 40°C than at 10°C; it is highly soluble (49 g/100 g at 40°C), so that substantial quantities are available for crystal growth when saturated solutions are evaporated; and its crystal habits are prismatic or acicular, so that forces of crystal growth are concentrated along one axis and are thus likely to be more effective in promoting disintegration in pores.

Sperling and Cooke (1985) showed that hydration of sodium sulphate in laboratory simulations was capable of breaking up rocks independently of, but not as successfully as, the process of crystal growth. The power of sodium sulphate has recently been stressed and reappraised (Rodriguez-Navarro et al., 2000), and it has become apparent that it is the crystallization of mirabilite from highly supersaturated solutions that is the crucial mechanism (Tsui et al., 2003; Steiger and Asmussen, 2008) rather than the hydration pressure caused by the increase in molar volume as the thenardite transforms to mirabilite (Scherer, 2004). Sodium sulphate is especially effective at causing rock scaling, whereas magnesium sulphate, because of its crystallization behaviour, is more likely to cause cracking (Ruiz-Agudo et al., 2007).

Sodium nitrate is another potentially effective salt (Goudie et al., 2002); it is highly soluble in comparison with many others. At 35°C, its solubility in water is 49.6 per cent, whereas that of sodium sulphate is 33.4 per cent, of sodium chloride 26.6 per cent and of calcium sulphate just 0.21 per cent. Thus large volumes of the salt can be precipitated within rock pores, where it can cause severe rock disintegration and brecciation (Searl and Rankin, 1993). Sodium nitrate is also hygroscopic (it can absorb moisture from the air) and deliquescent (it can dissolve in that moisture). When the relative humidity exceeds a critical level, water condenses on the surface of the mineral and dissolves it in that water. Then, if the critical relative humidity decreases below that critical level, the nitrate will effloresce or crystallize again. Thus cycling across this critical relative humidity is a major control of the frequency of salt crystallization cycles in the salt weathering process.

In general there appears to be some discrepancy between the efficacy of particular salts as revealed by field observations and their efficacy in laboratory simulations. Sodium chloride has been shown to cause rapid rock block disintegration in the field (Goudie and Watson, 1984; Goudie et al., 1997) but is generally not very effective in laboratory simulations. An exception to this was an experiment to simulate the role of sodium chloride in a foggy, coastal environment, which showed that when

Table 2.11 *Examples of the geomorphological effects of salt weathering*

Cliff sapping	Jutson (1918); Laity (1984)
Cobble development	Beaumont (1968); Kirchner (1996)
Desert varnish suppression	Dragovich (1994)
Enhancement of deflation potential	Tricart (1953)
Ground heaving and pseudo-anticline formation	Watts (1977); Horta (1985)
Loess formation and silt generation	Goudie et al. (1979); Wright (2001)
Pan formation and development of 'billiard-table' rock floors	Goudie and Thomas (1985); Aref et al. (2002)
Pedestal rocks	Cooke (1981)
Planation surfaces (haloplanation)	Abele (1983)
Rock meal formation	Wellman and Wilson (1965); Goudie and Viles (1995)
Tafoni development	Bradley et al. (1978); Abu Ghazleh and Kempe (2009)
Tor formation	Watts (1981)

critical amounts of fog moisture were applied, sodium chloride could cause rapid breakdown of rock samples (Goudie and Parker, 1998). It is also possible, as Gómez-Heras and Fort (2007) have pointed out, that many experimental simulations with sodium chloride have used convective heating, and that this has produced modest weathering, whereas experiments that have used radiative heating have been more effective. The reason for this discrepancy is that the latter treatment produces damaging subefflorescence rather than the less damaging efflorescences.

Recently, evidence has mounted for salt weathering's diverse geomorphological roles (Table 2.11). It has been held responsible for rock flaking, granular disintegration, splitting and for the production of silt-sized particles that may subsequently be deflated and then deposited as loess (Goudie et al., 1979, Smith et al., 1987; Goudie and Viles, 1995). It is probably crucial in forming cavernous weathering features (e.g. tafoni) and rock pedestals. Mass segregation of salts may create mounds and ground heaving (haloturbation). Salt weathering may create pitted cobbles and help the disintegration of alluvial fan gravels in Death Valley (Figure 2.5) (Goudie and Day, 1980; Butler and Mount, 1986) and the generation of debris in the Karakorams of Pakistan (Goudie, 1984b) and around Tunisian sabkhas (Goudie and Watson, 1984). It is certainly a major process of debris comminution. It may also help pan development in southern Africa (Goudie and Thomas, 1985) and Egypt (Aref et al., 2002). It is also likely that salt weathering is important in preparing bedrock for deflation. Woodward (1897, p. 365) recognized this in Australia and talked of 'salt wind-planed flats':

> The salts deposited upon these surfaces consist very largely of gypsum, which upon crystallizing out in the clay causes it to split up, thus allowing the weather to act upon it, which quickly

Figure 2.5 Alluvial fan gravels broken up by salt shattering at Badwater, Death Valley, USA. (ASG)

converts it into an almost impalpable powder, which the wind distributes far and wide over the surface of the country.

This was a view also adopted by Jutson (1918) to account for the development of over-steepened cliffs ('breakaways') and flat ('billiard-table') rock floors. He also believed that salt weathering helped to detach small particles, and that they might also be lifted up by crystal growth so that they would be exposed to wind removal. Jutson (1950) provided an impressive plate of a greenstone cliff being thus undermined at Yelladine Road in the Yilgarn Goldfield. More recently, Clarke (1994) has drawn attention to the role of halite efflorescences in causing cliff sapping in felsic and sedimentary rocks on the north and west shores of lakes in the same area.

Salt weathering can be rapid, as shown by laboratory simulations (Fahey, 1983), field observations and monitoring (Goudie, 1985b; Smith and McAlister, 1986; Goudie et al., 1997) and studies of the decay of ancient (Goudie, 1977) and modern engineering structures (Doornkamp et al., 1980; Bulley, 1986). Indeed, salt weathering has a fundamentally important applied context, because it damages buildings; roads; cultural heritage, including paintings and frescoes (Moussa et al., 2009); and engineering structures, especially as a result of groundwater ingress. This problem has been studied extensively (Fookes and Collis, 1977; Cooke et al., 1982; Horta, 1985). Further details are given in Section 6.5.

2.7 Chemical Weathering

Although chemical weathering has often been judged as relatively unimportant in deserts, moisture may be available to play a role in rock decomposition. As we saw in our discussion of wetting and drying weathering, desert rainfall occurs on a greater number of occasions and in a less intense way than is often supposed, coastal deserts receive moisture from fogs and moisture from dew can dampen rock outcrops with great frequency (Evenari et al., 1982). Appreciable moisture also occurs in specific geomorphological situations as in proximity to rivers, springs, playas, sabkhas and cliff-base seeps.

In mature, arid soils there may be different clay minerals from those found in humid soils, for there is less leaching, but it is not easy to distinguish between the effects of climate and those of parent material, amount and character of added dust, topographic position and environmental history. When these other factors can be controlled, it appears that there is more smectite (montmorillonite), illite, chlorite and palygorskite in arid than in humid soils (Reheis, 1990). Some palygorskite may be inherited from dust, itself derived from playas and sabkhas where it forms in large quantities (Coudé-Gaussen et al., 1984), but palygorskite may also be an authigenic transformation of montmorillonite in the presence of an enhanced supply of magnesium and in moderately saline (high pH) and perhaps temporarily waterlogged or brackish conditions (Singer, 1979). In any event, palygorskite has been widely reported from arid areas, not least in calcretes (Bouza et al., 2007). The presence of smectite is also common, and Singer (1979) demonstrated very clearly that on basaltic soils in Israel, smectite formed about 70 per cent of the clay fraction where the mean annual rainfall is around 350 mm, but that this had fallen to less than 20 per cent where the mean annual rainfall was around 750 mm. Illite is relatively rare in the arid soils of the Middle East and North Africa (Singer, 1988), but it is important in central Asia.

In the Sahara, across a transect from 35°N to 19°N, Paquet et al. (1984) identified four sectors. In northern Algeria, illite and chlorite accounted for around 7–75 per cent of the clay content, kaolinite about 15 per cent and attapulgite 10–15 per cent. Further south, around Beni Abbes and In Salah, attapulgite reached levels of 20–25 per cent. Still further south, illite and chlorite were dominant (60–70%), attapulgite was only 5–10 per cent and kaolinite 20–25 per cent. South of the Hoggar and in the Tanezrouft, smectites were dominant, followed by kaolinite (20–25%), illite (10–25%), attapulgite (10–15%) and chlorite (5%). Paquet et al. attributed this variability to the nature of the Quaternary sediments and bedrocks of the sectors concerned. For example, the sediments of the northernmost zone gain some of their characteristics as a result of deflation of inland basins (chotts), whereas high kaolinite contents may be derived from Tertiary weathering profiles.

In the south-west United States, smectite is the main clay mineral on alluvial fans in the western Mojave (Eghbal and Southard, 1993), and a general survey across a range of Californian climatic zones (Graham and O'Geen, 2010) showed that although kaolinite was almost ubiquitous, palygorskite, smectite and vermiculite were widely present in hot, dry locations.

2.8 Weathering by Biological Agencies

Lichens, cyanobacteria, fungi, bacteria and green algae occur widely in deserts, where they create weathering features (e.g. Krumbein, 1969; Friedmann, 1980; Danin et al., 1982; Viles, 2008). Lichens are especially widespread in areas with dew (Termina and Kidron, 2011) and in foggy coastal deserts, unless the rock surfaces are rendered too unstable by salt attack or sandblasting (Viles, 1995). In general, these organisms embroider or modify the micro-morphology of rock surfaces. Chemical processes include the aqueous dissolution of respiratory carbon dioxide, formation of water-soluble lichen compounds which are able to form soluble metal complexes, alkalization (Büdel et al., 2004) and excretion of oxalic and other acids by growing lichens (Jones et al., 1980). Of these processes, the last appears to be effective in etching feldspar, ferromagnesian minerals and quartz and in degrading hydrothermally altered minerals (Hallbauer and Jahns, 1977; Jones et al., 1980). Biophysical processes are associated with the volume changes in thallial tissues, which contract on withdrawal of water and under certain circumstances exert a pressure on the rock to which they are attached. Disintegration may also be caused by the wetting of thalli lodged in confined spaces, which swell and may exert sufficient pressure on the rock for it to fracture.

Endolithic organisms may dissolve limestones. Working in Israel, Danin et al. (1982) showed that dissolution tends to be greatest where two lichen colonies meet. Here, grooves 0.10–0.2 mm deep are formed at the line of contact and make a complex fretwork pattern. This is revealed when the lichens die or are removed. They also observed weathering pits that appeared to be formed by cyanobacterial colonies. It needs to be remembered, however, that lichens can in some situations protect rock from wind abrasion (Langston and McKenna Neuman, 2005), change its thermal response to heating and cooling stresses and prevent water from making direct contact with the surface (thus limiting dissolution) (Viles and Goudie, 2000).

Recent research by Puente et al. (2004) has suggested that bacteria associated with the roots of cacti produce organic acids that cause the weathering of lavas, marbles and

Figure 2.6 Solutional rillen on marble outcrops in the central Namib probably created by fog precipitation. (ASG)

limestones. They concluded (p. 639) that their study 'showed a massive population of rhizoplane microorganisms on desert cactus roots growing on rocks in the absence of soil. These microbes, together with the roots they colonize, may perform significant weathering of volcanic and sedimentary rocks in a hot desert'. Similarly, Lopez et al. (2012) found that endophytic bacteria of rock-dwelling cacti can promote mobilization of elements from rocks in Mexico.

2.9 Solution Processes and Limestone: Karst

Undoubtedly, solutional karstic phenomena occur in deserts (Smith et al., 2000) despite the fact that lack of water and vegetation might lead to a general expectation that such features would be relatively scarce (Jennings, 1985). Although they are not common, most solutional features, from the micro to the macro, have now been described in deserts, including facetted pebbles with bosses and rims (Bryan, 1929), polygonal cracks and concretionary forms (Soleilhavoup, 2011), hollows and pits, pavements, caves, dolines and dayas (shallow depressions) and tafoni (e.g. Jennings, 1985; Smith, 1987). Solutional flutes and grooves have been reported from Australia (Dunkerley, 1979), the Namib (Sweeting and Lancaster, 1982) (Figure 2.6) and the

Table 2.12 *Selected examples of karst in drylands in the Middle East and North Africa*

Location	Source
Saudi Arabia	Hoetzl (1995)
	Hussain et al. (2006)
	Amin and Bankher (1997)
Egypt, Eastern Desert	Aref et al. (1986)
Egypt, Western Desert	Kindermann et al. (2006)
	Aref and Refai (1987)
	Wanas et al. (2009)
Iran	Atapour and Aftabi (2002)
Israel	Frumkin et al. (2009)
	Buchbinder et al. (1983)
Libya	Halliday (2003)
Kuwait	Shaqour (1994)
Qatar	Sadiq and Nasir (2002)
United Arab Emirates	Fogg et al. (2002)

Sahara (Smith, 1987). Some studies of karst in the Middle East and North Africa are listed in Table 2.12.

Some karst may be inherited from more humid conditions. This may be the case with the Nullarbor karst in Australia, parts of which formed in the mid- to late Tertiary (Webb and James, 2006; Miller et al., 2012). Deep cave systems concurrent with low sea levels formed during the late Miocene and early Pliocene. Some of the karst in Israel (Buchbinder et al., 1983) also formed in the Neogene. Late Miocene karst systems are superbly developed in the Western Desert of Egypt, between Bahariya and Farafra (Wanas et al., 2009). Karst features of tropical morphology have been found under the Red Sea and off the United Arab Emirates and may have formed at times of low sea level when the shelf was exposed to solution under moist conditions (Purkis et al., 2010). It is also possible that karst development becomes greater in semi-arid areas rather than in truly arid areas, and notable karst has been recorded from such areas as the Carlsbad Caverns of New Mexico (Brook et al., 2006), north-east Brazil (Auler and Smart, 2003) and Queensland (Gale et al., 1997). Speleothems provide a good record of former hydrological activity in cave systems (Fleitmann et al., 2011).

In the hyperarid heart of the Sahara there are extensive areas of karst developed in silicate rocks (Busche and Erbe, 1987; Busche and Sponholz, 1992; Baumhauer, 2010). These include sinkholes, uvalas, poljes and caves, and they may be of early Miocene age.

Karst also occurs in and on rock salt outcrops. Under humid conditions such exposures are rapidly dissolved, but in drylands, although solutional attack occurs, it is not always sufficient to cause their complete destruction. Fine examples of salt karst are known from the salt domes of the Zagros Mountains in Iran (Zarei et al., 2012) and also from the Dead Sea area (Bowman, 2010), where there are shafts and caves.

2.10 Dayas

Some of the most extensive types of karst in drylands are dayas. These are 'Depressions not more than a few metres deep, but measuring from tens of metres up to several kilometres in diameter, which are sometimes found scattered over the surface of limestone, chalk or gypsum hamadas' (Mitchell and Willimott, 1974). They tend to occur at densities between 0.88 and 11.44 per km^2 (Goudie, 2010). They act as basins for runoff and also for groundwater recharge, have bottoms filled with silty soil, may have more vegetation than the surrounding landscape and may sometimes be sites of cultivation and grazing. They have ecological importance as wetlands. Their presence on calcareous rocks, including calcretes, suggests that karstic processes may be involved in their development and that they may therefore be a type of doline (Capot-Rey, 1939).

Dayas have been found in many semi-arid locations. Their small size and shallow dimensions may be a consequence of the relative paucity of solutional activity in comparison with more humid, soil-mantled areas. Most dayas are developed on calcretes and/or on relatively young limestones. This may be either because there has been insufficient time for the development of larger features or because the limestones are thin and less massive than classic karstic limestones and so cannot develop deep features.

Dayas have a range of morphologies which can be categorized into a series of types, although they are not mutually exclusive, and in some locations all types may occur:

1. *Simple circular depressions.* These are probably the classic daya and may represent an early stage of development.
2. *Centripetal.* These are dayas which act as a focus for drainage and which are fed by occasional overland flow. Examples of this type are from Bahrain and Namibia and very similar types are illustrated in Melton (1934).
3. *Structurally controlled.* These are dayas where the form, distribution and alignment appear to be related to control by the structural properties (e.g. joints, fractures, bedding planes, etc.) of the rocks in which they are developed. A good example of this is provided by sites in Western Sahara.
4. *Drainage channel.* These are dayas where development appears to have taken place along pre-existing drainage lines. Good examples of this type can be found in northern Kuwait and southern Iraq.

5. *Wind aligned.* These are dayas that show subparallel to parallel alignments that appear to be related to the presence of an actual or stripped dune cover. Examples are known from Algeria, Australia, the Namib and the High Plains of the United States.

The classic area for dayas is North Africa. In the far west, there are large expanses of closed depressions on Palaeogene outcrops either side of the border between Senegal and Mauritania. Further to the north, they occur widely in the west of Western Sahara, on Neogene and Oligocene limestones. They extend in a belt that extends about 200 km southwards from the Moroccan border. In Morocco itself, the Hamada du Guir, surfaced with limestones of probable Pliocene or lower Pleistocene age (Castellani and Dragoni, 1986), has thousands of depressions, some linked up to a network of drainage channels, which have depths of 3–4 m and diameters that range up to 1–2 km. Algeria also has extensive fields of dayas on the Hamada du Guir (Benhouhou et al., 2003) and elsewhere on the piedmont to the south of the Saharan Atlas (Taibi et al., 1999). This particular field of dayas occurs in parallel chains that run from c WNW to ESE. Dayas have also been identified in northern Libya and in the Western Desert of Egypt, between Kharga and Dakhla. In west-central Somalia, west of Gaalkacyo, and in neighbouring parts of south-east Ethiopia, there are many small depressions. Other extensive areas of dayas occur in eastern Kenya, around Wajir and Garissa, where Quaternary limestones are exposed. In the Middle East, dayas are known from Qatar (Babikir, 1986), Bahrain (Doornkamp et al., 1980), Kuwait, Iraq and Saudi Arabia.

In the Weissrand plateau of south-eastern Namibia (Figure 2.7), there is an area of broadly parallel drainage lines along which there is a multitude of small closed depressions. These have developed on calcretes which cap Ecca (Permian) sedimentary rocks (shales, mudstones, etc.) and Kalahari Bed sediments. The spacing and alignment of the drainage lines implies (a) that they may have been superimposed onto the calcrete by a formerly more extensive dune cover and (b) that the closed depressions may have developed along the drainage lines because of localized solution of the underlying calcrete, assisted perhaps by other pan-forming processes such as deflation and animal activity. There is a striking near correspondence between the alignment of the Weissrand drainage and the alignment of the linear dunes and inter-dunal depressions of the Kalahari. There is also an area of small dayas in the Ubib embayment of the central Namib (Marker, 1982). In South Africa, there is an extensive area of dayas on the Ghaap Plateau to the north-west of the Vaal River and to the north of Campbell. This is an area underlain by ancient Transvaal Dolomites but capped by calcretes.

In the High Plains of the United States, aligned drainage and associated small depressions are developed on the calcrete caprock of the Ogalalla Formation (Melton, 1934). In addition, remote sensing studies demonstrated the presence of numerous basins showing clearly preferred alignments in Lea County and around Lamesa, Texas (Goudie and Wells, 1995). Other notable aligned basin patterns occur surrounding the

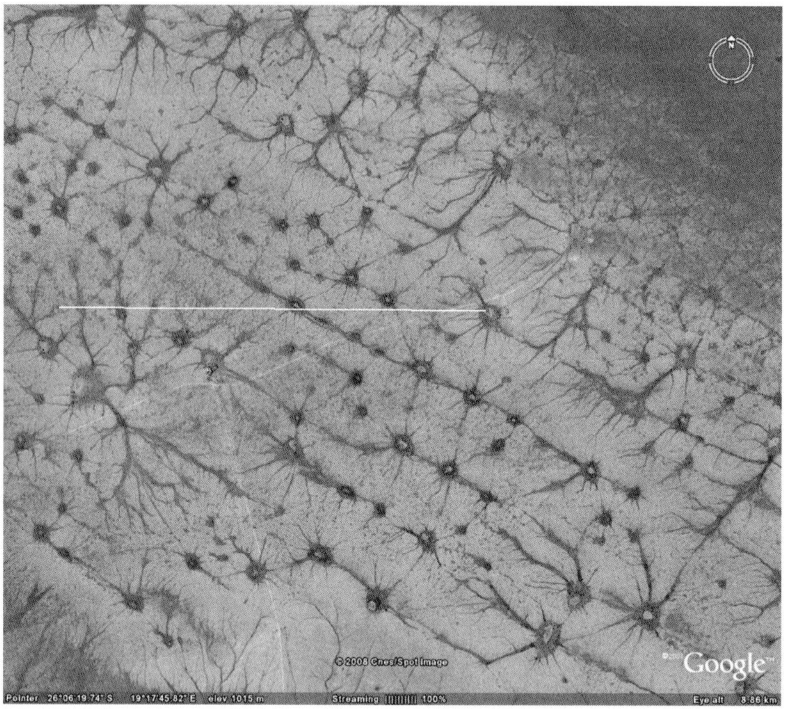

Figure 2.7 Parallel rows of small solution depressions, or dayas, developed in calcrete in the Weissrand plateau of Namibia. Scale bar is 5 km. (©Google Earth 2008, ©Cnes/Spot Image 2008)

Portales Valley on the Texas–New Mexico border. Lorenz (2002) has described dayas from calcrete deposits of the Pampas and Buenos Aires districts of Argentina.

In South Australia the Nullarbor Plain, covering about 200,000 km^2, developed on a very flat land surface formed on gently dipping upper Eocene and Miocene limestones. Although there are sinkholes and caves, the most widespread surface features are small ($<$200 m wide) hollows of limited depth ($<$4 m) (Doerr and Davies, 2007). These exist in the tens of thousands and, very remarkably, many of them occur in parallel alignments that run from WNW to ESE.

2.11 Weathering Forms

Flaking, spalling, splitting and granular disintegration are common. The small-scale exfoliation of individual silicate-rock boulders (onion skin weathering) is a form of rock spalling that was analysed by Blackwelder (1925). It can also occur at a large scale, involving rock sheets several metres thick and curved failure planes. Exfoliation may play a role in the development of natural sandstone arches, a distinctive feature in Utah (Blair 1987). Such weathering has been attributed to both mechanical and chemical processes.

Figure 2.8 Flat-bottomed weathering pit in the granite of the Erongo Mountains of Namibia. (ASG)

2.12 Weathering Pits

Small closed depressions – also called gnammas, Opferkessel or pias – are common on horizontal and gently inclined rock surfaces (Figure 2.8) and on a range of silicate rock types, most frequently granites (Twidale and Vidal Romani, 2005) and sandstones (Young et al., 2009). They may be similar in their broad morphology to solutional pits developed in carbonate rocks, to which the term 'kamenitza' is often applied. Goudie and Migón (1997) (Table 2.13) provide a list of references on weathering pits from a diverse range of morphoclimatic regions that range from polar to desert and humid tropical. The largest examples may be between 10 and 20 m long (Twidale and Corbin, 1963), but some may be even larger and be visible on remote sensing images (Figure 2.9). In Utah, some are as wide as 38 m and 16.5 m deep (Netoff et al., 1995). Pits may have a variety of forms, including pans, bowls, cylinders and armchairs. One particular type of pit is the rock doughnut, in which the pit occurs in the centre of a conical pipe (Netoff and Shroba, 2001).

There is uncertainty concerning the processes involved in the development of pits. Chemical processes of solution are usually invoked (Domínguez-Villar et al., 2007), but other processes may include hydration, the mechanical action of frost and salt

Table 2.13 *Studies of weathering pits on silicate rocks in drylands*

Source	Location	Rock type
Dzulynski and Kotarba (1979)	Mongolia	Granite
Goudie and Migón (1997)	Namibia	Granite
Jutson (1950)	Western Australia	Granite
Netoff et al. (1995)	Utah, USA	Sandstone
Robinson and Williams (1992)	Morocco	Sandstone
Schipull (1978), Howard and Kochel (1988)	Colorado Plateau, USA	Sandstone
Twidale (1978)	Uluru, Australia	Arkose
Domínguez-Villar et al. (2007)	Central Spain	Granite
Cui et al. (1999)	Inner Mongolia and Hebei, China	Granite

and biochemical weathering. Complex biofilms may accumulate at the base of rock basins; these are known to dissolve the cement between sandstone grains and also act as biological sealants to water infiltration (Chan et al., 2005). Positive feedback mechanisms related to the ever-growing amount of water available as a pit enlarges may account for a localized high intensity of weathering (Schipull, 1978).

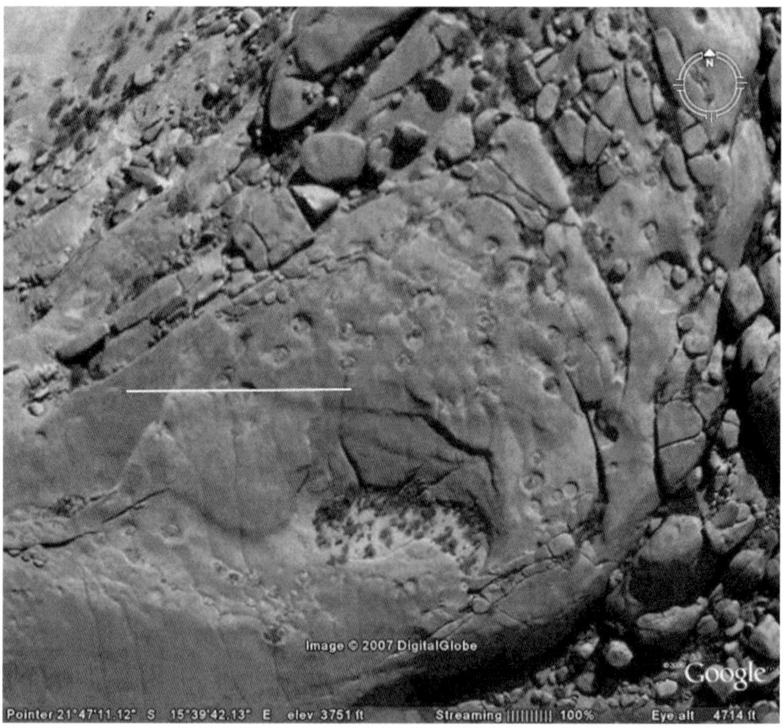

Figure 2.9 Satellite image of weathering pits in the Erongo Mountains, Ameib, Namibia. Scale bar is 100 m. (©Google Earth 2007, ©Digital Globe 2010)

Figure 2.10 A grotesque tafone (left) developed in volcanic rocks in the Atacama Desert near Arica, Chile. (ASG)

Many weathering pits are partially infilled with debris and organic matter (including algal coatings), yet there are also many which are empty, with bare flaky bedrock exposed on the floor. It seems that relatively little attention has been paid to the question of how the debris gets evacuated from the pit. Three possible ways have been suggested, however (Smith, 1941) – namely solutional transport, washing out during excessive rainfalls and deflation. Flotation is another possible mechanism, but no direct observations have been made. The occurrence of deep closed pits devoid of any sediment (cf. Watson and Pye, 1985) remains puzzling, and no satisfactory explanation of their emptiness has been offered (Netoff et al., 1995).

2.13 Tafoni

Tafoni (singular, tafone) are cavernous weathering forms which typically are several cubic metres in volume and have arch-shaped entrances, concave inner walls, overhanging margins (visors) and fairly smooth, gently sloping debris-covered floors (Mellor et al., 1997) (Figure 2.10). First described from Corsica (Brandmeier et al., 2010), they occur in many parts of the world (Table 2.14) (Goudie and Viles, 1997), including polar regions, but have been described in many deserts, including the Mojave (Blackwelder, 1929), the Sonoran (Bryan, 1925), the Colorado Plateau (Mustoe, 1982; Howard and Kochel, 1988), the Sahara (Smith, 1978), Saudi Arabia (Chapman, 1980),

Table 2.14 *Examples of dryland tafoni*

Location	Rock types	Reference
Algeria, Sahara	Rhyolites, granodiorites, etc.	Klaer (1993)
Andalucia, Spain	Sandstone	Mellor et al. (1997)
Argentina	Granite	Uña Alvarez and Vidal Romani (2008)
Dead Sea, Jordan	Sandstone	Abu Ghazleh and Kempe (2009)
Aravalli Hills, India	Granite	Goudie and Viles (1997)
Arizona, USA	Siltstone, fine-grained	Norwick and Dexter (2002)
Atacama, Chile	Sandstone, sandy dolomite Conglomerates	Segerstrom and Henriquez (1964)
Bahrain	Granitoid rocks	Doornkamp et al. (1980)
Gran Canaria	Phonolitic lavas	Goudie and Viles (1997)
Karakorams, Pakistan	Granite	Goudie (1984b)
Luderitz, Namibia	Granites	Goudie and Viles (1997)
Magadi, Kenya	Trachyte lavas	Smith and McAlister (1986)
New South Wales, Australia	Hawkesbury Sandstone	Young and Young (1992)
Petra, Jordan	Palaeozoic sandstones and granites	Goudie et al. (2002)
Ras Al Khaimah, UAE	Limestones	Goudie and Viles (1997)
South Australia	Granite	Dragovich (1969)
SW USA	Igneous, sandstone, sandy shale, conglomerates	Blackwelder (1929)
SW USA	Dolomitic limestones	Kirchner (1996)
Tenerife	Volcanic tuffs, etc.	Höllerman (1975)
Utah, USA	Calcareous sandstone	Mustoe (1982)
Utah, USA	Ashfall tuff	McBride and Picard (2000)
Baja California, Mexico	Granitoid clasts	Spelz et al. (2008)
Arizona, USA	Fanglomerate	Mills (1998)
Chile	Diorites, tonalities, granodiorites, amphibolites	Lopez and Johnson (1987)

South Australia (Dragovich, 1969) and southern Jordan (Goudie et al., 2002). As Table 2.14 shows, they occur in a wide range of rock types, but especially in medium- and coarse-grained granites, sandstones and limestones. Indeed, it is only rocks with relatively closely spaced discontinuities (bedding planes, foliation, joints), such as shales and slates, that seem to be relatively unaffected by these cavernous weathering forms. Most studies assume that tafoni develop on the surfaces of rock outcrops, but examples are known of their formation in subsurface situations (Roqué et al., 2012).

The cavernous hollows of tafoni are believed to result largely from flaking and granular disintegration caused by a range of possible weathering processes that include hydration, salt crystallization, lichen growth and chemical attack by saline solutions

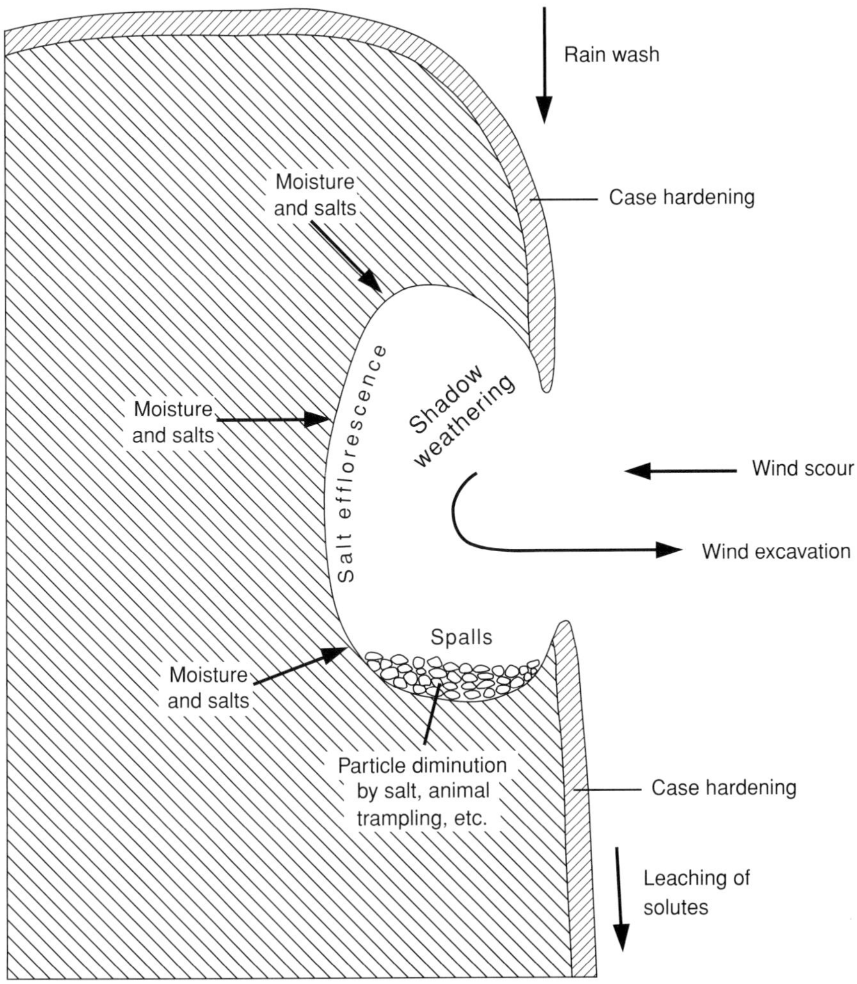

Figure 2.11 A model of tafoni development. (ASG)

(Figure 2.11). Some workers have found clear evidence of salt weathering being involved, whereas others have not. The role of case hardening in tafoni formation is also the subject of debate, but it can help to explain the formation of the visor. For a cavity to grow, there needs to be a mechanism to remove flakes and spalls. Wind may play a part, as do organisms such as pack rats. Although some early workers thought that the actual excavation of a cavity might be achieved by wind abrasion, many tafoni occur in environments where sandblasting does not occur, or they may have an aspect (e.g. the leeward side of a boulder) or a height up a cliff face that precludes such a mechanism.

Bradley et al. (1978) studied tafoni in South Australia and found that total salt content was two to thirteen times more abundant in tafoni flakes than in massive granite and accepted the association between salts and flaking granite as a genetic one. They believed that the saline material was derived from fluid inclusions within the granite and that this material dissolves when moisture seeps into the rock, to be precipitated as halite and gypsum when evaporation pulls the fluid back to the surface and renders it supersaturated. They undertook a study of the salt content of different portions of a boulder with tafoni development and remarked (p. 653): 'The sheltered upper parts of tafoni are most favourable for the retention of moisture, and this is where salts are concentrated and flaking most active. Salts either do not move to the more exposed rock surfaces, due to inadequate moisture, or they are flushed away from such locations by rainwater.' The amount of salt involved in the process was low, however – only 0.4 per cent of salt was present in the tafoni flakes.

Höllerman (1975) investigated tafoni on Tenerife in the Canary Islands and argued that they were associated with case hardening. He also found that the primary salts involved were sodium chloride (dominant), sodium sulphate and calcium sulphate. Another attempt to quantify the presence of salt associated with tafoni was made by Mustoe (1982). Working in the Capitol Reef desert of Utah, United States, he found that efflorescences were present, including gypsum, hexahydrate, bloedite and natron. Like Höllermann, he argued that salt crystallization was the most important cause of disintegration within the tafoni.

An ingenious approach to the role of salt crystallization in tafoni development was made at Machtesh Hagdol in the Negev by Rögner (1987), who monitored temperatures within a tafoni and found anomalous variations which he attributed to the process of salt crystallization out of highly concentrated solutions. He also recorded halite, gypsum, epsomite, hexahydrite and sylvite as being present behind flakes and scales. He believed that the presence of these salts and the fact that temperature observations showed that crystallization occurred within the tafoni were both reasons for believing that salt weathering was the main explanation for their formation.

It may be that once formed, tafoni develop through a positive feedback. This has been expressed thus by Smith and McAlister (1986, p. 456):

> Once a hollow is initiated it creates an environment in which weathering is favoured, weathering in turn extends the hollow to produce an optimum form in which weathering is further enhanced, and so on. Such re-enforcement cannot continue indefinitely and a condition must eventually be reached where the weathering rate is reduced. This could occur, for example, where a cavern becomes so deep that it either prevents the ingress of moist air, or rock temperature variations are reduced to the point where precipitation and evaporation no longer occur.

They argue that on exposed cliff faces, salts would be deposited by the outward migration of salts derived from within the rock, but that they would be removed by subsequent rainwash, whereas salts precipitated in hollows would be protected from such leaching and so could cause salt weathering to occur, which would progressively expand the hollow. The same mechanism has been proposed for the spectacular niches and columns of sandstone cliffs in South Jordan (Goudie et al., 2002). Oberlander (1977) described similar features, called 'colonnade walls' or 'alveolar walls', from sandstone terrains in south-eastern Utah.

Many tafoni show both inward and upward growth. This may be partly because downward growth is restricted by the fall of spall debris onto tafoni floors, but it may also be related to the receipt of warmth from the sun. Segerstrom and Henriquez (1964), following Blackwelder (1929), suggested that dampness is a key feature of hollows and promotes weathering. Given, it is argued, that the lower part of a cavity is more likely to be dried out by sunshine than the more shaded upper part, there will therefore be a tendency for the cavity to grow inward and upward. The flow vectors of moisture within boulders may also be a powerful control of tafoni form (Conca and Rossman, 1985; Conca and Astor, 1987). For similar reasons, tafoni may show a preferred orientation. For example, on the terraces of the pluvial Lake Lisan in Jordan, most of the cavities face west, south-west and north-west, which is in the direction of longest shadow duration (Abu Ghazleh and Kempe, 2009). This is an example of what German observers have called shadow weathering (*Schatten Verwitterung*).

It is evident that although salt weathering is not invariably the cause of tafoni formation, salts are often present in tafoni back walls and flakes and cause disintegration both through physical and chemical mechanisms (Young, 1987; Howard and Kochel, 1988; Brandmeier et al., 2011). The cavernous form may result either from the breaching of a case hardened exterior or from positive feedback effects leading to the enlargement of an initially small hollow. In coastal locations, where rates of tafoni development seem to be quicker than in deserts (Sunamura and Aoki, 2011) the salts may be derived from spray, and the same may apply in the vicinity of desert lake basins (Butler and Mount, 1986), but elsewhere the salts may be derived from within the rock mass by seepage.

One research area that needs to be addressed is why tafoni do not develop in some desert areas on suitable rock types.

2.14 Alveoles

Alveoles (Mustoe, 1982) are a second, extremely widespread cavernous form. They are small hollows (e.g. c 5–50 cm in diameter) that occur in clusters in apparently homogenous rocks to create a 'honeycomb' texture (Figure 2.12). They are morphologically similar to small tafoni, and occasionally the two forms occur together, although this does not necessarily imply a common origin. The origin of alveoles is similarly controversial, but there is also a convergence in the literature towards recognition of the importance of salt weathering, particularly because the process causes

Figure 2.12 Alveoles developed in granite in the Swakop Valley, central Namibia. (ASG)

the disaggregation of mineral grains without chemical decomposition (Mustoe, 1982). Individual alveoles are commonly separated by thin partitions, which may be relatively resistant to weathering if they are strengthened by 'case hardening' features or protected by algal growths. Other processes invoked to explain alveoles include wind erosion, frost shattering, exfoliation and solution of cements.

2.15 Amphitheatres and Alcoves

Another type of weathering-related development is shown in the formation of cavernous amphitheatrical valley heads. Such features result at least in part from a process called groundwater sapping, and recent research has shown that seepage processes are important in producing them in areas such as the Kalahari, the Wadi Rum area of Jordan and the sandstone terrains of the Colorado Plateau. In arid regions, seepage produces salt exudation, so that salt weathering may contribute to the undercutting processes that cause valley head recession. This is certainly what has been suggested by Laity and Malin (1985) and Howard and Kochel (1988) in the context of the Colorado Plateau.

Certainly, extreme cavernous weathering can cause cliffs to be over-steepened, and Hume (1925, p. 214) believed that on the Eocene Ma'aza Limestone Plateau in Upper Egypt, this had progressed sufficiently to cause them to become 'absolutely unscaleable'.

Figure 2.13 A natural arch developed in Cambro-Ordovician sandstone near Wadi Rum, Jordan. (ASG)

2.16 Natural Arches

Natural arches form when weathering, together with mass collapse, creates a tunnel through a slab of rock (Figure 2.13). In arid areas these processes may be aided by wind attack. They are not restricted to arid areas, but most of the world's finest examples occur in drylands. There are literally hundreds of natural arches in the south-west United States (Gregory, 1938; Barnes, 1987, Moore, 1999; Munthe, 2002) and also in the Tassili and Jebel Acacus areas of the central Sahara. Other important locations are the Wadi Rum area of southern Jordan (Goudie et al., 2002) and the Ennedi region of Chad. Arches most commonly occur in sandstones, which have sufficient permeability to provide the groundwater seepage that promotes weathering and yet which have the necessary cohesion for an arch to develop. They are best developed where the sandstones are massive and horizontal or gently dipping (Blair, 1987). That said, they can also develop in granites (as in the central Namib) and in chalk (as in the Farafra region of Egypt). Some of the arches have large dimensions: there are 400 known that are longer than 50 m (Young et al., 2009). The broadest is likely the Kolob Arch of Utah, which has a width of 94.5 m. Outside North America, the largest arch is probably that of Aloba in the Ennedi mountains of Chad, which has a span of 77 m but is also 120 m tall. There is considerable variety in the morphology of arches: some are in the form of horizontal lintels, others are round holes, while others have a classic arch shape. Most natural arches form by rock falls that occur

Figure 2.14 Polygonal cracking of granite in the Erongo Mountains of Namibia. Ruler is extended by 30 cm. (ASG)

when massive sandstones are undermined by seepage of groundwater and associated weathering processes, including salt weathering. Arches are most numerous, as in the Arches National Park of Utah, where long and closely spaced joints or fractures have been eroded to form narrow 'fins' of rock that are readily pierced by weathering (Cruikshank and Aydin, 1994). These fins, associated with salt tectonics, show up magnificently on air photographs of the region.

2.17 Polygonal Cracking (Tessellation or Alligator Cracking)

Polygonal fracture patterns (Figure 2.14) consist mainly of pentagonal or hexagonal cracks meeting at $\sim 120^\circ$ where they make tri-radial junctions on predominantly curved rock surface pavements (Robinson and Williams, 1989, 1992; Young et al., 2009). Polygonal crack diameters may vary from $\sim$5 cm to more than 50 cm, and in some instances micro-polygonal cracking has been observed within larger tessellation plates (Robinson and Williams, 1989). Although understanding the origin of tessellation has caused some debate (Young et al., 2009) and may differ between sites, it is a particularly common phenomenon in case-hardened rocks associated with regions of seasonal precipitation patterns (Robinson and Williams, 1989, 1992; Twidale and Campbell, 1993). Some of the likely causative mechanisms suggested include shrinkage of silica gel due to changing rock thermal and/or moisture conditions (Robinson and Williams, 1989, 1992), thermal changes which set up differential surface stress

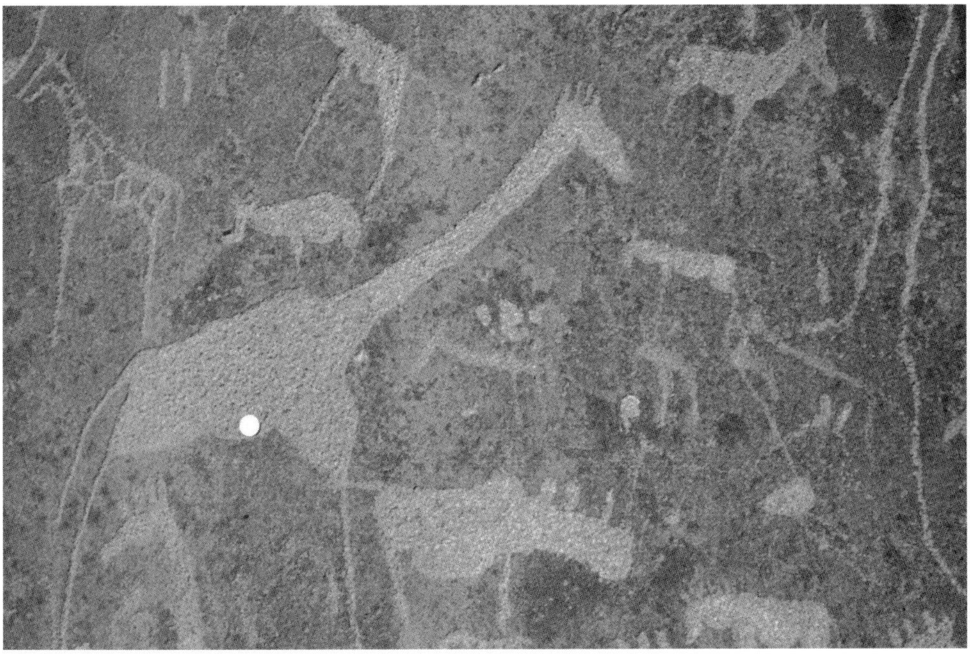

Figure 2.15 Rock engravings in sandstone at Twyfelfontein, Namibia, indicate the colour differences between the weathering rind and the sandstone rock beneath. Coin for scale. (ASG)

(Branagan, 1983; Croll 2009), expansion to accommodate precipitating solutions of iron, and thermal stresses or geochemical changes caused by emplacement of dolerite dykes (Velázquez et al., 2008).

2.18 Rock (Desert) Varnish and Other Rock Coatings

Rock or desert varnish is widespread (Figure 2.15). Some people consider that it has generated a literature whose size is out of all proportion to the importance of the topic. In the deserts of the south-western United States, it covers 75 per cent of all bare rock (Engel and Sharp, 1958), forming a hard rind, around 5–100 μm thick (rarely 500 μm), above and sometimes just below the soil surface. Although paper thin, it can completely alter the appearance of a landscape. Early artists scratched it away to make their rock paintings and petroglyphs (Dietzel et al., 2008). Varnish is rare or patchy on unstable surfaces, as, for example, where salt attack is active or wind abrasion severe. It grows at rates that have been estimated to be 1–15 μm per millennium (Fleisher et al., 1999) and $<$1–40 μm per millennium (Liu and Broecker, 2008), with rates being highest in micro-topographic lows (Reneau et al., 1992). Rock varnish becomes darker with age, but this is also affected by geomorphological position and the aeolian abrasion history of surfaces. Most varnishes (Table 2.15) consist of approximately

Table 2.15 *Examples of elemental variation exhibited in bulk chemical analyses of rock varnishes found in desert regions*

Site:	Salt Springs, Mojave Desert*	Trail Fan, Death Valley	Manix Lake, Mojave Desert	Makanak a Till, Hawai'i	Sinai Peninsula, Egypt	Petroglyph, South Australia	Ingenio, Peru Desert	Ayers Rock (Uluru), Australia
Position	Unknown	Former rock fracture	1 > m above soil	With silica skin	1 > m above soil	1 > m above soil	At soil surface	From rock fracture
Al	25.84	23.74	25.77	21.13	22.94	22.81	20.45	28.77
Si	37.49	39.09	32.35	29.77	32.81	33.34	45.88	35.69
K	2.35	3.45	2.11	3.3	2.42	2.79	2.91	2.11
Ca	0.8	4.87	1.35	4.89	2.91	2.18	6.22	1.45
Ti	0.74	1.52	10.84	0.73	0.68	0.65	0.85	1.19
Mn	11.77	10.87	12.47	13.6	11.97	21.7	4.94	11.91
Fe	14.5	13.47	18.09	21.13	22.94	13.26	12.03	16.57

* Results are normalized to 100%.
Source: Modified from data in Dorn (1998, table 10.2).

30 per cent Mn and Fe oxides and 70 per cent mixed layer illite/montmorillonite clay minerals (Potter and Rossman, 1977). Although many varnishes are Mn rich and Fe poor, this is not invariably the case (Jones, 1991). Layering at the scale of nanometres (Krinsley, 1998) is common and may reflect environmental changes (Cremaschi, 1996), including the transition from humid to dry climates (Dietzel et al., 2008; Zerboni, 2008). The increasing use of high-resolution geochemical techniques and microscopy has revealed just how complex the composition of varnishes can be and how many processes contribute to their evolution (Garvie et al., 2008).

Broecker and Liu (2001) observed pronounced temporal variations in Mn and Ba concentration within varnish layers, and argued that these variations are similar over large regions and could relate to variations in past moisture conditions. Together with Liu and Dorn (1996), they postulated that these can be correlated with Heinrich Events and the Younger Dryas. Furthermore, by studying the MnO_2 content of the Holocene layers, they found that manganese contents increase with precipitation. They therefore argued that layers rich in manganese record relatively moist conditions, as did Dietzel et al. (2008). In the western United States, Mn-poor yellow layers (usually containing 5–15% MnO) formed during dry periods of the Holocene, whereas Mn-rich black layers (usually containing 25–45% MnO) were deposited during wet phases (Liu and Broecker, 2008).

Early workers believed varnishing to be a physicochemical process, associated with high Eh (dry, oxidizing) and pH (unleached, alkaline) conditions. The occasional incoherence of the underlying rock suggested to some that iron and manganese had been drawn in solution from the rock beneath and precipitated by evaporation. Early models saw varnish constituents as being from the underlying rock (Dorn, 1998). More recently, it has been demonstrated that most varnish constituents could not have come from underlying rocks, many of which have little iron or manganese. Engel and Sharp (1958), while still assuming physicochemical fixation, discovered that many of the elements in varnishes were derived from an external source: dust. The dominant view now is that varnish is an external accretion (Potter and Rossman, 1977) derived not only from dust but also from direct aqueous atmospheric deposition (Thiagarajan and Lee, 2004). This view is supported by the presence of radioisotopes (^{137}Cs and ^{210}Pb) and anthropogenic metals in varnish samples (Fleisher et al., 1999). Moreover, most varnishes overlie the rock with a sharp boundary (Potter and Rossman, 1977).

The physicochemical model was challenged by biological models of manganese fixation (Whalley, 1983), and this explanation is now widely adopted (Nagy et al., 1991; Kuhlman et al., 2008; Wang et al., 2011). Microorganisms (lichens, fungi and bacteria) (Dragovich, 1993) play the major role in fixing manganese.

Many attempts have been made to date varnishes using techniques that include uranium series dating, cation-ratio dating, microlamination analysis (Liu and Broecker, 2008; Stirling et al., 2010) and AMS radiocarbon analysis. The nature and multiple problems of these techniques have been reviewed by Watchman (2000). Such techniques have recently been applied to establish the ages of landforms in areas of seismic activity and hence to obtain an idea of the frequency and magnitude of past ground motions (Stirling et al., 2010). More generally, varnish dates have been produced for a range of landforms, including stone pavements, colluvial accumulations and alluvial fans (Liu and Broecker, 2008).

In addition to rock varnish, there are various other types of rock coatings that have been identified (Dorn, 2009). These include silica glaze, a broad category of rock coating dominated by amorphous aluminium and iron and various types of iron film and oxalate crusts.

The presence of rock coatings means that case hardening can be conspicuous in deserts. For example, it occurs on sandstones in the south-western United States, where it varies in thickness from 0.5 to 5 mm (Conca and Rossman, 1982). Thus it is thicker than rock varnish, paler in appearance and much harder than the underlying rock from which it is separated by an abrupt boundary, and it has a calcareous surface horizon over a zone of kaolinite enrichment, which is poorer in quartz and haematite than the underlying rock. Conca and Rossman believed that the kaolinite had been brought to the surface by evaporating water and later cemented by dew and rain. Elsewhere, however, iron films may be the case-hardening agent, as on the sandstones of south Jordan (Goudie et al., 2002). Abiotic processes may be dominant. It is widely held that solutions are mobilized from the rock, drawn out by evaporation and re-precipitated on the rock's outer shell. External sources (e.g. dust) may also be important, however – particularly where the case hardening involves accumulation of material such as manganese and oxalate, concentrations of which in the host rock are minimal (Dorn, 1998).

2.19 Duricrusts

The word 'duricrust' was introduced by Woolnough (1927), who subsequently defined the term thus (Woolnough, 1930, pp. 124–5):

> The widespread chemically formed capping in Australia, resting on a thoroughly leached substratum.... The nature of the deposit varies from a mere infiltration of pre-existing surface rock, to a thick mass of relatively pure chemical precipitate.

As a result of subsequent work on the individual duricrust types, the *crete*-based terminology of which had been laid down by Lamplugh (1907), Goudie (1973, p. 5) proposed a modified definition which resulted from a synthesis of a various definitions that had already been developed for the individual types, stressing their essentially subaerial and near-surface origin and nature:

> A product of terrestrial processes within the zone of weathering in which either iron and aluminium sesquioxides (in the case of ferricretes and alcretes) or silica (in the case of silcrete) or calcium carbonate (in the case of calcrete) or other compounds in the case of magnesicrete and the like have dominantly accumulated in and/or replaced a pre-existing soil, rock, or weathered material, to give a substance which may ultimately develop into an indurated mass.

As this definition suggests, deserts have various duricrusts, including nitrates, calcretes, gypcretes, silcretes and ferricretes (Goudie, 1973). These have a profound effect on the land surface and provide the main relief features of low-relief deserts. In general, duricrust caprocks preserve ancient landforms; an example is the extensive Ogallala calcrete caprock in the High Plains of the United States (Brown, 1956). In high-relief deserts such as those of the south-western United States and south-east Spain, calcrete preserves alluvial fans (Gile and Hawley, 1966; Stokes et al. 2007). Because most duricrusts preserve the deposits of valleys rather than of deeply weathered (and weakened) interfluves, incision or deflation may create relief inversion, whereby the ancient valleys are preserved as elongated plateaux or mesas, while the divides become valleys (Ollier, 1988; Oviatt et al., 2003). In Australia, there are excellent examples of inverted relief from the Pilbara Region involving ferricretes and from the Painted Desert of South Australia involving silcrete (West et al., 2010). In Arabia, the central Sahara and parts of China, narrow, often markedly meandering channels, were duricrusted and have survived as 'raised channels'. (Maizels, 1987; 1988).

Other landforms are produced by incision into duricrusts. These include steep vertical scarps or 'breakaways', caverns in the softer material beneath, and blocky talus slopes onto which fragments of the duricrust collapse (Blume and Barth, 1979).

Figure 2.16 The calcrete caprock on the incised cliffs of the Molopo River on the border between South Africa and Botswana. (ASG)

This is, for example, evident along the Molopo Valley between South Africa and Botswana (Figure 2.16). In some places, groundwater seepage may undermine duricrusted scarps, as in Botswana (Shaw and De Vries, 1988). Where duricrusts cover an undulating landscape, incision produces complex flatirons (Everard, 1963). Remarkable karstic landforms develop in calcretes, ferricretes and silcretes (Goudie, 1973; McFarlane and Twidale, 1987); solution hollows (including dayas), collapse features and caves. It has been claimed that some of the large depressions in the central Sahara are karstlike dolines and poljes dissolved in an ancient silcrete (Busche and Erbe, 1987).

2.20 Nitrates

Sodium nitrate deposits are less widespread than the other major types of chemical crust (Goudie and Heslop, 2007). Indeed, the only deposits of any great spatial extent and thickness are those of portions of the hyperarid Atacama Desert in South America. These materials are, in the words of Ericksen (1981, p. 366), 'so extraordinary that, were it not for their existence, geologists could easily conclude that such deposits could not form in nature'.

In addition to the Atacama, Sodium nitrate is known from other dryland situations, although nowhere – except perhaps in the Turpan-Hami area of China (Qin et al.,

2012) – does it attain the same significance: from numerous caves in the south-western United States (Hill, 1977) and Argentina (Forti and Buzio, 1990), and on open sites from the Mojave Desert of California (Ericksen et al., 1988; Böhlke et al., 1997), the Basin and Range Province of the United States (Kirchner, 1996) and numerous locations in Antarctica (Keys and Williams, 1981; Matsuoka, 1995). It has also been recorded from various buildings surfaces in Bukhara in Uzbekistan (Cooke, 1994).

The sodium nitrate deposits of the Atacama are called caliche. This material occurs primarily in the provinces of Tarapacá and Antofagasta in northern Chile (Figure 2.17), although it also extends northwards into Peru. The nitrates occur as a band, up to 30 km wide, along the eastern (inland) side of the Coastal Range (Ericksen, 1981). They extend from c 19°30°S to 26°S, a distance of about 700 km. The lower-grade deposits are more extensive than this, and much of the Coastal Range is encrusted with nitrate-bearing saline-cemented regolith. The deposits extend over a considerable altitudinal range (some up to 4,000 m above sea level but most below 2,000 m). They occur in a wide range of topographic situations, from tops of hills and ridges to the centres of broad valleys, many of which are occupied by closed salt lakes (*salars*). The deposits also occur in and on all types of rock and superficial sediment in the area, and there appears to be little lithological control of their different types and mineral assemblages.

While some of the Atacama sodium nitrate deposits are relatively pure (Ericksen and Mrose, 1972), most are impure and contain substantial amounts of other salts, including a range of sulphates, chlorides, nitrates, borates, iodates, perchlorate and chromates. Ericksen (1981), Pueyo et al. (1998) and Searl and Rankin (1993) provide a detailed list of such minerals, although some of Searl and Rankin's identifications have been challenged by Ericksen (1994).

Sodium nitrate is highly soluble in water in comparison with many other salts. At 35°C its solubility in water is 49.6 per cent, whereas that of sodium sulphate is 33.4 per cent, of sodium chloride 26.6 per cent and of calcium sulphate just 0.21 per cent. Thus it can only persist in extremely arid environments. Because of its ready solubility, it has been argued that it exists to such a great extent in the Atacama as it is drier than any other desert in the world. Another reason that has been given for the development of nitrates in the Atacama is that the desert is one of the world's oldest, so that there has been an extended period for large amounts of nitrate to build up, even from modest rates of inputs.

Many theories have been developed over the last century and a half to account for the development of the Chilean nitrate deposits, and the issue is still far from resolved. Ericksen (1981) provides a review of some of the early ideas about the derivation of the nitrate, including the idea that it is drawn from guano or seaweed.

That nitrates and related salts are the result of atmospheric deposition now receives relatively widespread support. Ericksen (1981) argued that some of the nitrates could have an atmospheric origin, with their source originating from the Pacific. He also

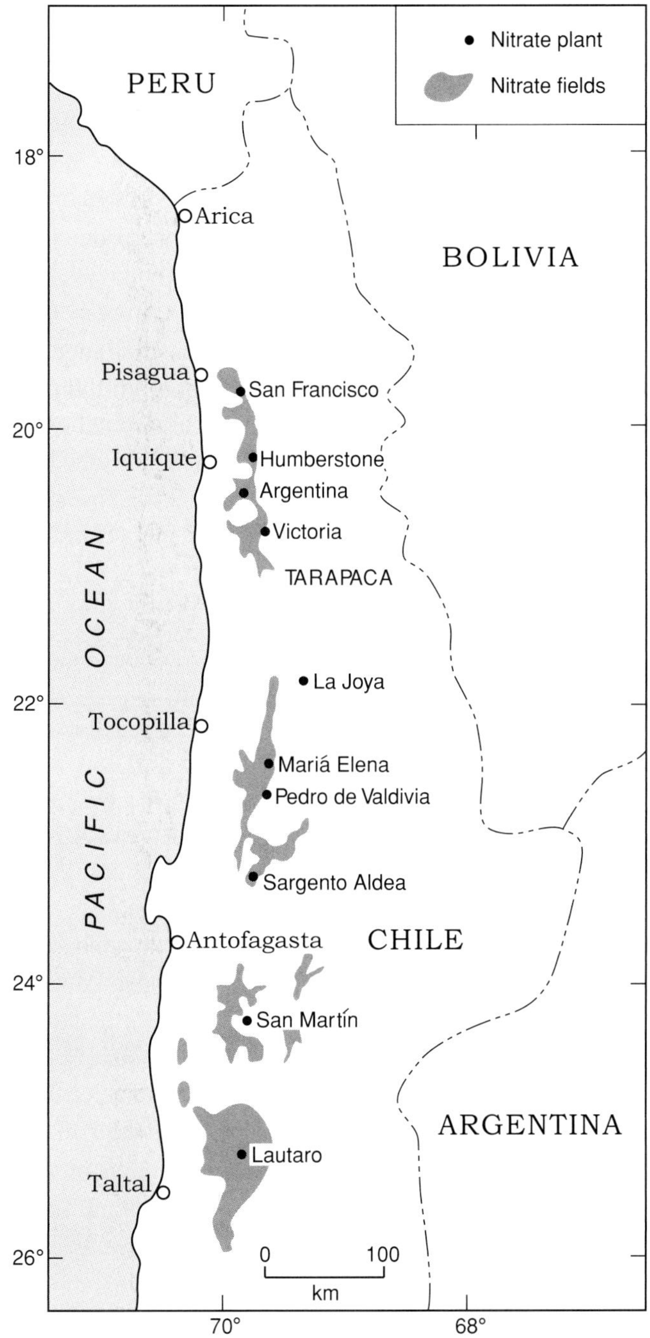

Figure 2.17 The distribution of nitrate deposits in northern Chile. (ASG)

recognized, however, that volcanic emissions from the Andes could have played a role, as could biological activity and bedrock weathering. He recognized atmospherically derived materials would be leached, redistributed and enriched in their highly soluble components and would accumulate on old land surfaces that have had little or no modification since the Miocene, on lower hillsides and at breaks in slopes as the result of leaching and redistribution by rainwater and in saltpans. In 1981, however, Ericksen admitted that his ideas had changed, and argued that most of the nitrate was formed 'by fixation of atmospheric nitrogen by microorganisms in playa lakes and associated moist soils' (p. 372). Subsequent stable isotopic studies – using N, O and S isotopes – by Ericksen and co-workers (see Böhlke et al., 1997, p. 135), 'support the hypothesis that some high-grade caliche-type nitrate-rich deposits in some of the Earth's hyperarid deserts represent long-term accumulations of atmospheric deposition . . . in the relative absence of soil leaching or biologic recycling'. This was confirmed by Michalski et al. (2002). Böhlke and Michalski (2002) investigated the oxygen isotopic composition of nitrates from both the Atacama and from the Mojave and argued (p. 1):

> The magnitude of the non-mass-dependent isotope effect in the Atacama Desert nitrate deposits is consistent with the bulk of the nitrate having been derived from atmospheric nitrate deposition, with relatively little having formed by oxidation of reduced N compounds on the earth's surface. The proportion of the microbial end member is larger in the Mojave deposits, possibly because of slightly higher rainfall and more biological activity.

Arias (2003) also supported an atmospheric origin for the Atacama deposits, and suggested that the nitrates resulted from the decay of marine algae concentrations and inland transport of aerosols in sea spray and fog.

Ericksen (1981, p. 9) argued that the coastal fogs (camanchaca) of northern Chile are saline and so may have been important sources of nitrate deposit constituents. The purity of fog water and the limited amounts that are deposited away from the coast, however, suggest that it can only be a minor contributor to the nature of the nitrate deposits. Indeed, S and Sr isotopic studies of Atacama aerosols and sediments confirm this, and Rech et al. (2003) suggest that marine or local salar salts are the main source of not only the nitrates but also of the perchlorate and iodate that the caliche contains. The large Chinese nitrate deposits also appear to have an atmospheric origin (Qin et al., 2012).

2.21 Gypsum Enrichment

Gypsum-rich materials may cover around 207 million ha of Earth's surface, and the majority occur where the mean annual rainfall is less than 200–250 mm. This is because gypsum ($CaSO_4.2H_2O$) is semi-soluble ($\sim$2.6 g/l at 25°C) and is normally leached out under higher rainfall conditions (Herrero and Porta, 2000). With respect to soils, the depth to the gypsum horizon increases with mean annual precipitation

Table 2.16 *Selected studies of gypcretes*

Location	Author(s)
Algeria	Horta (1980)
Australia	Chen (1997)
Chile	Pueyo et al. (2001)
Egypt	Aref (2003)
Iraq	Tucker (1978)
Kuwait	El-Sayed (1993)
Namibia	Eckardt et al. (2001)
Syria	Florea and Al-Joumaa (1998)
Tunisia	Watson (1988)
UAE (Abu Dhabi)	Shahid and Abdelfattah (2009)
USA (New Mexico)	Buck and Van Hoesen (2002)
USA (Wyoming)	Reheis (1987)

(Retallack and Huang, 2010). Gypsum crusts have been reported from many drylands (Table 2.16) (see Watson, 1983, fig. 5.1, for a global distribution map), but it is probably those of Australia, Namibia and Tunisia that have received the greatest attention.

Gypsum-rich horizons are up to 5 m thick (Watson, 1983) and have a gypsum content that is usually 75–97 per cent. Gypsum cements are never as tough as calcretes or silcretes. Hardened horizons are known as gypcrete, and they have formed in four distinct conditions: (1) in well-drained soils (as a result of aeolian deposition), (2) as buried evaporites, (3) in hydromorphic soils (*croûte de nappe*) and (4) by the exposure of subsurface horizons by erosion (Watson, 1988).

Reheis (1987) recognized four stages of pedogenic gypcrete development. The first stage is characterised by thin, discontinuous gypsum coatings on the undersides of stones. The second stage has a greater abundance of gypsum pendants under stones as well as gypsum, crystals or nodules scattered through the matrix. In stage three, the gypcrete is characterised by the presence of large gypsum pendants beneath stones but above all by the presence of continuous gypsum through the soil matrix. In stage four, the gypcrete consists of a continuous gypsum-plugged matrix, with stones and small debris floating in the gypsum matrix.

The development of gypsum-enriched layers (Figure 2.18) can be seen in terms of three main stages: (1) primary crystallization, (2) transportation and redeposition and (3) post-depositional alteration (Chen, 1997). The first of these involves crystallization from surface brines and from groundwater, the second involves transport by wind or water and the third involves alteration above or below the capillary fringe. Gypsum crust formation involves a complex interplay of mobilization, depositional and translocational phases (Eckardt et al., 2001; Drake et al., 2004).

Figure 2.18 Polygonal structures in a gypsum crust in the Chott region of southern Tunisia. (ASG)

In Namibia, sulphur isotope studies indicate the importance of initial input of sulphate-rich marine aerosols (Figure 2.19). These are transported into the desert by wind and rain and accumulate in depressions (inland playas and coastal sabkhas). Here, gypsum is precipitated and is then subjected to aeolian dispersal and deposition on stone pavements and other surfaces. Some gypsum enriches groundwater and then may be precipitated in zones of groundwater discharge.

One particularity of the Namib is the eruption of hydrogen sulphide from the coastal fringe (Logan, 1960) and, more important, from offshore. Remote sensing studies suggest that such eruptions are large in extent, of frequent occurrence and of long duration (Ohde et al., 2007; Brüchert et al., 2009). It has often been suggested that the sulphur from the eruptions has contributed to the widespread development of the gypsum crusts over the Namib plains (see, for example, Martin, 1963; Wilkinson et al., 1992). However, stable isotopic work by Eckardt and Spiro (1999) does not support this theory and indicates that the sulphur is largely derived from primary marine production of dimethylsulphide (DMS) in the upwelling offshore waters. In addition, $\Delta^{17}O$ studies by Bao et al. (2000, 2001) point in the same direction. Furthermore, the ionic content of the fogs that have been proposed as one of the pathways for bringing sulphate into the desert is much lower than previously thought (Eckardt and Schemenauer, 1998, p. 2598) and is thus "unlikely to be a major vector

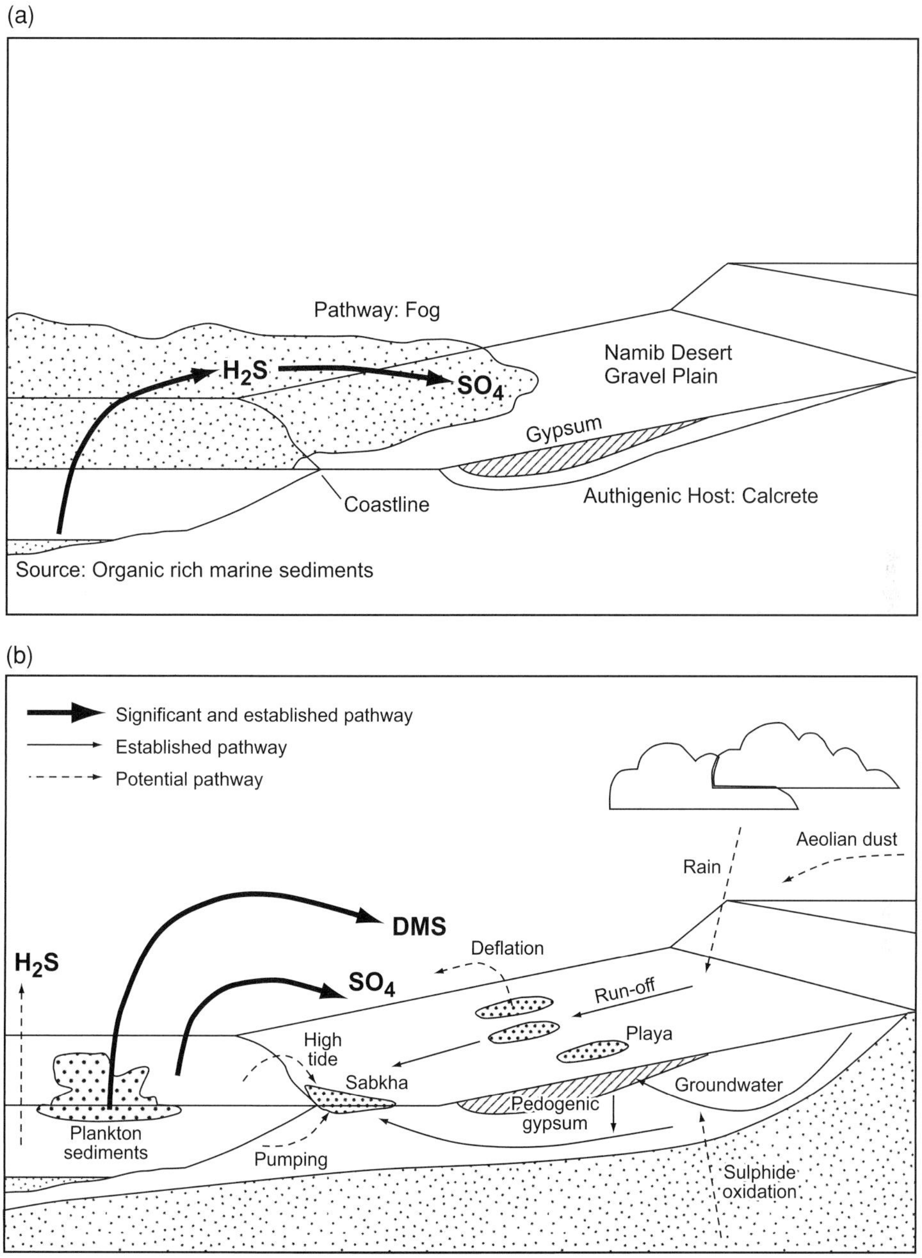

Figure 2.19 Two models of sulphate movement into and within the Namib: (a) Martin's model, (b) Eckardt's model. (From Eckardt, 1996, fig. 8.1)

of marine biogenic sulphur compounds and unlikely to contribute directly to gypsum accumulation".

It is also possible that a small proportion of the sulphate is derived from the weathering of bedrock, as some outcrops of gneiss, for example, contain primary anhydrite (Nash, 1972) or sulphide minerals. Cagle (1975) proposed that such bedrock sources could be significant, but the isotopic work of Eckardt and Spiro (1999, p. 267) suggested that they are of limited importance in the formation of gypsum. Whatever its primary source, some of the Namib gypsum may be distributed across the desert surface by dust storms that have deflated evaporate material from the many small pans and sabkhas that have been identified on satellite images (Eckardt et al., 2001).

Similar sulphur isotope techniques have also been employed in southern Tunisia (Drake et al., 2004). Here, it has been found that gypsum recycled from bedrock of various ages is the ultimate source, but that the gypsum is then transported into playas (chotts), from which it is widely redistributed by aeolian processes. Once the gypsum is deposited by the wind, it is then subjected to illuvial and pedogenic processes that cause consolidation.

2.22 Calcretes

Carbonate-enriched horizons, sometimes called petrocalcic horizons, may underlie as much as 13 per cent of Earth's total land surface (Goudie, 1983b, quoting Yaalon, 1981). In Australia, they occupy some 21 per cent of the land surface (Chen et al., 1997. (In some parts of the world, calcretes are called caliche, but given the use of this term for nitrate deposits, this only causes confusion [Goudie, 1972c]). When rapid evaporation produces supersaturated solutions of calcite and where there is also a high magnesium to calcium ratio in the waters, crusts may be dolomitic (Watts, 1980; Maizels, 1988); when hardened, these are termed dolocrete (El-Sayed, 1991a; Dixon, 2010). Calcretes can also be enriched in silica – calsilcretes or silcalcretes have been reported (Nash and Shaw, 1998; Nash et al., 2004; Shaw and Goudie, 2004; Kampunzu et al. 2007; Ringrose et al., 2009). Alonso-Zarza and Wright (2010) provide detailed information on the micromorphology of calcretes, and Wright (2009) on their macromorpohology. Many calcretes are characterised by the presence of palygorskite as a predominant clay mineral (Kadir and Eren, 2008) Gallala et al., 2010), and in some parts of the world (e.g. Namibia, Botswana, Australia, Mauritania, Somalia, Tanzania, India (Misra et al., 2011) and the south-west United States), calcrete is a host for uranium. It may also be a host for gold (Lintern et al., 2006, 2011).

Goudie (1973, p. 18) reviewed around 300 bulk analyses of calcretes from around the world and found that the mean percentages of the main constituents were:

$$CaCO_3 - 79.3\%,$$
$$SiO_2 - 12.3\%,$$
$$Al_2O_3 - 2.1\%,$$
$$Fe_2O_3 - 2.0\%,$$
$$MgO - 3.05\%,$$
$$CaO - 42.6\%$$

Table 2.17 *A morphological classification of calcretes*

Calcrete type	Characteristics	Occurrence
Calcified soils	Weakly cemented soil	Soil horizons
Powder calcrete	Fine powder, some carbonate replacement	Pans and playas
Nodular calcrete	Concretions or nodules in a calcareous matrix	Various
Honeycomb calcrete	Honeycomb texture of coalesced nodules. May be conglomeratic	Various
Hardpan calcrete	Hard layer often composed of cemented honeycomb or nodular horizons. Includes calcretised gravels	Above or between nodular or powder calcretes. Frequent as a surface horizon
Laminar calcrete	Laminated crust or layers <25 cm thick	Frequent cap to hardpan exposures
Boulder calcrete	Discrete to coalesced boulders. Re-solution often apparent	Secondary calcrete formed from other types

Source: After Goudie (1983b).

Calcic horizons develop in soils where there is a net moisture deficit, such that carbonate produced in a drier season is not leached away in a wetter season. Most occur in areas where mean annual rainfall is between 100 and 1,000 mm (Wright, 2009). They commonly occur in ancient, complex profiles more than 200 m thick, which show signs of multiple phases of sedimentation: carbonate accumulation, replacement, solution, break-up and re-cementation (Gile and Hawley, 1966; Goudie, 1983b). The accumulation of calcrete in pre-existing host material can lead to the development of folded structures called pseudo-anticlines (Price, 1925; Watts, 1977) or tepees (Eren, 2007). The calcrete pseudo-anticlines described from the Kimberley District of Western Australia occur as low ridges 1.7–3.3 m high, 10–45 m across and several hundreds of metres in length (Jennings and Sweeting, 1961).

Netterberg (1967) developed a useful classification of carbonate enrichment types, which included calcified soils, powder, nodular, honeycomb, hardpan, laminar and boulder calcretes. The Netterberg scheme (Table 2.17) sees calcified soils as the least developed form. With progressive calcification a powder calcrete develops, which is characterised by loose carbonate silt or sand particles with few or no host soil particles or nodules present. Nodular (or glaebular) calcrete is the next stage. As calcrete nodules increase in size and number they coalesce to form a honeycomb calcrete, the voids of which are still filled by soil material. Next, when the voids of the honeycomb calcrete become infilled or cemented, a hardpan calcrete develops. Some of these are indurated layers up to 3 m thick. This may often be capped with a laminar horizon (Gallala et al., 2010). Finally, the calcrete may become degraded, producing a boulder calcrete or brecciated masses.

Most calcretes are pedogenic, although by no means all (Nash and McLaren, 2003). In the case of pedogenic calcretes, carbonate is mobilized in surface horizons with high CO_2 contents, through which rainwater infiltrates, and deposited lower down the profile where CO_2 levels are less or from which water is lost by evaporation. This is the *per descensum* hypothesis. Until the late 1970s, most models of calcrete formation were physiochemical. Since then, the role of organic processes has become clearer (Goudie, 1996a). Indeed, Wright and Tucker (1991) draw a distinction between two classes of calcrete fabric: beta fabrics exhibiting micro-scale features attributable to the existence of macro- and microorganisms, and alpha fabrics, showing the products of mainly physicochemical processes (e.g. evaporation, degassing, desiccation and expansive growth). They suggest that alpha calcretes occur in areas with a more arid climate and less biological activity, whereas beta fabric calcretes are best developed in semi-arid and subhumid regions where there is an extensive vegetation cover. In Texas, the importance of biogenic structures, which is considerable, decreases as the climate becomes more arid (Zhou and Chafetz, 2009). Ant nests may also be preserved in the fossil form in calcretes (Smith et al., 2011).

Many calcretes, including those of the Nullarbor Plain in Australia (Miller et al., 2012), have various organic structures, and a good study of biogenic calcrete on alluvial fans in Spain is provided by Alonso-Zarza et al. (1998). Among the biogenic fabrics that have been identified are rhizoliths (root structures), calcified filaments produced by bacteria, lichens, fungi and algae, needle fibres associated with fungi or root hairs, fecal pellets, stromatolitic structures, and tree throw brecciation (Goudie, 1996a). However, although organic structures have been found in calcretes, the role of the organism may in some cases have been passive rather than active. In other words, the carbonate deposition may have incorporated organic material rather than necessarily having been promoted by the organisms themselves.

Much calcium comes from dust or rain. This hypothesis (the *per descensum* hypothesis) is corroborated by measurements of carbonate inputs (Machette, 1985; Mayer et al., 1988). Rates of calcium deposition from dust and rain are consistent with the age and amounts of calcium in calcic horizons throughout the south-western United States (Schlesinger, 1985). In New Mexico, strontium isotope studies indicated that at least 98 per cent of the calcium in calcretes originated from atmospheric inputs (Capo and Chadwick, 1999; Van der Hove and Quade, 2002), and similar values were obtained for South Australian calcretes by Lintern et al. (2006). Likewise, studies of calcretes developed on granites in Spain showed the dominant role of atmospheric inputs (Chiquet et al., 1999; Chiquet et al., 2000); very little was derived from weathering of bedrock. Naiman et al. (2000), however, working in the south-west United States, found from isotopic studies that both dust and bedrock weathering were important sources of soil carbonates.

Various factors influence development of calcretes (McFadden and Tinsley, 1985), including the texture of the host material; dense accumulations cannot develop in soils with little void space. Also important are soil carbon dioxide pressures, dust input rates, the amount of infiltrating water available for leaching and the age of the profile. In the classic model of Gile et al. (1966), horizons proceed through a sequence from light cementation to the development of the hardpan calcrete, which prevents further infiltration and so allows the development of the laminar layer above. Following this model, one can view the total amount of carbonate in a single-stage profile as a measure of age (Machette, 1985). It is probable that many well-developed horizons are relics of former climates. In the south-western United States, calcic horizons are too deep to have formed under present rainfalls (Marion et al., 1985), and in Texas, there is widespread evidence for calcrete degradation by solution processes (Hirmas and Allen, 2007). In the southern Negev, Amit et al. (2011b) suggested that conditions are now too dry for extensive and mature calcretes to form, instead suggesting that most formed between 1 and 2 Ma ago before hyperaridity was established. In the Mersin district of Turkey, hardpan calcrete surfaces have a number of karren features, including pits (kamenitza), rills (rillenkarren) and solutionally enlarged fractures (kluftkarren) (Eren and Hatipolgu-Bagci, 2010). As was shown in Section 2.10, solutional hollows – dayas – are widespread on calcrete surfaces. The stable isotope composition of calcretes can also indicate that they formed under different conditions (Andrews et al., 1998). Equally, it is possible that difference types of enrichment relate to different climates and palaeoclimates (Khadkikar et al., 2000; Gunal and Ransom, 2006).

Field relations and characteristics show that many calcretes are non-pedogenic, having been deposited near a water table (the *per ascensum* hypothesis) (Alonso-Zarza and Wright, 2010). The main mechanism may be evaporation from the capillary fringe, but carbonate deposition can be induced below a water table by the changing CO_2 contents of the water. Water-table calcretes develop especially in alluvium, where water tables are near the surface, and where the matrix is coarse. In many valleys, both pedogenic and groundwater calcretes occur, their presence varying both vertically within a profile and laterally across the landscape (McQueen et al., 1999).

Water-table calcretes may be difficult to distinguish from calcretes formed by infiltration of carbonate-rich floodwaters. Individual horizons formed in this way are usually thicker (often >10 m) than other forms of calcrete, occur only in alluvium close to stream channels and show evidence that the alluvium was fresh when cemented (Arakel and McConchie, 1982). Floodwater calcretes and dolocretes (Maizels, 1988), also called channel calcretes (McLaren, 2004), are commonly interstratified with (and in turn hard to distinguish from) tufas or lacustrine marls (Goudie, 1983b). Many channel calcretes are ancient: in Oman and Western Australia, some date to the early Tertiary (Arakel and McConchie, 1982).

2.23 Silcretes

Silcrete is 'a very brittle, intensely indurated rock composed mainly of quartz clasts cemented by a matrix which may be well-crystallized quartz or amorphous (opaline) silica' (Langford-Smith, 1978, p. 2). It commonly consists of brittle masses or nodules of hard, silica-cemented sand with a conchoidal or subconchoidal fracture. The silcrete matrix consists of opal, chalcedony, cryptocrystalline silica and quartz. Webb and Golding (1998) provide details of silcrete petrology, geochemistry and isotopic characteristics. Silcretes have a diverse range of colours, including grey, green, brown, red and white. They are chemically rather simple, generally containing in excess of 90 per cent silica. Horizons are 1–5 m thick, and many form prominent caprocks. They occur in weathered profiles, in sediments and even in unweathered rock. There is commonly more than one silcrete in a profile; they may occur in the same profile with, and superimposed on, calcrete, ferricrete or gypcrete (Summerfield, 1983a). Silcretes have developed particularly extensively in Australia and southern Africa, and examples are known from the Middle East (El-Sayed, 1991b). They are relatively rare in North America but are known from the Ogallala Formation of the High Plains (McCoy, 2011).

Nash and Ullyott (2007) classify silcretes into a variety of genetic types: those which are essentially pedogenic and those which are non-pedogenic (groundwater, drainage-line and pan/lacustrine types). Precipitation from groundwater has been proposed for some of the silcretes in the Lake Eyre Basin of Australia (Alexandre et al., 2004) and elsewhere in inland Australia (Thiry et al., 2006). Modern silcretes may occur on playa margins, as in the Kalahari (Summerfield, 1982; Ringrose et al., 2009), where silica (some derived from diatoms) has been released by high pH conditions and precipitated when silica-rich solutions come into contact with saline solutions.

Where massive, extensive silcretes now occur in deserts – as they do in parts of Australia, southern Africa and the Sahara – most are relics of former climates. In the Sahara, silcrete is thought to date from wet periods in the Oligocene (Busche, 1983), while in southern Africa it may date to the Cretaceous (Summerfield, 1983a). Such ancient silcretes may have been the products of warm, humid climates, although just how wet is a matter for discussion (Twidale and Hutton, 1986). Only a climate considerably wetter than semi-arid would permit the thorough weathering necessary to release silica from silicate minerals in the quantities required. In many silcretes, the silica must have migrated long distances to its place of deposition – another indication of derivation in a wet climate – for migration would also need more water than could have been supplied in a semi-arid climate (Stephens, 1971).

Silcrete is widespread in Australia (Thiry et al., 2006) because of the presence of extensive low-relief surfaces, favourable climatic conditions, large areas of centripetal drainage, geomorphological stability associated with low rates of denudation and long periods undisturbed by tectonic action. Silcretes have formed a number of different surfaces at a number of different times from the Palaeozoic to the Holocene (Ollier, 1991). Silcretes form on diverse lithologies. In central Australia, many silcretes have developed on sedimentary rocks but elsewhere are developed on granites (Butt, 1985) and basalts. They may also develop on terrestrial sediments, including lakebeds (Ambrose and Flint, 1981).

The distribution of Australian silcrete has been the subject of controversy (Stephens, 1971; Young, 1985). Even if climate is not a paramount control (Webb and Golding, 1998), there is a clear tendency for silcretes to be a feature of the continental interior, where their distribution corresponds broadly to that of some of the basin areas extant in the Permian and the Early Cretaceous, including the Canning, Officer and Great Artesian Basins.

2.24 Ferricretes

A ferricrete is a duricrust that has been cemented by iron and is often referred to as laterite. Iron concentrations may be up to 80 per cent. Individual ferricretes are between 1 m and 10 m thick. In deserts, they are undoubtedly relics of a former climate – probably from the Cenozoic – that was warm, wet, although probably highly seasonal. Ferricretes are best developed in the Sahel of West Africa and become thinner – and more discontinuous – northwards into the Sahara (Nahon, 1986). They are also developed on the moister margins of the Australian deserts (Twidale and Hutton, 1986), in the Thar Desert of India (Ramakrishnan and Tiwari, 2006) and in parts of Arabia (Edgell, 2006, p. 17).

2.25 Desert Tufas and Stromatolites

A further type of carbonate accumulation that occurs in desert areas are freshwater meteogene carbonate deposits called tufas or travertines (Viles et al., 2007) (Figure 2.20). These are freshwater carbonates formed from ambient, non-thermal waters. Some are deposited in fluvial channels and form a series of dams or barrages, some are deposited as cascades as streams debouch over mountain fronts, others are deposited by springs issuing on slopes, others are formed as algal bioherms on the margins of lakes, and still others are laid down in swamps and other poorly drained areas (paludal tufas) (Johnson et al., 2009). Ford and Pedley (1996) provide a general discussion of the nature and origin of tufas on a global scale, while Stone et al. (2010) discuss some of the problems of obtaining reliable dates on these materials.

Some of the most notable desert tufas occur in the hyperarid Western Desert of Egypt, especially in the Kharga, Dakhla and Kurkur Oases (Nicoll et al., 1999; Smith et al., 2004). Some were deposited as spring deposits, but many show fine barrage structures. Stromatolitic tufas occur on the margins of the Red Sea Hills in eastern Egypt (Freytet et al., 1994). Extensive Holocene tufas occur in south-west

Figure 2.20 A large mound of calcareous tufa in the West Kimberley District of Western Australia. (ASG)

Libya (Cremaschi et al., 2010), and great tufa cascades are known from Morocco (Weisrock, 1986). The Atacama also has tufas, with huge lacustrine forms known from depressions such as the Salar de Uyuni (Rouchy et al., 1996; Rech et al. 2002). The lake basins of North America are also notable for their tufa pinnacles, formed by springs discharging on the floors of pluvial lakes, as in Searles Lake (Smith, 2009; Guo and Chafetz, 2011), Lake Tecopa (Nelson et al., 2001), Big Soda Lake (Rosen et al., 2004), Owens Lake (Bradbury, 1997), Mono Lake (Scholl and Taft, 1964) and Lake Lahontan (Benson, 1994; Lin et al., 1996). Tufas also formed in the swash zone of some pluvial lakes (Nelson et al., 2005; Felton et al., 2006). Tufa pinnacles on lake floors also occur in the Badain Jaran Desert of China (Arp et al., 1998). Other travertine deposits in the United States are associated with waterfalls at Havasu in the Grand Canyon (Black, 1955; Szabo, 1990; O'Brien et al., 2006) and in Fossil Creek, Arizona (Fuller et al., 2011). There are also deposits associated with springs in Death Valley (Miner et al., 2007).

In the Middle East, spring tufas are known from the Negev and elsewhere in Israel (Weinstein-Evron, 1987; Kronfeld et al., 1998), along the Dead Sea Rift in Jordan and in Oman (Clark and Fontes, 1990). Tufa deposits near Salalla in Dhofar consist of cascades and tufa-dammed lakes. In Wadi Darbat, one tufa dam is more than 70 m high (Hoorn and Cremaschi, 2004). In the drier parts of the Hindu Kush in Afghanistan, major tufa barrages impound the Band-e-Amir lakes. Tufa barrages and large spring tufas are also a feature of the Naukluft Mountains in Namibia (Brook et al., 1999;

Viles et al., 2007; Stone et al., 2010), where a waterfall/cascade tufa at Blasskrantz is 80 m high and more than 400 m across. Massive quantities of tufa, with a great range of morphologies, occur in the semi-arid and subhumid Oscar and Napier Ranges of the Kimberley District of north-west Australia (Viles and Goudie, 1990; Wright, 2000): drapes formed on cliff faces, cones where water issues from caves, barriers in channels and rimstone pools on gentle pediments. Some of the older tufas were karstified (to produce caves and karren) or altered into calcrete crusts.

Although it is possible that under strongly evaporative conditions tufas are forming today in some desert regions, many of them may be the result of enhanced hydrological activity under pluvial conditions, including the early Holocene moist phase (Cremaschi et al., 2010).

Tufas also show the imprint of highly variable flow regimes that cause differing facies associated with deposition, quiescence and erosion, which is brought out in Viles et al. (2007). They suggest the following model of tufa formation to explain the different facies that occur in the Naukluft Mountains of Namibia:

1. An initial irregularity in the stream-long profile creates turbulence in the stream which leads to carbonate deposition if water supply conditions are adequate. Moss also contributes to tufa deposition. A barrage gradually builds up. Pools develop behind the barrage so that laminated and reed facies develop.
2. A phase of episodic high flows causes incision of the softer tufa barrages by large transported boulders. Total obliteration may sometimes occur but normally remnants are preserved.
3. Subsequent lower-flow conditions cause boulder deposition in the stream bed, and tufa is deposited in their interstices.

These three stages may be repeated again and again.

Not all desert tufas are formed of calcium carbonate. For example, in Namibia, there are some deposits in the Swakop Valley that are formed of halite (sodium chloride), which encrusts bird feathers, while in the Layla Lakes region of Saudi Arabia there are what have been described as the largest lacustrine gypsum tufas in the world (Kempe and Dirks, 2008).

Related to such tufa deposits are extensive accumulations of biogenic freshwater carbonates (stromatolites) associated with salt lakes and their strandlines (Casanova, 1991). Examples include those from Walker Lake, United States (Petryshyn and Corsetti, 2011; Petryshyn et al., 2012); Mexico (Winsborough et al., 1994; Kazmierczak et al., 2011); the Turkana Basin, Kenya (Ekdale et al., 1989); various other lakes from the East African Rifts (Grove et al., 1975; Hillaire-Marcel and Casanova, 1987; Casanova and Hillaire-Marcel, 1992; McCall, 2010); the Etosha Basin of Namibia (Smith and Mason, 1991); the Yalgorup lakes of Western Australia (Wood et al., 1991); the Coorong region of South Australia (McKirdy et al., 2010); Lake Bugunnia in south-eastern Australia (McLaren et al., 2012); southern Transbaikalia (Namsaraev et al.,

Figure 2.21 A large spring mound formed by silt accumulation around an artesian spring orifice, Bahariya Oasis, Western Desert of Egypt. (ASG)

2010); the Dead Sea (Lisker et al., 2009); and north-western Argentina (Valero-Garces et al., 2001).

2.26 Spring Mounds

Sometimes composed of carbonate tufa, mound springs – often with a conical morphology – form when artesian water discharges at the ground surface and builds up a mound of deposited material. Some of this material is caused by precipitation of solutes from the escaping groundwater, but some of it is formed by aeolian deposition of silt and sand that is trapped on the moist and vegetated surfaces that surround the spring orifice. Mound springs are widespread in, inter alia, the Great Artesian Basin of Australia, the western United States (e.g. Nelson et al., 2001), Syria (Le Tensorer et al., 2007), parts of Tunisia, the Etosha Pan region of Namibia and the Libyan Desert of Egypt (Brookes, 1993) (Figure 2.21). The mounds found in southern Tunisia (Roberts and Mitchell, 1987) are 25–30 m high, and those in Egypt are up to 20 m high (Caton-Thompson and Gardner, 1932). In Central Australia, the Dalhousie Mound Spring Complex includes some mounds that are as much as 5 km long and 20 m high (Clarke and Stoker, 2003; Keppel et al., 2011), although a height of 8 m may be more normal (Mudd, 2000). The mounds of the Dakhla Oasis in Egypt, fed by the iron-rich waters of the Nubian Aquifer (Adelsberger and Smith, 2010) are capped with a ferruginous sandstone crust.

2.27 Surface Types

In addition to the surfaces dominated by the presence of the various types of geochemical crust outlined in earlier sections deserts have a number of other soils and surface material types. These include stone pavements, biological and rain-beat crusts, gilgai, polygonal ground and features produced by organic activity (such as banded vegetation and mima mounds).

2.28 Stone (Desert) Pavements

Stone (or desert) pavements are armoured surfaces composed of a mosaic of fragments, usually only 1 or 2 stones thick, set on or in matrices of finer material comprising varying mixtures of sand, silt or clay (Wood et al., 2005). They are generally underlain by a vesicular layer (McFadden et al., 1998). Pavements occur on weathered debris mantles and alluvial terraces and fans in many deserts. In Australia, these stony surfaces are often called gibber plains, and their significance was recognized early on by Gregory (1906).

They are formed by a range of processes that cause coarse particle concentration at the surface: (a) the classic mechanism of deflation of fine material by wind (Figure 2.22), (b) removal of fines by surface runoff and/or creep and (c) processes causing upward migration of coarse particles to the surface. In addition, it has become increasingly clear that pavements may evolve in close association not only with aeolian erosion but also with dust deposition and soil-profile differentiation caused by weathering. Their stability, antiquity and lack of leaching means that they favour the retention and accumulation of salts, including nitrates (Graham et al. 2008). The nature of the clasts that form the pavements is also affected by various cracking processes (Adelsberger and Smith, 2009). Lateral surface process such as unconcentrated overland flow and creep also contribute to pavement formation and to the orientation of their surface clasts (Dietze and Kleber, 2012).

An interesting question is just how capable stone pavements are of recovering from disturbance (Figure 2.23). Sometimes, once damage has been done, recovery may be slow and difficult. In Egypt's Western Desert, tracks of vehicles from the First World War are still visible in desert surfaces. Off-road vehicle tracks in the Mojave Desert, in contrast, seem to have recovered in places within a few decades (e.g. Elvidge and Iverson, 1983). Much depends on the availability of stones and the effectiveness of compaction, which may restrict pavement-producing processes. In other localities, pavements have been seen to heal relatively rapidly following deliberate local disruption of their surface cover of stones (Dietze and Kleber, 2012).

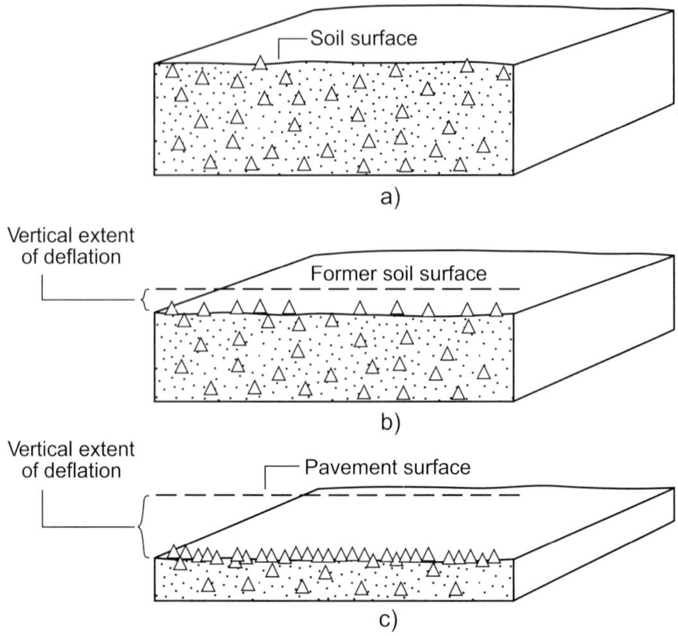

Figure 2.22 The classic deflation model of stone pavement development with an initial alluvial sediment containing both fine and coarse materials (a) being subjected to deflation (b) until such time as the surface is lowered to such an extent that (c) a coarse gravel lag or armour is left at the surface, the fine material having blown away. (ASG)

Haff and Werner (1996), working in California, found that gaps healed in around five years and that displacement of surface stones by small animals was a major component of the healing process. Similarly, Wainwright et al. (1999), using rainfall simulation experiments at Walnut Gulch in Arizona, found raindrop erosion processes resulted in rapid surface recovery. They reported (p. 1034) that 'significant accumulation of coarse particles formed within five 5 min events following the disturbance. Given that more than five such events typically occur every year at Walnut Gulch, pavements can generally be expected to recover on an annual cycle after disturbance.'

Pavements are important because of their effects on infiltration and runoff (Meadows et al., 2008); their tendency, once formed, to inhibit aeolian deflation of fine material; and their use as relative dating tools (although see Pelletier et al. [2007], for a criticism of this approach). Some stone pavements are undoubtedly of great antiquity and associated with surfaces that have only undergone very limited modification since they were formed. Those in the Negev, according to studies using cosmogenic radionuclides, have retained their original topography for more than 2 million years (Matmon et al., 2009), whereas those in the Capitol Reef National Park in Utah gave dates ranging from c 98–159 ka (Marchetti and Cerling, 2005).

Many hypotheses have been put forward to account for pavement formation, and these are considered in the following sections.

Figure 2.23 A stone pavement in Egypt's Western Desert which has been damaged by an unthinking driver. (ASG)

Deflation

Pavements have usually been explained as being produced by deflation of fine material from the surface, which leaves a residue, lag or armour of coarse particles. The concentration of coarse particles is seen as a function of their distribution in the original sediment and the extent of deflation.

Loose, fine material can indeed be removed by wind, as the frequency of dust storms – together with experimental work – has demonstrated (Symmons and Hemming, 1968). Arguments can be advanced against the ubiquitous application of the deflation hypothesis, however. Although desert winds are probably strong enough to move most loose, fine material, in semi-arid areas scrubby vegetation may reduce winds to below the requisite threshold velocities, and surface vegetable litter may protect loose fines. Even more important is the fact that 'loose' fine material may be bound at the surface into a thin biological or rain-splash crust. Deflation will occur only if this carapace is broken by, for example, prior wind abrasion, animal activity (Haff, 2001) or vehicular traffic.

Water Sorting

Experimental observations show that some pavements are often composed – at least in part – of coarse particles that remain after finer materials have been dislodged and removed by raindrop erosion and running water (Wainwright et al., 1995). Sharon (1962) demonstrated that surface runoff is the dominant process of pavement formation in Israel. At cleared pavement sites in the Mojave, Cooke (1970) showed that surface fines and some buried coarse particles were removed by surface runoff and collected in sediment traps downslope. Plainly, the role of sheetfloods should not be ignored as a horizontal transport mechanism (Williams and Zimbelman, 1994; Dietze and Kleber, 2012).

Upward Clast Migration

The concentration of coarse particles at the surface and at depth, and the relative scarcity of coarse particles in the upper soil profile (Cooke, 1970, Mabbutt, 1965, Springer, 1958), suggest that stones may have moved upwards through the soil to the surface by cycles of freezing and thawing, wetting and drying or salt heave. Indeed, laboratory experiments have shown that alternate freezing and thawing of saturated mixed sediments causes coarse particles to migrate towards the ground surface (Corte, 1963; Mackay, 1984). Although sorting by freeze/thaw action may seem improbable in hot deserts today, it cannot be ruled out, especially in high-altitude deserts. It may also have been more effective in some places during cooler, moister periods of the Quaternary. Nonetheless, a much more effective and widespread migration mechanism in deserts is thought to be associated with wetting and drying of the surface soil (McFadden et al., 1987), and various experiments have shown this to be feasible (Springer, 1958; Cooke, 1970). Springer (1958) suggested that when a soil containing expanding clay minerals is wetted, it expands and a coarse particle is lifted slightly. As the soil shrinks on drying, cracks are produced around the particle and within the soil. Because of its large size, the coarse particle cannot move down into the cracks, whereas finer particles can. The net effect is an upward displacement of the coarse particle. Subsidiary mechanisms of upward migration may be salt heave (Horta, 1985; Searl and Rankin, 1993) and the activity of soil fauna, including ants, termites, and burrowing mammals. Whether or not bioturbation causes stone pavement formation or disruption is, however, still a matter of debate. On the one hand, churning and burrowing may bring fine material to the surface, where it can be deflated, whereas on the other hand, the process may cause coarse particles to sink and for homogenisation to occur. Under higher rainfall conditions (e.g. during pluvials), it is probable that pavement disruption predominates.

The Role of Dust Deposition

In recent years, it has become accepted (see Section 3.4) that significant amounts of dust are delivered to desert surfaces by dust storms, and it is therefore inevitable that such dust contributes to the development of stone pavements. McFadden et al. (1987), followed by Anderson et al. (2002) and Valentine and Harrington (2006), have argued that on the basalt lavas of the Cima Volcanic Field in the Mojave Desert, pavements are born at the surface, and that the individual pavement stones were never buried. The pavements in this area were said to evolve as follows. First, volcanic rocks are broken up by weathering into clasts, a process accelerated by the ingress of aeolian silty dust into cracks and subsequent volume changes with wetting and drying. Such clasts move by colluvial and alluvial processes from higher to lower areas that are already filled with silt and clay. Deposition of aeolian material from dust storms continues. The aeolian mantle is altered pedogenically by, for example, leaching of soluble materials (Ikeda et al., 1998) to create a vesicular A_v horizon. Finally, the clasts are maintained at the surface, while aeolian deposition and pedogenesis continue. In this explanation, therefore, no aeolian deflation or upward migration is necessary, although it is probable that both may occur from time to time.

Amit and Gerson (1986) worked on a series of fifteen Holocene alluvial terraces in the Dead Sea region, where the degree of pavement development (measured by per cent surface cover of stones and sorting, for example) is most pronounced on the oldest surface and weakest on the youngest. Such changes are associated over around 14,000 yr with surface weathering by salt and other processes, and contemporaneous soil-profile differentiation. In this area, as in the Mojave, the aeolian input of silt and salt is fundamental, and the creation of a full stone cover helps to reduce infiltration capacity and promote runoff. Indeed, many of the characteristics of pavement surface clasts, including their degree of splitting and their grain sizes, may be affected by weathering processes (Al-Farraj and Harvey, 2000; Al-Farraj, 2008), the significance of which may increase with land-surface age.

So then, deflation may create stone pavements, although its operation is reduced by their presence. Other processes contribute to their development, however, including surface runoff, weathering, heave and dust accretion. Stone pavements cannot, therefore, be thought of solely as products of aeolian action.

2.29 Takyrs

Takyrs are 'flat or slightly sloping dense clay surfaces sited in closed drainage areas of accumulative landscapes on alluvial, proluvial and deluvial deposits, and in areas of ancient irrigation in the arid zones of Central Asia' (Fleskens et al., 2007). They are characterised by compact, polygonally cracked surface crusts with little or no vegetation. Waterlogging and the activity of algae and invertebrates play an important role in their development (Lebedeva-Verba and Gerasimova, 2010). They have high clay contents, are sodic, poor in humus and of low permeability. Thus they can generate a large amount of runoff. When abraded by saltating sand, they also appear to be a major source of dust, as in the southern Aral Sea Basin (Singer et al., 2003).

2.30 Biological Crusts

In recent years, it has come to be recognized that organic (also called microphytic, microbiotic, cryptogamic, cyanobacterial or biological) crusts in and on the surfaces of desert soils and sediments play an important hydrological and geomorphological role (Eldridge and Rosentreter, 1999; Viles, 2008; Maestre et al., 2011). They are composed of mosses, algae and lichens. Their distribution varies with surface conditions, and in the western United States, for example, they are especially well developed on surfaces developed in granite grus (Pietrasiak et al., 2011). In the Kalahari, Thomas and Dougill (2007, p. 23) found that in the absence of livestock, grazing crust cover was 84–95 per cent, compared to 44–68 per cent in grazed locations in proximity to boreholes. Crusts are typically 2–15 mm thick (Zaady and Offer, 2010). In Qatar, Richer et al. (2012) found that mean percentage cover varied between 0 and 87 per cent, with the greatest cover being in the north of the country, where there are many closed depressions (dayas), where rainfall is slightly higher than in the south and where surfaces tend to be more stable.

Unlike vascular plants, the cover of organic crusts is not reduced in drought years, and it is present the whole year round. They are, however, very susceptible to anthropogenic disturbance (Belnap and Gillette, 1997; Thomas and Dougill, 2007; Liu et al., 2009), although they are also capable of reforming rapidly once the disturbance is removed (Zaady and Offer, 2010).

Organic compounds, including plant waxes, can produce hydrophobic (water repellent) substances, as can a range of fungi and soil microorganisms. Although water repellent soils occur in more humid environments, many examples of them also have been reported from semi-arid areas (Doerr et al., 2000). These hydrophobic surfaces tend to be zones of reduced soil infiltration capacity and thus of increased overland flow. Following from this is the likelihood that enhanced soil erosion also occurs. Removal of the crusts has indeed in some cases been shown to have a dramatic effect on infiltration rates and runoff (Eldridge et al., 2000), but the picture is not consistent (Issa et al., 2009), and there are cases where lower infiltration rates have been observed for soils with microbial cover compared to soils with low levels or no microbial cover.

The filaments and extracellular secretions of cyanobacteria also are important in the formation of water stable aggregates that help soils to resist water erosion and raindrop impact effects (Issa et al., 2001). It also needs to be appreciated that not all organic crusts are hydrophobic, and that by eliminating the effect of raindrops, they prevent the rapid development of a sealed layer (rain crust) conducive to runoff generation (Eldridge and Greene, 1994; Kidron and Yair, 1997). Desert shrubs may produce a litter layer that reduces rain splash (Geddes and Dunkerley, 1999).

Likewise, biological soil crusts have an influence on aeolian processes (Goossens, 2004). A cover of cyanobacteria, green algae, lichens and mosses is important in stabilizing soils in drylands and protecting them from wind erosion (Belnap and Gillette, 1997 a,b) Kurtz and Netoff, 2001; Eldridge and Leys, 2003). Filamentous cyanobacteria mats are especially effective against wind attack (McKenna-Neuman et al., 1996). They also play a role in dune stabilization (Kidron et al., 2000) and can improve soil quality by promoting dust accumulation (Zaady and Offer, 2010).

2.31 Inorganic (Rain-Beat) Crusts

The surfaces of some deserts, especially those with silt and clay-rich materials, are covered with crusts that develop when physico-chemical changes occur at the interface between the air and the soil during and after rainfall events These rain-beat crusts develop in dry regions because a protective plant canopy is seldom present to intercept the impact of falling raindrops. Direct raindrop impact on such unprotected surfaces causes speedy disruption of soil aggregates, compaction of the surface, slaking (see Section 2.5) and the filling of cracks and pores by wash-in of fine material (Ben-Hur and Lado, 2008). Simple calculations suggest that an enormous number of drop collisions occur on exposed soil surfaces during rain (Dunkerley, 2011, p. 106):

> If we consider a rain event delivering 10 mm and suppose for simplicity that all of the raindrops are spheres uniformly 1.5 mm in diameter, then over each square metre during the storm there are about 5.65×10^6 drop impacts. Every point on the soil surface would be struck multiple times.

Physical crusts are best developed on fine-textured soils (i.e. with high proportions of silt and clay), particularly those with a high Exchangeable Sodium Percentage.

2.32 Patterns

Gilgai

Gilgai is a type of micro-relief consisting of mounds and depressions arranged in random to ordered patterns (Verger, 1964). It is an Australian aboriginal word meaning 'small waterhole' (Hubble et al., 1983), and seasonal ponding of water does occur in some of the closed depressions of the larger forms. There is a great variety of forms, and they occur on a range of swelling clay and texture-contrast soils that have thick subsoil clay horizons, especially vertisols. They tend to occur on level or gently sloping plains in areas subject to cycles of intense wetting and drying, but they are by no means restricted to dry regions. They have been reported from a number of arid and several semi-arid areas, including Coober Pedy (Ollier, 1966) and New South Wales (Hallsworth et al., 1955) in Australia, the central Sahara (Meckelein, 1959), the Middle East (Harris, 1959; White and Law, 1969; Khresat and Taimeh, 1998) and on tropical black earths in East and central Africa (Stephen et al., 1956), South Dakota (White and Bonestall, 1960) and elsewhere (Verger, 1964).

Some gilgai consist essentially of mounds and depressions; however, others are linear forms (Hallsworth et al., 1955). Beckmann et al. (1973, p. 365) see surface runoff and soil heaving as working together to produce such features, particularly on pediment slopes. In the Kimberley of north-western Australia there are individual linear gilgai up to 2 km long, and it is possible that in their case aeolian processes have contributed to their development (Goudie et al., 1992). Round gilgai and some network gilgai occur on flat ground; lattice, wavy and some network gilgai occur on gently sloping ground, and their puffs or depressions are generally oriented parallel to slope contours. The dimensions of the different elements are extremely variable. The vertical relief between mound crest and depression floor may vary from a few centimetres to about 3 m; the diameter of non-linear positive or negative features may be up to about 50 m; linear features may be many metres or even some kilometres long and as much as 12 m wide.

The small steps which interrupt generally smooth but sloping stone pavement surfaces are a type of stepped gilgai. They are characterised by risers of pebbles and treads of finer material which is often capped by a thin veneer of stone pavement. In some areas, such as in Panamint Valley, California, the steps appear to be produced by the accumulation of fine material upslope of large boulders that extend through the soil profile. In Death Valley, California, (Denny, 1965, 1967), downslope creep of fine, relatively stone-free material beneath the pavement when it is saturated may lead to the development of steps. Hunt and Washburn (1960) indicated that the treads of some steps had up to ten times as much salt as stable ground around the steps, which might indicate that the salts play a role in developing them.

Most mechanisms of gilgai development involve swelling and shrinking of clay subsoils under severe seasonal climate. A widely adopted hypothesis for their formation is as follows (Hubble et al., 1983, p. 31):

> When the soil is dry, material from the surface and the sides of the upper part of major cracks falls into or is washed into the deeper cracks, so reducing the volume available for expansion on rewetting of the subsoil. This creates pressures which are revealed by heaving of the soil between the major cracks which, once established, tend to be maintained on subsequent drying. This process is repeated, with the result that the subsoil is progressively displaced, a mound develops between the cracks, and the soil surface adjacent to the cracks is lowered to form depressions.

Pioneer studies by Hallsworth and Beckmann (1969) and Hallsworth et al. (1955) suggest that the nature of gilgai depends on the presence of swelling clays (especially montmorillonite) and on the sodium saturation of the exchange complex (the sodium ions adsorbed to the clay complex produce larger, more resistant clods, and the larger the clods moved by swelling, the greater the amplitude of the undulations). In addition to soil properties, gilgai formation depends on climatic conditions (controlling the nature of wetting and drying) and soil moisture. For instance, if the swelling layers

Figure 2.24 Desiccation cracks developed in flood silts, Tumas Flats, Namibia. (ASG)

never dry out, movement of the wetting front will be minimal and gilgai would be unlikely to form. Gilgai formation is most likely in areas with a marked alternation of wet and dry seasons.

Desiccation Cracks

As a saturated, fine-grained, cohesive sediment dries out due to evaporation, sufficient tensional stress for rupture often occurs and cracks are formed (Figure 2.24) (Lachenbruch, 1962; Maizels, 1987). In general, the morphology of the rupture patterns depends mainly on the intrinsic conditions of the material (such as moisture content, structure and degree of packing) and on extrinsic conditions of the environment (temperature, humidity, rate of desiccation, etc.) (Corte and Higashi, 1964). Research by Kindle (1917) suggested that the spacing of cracks may increase with the rate of desiccation and the proportion of clay present in the material. The type of clay present (Chico, 1963) and the cohesive properties of the material may also influence the extent of contraction. For instance, montmorillonite-rich sediment is likely to contract more than sediment with a comparable proportion of kaolinite.

Giant desiccation fissures, sometimes called earth fissures, are found in some alluvial sediments and especially on playas. Their suggested causes have included seismic events, piping erosion, tensional cracking due to subsidence resulting from differential compaction caused by hydrocompaction or groundwater withdrawal for

Table 2.18 *Main characteristics of banded vegetation patterns as indicated in the literature*

Country	MAR* (mm)	MSG† (%)	WL‡ (m)	Band type
Australia	190	1.833	45	Chenopod shrub
Australia	225	0.350	80	Trees/shrubs
Australia	240	0.400	20	Perennial grass/shrubs
Australia	240	1.320	38	Perennial grass/shrubs
Australia	250	0.200	160	Trees/grass
Jordan	75	NA§	68	Shrubs/grass
Mali	200	2.100	60	Trees/grass
Mali	550	1.300	250	Trees/grass
Mauritania	250	0.300	94	Perennial grass
Mexico	260	0.372	120	Shrubs
Niger	310	0.400	55	Trees/grass
Niger	560	0.524	74	Trees/grass
Niger	641	0.480	55	Trees/grass
Somalia	150	0.602	NA	Perennial grass
Somalia	213	0.222	133	Grass, woodland
Sudan	250	0.123	233	Tree
Sudan	250	0.357	32	Grass
Sudan	350	0.500	NA	NA
Sudan	450	0.500	90	Trees

* MAR: Mean annual rainfall.
† MSG: Mean slope gradient.
‡ WL: Wavelength.
§ NA: Not available.
Source: Modified after Valentin et al. (1999, table 2)

irrigation (Wehmeier, 1998; Hernandez-Marin and Burbey, 2010), salt mobilization within playa sediments, shrinkage due to desiccation or some combination of these. On playas, which afford flat surfaces for vehicular movement, these giant fissures can prove to be dangerous, so that considerable effort has been expended in studying them (e.g. Neal, 1965, 1968). The sudden creation of giant fissures in playas can seriously damage houses, runways and roads, as in the western United States (Harris and Allison, 2006).

Banded Vegetation

From the air, many dryland surfaces can be seen to have a surface characterised by alternating light and dark bands (Valentin and Poesen, 1999; Tongway and Ludwig, 2001; McDonald, et al., 2009). The bands, which are often arcuate (Figure 2.25a and b), tend to occur in regions with between 50 and 750 mm mean annual rainfall and on slopes that are gentle and uniform (0.2–2%) (Table 2.18). The great majority of

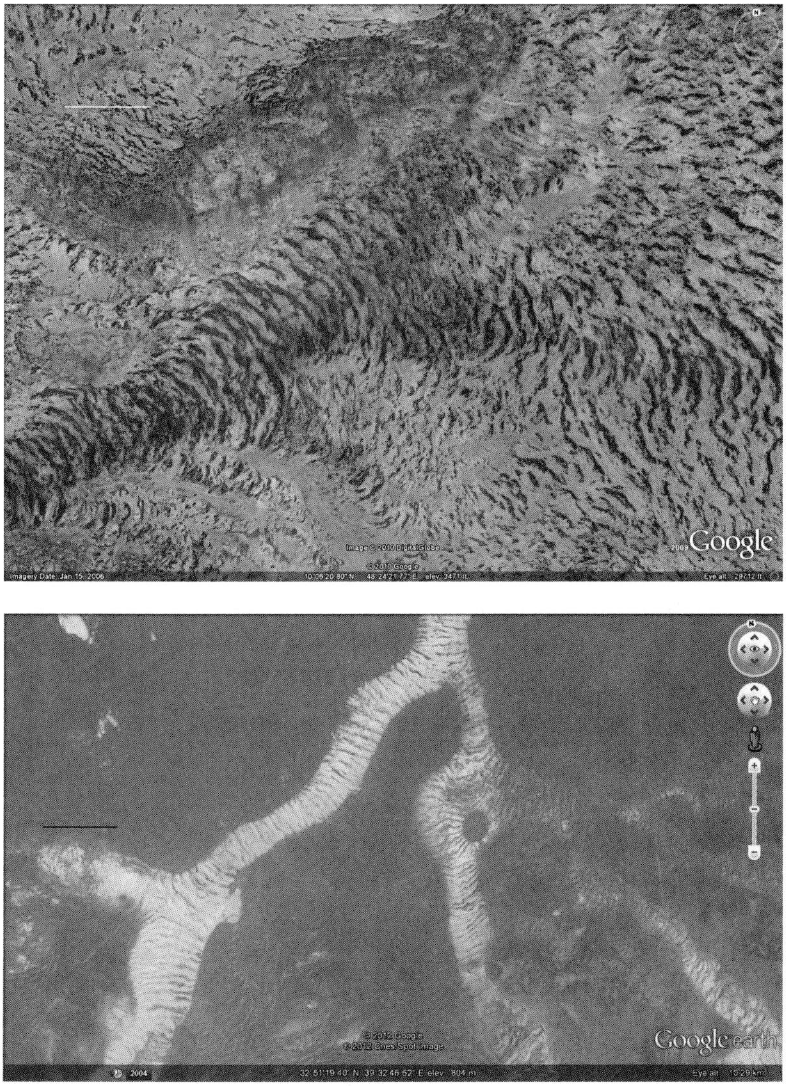

Figure 2.25 Tiger Bush (a) Northern Somalia. Scale bar 1 km. (©Google Earth 2010, ©Digital Globe 2010). (b) Syrian Desert, Iraq. Scale bar 1 km. (©Google Earth 2012, ©Cnes/Spot Image 2012)

the studies on banded vegetation in Africa have been carried out in eastern or central-western Africa, starting from the pioneering research of Gillet (1941) and MacFadyen (1950) in the former British Protectorate of Somaliland. Now, however, it is recognized that they are of much wider extent, with locations identified in, inter alia, Australia, Namibia, Bolivia, Iran, Iraq, Syria, Kazakhstan, Madagascar, Turkmenistan, Yemen, Zambia and the United States (Paron and Goudie, 2007).

The stripes tend to be perpendicular to the slope (Valentin et al., 1999) although this is not invariably the case (Dunkerley and Brown, 2002). They develop in a range of geomorphological situations, including gentle slopes subjected to surface runoff, ephemeral drainage lines and inter-dune corridors (Goudie, 2007b). The patterns are formed by vegetation differences and are often termed *brousse tigreé* (tiger brush). The banded vegetation can consist either of grass, trees, shrubs or trees and grass. The wavelength of the features ranges between c 20 and 250 m, with the wavelength being the length of a cycle including a band and an interband.

The differences in vegetation relate to differences in soil depth and character, with the vegetated bands occurring on deeper soils with higher infiltration capacities than the bare ground in between. Monitoring has shown that the widths of the bands vary with rainfall, and that the bands can migrate upslope at rates of c 0.2–0.75 m yr^{-2} (Leprun, 1999).

The prime reason for the development of soil and vegetation banding is sheetflow (Mabbutt and Fanning, 1987). A mechanistic explanation for these features (Dunkerley and Brown, 2002) sees this phenomenon as being the product of the different properties of bare and vegetated soils, in which bare surfaces develop crusts and become impermeable, thereby directing runoff downslope to a neighbouring grove of plants where the water is absorbed into more permeable soils, thereby supporting a greater plant cover. The partitioning of the surface into those zones which are water scarce and those which are water rich is regarded as the fundamental cause of the banded vegetation (Saco et al., 2007). Some field studies have reported that infiltration rates under vegetation patches or bands is up to ten times that of inter-band areas. The water also acts as a conveyor sheet and leads to further accumulation of litter, sand and other organic and inorganic particles at the vegetated band, which thus acts as a dam. As a consequence, the bands tend to migrate upslope and to contract downslope (Deblauwe et al., 2012).

The circular 'fairy rings' of the pro-Namib in Namibia (Albrecht et al., 2001) (Figure 2.26) are another type of intriguing, but little understood, vegetation pattern. These occur in the thousands in areas where the mean annual rainfall is 100–150 mm, and consist of bare patches c 2–12 m in diameter. They are widespread between the Orange River and southern Angola (Becker and Getzin, 2000; Picker et al., 2012). It is possible that their origin owes something to the foraging action of termites (Gruber, 2002) or ants (Picker et al., 2012) or to growth inhibition as a result of allelopathic compounds released by dead *Euphorbia damarana* plants, but as yet there is no entirely satisfactory explanation for their origin (van Rooyen et al., 2004). It has also been proposed that they may be the result of micro-seepage of gases and hydrocarbons (Naudé et al., 2011).

Mima and Other Mounds

Also called prairie mounds and pimple mounds, mima mounds take their name from Mima Prairie, Thurston County, Washington, United States. Such mounds are

Figure 2.26 'Fairy circles' in the grasslands of the Namib Desert near Rostock Ritz. Scale bar 100 m. (©Google Earth 2011, ©GeoEye 2011)

characteristically up to around 2 m in height, 25–50 m in diameter and occur at a density of 50–100 or more to the hectare. There are many hypotheses for their origin (Cox and Gakahu 1986), including that they are erosional residuals, result from depositional processes around vegetation clumps, are the product of frost sorting, have been formed by communal rodents, are degraded termitaria or have been created by seismic activity or groundwater vortices (Reider et al., 1996). In the United States, they are found from the Gulf of Mexico to Alberta, and some of them may be relict nebkhas (dunes that accumulate around plants) that formed during past drought phases (Seifert et al., 2009). Certainly, some of the mounds occur in quite dry locations with less than 300 mm of mean annual rainfall, as in southern California and in the Columbia Plateau of Oregon (Cox and Allen, 1987). Similar forms are also known from Argentina and Kenya. These mounds probably have many different origins, but the role of such burrowing beasts as mole-rats, gerbils, wombats, prairie dogs and gophers should not be underestimated (Cox, 1990; Whitford and Kay, 1999).

In southern Africa, the mounds are called heuweltjies, and they are widely distributed in the drier, western parts (Lovegrove and Siegfried, 1989). Evidence suggests that they are the large and mature, long-lived mounds of the southern harvester termite *Microhohotermes viator*. The highest densities of these mounds (>400 km^{-2}) occurs

Figure 2.27 Mounds formed beneath *Larrea tridentata* bushes in the Mojave Desert, California, USA. (ASG)

in the zone where mean annual rainfall is between 150 and 350 mm per year. The mounds average 17 m in diameter and have a height of 1.45 m (Picker et al., 2007).

Not all mounds are the result of the activities of burrowing animals. In sea-arid eastern Australia, for example, there are mounds associated with mulga (*Acacia aneura*) bushes and logs, which although partially the result of termites, are also the result of aeolian and fluvial processes (Tongway et al., 1989). Indeed, in the south-western United States there are huge numbers of creosote bushes (*Larrea tridentata*) that have mounds beneath them that are some tens of centimetres high (Figure 2.27) (Soulard et al., 2012). Many of these mounds are formed by a combination of rainsplash, raindrop detachment and overland flow across the bare patches between the bushes (Parsons et al., 1992). In other words, they are largely residual features. This has also been seen as true of mounds found in Patagonia (Rostagno and del Valle, 1988) and in the Negev, where inter-shrub areas have biological crusts which generate runoff and are absent under shrubs (Wright et al., 2006). Mounds may be areas of relative accumulation, however, catching as they do detritus transported by wind and water (Furbish et al., 2009). A positive feedback may be involved in that the mounds accumulate material, so that they receive nutrients and also have deeper soils that can retain more moisture. Both factors encourage bush or patch growth. Furthermore, they may have low-infiltration capacities and so will be less subject to water erosion (Bedford and Small, 2008). The mounds are in effect 'islands of fertility' (Tongway et al., 1989; Wright et al., 2006) with a self-reinforcing tendency (Bartley et al., 2006). Many mounds are probably the result of a combination of inter-shrub erosion and under-shrub accumulation (Buis et al., 2010).

3
Aeolian Geomorphology

3.1 Introduction

It is in the world's deserts that wind action is at its most potent as a geomorphological agent, although as we have seen in Chapter 1 there have been great debates about its relative importance with respect to water action. In reality there are great differences in the roles of wind in different deserts, and there are also many ways in which wind and water work together. Wind is more effective in deserts not because it is stronger than in wetter environments – for in general it is not – but for two other reasons: dry surfaces and sparse vegetation.

In this book, the role of wind is divided into two parts; dunes have their own section. In this chapter, we consider a range of other aeolian phenomena, including the role of wind as an agent of erosion and deflation and its contribution to the development of desert loess, dust storms, pans, yardangs and ventifacts. Research into aeolian phenomena is now very active, and so in 2009 a new journal appeared, *Aeolian Research*.

3.2 The Spatial Variability of Wind Power

The significance of wind erosion varies enormously between and within deserts because of the great variability that exists in wind power. This becomes evident when one examines data for annual sand drift potentials based on the method of Fryberger (1979) (Table 3.1). These range from values as low as 11 vector units to as high as 1,401 vector units. Fryberger (1979) classified locations with values greater than 400 as high-energy wind environments, those between 200 and 400 as intermediate-energy environments and those with less than 200 as low-energy environments. On this basis it appears, for example, that the Great Plains of the United States are a high-energy environment, as are parts of the Libyan Desert in Libya and Egypt, whereas the Kalahari, the Thar and many of the Chinese deserts are relatively

Table 3.1 *Annual drift potentials*

Vector units	Location	Source
1,401	El-Khanka, Egypt	Moursy et al. (2002)
1,155	Akron, Colorado, USA	Muhs et al. (1996)
1,090	Hutchinson, Kansas, USA	Arbogast (1996b)
750	Northern Great Plains, USA	Muhs and Wolfe (1999)
687	Milford, Utah, USA	Jewell and Nicoll (2011)
655	Sabha, Libya	Linsenbarth (1991)
655	Wadi Araba, Jordan	Saqqa and Atallah (2004)
595	Guaizihu, China	Yang et al. (2011)
588	Ghadamis, Libya	Linsenbarth (1991)
540	Bahrain	Fryberger (1980)
518	Walvis Bay, Namibia	Fryberger (1980)
489	An Nafud, Saudi Arabia	Fryberger (1979)
468	Kufra, Libya	Linsenbarth (1991)
457	Alashan Right Banner, China	Yang et al. (2011)
414	Jalu, Libya	Linsenbarth (1991)
392	El Centro, California, USA	Muhs et al. (1995)
405	Qatar	Embabi and Ashour (1993)
391	Simpson Desert, Australia	Fryberger (1979)
384	Mauritania	Fryberger (1979)
366	Karakum/Kyzlkum	Fryberger (1979)
362	Hun, Libya	Linsenbarth (1991)
354	Kuwait	Al-Awadhi et al. (2005)
335	Bayinmaodao, China	Yang et al. (2011)
321	Algodones, California, USA	Muhs et al. (2003)
293	Algeria	Fryberger (1979)
281	Ejina Banner, China	Yang et al. (2011)
270	Al Jaghbub, Libya	Linsenbarth (1991)
266	Dunhuang, China	Wang et al. (2005a)
237	Namib	Fryberger (1979)
203	Blythe, California, USA	Muhs et al. (1995)
201	Rub 'Al Khali, Saudi Arabia	Fryberger (1979)
191	Kalahari	Fryberger (1979)
178	Dingxin, China	Yang et al. (2011)
160	Badain Jaran, China	Z. Wang et al. (2005b)
144	Winnemucca, Nevada, USA	Jewell and Nicoll (2011)
147	Ashdod, Israel	Tsoar and Blumberg (2002)
139	Mali	Fryberger (1979)
134	Kara-Kum, Kyzyl-Kum	Maman et al. (2011)
127	Gobi, China	Fryberger (1979)
116	Ghat, Libya	Linsenbarth (1991)
114	Indio, California, USA	Muhs et al. (1995)
86	Tazirbu, Libya	Linsenbarth (1991)
82	Thar Desert, India	Fryberger (1979)
75	Northern Kalahari	Thomas (1984)
41	Taklimakan, China	Z. Wang et al. (2005b)
37	Mu Us, China	Z. Wang et al. (2005b)
11	Tengger, China	Ha (2002)

low-energy environments. That said, there is often great variability in wind energy within deserts, in part because of topographic control. Values in the Taklimakan, for example, range between 2 and 112 (Wang et al., 2002; Wang et al., 2005), and for northern China as a whole between 0.9 and 1,052 (Wang et al., 2006), whereas, those in Libya range between 86 and 655 (Linsenbarth, 1991). In the Western (Libyan) Desert of Egypt values range between 27 and 571 (Hereher, 2010). In the Great Basin of the United States, values range between 144 and 687 (Jewell and Nicoll, 2011). There is also a substantial degree of temporal variability (see, for example, Bullard et al., 1996; Jewell and Nicoll, 2011). Saqqa and Atallah (2004), for instance, analysed annual drift potentials in the Wadi Araba of Jordan for each of 23 years, and found that they ranged from 141 to 1,070.

However, the significance of wind erosion is also controlled by small-scale wind events with high vertical velocities and gustiness, something which is hidden in the general wind velocity characteristics discussed in preceding paragraphs.

Traditionally, our knowledge of the wind characteristics of deserts has been based on analysis of data from weather stations, but there are relatively few stations in the heart of deserts for which there are reliable, long-term wind records. This limitation is now being partially resolved by the use of global wind data obtained from the European Centre for Medium Range Weather Forecasts (ECMWF) ERA-40 reanalysis project (see, for example, Livingstone et al., 2010), but the spatial resolution of this is still quite coarse, and the influence of local topographic conditions is inadequately represented.

3.3 Past Wind Velocities

The strength of the trades may have intensified in the Pleistocene as a whole and also during particular phases of the Pleistocene. In the late Pliocene (3.2–2.1 Ma), there may have been an increase in atmospheric circulation driven by a steeper pole-equator temperature gradient owing to the development of the bipolar cryosphere (Marlow et al., 2000). Within the Pleistocene, analysis of ocean core sediments has shown variability in pollen, diatom, phytolith and sediment influx to the oceans which may be explained by variations in wind velocities (Parkin, 1974; Kolla and Biscaye, 1977; Sarnthein and Koopmann, 1980). Hooghiemstra, 1989). Moreover, changes in upwelling intensity and oceanic productivity, established by analyses of benthic foraminifera, have been linked to changes in wind intensity (Loubere, 2000).

Studies based on these lines of evidence have indicated that north-east trade velocities were higher during glacials. This was probably because of an intensified atmospheric circulation caused by an increased temperature gradient between the North Pole and the equator, which resulted from the presence of an extended Northern Hemisphere ice cap (Stein, 1985; Ruddiman, 1997; Kim et al., 2003). Work off Namibia, however, suggests that the south-east trades were also intensified during

Figure 3.1 A dust storm at Jazirat Al Hamra, United Arab Emirates. (ASG)

glacials compared to interglacials (Stuut et al., 2002). Off north-west Africa, the highest wind velocities may have occurred during the last deglaciation rather than at the times of maximum ice conditions (Moreno et al., 2001). Moreno and Canals (2004) attribute this to the lowering of North Atlantic sea surface temperatures during deglaciation because of glacial meltwater releases. This in turn strengthened the North Atlantic high-pressure system, caused a high temperature difference between land and sea and enhanced the trade-wind system.

3.4 Dust Storms

Dust storms (Figure 3.1) are an important manifestation of geomorphological activity in deserts (Goudie and Middleton, 2006) and play an important role in the Earth system (as reviewed by Shao et al., 2011). On the one hand, they are a consequence of wind deflation, and on the other, they lead to deposition of substantial amounts of material, much of which consists of silt. Rates of dust deposition range from almost 0 to greater than 450 g m^{-2} yr^{-1}, with distance from source being a primary control (Lawrence and Neff, 2009). With regard to the Sahara, the coarsest dust (more than 70 μm) occurs in or close to the Sahara itself, whereas dust that has travelled further tends to be finer silt, between 5 and 30 μm in diameter. Aeolian dust is dominated by SiO_2 and Al_2O_3, but other significant components are Fe_2O_3, CaO and MgO.

It may also have an organic content (Zaady et al., 2001) and contain large quantities of soluble salts and carbonates (Hirmas et al., 2011).

Standard World Meteorological Organization (WMO) definitions for dust events are given by McTainsh and Pitblado (1987): (a) *Dust storms* are the result of turbulent winds raising large quantities of dust into the air and reducing visibility to less than 1,000 m. (b) *Blowing dust* is raised by winds to moderate heights above the ground reducing visibility at eye level (1.8 m) but not to less than 1,000 m. (c) *Dust haze* is produced by dust particles in suspended transport which have been raised from the ground by a dust storm prior to the time of observation. (d) *Dust whirls* (or *dust devils*) are whirling columns of dust moving with the wind and are usually less than 30 m high (but may extend to 300 m or more) and of narrow dimensions.

There is some confusion in the literature between 'sand storms' and 'dust storms'. The former tend to be low-altitude phenomena of limited areal extent, composed of predominantly sand-sized materials. Dust storms reach higher altitudes, travel longer distances and are mainly composed of silt and clay. In this book, the term 'dust storm' refers to an atmospheric phenomenon in which the horizontal visibility at eye level is reduced by mineral dust to less than 1,000 m.

The passage of low pressure fronts with intense baroclinal gradients that are accompanied by very high-velocity winds is the dominant dust-generating mechanism in many of the world's dusty regions, including Australia (Strong et al., 2011), Iran (de Villiers and van Heerden, 2011), northern China and Mongolia, central Asia, the Middle East (Vishkaee et al., 2012), the Mediterranean coast of North Africa, the Sahel of West Africa, the High Plains of the United States, the Chihuahuan Desert (Rivera et al., 2009) and the plains of the Argentine Pampas. Surface cyclones themselves may sweep out gyres of dust when circulation around the low pressure becomes very intense. More localized dust storms occur when katabatic winds deflate mountain foot sediments, as on the northern slopes of Kopet Dag on the Iran-Turkmenistan border, or in California (the Santa Ana wind). The high Andean Altiplano experiences strong dust raising from the upper westerlies, and similar upper airflow deflates sediments from the arid Tibetan Plateau. The cold downburst wind of a dry thunderstorm, the classic haboob (Miller et al. 2008; Knippertz and Todd, 2012), is perhaps the most common mesoscale dust-raising system, which raises dust at the gust front some kilometres in advance of the towering convective clouds. Dust can also be raised and transported by dust devils, which are called willy-willies by the Australians (Oke et al., 2007a and b), although their quantitative significance is not clear.

Source Areas and Frequency

Dust storm incidence is highly variable in time and space (Wang et al., 2011). Some areas are major generators of dust (Figure 3.2) and may have dust storms on some tens of days in the year. Other areas are much less active. The frequency of dust

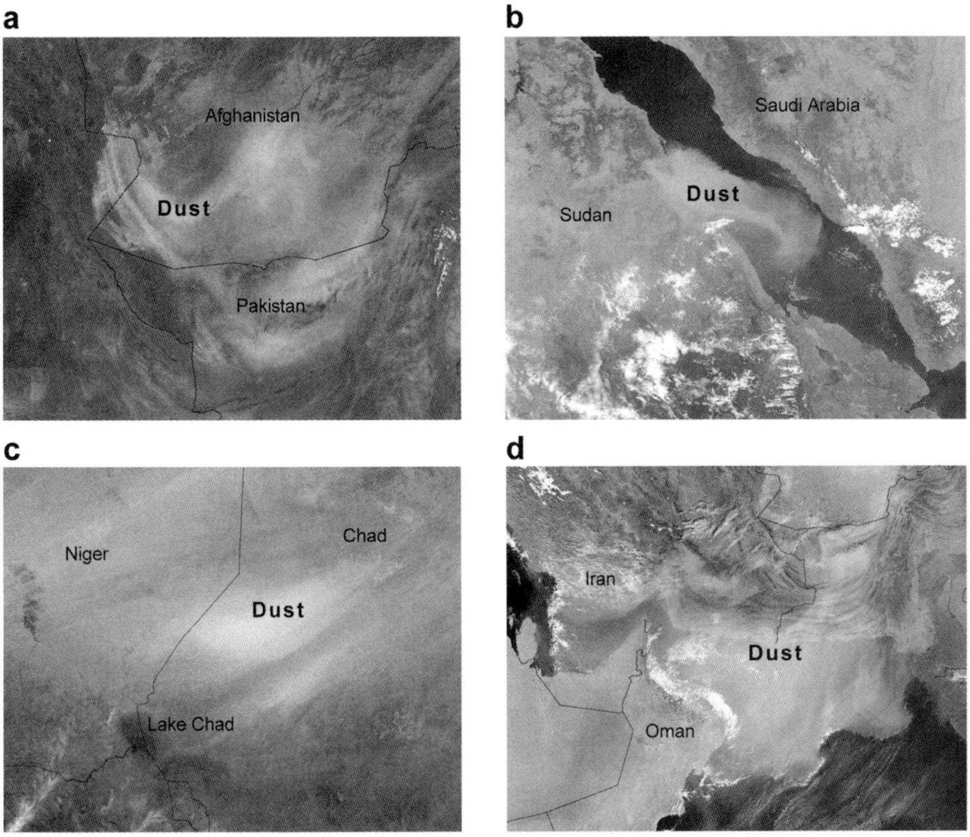

Figure 3.2 Satellite images of dust plumes. (Courtesy of NASA): (a) Seistan Basin (MODIS, June 16, 2004); (b) Tokar Delta, Sudan and Red Sea (MODIS, September 1, 2004); (c) Bodélé Basin (MODIS, January 5, 2005); (d) Gulf of Oman and Makran Coast (SEAWIFS, December 13, 2003)

events can be determined from standard meteorological records, although satellite sensors are the best means of determining source areas (Schepanski et al., 2012). The Total Ozone Mapping Spectrometer (TOMS) has proved to be an effective instrument for detecting atmospheric mineral dust (Prospero et al., 2002; Washington et al., 2003), but we also have global or near-global maps of aerosol optical thickness (a measure of aerosol column concentration) derived from satellites such as the NOAA Advanced Very High Resolution Radiometer (AVHRR), the Moderate Imaging Spectroradiometer (MODIS) (see, for example, Yu et al., 2003; Chin et al., 2004; Ginoux et al., 2004; Baddock et al., 2009), the Spinning Enhanced Visible and Infrared Imager (SEVIRI), the Multiangle Imaging Spectroradiometer (MISR) and the Cloud-Aerosol Lidar with Orthogonal Polarization (CALIOP) (Ridley et al., 2012). Different sensors do not always provide the same picture of dust storm activity, and Brooks et al. (2005), for example, point to the difference in results obtained from TOMS compared to those derived as an Infra-Red Difference Dust Index (IDDI) from METEOSAT.

Table 3.2 *Maximum mean Aerosol Index (AI) values for major global dust sources determined from TOMS*

Location	AI Value	Average annual rainfall (mm)
Bodélé Depression of south-central Sahara	>30	17
West Sahara in Mali and Mauritania	>24	5–100
Arabia (southern Oman/Saudi border)	>21	<100
Eastern Sahara (Libya)	>15	22
South-west Asia (Makran coast)	>12	98
Taklamakan/Tarim Basin	>11	<25
Etosha Pan (Namibia)	>11	435–530
Lake Eyre Basin (Australia)	>11	150–200
Mkgadikgadi Basin (Botswana)	>8	460
Salar de Uyuni (Bolivia)	>7	178
Great Basin (USA)	>5	400

Schepanski et al. (2012) also compare the results of different satellite products in the context of the Sahara.

TOMS data include an Aerosol Index (AI), values for which are linearly proportional to aerosol optical thickness. The world pattern of annual mean AI values has certain clear features (Table 3.2). First, the largest area with high values extends eastwards from West Africa through the Sahara to Arabia and south-west Asia (Figure 3.3). In addition, there is a large zone with high AI values centred over the Tarim Basin in central Asia. Australia has a relatively small zone, located in the Lake Eyre Basin, while southern Africa has two zones, one centred on the Mkgadikgadi Basin and the other on Etosha Pan. In Latin America, there is one easily identifiable zone in the vicinity of one of the great closed basins of the Altiplano – the Salar de Uyuni. North America has only one relatively small zone with high values – the Great Basin.

The importance of these different dust 'hotspots' can be gauged by looking not only at their areal extents but also at their relative TOMS AI values (Figure 3.4); Table 3.2 lists the latter. This again brings out the dominance of the Sahara in particular and of the Old World deserts in general. The Southern Hemisphere and the Americas both have relatively low AI values. So, for example, the AI values of the Bodélé Depression of the Sahara are around four times greater than those recorded for either the Great Basin of the United States or the Salar de Uyuni of Bolivia.

TOMS data have demonstrated the primacy of the Sahara, and the importance of large basins of internal drainage (Bodélé, Taoudenni, Tarim, Seistan, Eyre, Etosha, Mkgadikgadi, Etosha, Uyuni and the Great Salt Lake) as dust sources (Engelstaedter, 2001), although there will be great variability in the importance of individual basins depending on their surface state (e.g. whether they are wet or dry, compacted or

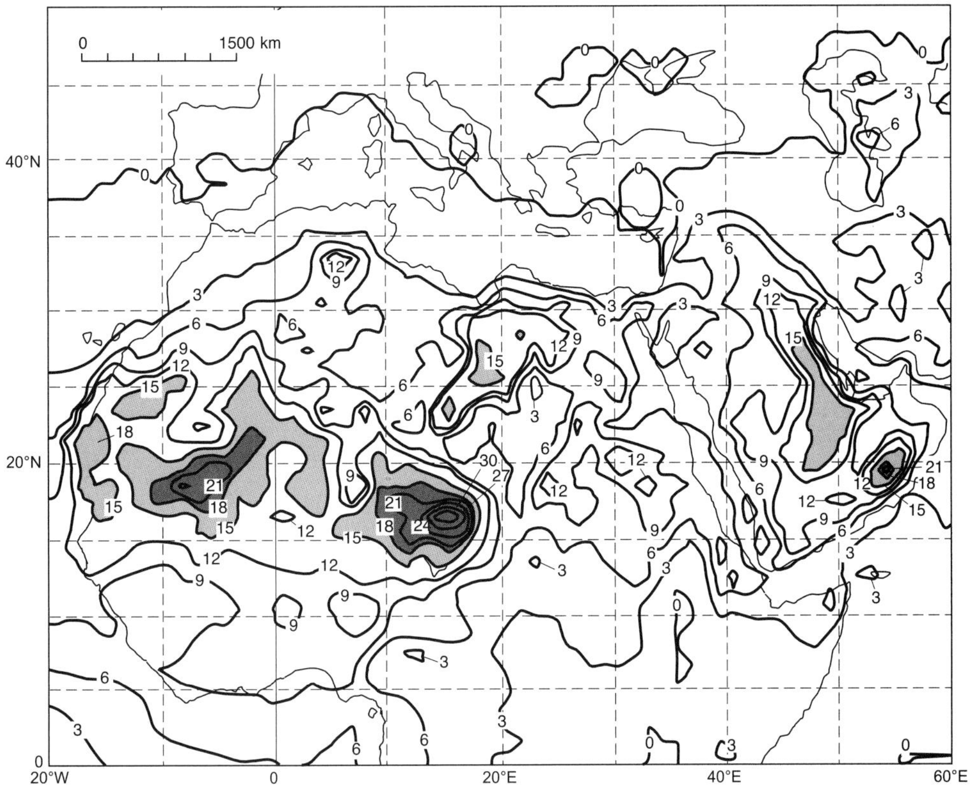

Figure 3.3 Annual mean Aerosol Index (AI) values from TOMS data (1980–93, 1997–2000). (From Middleton and Goudie, 2001, fig. 4)

puffy, etc.). As Elmore et al. (2008, p. 1754) remarked, 'A global map of endorheic basins is nearly indistinguishable from a map of global dust.' Also, many sources are associated with deep and extensive alluvial deposits (Prospero et al., 2002) or with extensive piedmont alluvial fans (Wang, Zhou and Dong, 2006; Tegen and Schepanski, 2009). Sand dune systems and sandy deserts are not good sources of fine grained dust, however, unless they contain an appreciable content of fines. Furthermore, TOMS data indicate that many of the world's major dust sources are very arid. The prime global source, Bodélé, has a mean annual rainfall of 17 mm, whereas the large west Saharan source has annual precipitation levels between 5 and 100 mm. In Arabia, dust storms are most prevalent where the mean annual rainfall is less than 100 mm (Goudie and Middleton, 2001), and the Taklamakan dust source in north-west China has large areas where the annual rainfall is less than 25 mm.

With regard to individual source regions, in the United States the greatest frequency of dust events occurs in the panhandles of Texas and Oklahoma, Nebraska, western Kansas, eastern Colorado, the Red River Valley of North Dakota, northern Montana and the Great Basin. These areas combine erodible materials with a dry climate and high wind energy (Gillette and Hanson, 1989). Dust emissions may also increase after

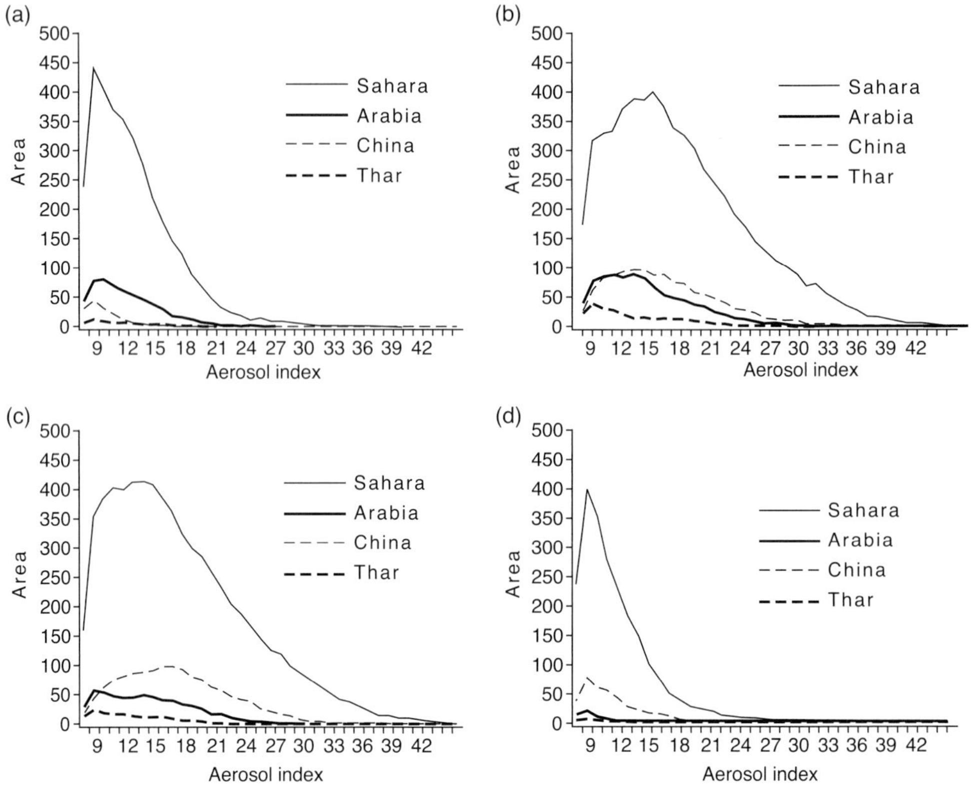

Figure 3.4 The extent and intensity of the Aerosol Index (AI) derived by TOMS for four main desert areas over the four quarters of the year, starting in January. The figure shows the areas (in $km^2 \times 10^3$) covered by different intensities of the AI. (From Goudie and Middleton, 2001, fig. 2)

fires (Stout, 2012). The dust has a major impact on surface materials in the south-west United States (Reheis et al., 2009; Quick and Chadwick, 2011). A discussion of the spatial and temporal variability of dust storms in the Mojave and Colorado Deserts is provided by Bach et al. (1996), who identify the Coachella Valley as being the dustiest region. Dry washes, playa margins and distal portions of alluvial fans are major dust sources in the Mojave (Sweeney et al., 2011). In the High Plains of Texas and New Mexico, anthropogenically disturbed sandsheets produce more dust sources than any other geomorphic categories, but playas produce the most sources per unit area (Lee et al., 2012). In the Chihuahuan Desert, almost one-half of dust plumes are derived from ephemeral lakes, even though these only cover 4 per cent of the area (Baddock et al., 2011b). Indeed, dust storms are a major phenomenon in the semi-arid parts of central-northern Mexico (Pineda-Martinez et al., 2011).

The main source areas for Saharan dust, which used to be unclear (Goudie and Middleton, 2001), include Bodélé; an area that comprises southern Mauritania, northern Mali and central-southern Algeria (Knippertz and Todd, 2010); southern Morocco and western Algeria; the southern fringes of the Mediterranean Sea in Libya (O'Hara et al., 2006) and Egypt (Koren et al., 2003); and northern Sudan (Brooks et al., 2005). Bodélé is the largest dust source in the world (Ridley et al., 2012), and this is because of the topographically controlled Low Level Jet, channelled between the Tibesti and Ennedi Mountains, which deflates low-density diatomites and other susceptible materials from a desiccated lake floor (Washington et al., 2006a and b; Engelstaedter et al., 2006; Koren et al., 2006; Bristow et al., 2009). The synoptic conditions associated with dust events in the Sahara and Sahel are discussed by Schepanski et al. (2009), Klose et al. (2010) and Knippertz and Todd (2010, 2012). Much dust is moved by the north-easterly trades over Nigeria and the Guinea zone to give the Harmattan haze (Breuning-Madsen and Awadzi, 2005). The Tokar Delta of the Sudan, an arid, silty region across which high-velocity winds are funnelled by a gap in the Red Sea Hills, is a frequent source of dust over the Red Sea (Hickey and Goudie, 2007) – dust which often moves into Arabia.

Dust storms in the Middle East itself have been analysed by Middleton (1986a). They are frequent on the alluvial plains of southern Iraq and Kuwait. On the basis of the study of aerosol geochemistry over the Arabian Sea, Pease et al. (1998) have suggested that the Wahiba Sands of Oman is also a major source. Analysis of TOMS data indicates that the Oman-Saudi Arabia border is a large dust-generation area that has not been picked up from ground meteorological observations (Middleton and Goudie, 2001). Also important is the eastern part of Saudi Arabia to the north of the Rub 'Al Khali sand sea. Some dust in the United Arab Emirates may be derived from Iran and central Asia (de Villiers and van Heerden, 2011). Another major region of dust storms occurs in Iraq and Iran (Middleton, 1986b; Zarasvandi et al., 2011; Vishkaee et al., 2012) and at the convergence of the common borders between Iran, Pakistan and Afghanistan. At Zabol, in Iranian Seistan, there are on average eighty-one dust storm days per year, making it one of the world's dustiest locations. This is a closed basin fed by the silt-laden Helmand River and through which high-velocity winds are channelled by the high mountains that bound it (see Section 7.11). The frequency of dust storms in that region varies greatly from year to year depending on whether the Helmand lakes are wet or dry (Miri et al., 2010), but in dusty years very high particulate concentrations have been recorded (Rashki et al., 2011, 2012).

Middleton (1986c) has mapped dust storm activity in the Indian subcontinent and has demonstrated that the greatest number of dust storms occurs in the Thar. The Makran coast is also an area of significant dust storm occurrence, and plumes are often identifiable on satellite images, showing dust being blown off dry riverbeds.

In the southern former Soviet Union, the number of dust storms exceeds forty per year, and some locations have more than eighty, one of the highest occurrences in the world (Orlovsky et al., 2005). The largest source area has been the Karakum Desert, but there is a large dust belt that extends from west to east, lying north of the Caspian Sea, south of Lake Balkhash and in the Aral Sea region (Indoitu et al., 2012). The dust storms can cover immense areas and may transport dust particles to the more humid parts of China (Zhao et al., 2010), Korea, Japan and beyond.

Dust storms are of particular importance in China because of their role in the formation of Earth's greatest loess deposits (Derbyshire et al., 1998; Kar and Takeuchi, 2004) (see Section 3.5). Moreover, according to Kes and Fedorovich (1976), the Tarim Basin seems to have more dust storms than any other location on Earth, with 100 to 174 per year. Studies of dust loadings and fluxes suggest that there are two main source areas: (1) the Taklamakan (Gao and Washington, 2009) and (2) the Badan Jarain (Zhang et al., 1998; Xuan, 1999; Shao and Wang, 2003), although in glacials the Qaidam Basin may have had a greater significance than today because of a southwards shift in dust-generating winds (Pullen et al., 2011). Dust storms also occur on the Mongolian Plateau, especially at times when the ground is not covered by snow or is frozen (Han et al., 2011).

Southern Africa is not a major area of dust production. Nevertheless, satellite images show plumes blowing off the Namib and the Kalahari towards the South Atlantic (Eckardt et al., 2002). TOMS analyses indicate that there are two relatively small but clearly developed dust-source areas: the Etosha Pan in northern Namibia (Bryant, 2003) and the Mkgadikgadi Depression in northern Botswana (Resane et al., 2004). Ephemeral river channels also appear to be significant sources for dust that is blown out over the Atlantic (Eckardt and Kuring, 2005), however, and the desiccated Kuiseb Delta near Walvis Bay is especially powerful.

In South America to the west of Buenos Aires in Argentina, there are more than eight dust storms per year (Middleton, 1986b). Dust is also generated from Patagonia, a dry environment with high-velocity winds (Gassó and Stein, 2007). The presence of extensive areas of closed depressions and of wind-fluted topography, combined with the probable importance of salt weathering in the preparation of fine material for deflation (Goudie and Wells, 1995), suggest that the dry areas of the Altiplano should be major source areas for dust storms. TOMS identifies one area where aerosol values are relatively high: This is the Salar de Uyuni, a large closed basin in Bolivia which is located in an area with 200–400 mm of annual rainfall. In the late Pleistocene, it was the site of the huge (600 km long) pluvial Lake Tauca (Placzek et al., 2006). It is possible that the deflation of fine sediments from its desiccated floor is one of the reasons for high aerosol values in this region.

Australia, although not as dusty as some regions of the world (McTainsh, 1989), is the largest dust source in the Southern Hemisphere. Moreover, at the Late Glacial Maximum (LGM) it contributed three times more dust to the south-west Pacific than it does now (McTainsh and Pitblado, 1987; Hesse and McTainsh, 1999). Dust departs Australia in two main plumes: one that crosses the Tasman Sea to New Zealand (Marx et al., 2005a, 2005b) and another that heads westwards out into the Indian Ocean (Hesse and McTainsh, 1999). Within the Lake Eyre Basin, using MODIS data, Bullard et al. (2008) found that 37 per cent of dust plumes originated in areas of

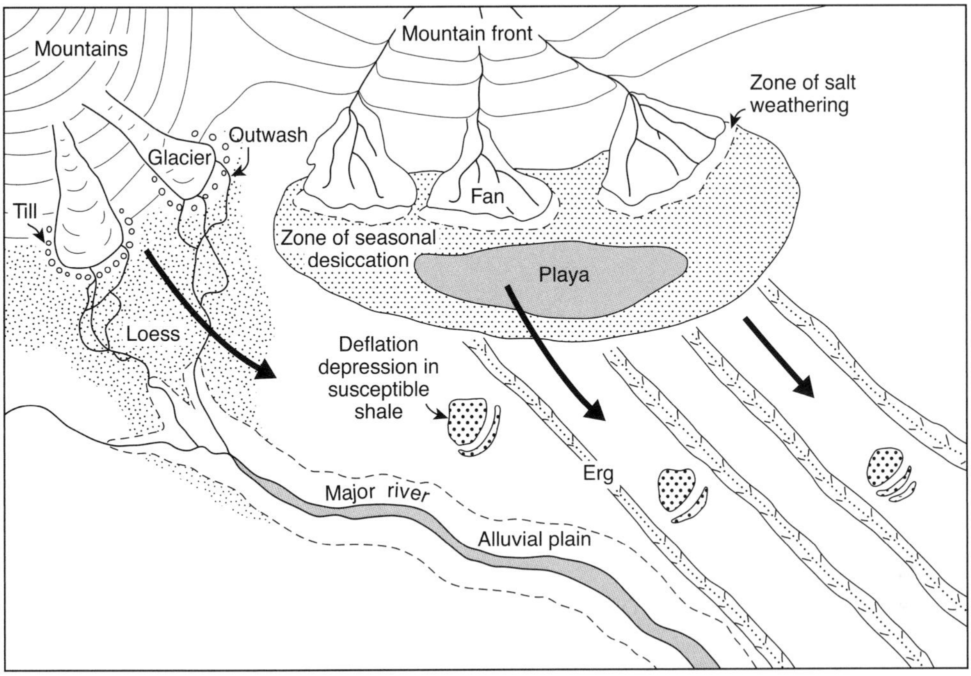

Figure 3.5 A model of geomorphological environments from which substantial deflation occurs.

aeolian deposits, 30 per cent from alluvial deposits and flood plains and 29 per cent from ephemeral lakes. Such lakes make some dust rich in NaCl (Shiga et al., 2011). The importance of dunes destabilized by fire as dust sources has been stressed by Strong et al. (2010). A full account of the history and distribution of dust storms in Australia is given by McTainsh et al. (2010).

What emerges from this regional picture is that dust storms are generated from certain key hotspots, most of which are closed depressions in hyperarid areas, especially in the Northern Hemisphere, but a number of landform types are involved (Figure 3.5).

The Global Picture

Estimates of the total soil dust emissions to the atmosphere on a global scale (Tables 3.3 and 3.4) show a large range (Prospero, 1996b) but seem to be between 1,000 and 3,000 millions of tonnes per year (Cakmur et al., 2006). The largest source is the Sahara which probably contributes around half of the global total. Estimates of its annual dust emissions are shown in Table 3.5.

One of the reasons dust emissions have global impacts is because of the huge distances over which dust plumes can travel (Zhu et al., 2007). Thus dust from the Lake Eyre Basin of Australia may accumulate in New Zealand and East Antarctica

Table 3.3 *Estimates of dust emissions to the global atmosphere*

Author(s)	Annual quantity (millions of tonnes)
Tegen and Fung (1994)	3,000
Andreae (1995)	1,500
Duce (1995)	1,000–2,000
Mahowald et al. (1999)	3,000
Luo et al. (2003)	1,654
Zender et al. (2003a)	1,490
Ginoux et al. (2004)	1,950–2,400
Liao et al. (2004)	1,784
Miller et al. (2004)	1,019
Cakmur et al. (2006)	1,000–3,000
Tanaka and Chiba (2006)	1,877

(Revel-Rolland et al., 2006; De Deckker et al., 2010), and dust from Patagonia may also reach Antarctica (Bory et al., 2010). Saharan dust reaches North America by way of the Pacific (McKendry et al., 2007), and some is carried thousands of kilometres to the Americas (Ben-Ami et al., 2010), Europe and the Near East (Doherty et al., 2008; Thevenon et al., 2011) and also to the Arctic (Barkan and Alpert, 2010). Large amounts of Saharan dust are blown southwards into the Gulf of Guinea (Resch et al., 2007). The Sahara is also a major source of dust deposition into the Mediterranean Sea and neighbouring countries (Santese et al., 2007; Israelevich et al., 2012). Dust from the Sahara and Thar reaches Mount Everest and Hong Kong in substantial quantities (Kaspari et al., 2009; Lee et al., 2010). It also contributes nutrients to Amazonia (Ridley et al., 2012). Ukraine has recently delivered large quantities of dust to central Europe (Birmili et al., 2007). Middle Eastern dust reaches India (Badarinath et al., 2010), and Chinese dust may reach North America via the Pacific (Bennett et al., 2006; Fairlie et al., 2007; Zdanowicz et al., 2006; Uno et al., 2009). North American dust storms may deposit fine materials on the California Channel Islands and the eastern Pacific Ocean (Muhs et al., 2007a). Dust in a Greenland ice core has been traced by isotopic means to both the Sahara and to China (Lupker et al., 2010).

Quaternary Dust Loadings

Extensive deposits of loess (see Section 3.5), the presence of large amounts of dust in ice, ocean, lake and peat bog core sediments and in speleothems, indicate that during parts of the Quaternary the world was dustier than today. For example, Mahowald et al. (2006) have suggested that dust deposition rates at the LGM were 2.1–3.3 the present levels.

Table 3.4 *Estimates of global and regional dust emissions*
(a) % of global emissions

Sahara/Sahel	50.7
Central Asia	16.0
Australia	14.5
North America	5.2
East Asia	4.9
Arabia	4.2
Others	4.5

Source: Derived from data in Miller et al. (2004, fig. 1).

(b) *Dust emissions*

	(Tg per year)	(%)
North Africa	1,430	69.0
South Africa	322	1.1
North America	9	0.4
South America	55	2.7
Asia	496	23.9
Australia	61	2.9
Total	2,373	100

Source: Derived from data in Ginoux et al. (2001).

(c) *% of global emissions in 1998*

	(Tg per year)	(%)
North Africa	1,114	67.4
Arabian Peninsula	119	7.2
Asia	54	3.3
Australia	132	8.0
Miscellaneous	235	14.2
Total	1,654	100

Source: Derived from data in Luo et al. (2003).

(d) *Global dust emissions*

North Africa	1,087	57.9
Arabian Peninsula	221	11.8
Central Asia	140	7.5
Western China	68	3.6
Eastern China	146	7.8
North America	2	0.1
South America	44	2.3
Southern Africa	63	3.4
Australia	106	5.6
Total	1,877	100

Source: Derived from data in Tanaka and Chiba (2006).

Table 3.5 *Estimates of the strength of the Sahara dust source*

Author(s)	Annual quantity (millions of tonnes)
Jaenicke (1979)	260
Schütz et al. (1981)	260
D'Almeida (1986)	630–710
Prospero (1996a, b)	170
Swap et al. (1996)	130–460
Marticorena and Bergametti (1996)	586–665
Callot et al. (2000)	760
Ginoux et al. (2004)	1,400
Miller et al. (2004)	517
Tanaka and Chiba (2006)	1,087

The enhanced dustiness during glacials may relate to a range of factors: larger sediment sources (e.g. areas of glacial outwash), changes in wind characteristics both in proximity to ice caps and in the trade-wind zone (Ruddiman, 1997) and the expansion of low-latitude deserts. Not all cases of higher dust activity can be attributed to greater aridity in source regions, however, for as Nilson and Lehmkuhl (2001) point out, this is but one factor, albeit an important one. Also important are changes in the trajectories of the major dust-transporting winds, changes in the strength and gustiness of winds in source regions (McGee et al., 2010), the balance between wet and dry deposition (which may determine the distance of dust transport), the degree of exposure of continental shelves in response to sea-level changes and the presence of suitable vegetation to trap dust on land. Broecker (2002) suggested that the increase in dust production and deposition during glacials can be attributed to the steepened temperature gradients and associated aeolian activity related to the equatorward extension of continental glaciers and sea ice. Changes in the hydrology and vegetation cover of source regions, however, would also have been very important (Werner et al., 2002). The lower atmospheric carbon dioxide levels in glacial times may have caused a reduction in vegetation cover because of their impact on the ability of plants to withstand droughts (McGee et al., 2010).

Around the time of the LGM, the amount of dust transported from the Sahara into the Atlantic was augmented by a factor of 2.5 (Tetzlaff et al., 1989, p. 198), and Australia contributed three times more dust to the south-west Pacific (Hesse and McTainsh, 1999). By contrast, dust activity appears to have been very low during the African Humid Period (AHP). From 14.8 to 5.5 ka, the mass flux off Cape Blanc was reduced by 47 per cent (DeMenocal et al., 2000). This is confirmed by analyses of the mineral magnetics record from Lake Bosumtwi (Ghana), which suggest a high dust flux during the last glacial period and a great reduction during the AHP (Peck et al., 2004). On the basis of cores from the Arabian Sea, Sirocko et al. (1991) suggested that dust additions were around 60 per cent higher during glacials than in post-glacial times, although there was a clear ‘spike’ of enhanced dust activity at around 4000 years BP associated with a severe arid phase. Jung et al. (2004) also reported on Holocene dust trends in the Arabian Sea, and suggested that dry, dusty conditions were established by 3.8 kyr BP.

Another source of long-term information on rates of dust accretion is the record preserved in ice cores. The Epica Dome C dust core in Antarctica showed around a twenty-five-fold increase in glacial dust flux over the last eight glacial periods (Lambert et al., 2008). The Dunde Ice Core from high Asia (Thompson et al., 1990) also shows high dust loadings in the Late Glacial and a sudden fall-off at the transition to the Holocene. The Little Ice Age, however, was also a time of relatively high dust activity (Yang et al., 2006).

Within the last glaciation, dust activity both in Europe and in Greenland appears to have varied in response to millennial-scale climatic events (Dansgaard-Oeschger Events and Bond Cycles) (Rousseau et al., 2002). For China, Kohfeld and Harrison (2003) indicated that in glacial phases (e.g. OIS 2) aeolian mass accumulation rates were c 310 g m^{-2} yr^{-1} compared to 65 g m^{-2} yr^{-1} for an interglacial stage (e.g. OIS 5) – a 4.8× increase. For Europe, Frechen et al. (2003) found large regional differences in loess accumulation rates but suggested that along the river Rhine and in eastern Europe, they were from 800–3,200 g m^{-2} yr^{-1} in OIS 2. Loess accumulation rates over much of the United States during the LGM were also high, being c 3,000 g m^{-2} yr^{-1} for mid-continental North America (Bettis et al., 2003). From 18,000 to 14,000 years ago, rates of accumulation in Nebraska were remarkable, ranging from 11,500 g m^{-2} y^{-1} to 3,500 g m^{-2} y^{-1} (Roberts et al., 2003).

Decennial and Centennial Scale Fluctuations

Considerable progress has been achieved in reconstructing dust storm frequencies and trends at decennial and centennial scales using cores from ice caps, lakes and glaciers (e.g. An et al., 2011; Marx et al. 2011; Thevenon et al., 2011), tree rings and historical documents. Information for Asia is provided by Kang et al., 2001; Yang et al., 2006, 2007; Xu et al., 2007; and Kaspari et al., 2009. Lim and Matsumoto (2006) investigated sediments from a volcanic depression – a maar – from Cheju Island, Korea, which dated back 6,500 years. The coarse quartz flux showed considerable variability over that period, with a high flux from 4000 to 2000 years BP and 2000 to 1000 years BP. This was attributed to variation in the aridity of the Chinese source areas. Analysis of a lake core from China showed strong aeolian deposition at 11.8–11.1, 10.6–8, 6.1–4.9 and after 3.3 years BP (An et al., 2011). Wang (2005), working on the Malan ice core from the northern Tibetan Plateau, found a decreasing trend for the last 200 years. A core from the Aral Sea (Huang et al., 2011) showed dust deposition was at a very high level during the Little Ice Age (AD 1400–1780).

Deposits of dune sand and loess also have the potential to indicate changing dust storm activity. Miao et al. (2007) used a large number of optically stimulated luminescence dates from aeolian sections in the central Great Plains of the United States to identify when aeolian activity has been most intense over the past 10,000 years and gave dates of 1.0–0.7 ka, 2.3–4.5 ka and 6.5–9.6 ka.

Human activities may have had an important effect on dust storms in some parts of the world. Von Suchodoletz et al. (2010) have even speculated that humans intensified dust storm activity in the northwest Sahara as early as 7–8 ka ago. The situation becomes less speculative as we move towards the present, and Neff et al. (2008), for instance, used analyses of lake cores in the San Juan Mountains of south-western Colorado, United States, to show that dust levels increased by 500 per cent above the late Holocene average following the increased western settlement and livestock grazing during the nineteenth and early twentieth centuries. The U.S. Dust Bowl of the 1930s was caused by a combination of a major drought and adverse land management, with the latter having a feedback effect on the drought itself (B.I. Cook, et al., 2009). A dust core from the Antarctic Peninsula (McConnell et al., 2007) showed a doubling in dust deposition in the twentieth century, and this is explained by increasing temperatures, decreasing relative humidity and widespread desertification in the source region – Patagonia and northern Argentina. In south-east Australia a two- to tenfold increase in dust deposition rates above Holocene norms is recorded in the last 200 years, following settlement by European farmers (Marx et al., 2011). Finally, analysis of a 3,200-year marine core off West Africa shows a marked increase in dust activity at the beginning of the nineteenth century, which was a time that saw the advent of commercial activity (including groundnut production) in the Sahel region (Mulitza et al., 2010).

Fluctuations in Dust Activity During the Period of Meteorological Observations

Both TOMS observations (e.g. Ogunjobi et al., 2012) and analysis of meteorological data have enabled the changing frequency of dust events to be established for the last six decades or so (Goudie and Middleton, 1992). Some areas have shown increasing trends (e.g. the Sahel zone of Africa and the eastern Mediterranean), whereas others have shown declining trends in the late twentieth century (e.g. central Asia, China and Australia) followed by a spike of activity in the early years of the present century (e.g. Mitchell et al., 2010). An exception to the decreasing trend in dust storm occurrence in central Asia occurs in the vicinity of the Aral Sea, where its desiccation has caused dust events to almost double since 1980 (Indoitu et al., 2012). With respect to China, both natural and anthropogenic factors are implicated in these trends (Xu, 2006). The general consensus, however, is that natural climatic fluctuations have played a greater role in explaining the observed trends than have changes in human pressures on the land (X. Wang et al., 2007). There is certainly no clear upward trend in dust storm activity when the past fifty years are considered (Guo and Xie, 2007; Yao et al., 2010), although the early years of the new millennium have seen some severe events (Yang et al., 2008; Kurosaki et al., 2011). Links have been established between dust emissions from the Tarim Basin and the Arctic Oscillation (AO) index, with dust activity being high during the negative phase of the AO (Gao and Washington, 2010; Mao et al., 2011).

In southern Africa, fluctuations in dust emission from the Mkgadikgadi Pans during the period 1980–2000 suggest that dust loadings are intermittently influenced by the extent and frequency of lake inundation, sediment inflows and surface wind speed variability. The variability of these is in turn influenced by the El Niño-Southern Oscillation and Indian Ocean sea surface temperature anomalies (Bryant et al., 2007). Lake flooding and desiccation is also an important control on dust storm frequencies in the Seistan Basin of Iran (Miri et al., 2010).

In the West African Sahel, where drought has been persistent since the mid-1960s, analysis of wind, precipitation and visibility data by Ozer (2003) showed that there have been remarkable changes in dust emissions since the late 1940s. Using a model developed by D'Almeida (1986), he indicated that during the pre-drought conditions that existed from the late 1940s to the late 1960s, yearly dust production was 126×10^6 tons. It rose to 317×10^6 tons during the 1970s and has been $1{,}275 \times 10^6$ tons since 1980, a tenfold increase over the whole period. Variability in Sahel dust emissions may be related not only to droughts and the degree of vegetation cover (Pierre et al., 2012) but also to an increase in the frequency by which wind exceeds the threshold wind velocity for dust entrainment. It also seems to be related to changes in the North Atlantic Oscillation (Engelstaedter et al., 2006), North Atlantic sea surface temperatures (Wong et al., 2008) and the Atlantic Multidecadal Oscillation (Foltz and McPhaden, 2008; Jilbert et al., 2010). The steadily increasing trend of dust storms from Africa in the eastern Mediterranean during the last five decades in association with changing synoptic conditions has been analysed by Ganor et al. (2010). The slope of the increase is 0.27 days per year.

Using a variety of data sources, Mahowald et al. (2010) have tried to estimate the global picture of changes in dust storm activity for the twentieth century. They suggest a doubling of desert dust took place over much of the globe.

3.5 Desert Loess

Loess has been the subject of an enormous literature, ever since Charles Lyell (1834) drew attention to the loamy deposits of the Rhine Valley in Germany. Many theories have been advanced to explain loess formation, and Smalley (1975) provides excerpts from the early literature and a commentary to go with them. It was, however, Ferdinand

Figure 3.6 F. von Richthofen, the scientist who discovered the aeolian origin of China's loess. (Photographer unknown. Accessed via Wikipedia, March 11, 2011)

von Richthofen (1882, pp. 297–98) (Figure 3.6) who cogently argued that these intriguing deposits probably had an aeolian origin and that they were produced by dust storms transporting silts from deserts and depositing them on desert margins:

> In regions where the rains are equally distributed through the year, little dust is formed, and the rate of growth of the soil covered with vegetation will be exceedingly small. But where a dry season alternates with a rainy season, the amount of dust which is put in motion and distributed through atmospheric agency can reach enormous proportions, as witnessed by the dust storms which in Central Asia and Northern China eclipse the sun for days in succession. A fine yellow sediment of measurable thickness is deposited after every storm over large extents of country. Where this dust falls on barren ground, it is carried away by the next wind; but where it falls on vegetation, its migration is stopped.
>
> In rainless deserts the wind will gradually remove every particle of fine-grained matter from the soil, though a new supply of this may constantly be provided by the action of sandblast. The sediments of desiccated lakes, the soil which is laid bare by the retiring of the sea, the materials which are carried down by periodical torrents from glaciated regions to desert depressions, the particles which on every free surface of rock are loosened by constant decay – all these will be turned over and over again by the wind.

Loess is largely non-stratified and non-consolidated silt, containing some clay, sand and carbonate (Smalley and Vita-Finzi, 1968). It consists chiefly of quartz, feldspar, mica, clay minerals and carbonate grains in varying proportions. The grain size distribution of typical loess shows a pronounced mode in the range 20–40 μm and is generally positively skewed towards the finer sizes, although a bimodal grain size distribution has also been identified (e.g. O'Hara-Dhand et al., 2010) with both fine and coarse silt fractions. It can, however, sometimes have a sand content of more than 20 per cent, in which case it is termed sandy loess, or a clay content in excess of 20 per cent, in which case it is termed clayey loess (Pye, 1987, p. 199). Grain size depends on distance from source, formative wind velocities and the granulometry of the materials from which it is derived.

Loess terrain is notable for a number of geomorphological hazards (Derbyshire, 2001), including debris flows, many types of landslide, piping and gullying. These are matters returned to in Chapter 6.

There are many mechanisms that could produce desert loess (Wright, 2001; Smith et al., 2002) (Figure 3.7). Among the processes (Table 3.6) involved in desert silt production are wind and water abrasion (Bullard et al., 2004; Bullard and White, 2005; Crouvi et al., 2008; O'Hara-Dhand et al., 2010) and weathering of particles by salt (Goudie et al., 1979) and frost (Smith et al., 2002). There is certainly no reason to believe that silt is solely the product of glacial grinding.

Loess has been recorded from various deserts (Table 3.7). In Arabia, Australia and Africa, where glaciation was relatively slight, loess is much less well developed, although increasing numbers of deposits in these regions are now becoming evident. Of all the world's loess deposits, China's are the most impressive for their extent and thickness, and loess reaches its supreme development in the Loess Plateau, a 450,000 km^2 area in the middle reaches of the Yellow River (Hwang Ho). North-west of Lanzhou, it attains a maximum thickness of 334 m, while in Jingyuan County, Gansu Province, a thickness of 505 m has been reported (Huang et al., 2000). Over most of the plateau a thickness of 150 m is more typical. Much of the Chinese loess was deposited by dust-laden winds blowing from the mountains, basins and deserts to the west (Stevens et al., 2010), and Lu et al. (2010) have mapped the increasing spread of loess deposits from 22 Ma until the present. There are also extensive loess deposits in Tajikistan, Uzbekistan and other parts of central Asia (Figure 3.8).

In the United States, some of the Peoria loess – including that in Nebraska – may not be glaciogenic, having been transported by westerly to northerly winds from parts of the Great Plains not directly influenced by the Laurentide ice sheet or alpine glaciers (Mason, 2001; Aleinikoff et al., 2008). Some of the loess in the Great Plains (the Bignell Loess) is of Holocene age (Mason and Kuzila, 2000; Mason et al., 2003; Jacobs and Mason, 2005), and Miao et al. (2005) believe that most of it, dating from 9,000–10,000 to 6,500 years ago, was produced in dry phases as a result of the winnowing of dunefields.

a)

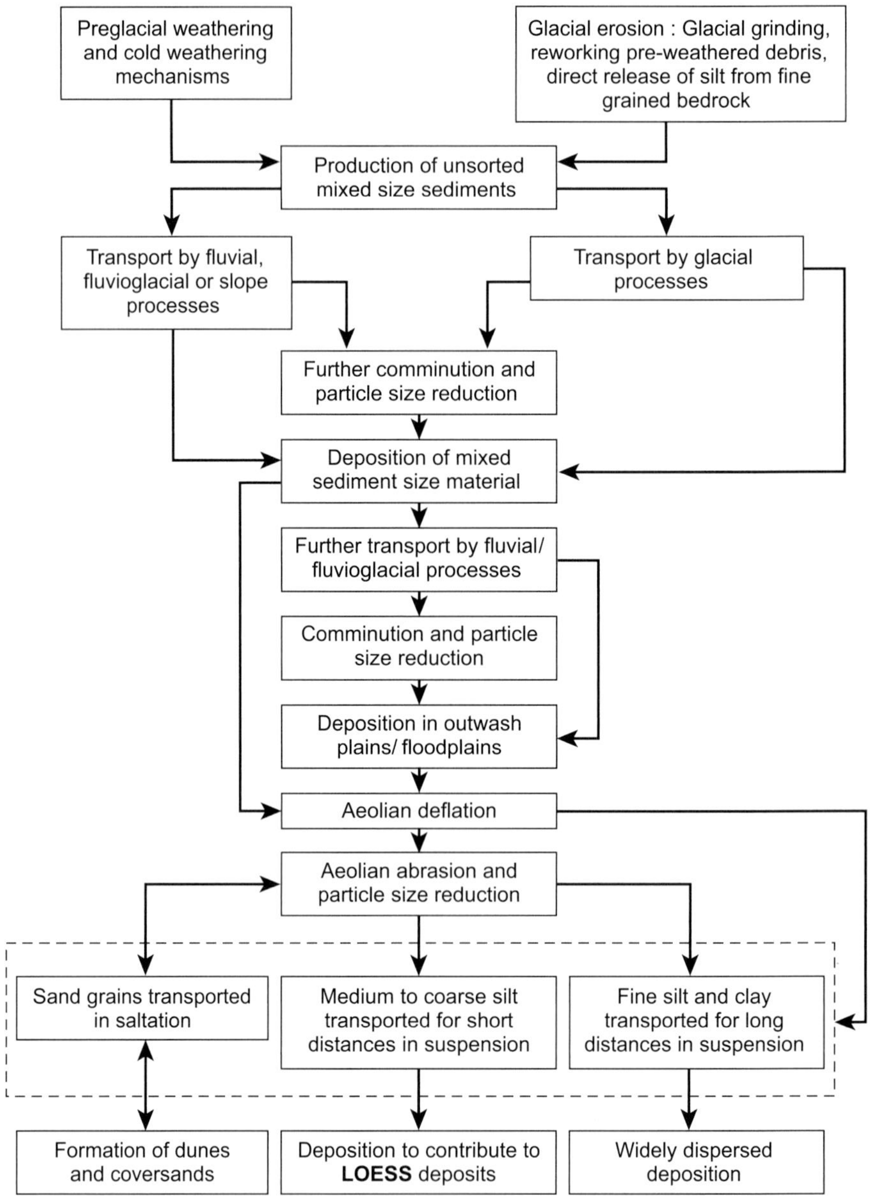

Figure 3.7 Events in the formation of loess deposits – a hypothetical pathway to explain the formation of loess deposits in (a) cold environments and (b) hot environments. (From Wright, 2001b, figs. 3 and 4)

b)

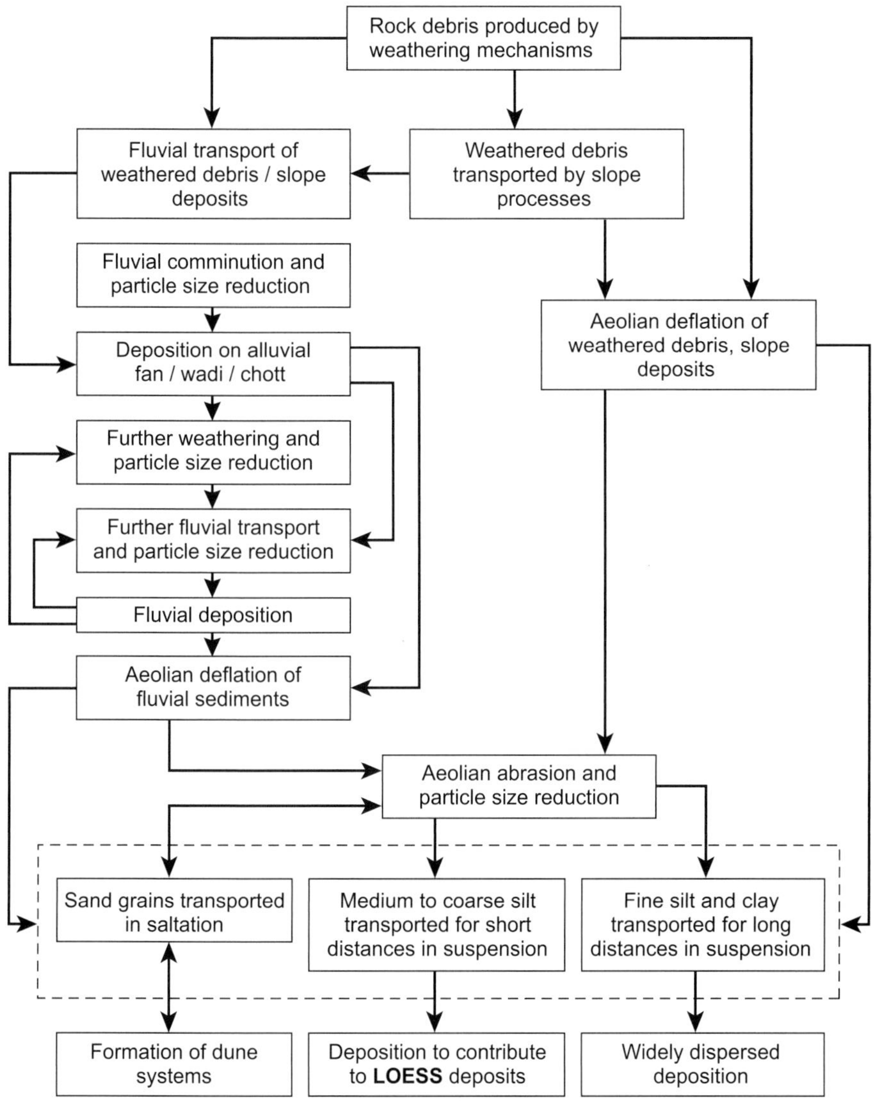

Figure 3.7 (*continued*)

In South America, where the Pampas of Argentina and Uruguay has thick deposits, a combination of semi-arid and arid conditions in the Andes rain shadow – combined with glacial outwash from those mountains – created near ideal conditions (Zárate, 2003). In Australia, where there has been a tendency to use the term ‘parna’ to describe dust deposits (Butler, 1974), dust additions to soils have undoubtedly been important (Cattle et al., 2009), and late Pleistocene dust contributed to the development of silty valley fills (Haberlah et al., 2010).

Table 3.6 *Mechanisms for formation of loess-sized silt particles*

Aeolian abrasion
Biotite weathering in granite profiles
Fluviatile abrasion
Fracturing and/or dissolution in duricrusts
Frost weathering
Inherited glacial grinding
Inherited silica dissolution of dune sands
Inherited tropical deep weathering
Inherited weathering from temperate soils
Microdilation in granitic profiles
Salt weathering of bedrock
Salt weathering of loose granular material
Silica dissolution in saline environments

Source: Modified from Smith (2009, table 4.5).

Although loess has been estimated to cover up to 10 per cent of the world's land area (Tsoar and Pye, 1987), its occurrence in Africa is limited. This appears surprising given that the Sahara is currently the world's largest area of contemporary dust storm activity and that evidence from ocean and ice cores suggests it produced more dust during the cold phases of the Pleistocene. The reasons for the relative lack of loess deposits around the Sahara are a subject for debate (see J.S. Wright, 2001). Some have argued

Table 3.7 *Examples of desert loess*

Location	Source
Bahrain	Doornkamp et al. (1980)
Eastern and north-eastern Afghanistan	Pias (1971); Shroder et al. (2011)
Iran	Lateef (1988); Karimi et al. (2009)
Matmata, Tunisia	Coudé-Gaussen et al. (1982); Dearing et al. (1996, 2001)
Namib	Blümel (1982); Brunotte et al. (2009)
Negev	Yaalon and Dan (1974); Crouvi et al. (2008, 2009)
Northern Nigeria	McTainsh (1987)
Pakistan	Rendell (1984)
Saudi Arabia	Al-Harthi and Bankher (1999)
Libya	Giraudi (2004)
Algeria	Nouaouria et al. (2008)
India	Jayant et al. (1999); Malik et al. (1999)
Australia	Haberlah et al. (2010)
Syria	Rösner (1989)
United Arab Emirates	Goudie et al. (2000)
Yemen	Nettleton and Chadwick (1996); Coque-Delhuille and Gentelle (1998)

Figure 3.8 Thick loess profiles with palaeosols at Lakuti, Tajik Republic, central Asia. (ASG)

that sufficient silt-sized material could only be produced in glacial environments, and that the Sahara lacks loess because it has few mountains and therefore receives insufficient material from mountain glaciers (Smalley and Krinsley, 1978). This is unlikely to be the full explanation because, as we have already seen, there are many mechanisms whereby silt is produced in deserts, and there is self-evidently plenty of silt in the Sahara at the present day to provide material for dust storm transport (McTainsh, 1987; Yaalon, 1987). Certainly much Saharan dust has been deposited over the oceans, but on land only certain desert margins appear to have been favourable for loess formation. Tsoar and Pye (1987) suggest that globally, the absence of more widespread peridesert loess is largely due to a lack of available vegetation traps for dust, an idea also put forward by Coudé-Gaussen (1990) in comparing loess deposits north and south of the Mediterranean. Another possible reason is the relative high intensity of rainfall (and therefore of water erosion) on the south side of the Sahara. The mean rainfall per rainy day in the drier parts of West Africa averages 9.75 mm, whereas in the drier parts (mean annual rainfall less than 400 mm) of the classic loess belts it is 4.51 (China) and 2.56 (former USSR). Several authors suggest that the current inventory of loess derived from the Sahara is incomplete (e.g. Coudé-Gaussen, 1987; Yaalon, 1987), but three areas have been studied in some detail: (1) southern Tunisia (Coudé-Gaussen et al., 1982), (2) northern Nigeria (McTainsh, 1987) and (3) the Negev (Yaalon and Dan, 1974). The Matmata plateau loess of southern

Figure 3.9 Desert loess at Matmata, Tunisia. (ASG)

Tunisia (Figure 3.9) reaches a thickness of 18 m at Téchine and contains up to five palaeosols typically rich in smectite and palygorskite. The loess probably derives from the Tunisian sabkhas (Figure 3.10) and from the Grand Erg Oriental. Silty loess in north-west Libya, which reaches a maximum thickness of 4–5 m and contains interbedded palaeosols and calcretes (Giraudi, 2005), is effectively an extension of the Matmata loess. Elsewhere in Libya, clay-rich loess has been documented in the Ghat area in the south-west (Assallay et al., 1996).

On the south side of the Sahara, material from the Chad Basin, transported by the Harmattan, has been the source of the Zaria loess mantle in northern Nigeria, displaying a clear decrease in grain size with distance from that basin (McTainsh, 1987). Other sparse deposits are catalogued by Coudé-Gaussen (1987): to the north of the Sahara in the Canary Islands, southern Morocco, south-western Egypt and to the south in Guinea and northern Cameroon. In the Negev of the Middle East, the Netivot loess section is up to 12 m thick and contains palaeosols of upper Pleistocene and Holocene age. Loess has also been identified in central Sinai (Rögner and Smykatz-Kloss, 1991). Some of these Near Eastern dust deposits have an African origin, and some of the Negev loess may have been derived from Nile sediments at times when delta sediments were exposed because of low sea levels during glacials (Amit et al., 2011a).

Figure 3.10 Blowing dust at Chott Ghasa, Tunisia. (ASG)

3.6 Pans

Arid regions are often zones of interior or centripetal drainage (Figure 3.11), and nearly all those places on the face of the Earth that lie below sea level occur in deserts (Table 3.8) (see Section 1.9). Desert depressions originate through four main types of process (Shaw and Thomas, 1997): (1) structural controls (e.g. faulting and rifting, downwarping), (2) erosional controls (e.g. deflation, solution, animal scouring), (3) ponding (e.g. in interdune troughs or ephemeral rivers) and (4) dramatic (e.g. meteorite impacts, volcanic cratering). It is likely, however, that many of the depressions, especially the smaller ones, called pans, have a partially or dominantly aeolian origin. Another important aspect of their development, which has considerable implications for the nature of their surfaces, is whether they are dominantly fed by groundwater or by surface runoff, and whether their sediments are dominantly limnogenous (i.e. formed in and under the lakes) or terrigenous (i.e. have an origin from the surrounding land).

Pans tend to be most prevalent in semi-arid areas, as the list of mean annual rainfalls suggests (Table 3.9). In addition to a rainfall control on their distribution, there is also a strong control exercised by surface materials. The pans of West Siberia occur on

Table 3.8 *Surface altitudes (m) below sea level of major depressions*

Depression	Country	Altitude
Death Valley	USA	−86
Salton Trough	USA	−76
Lake Eyre South	Australia	−16
Dead Sea	Israel, Jordan	−400
Qattara	Egypt	−133
Fayum	Egypt	−45
Siwa	Egypt	−25
Turpan (Turfan)	China	−154
Sea of Galilee	Israel	−210
Caspian Sea	CIS	−28
Vpadina Karagiye	Kazakhstan	−132
Karynzharyk Kendyrlisov	Kazakhstan	−70
Oz El'ton	Russia	−15
Oz Baskunchak	Russia	−21
Lake Assal	Djibouti	−150
Lake Dallal	Ethiopia	−116
Chott Melrhir	Algeria	−31
Chott et Gharsa	Tunisia	−23
Salina Grande	Argentina	−42
Salina Chica	Argentina	−12
Laguna del Carbon	Argentina	−105
Akdzhakaya	Turkmenistan	−81
Laguna Salada	Mexico	−10
Sabkha Teh	Morocco/W. Sahara	−55
Sabkha Ghuzayyil	Libya	−47

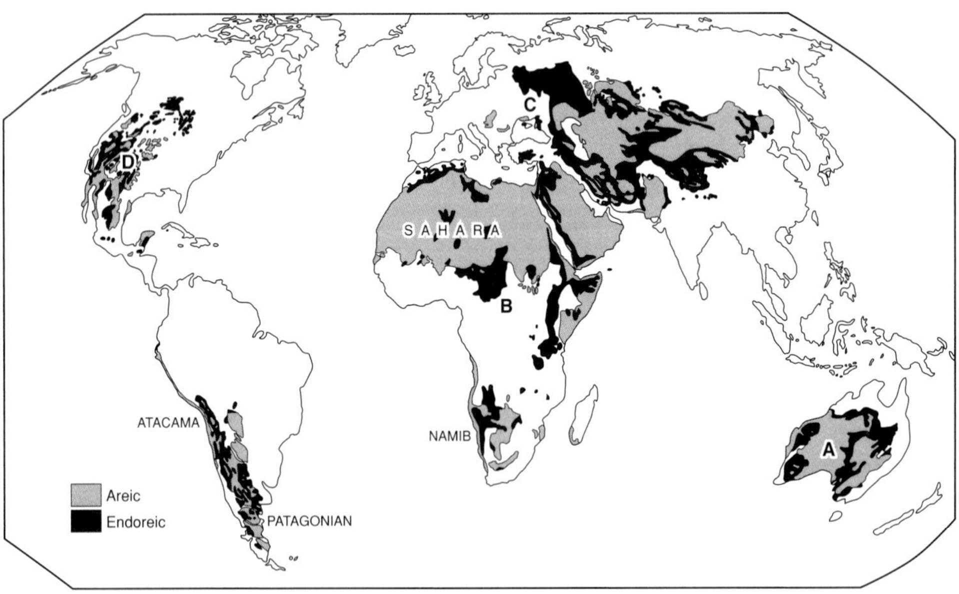

Figure 3.11 Areas of areic and endoreic drainage on a global basis. A is the Lake Eyre Basin, B is the Bodélé-Chad Basin, C is the Caspian Basin and D is the Great Basin. (From Goudie, 2002, fig.1.8)

Table 3.9 *Annual precipitation totals (mm) for major pan areas*

Atbasar, Kazakhstan	320
Bahia Blanca, Argentina	550
Duolan, Inner Mongolia, China	389
Katanning, W. Australia	480
Keetmanshoop, Namibia	143
Kimberley, South Africa	420
Lubbock, Texas, USA	450
Mildura, Victoria, Australia	260
Rio Grande, Tierra del Fuego	330

Neogene and Quaternary materials, those of southern Africa on Ecca shales and Kalahari Sands, those of the High Plains of the United States on Ogallala beds and various shales and sandstones and those of the Yorke Peninsula in Australia on Permian clays and sands. Elsewhere, pans have developed in dunefields, on coastal plains and in deflated lakebeds. In other words, they only develop in classic and abundant form on materials that are relatively lightly lithified and which are composed of fine-grained materials that are susceptible to deflation.

Pans occur widely, notably in the Pampas and Patagonia of Argentina (Mazzoni 2001), eastern Tierra del Fuego, the High Plains of the United States (Bowen et al., 2012), the interior of southern Africa (Goudie and Thomas, 1985; Holmes et al., 2008), south-eastern Mozambique, extensive tracts of Australia (Bourne and Twidale, 2010) (Figure 3.12), the West Siberian Plain, and Manchuria (Goudie and Wells, 1995). Some pan fields consist of tens of thousands of basins – for example, those of Western Australia (Boggs et al., 2006) and of Kansas and eastern Colorado (Bowen et al., 2010, Quillin et al., 2005). Many pans are modest in size. In the Great Plains, the mean area ranges from 0.57 to 7.58 ha (Bowen et al., 2010, table 2).

The pans of North America illustrate the diversity of views on pan origin and also the diversity of pans themselves; Gustavson et al. (1995) provided a good review. In the late nineteenth century, Gilbert (1895) speculated on the role of deflation in the creation of pans on shale in the Arkansas Valley of eastern Colorado, while Baker (1915) attempted to explain the depressions of the southern High Plains as a result of karstic collapse through calcrete (caliche) solution. The solutional hypothesis has received support from Osterkamp and Wood (1987), who argued that pan development and enlargement in the area occurs by carbonate dissolution through groundwater in the unsaturated zone, piping of water and fine clastic material towards the zone of saturation and eluviation by groundwater of dissolved and particulate material. Likewise, Paine (1994) has argued that pans in the Texas Panhandle are related to subsidence above Permian evaporite-bearing strata. Some other workers have postulated that bestial activities (e.g. wallowing) may have contributed to pan development.

Figure 3.12 Characteristically shaped pans developed along ancient drainage lines in Western Australia. (ASG)

American pans appear to have developed preferentially on certain lithologies in association with desiccated lakebeds and along relic drainage lines. Favoured rock types on which they have developed include the carbonate calcrete (caliche) caprock of the Ogallala Formation of Texas and New Mexico; the Pierre, Carlile and Steele shales of Colorado, Wyoming, Montana and the Dakotas; and the Mesaverde and Fox Hill sandstones of Wyoming and the Dakotas, respectively. Relatively fewer and smaller basins are found on the cover sands and loess blanket of the High Plains in Nebraska, Kansas and Colorado. Many deflation basins have formed on the Pierre Shale surface to the east of the Colorado Front Range from Denver to Fort Collins. Most of these are circular to oval in shape, with diameters between 0.5 km and 3.0 km (Colton, 1978). Other occurrences of pan development on the Pierre Shale include western South Dakota (Crandell, 1958) and eastern Wyoming and Montana. The largest deflation basin excavated into shale is in the Big Hollow in Wyoming (De la Montagne, 1953), which has a length of 6.7 km and an area of 11.4 km^2.

Other pans occur in association with desiccated pluvial lakes. Their drying up was accompanied by deflation, leading to the migration of dunefields into areas adjacent to the leeside shorelines of the larger playas and the construction of substantial lunette dunes during stages of lake-level recession. The floor of Pluvial Lake Estancia, on

Figure 3.13 Bestial activities such as wallowing may contribute to pan enlargement as here in the Etosha Park, northern Namibia. (ASG)

the eastern side of the Mozama Range in New Mexico, contains more than sixty such deflated basins, including Laguna del Perro and Salida Lake, which are bounded by lunette dunes approaching 45 m in height (Bachhuber, 1971).

Approximately fifty basins on the southern High Plains with areas of 5–90 km^2 are linked to regional palaeodrainage channels. The disruption of much of this drainage may be attributed to the beheading of major tributaries of the Brazos River by the northward extension of the Pecos, which pirated streams with source regions in the Rocky Mountains that formerly flowed eastwards across the southern High Plains. Frye and Leonard (1957) consider the drainage to have been developed during the Kansan Glaciation, and Reeves (1966) has noted the association of several playa basins south of Lubbock, Texas, with Cretaceous drainage valleys.

The origin of pans has generated a large literature, and hypotheses for their formation have included solution, excavation by animals (elephants, hogs, etc.) (Figure 3.13), karstic and pseudo-karstic solution and tectonic subsidence (Goudie and Wells, 1995). Many, as in Australia and the High Plains of the United States, occur along dismembered drainage lines. That pans are at least in part of aeolian origin, however, is indicated by their distinctive morphology (it has often been likened to a

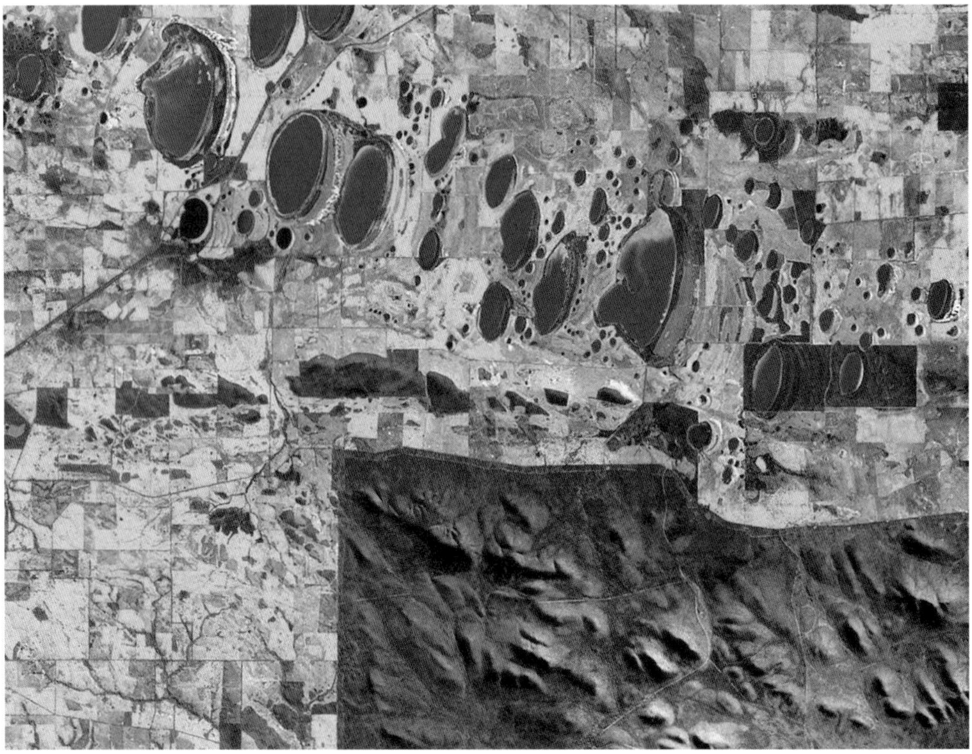

Figure 3.14 A Landsat image of pans developed in the Stirling Range area of Western Australia. (Courtesy of NASA)

pork chop), their orientation with regard to prevailing wind directions (Figure 3.14), the presence of lunettes (composed in part of sediment deflated from pan floors) on their lee sides and observations on the ground and from space of dust plumes blowing from their surfaces. Lunette dunes, composed of material excavated by the wind, frequently (although not invariably) occur on their lee sides (Sabin and Holliday, 1995). On the southern High Plains of Texas there are some 25,000 pans, only 1,100 of which have lunettes, but elsewhere the proportion of pans that have lunettes is considerably higher. In West Siberia, for example, of 8,900 pans studied, 71.4 per cent had lunettes (P. Kent, pers. comm.).

That said, processes other than – or in addition to – aeolian ones may contribute to the development of closed depressions in desert areas, including solution and animal excavation (Gustavson et al., 1995). Moreover, they have undergone periods of modification during more humid phases (Holliday et al., 1996; Holliday, 1997) and may contain detailed sedimentary evidence for environmental reconstruction (Bowen and Johnson, 2012). Dating of lunette dunes by luminescence techniques has great potential for establishing when deflational activity has been active (e.g. Lawson and Thomas, 2002; Telfer and Thomas, 2006; Bowen and Johnson, 2012).

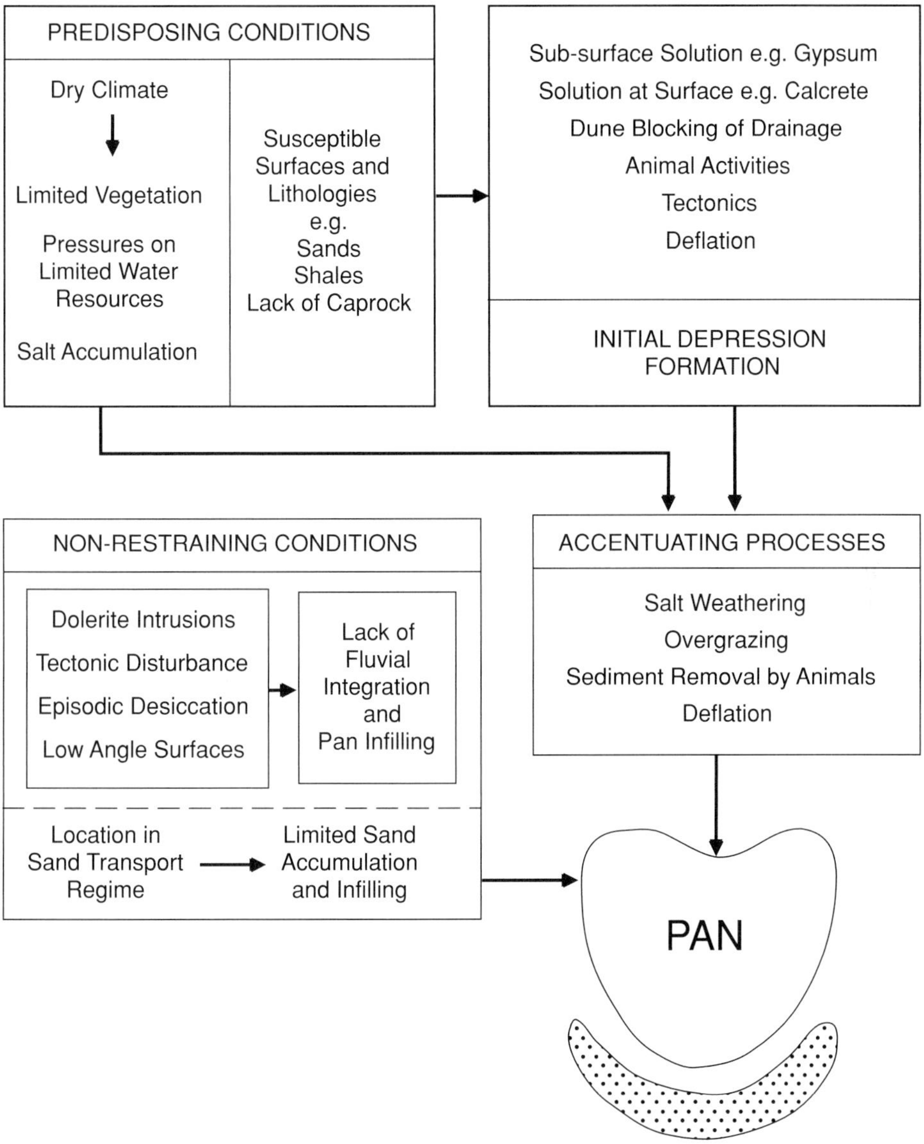

Figure 3.15 A model of pan formation. (ASG)

Goudie (1999b) developed a model of pan development (Figure 3.15) which both recognized the variety of formative influences and classified them into various categories. First of all, there is the *predisposing condition* of low precipitation which has various consequences: vegetation cover is limited so that deflation can occur; animals concentrate at pans, causing trampling and overgrazing which also promote deflation; and salt accumulation occurs so that salt weathering can attack the fine-grained bedrock in which the depression lies, producing rock flour which can then be evacuated by the wind (Goudie and Viles, 1997). These are *accentuating processes*, which serve to enlarge hollows, whether they are formed by other *initial formative processes*, such as solution of carbonate and gypsum beds or tectonics. It is also important if pans are to develop that the initial surface depression is not obliterated by the action of integrated or effective fluvial systems. *Non-restraining conditions* that limit fluvial activity are low-angle slopes, episodic desiccation and dune encroachment, the presence of dolerite intrusions and tectonic disturbance. In addition, it is important that pans do not lie in areas of active sand accumulation which might cause infilling of an existing hollow, although pans can and do develop in inter-dune depressions, particularly in linear and parabolic dunefields.

3.7 Yardangs

'Yardang' was introduced by Hedin (1903) as a term for wind-abraded ridges of cohesive material. They range in size from small centimetre-scale ridges (micro-yardangs), to forms that are some metres in height and length (meso-yardangs), to features that may be tens of metres high and some kilometres long (mega-yardangs) (McCauley et al., 1977b; Cooke et al., 1993, pp. 296–97; Al-Dousari et al., 2009). Mega-yardangs are ridge and swale features of regional extent (Mainguet, 1972) (Figure 3.16), and their curving trajectories in the Sahara very closely mirror the trajectories of the winter trade winds, both being largely north to south over Egypt and almost east to west in the lee of Tibesti.

Greeley and Iversen (1985, p. 140) believed that the shape of yardangs, like an upturned ship's hull, was an equilibrium shape, which typically would be 'an elongate hill of 1:4 width-to-length ratio, asymmetric in profile, and with the highest part in the upwind one-third of the hill'. Various morphometric studies of yardangs have revealed relationships between different parameters. Whereas Ward and Greeley (1984) found a 1:4 width-to-length ratio, Halimov and Fezer (1989) found that the ratios of length, width and height were 10:2:1; and Goudie et al. (1999) found volume, length, width, height ratios of 18.7:9:9:2.7:1.

The forms may go through a cycle of development and eventual obliteration (Halimov and Fezer, 1989; Dong et al., 2012) (Table 3.10). Although they are dominantly aeolian erosion features, there has been a considerable debate as to the relative importance of deflation, aeolian abrasion, fluvial incision and mass movements in moulding yardang morphology (McCauley et al., 1977a; Laity, 1994; Goudie, 1999b). That abrasion is important is indicated by polished, fluted and sandblasted slopes, and the undercutting of the steep windward face and lateral slopes. It is probably the dominant process in hard bedrock yardangs, whereas deflation may be important in the evolution of yardangs developed in soft sediments, such as old lakebeds. Excessive fluvial erosion, however, would tend to obliterate yardangs, although fluvial erosion

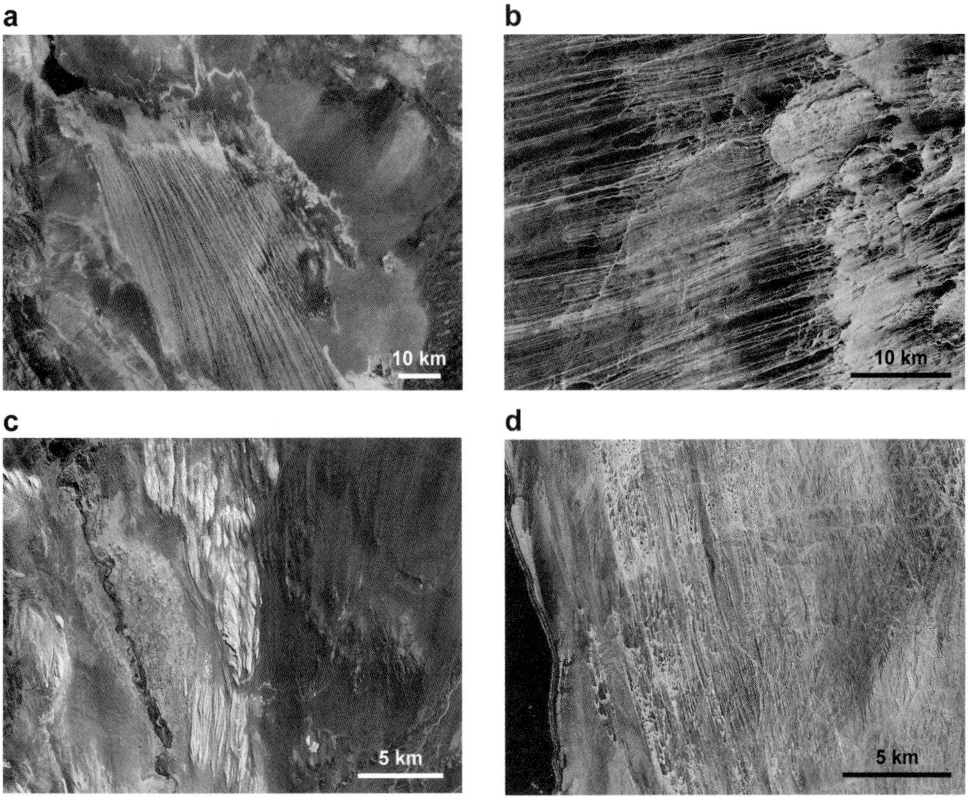

Figure 3.16 Landsat images of mega-yardangs in (a) the Lut Desert of Iran, (b) the central Sahara, (c) the Peruvian Desert, (d) northern Namibia. (Courtesy of NASA)

may provide an avenue along which wind erosion occurs, and water-formed rills are often present on their slopes. Mass movements may also be significant when their slopes have been over-steepened by wind erosion, while leaching of binding salts may contribute to their evolution (Dong et al., 2012).

Table 3.10 *An evolutionary model of yardang development*

1. Formation of lakebed, swamp deposits or cultivation deposits under humid conditions
2. Desiccation
3. Incision of beds by wind and/or water towards base of sediment
4. Formation of high ridges and mesas
5. Abrasion widens corridors and plains between ridges. Over-steepening causes mass movements. Salt weathering encourages deflation.
6. Ridges become transformed into cones, pyramids, sawtooth forms, hogbacks, etc.
7. After relief is reduced to less than 2 m, the whole surface is subjected to abrasion to give simple aerodynamic form.
8. Gradual removal of sediment until plain surface formed

Source: Modified from Halimov and Fezer (1989) and Goudie (1999d), table 8.4).

Table 3.11 *Mean annual rainfall (mm) in mega-yardang areas*

Al Muharraq, Bahrain	68
Chanaral, Chile	10
Dunhuang, China	40
Faya Largeau, Chad	15
Luderitz, southern Namibia	10
Lut, Iran	<50
Mut, Egypt	1
Northern Namibia	10–30
Pisco, Peru	0
Potrerillos, Chile	50
Tayma, Saudi Arabia	61

The significance of yardangs is now being re-evaluated, largely because analysis of satellite images has shown that mega-yardangs cover extensive areas in the coastal deserts of Peru and Chile, the high Andes of South America (de Silva et al., 2010), the Sahara and Libyan Deserts, the northern and southern Namib, the north-west of Saudi Arabia (Vincent and Kattan, 2006), the Lut Desert of Iran, the Seistan Basin of Afghanistan and the Taklamakan and Turpan Basins of high Asia (Rohrmann et al., 2010).

Goudie (2007a) attempted to analyse the key factors that determine the global distribution of mega-yardangs and came up with the following relationships: First, large yardangs occur in hyper-arid areas where deflation is at a maximum, vegetation cover is minimal and where sand abrasion can occur. Nearly all mega-yardangs occur where rainfall totals are less than 50 mm per annum, whereas pans, as we have seen, become more significant in areas where rainfall is between 150 and 500 mm per year (Table 3.11).

Second, yardangs do not occur in sites of active dune accumulation (e.g. sedimentary basins), although they do occur in former pluvial lake depressions. Basins are areas of sand-sea development rather than surface aeolian erosion. Yardangs do not occur in areas with massive alluvial fan accumulation, in truly mountainous areas or in areas with integrated drainage systems.

Third, mega-yardangs occur in trade-wind areas with unidirectional or narrow bimodal wind directions, as is made evident by their association in some cases with barchans (e.g. northern Namib, Peru, central Chile, Egypt) (Figure 3.17), a dune form that only occurs where winds are relatively constant in direction. It is only with such constant wind directions that forms which are parallel to the prevailing wind can develop. They sometimes occur upwind of sand seas, in areas where sand transport occurs (e.g. the Lut, northern and southern Namibia, Saudi Arabia).

Fourth, mega-yardangs occur in relatively homogeneous rocks without complex structures (e.g. sandstones), but with jointing along which incision can occur. They do

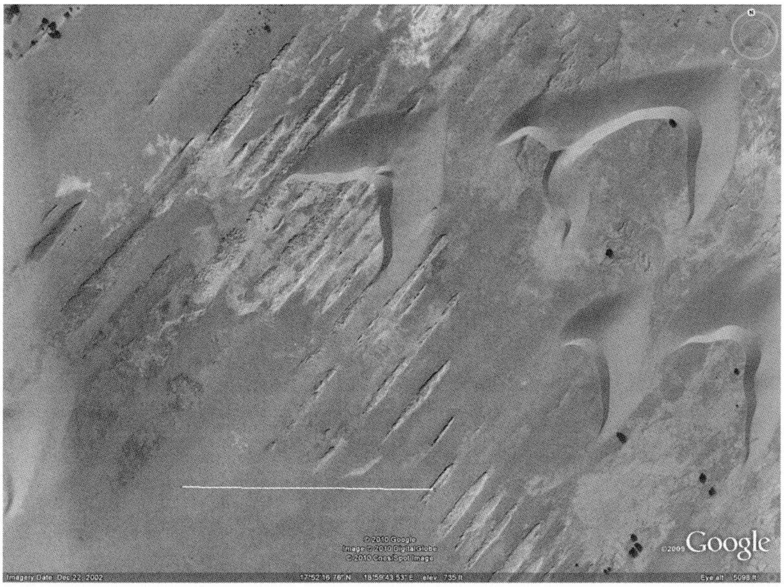

Figure 3.17 Yardangs in Chad being overwhelmed by barchans. Note the similar orientations of both phenomena. Scale bar 0.5 km. (©Google Earth 2010, ©Cnes/Spot Image 2010)

not solely occur in 'soft rocks', as has so often been asserted (see, for example, Cotton, 1942, p. 7), although chalk landscapes in the Libyan Desert have been dramatically scoured and moulded into mound and trough topography (Table 3.12). Indeed, small lake deposits only lead to small forms (Figure 3.18), although the large spreads of diatomite and other lake deposits in the Bodélé Depression have some features about 10 m high (Bristow et al., 2009)

Finally, there is very little evidence as to the age and rate of formation of mega-yardangs developed in hard rocks. Rates of vertical downcutting in weak ignimbrites in the Altiplano of Argentina are only 0.07–0.3 mm per year (de Silva et al. 2010). There is, however, some evidence that smaller yardangs can be excavated to depths of some metres in lake and swamp deposits of mid-Holocene age (Riser, 1985; Embabi, 1999; Goudie, 1999b; Goudie et al., 1999). Mega-yardangs may be old and persistent features that have been shaped over millions of years, not least by high-velocity glacial-age trade winds (Rea, 1994). The Atacama and the Namib, for example, are deserts that originated in pre-Pleistocene times, possibly in the Miocene or earlier (Goudie, 2002), so that there has been an extended time for yardang formation. Some yardangs, however, such as those in parts of the southern Namib, may be relict features, currently undergoing weathering rather than abrasion, because of past changes in sand-flow patterns associated with sea-level changes (Corbett, 1993).

Table 3.12 *Rock types of yardang fields*

Location	Rock type	Source
Meso-yardangs		
Syria	Consolidated dunes	Rösner (1998)
Ebro Depression, Spain	Miocene limestones and gypsum, unconsolidated lake deposits	Gutiérrez-Elorza et al. (2002)
Jaisalmer, India	Eocene limestones	Goudie (1999d)
Rogers Lake, USA	Lunette material	Ward and Greeley (1984)
South Dakota, USA	Shale	Baker (1951)
Mali	Holocene lacustrine beds	Riser (1985)
Mojave, California, USA	Holocene playa dunes	Clarke et al. (1996)
Kharga, Egypt	Lake and swamp beds (Holocene)	Goudie et al. (1999)
Farafra, Egypt	Holocene lakebeds	Embabi (2005)
Kara-Bura, Tajikiststan	Quaternary alluvium	Goudie (2008a)
Hamoun, Afghanistan	Quaternary alluvium and lakebeds	Goudie (2008a)
Northern Peru (Talara)	Upper Eocene to Palaeocene shales and sandstones	McCauley et al. (1977b)
Mongolia	Upper Cretaceous sandstone and mudstone	Ritley et al. (2004)
China	Sandstones, claystones and loess	Wang et al. (2011)
China	Lake, fluvial and aeolian strata	Dong et al. (2012)
Mega-yardangs		
Dakhla region, Egypt	Palaeogene limestones	Brookes (2001)
Southern Namibia	Gariep Complex rocks (volcanic intrusives, schists, rhyolite, etc.)	Goudie (2008a) and Corbett (1993)
	Nama System dolomites	Krenkel (1928)
Cunene erg, Namibia	Damara rocks (schists, marbles, phyllites, etc.)	Goudie (2008a)
Bahrain	Aeolianite and dolomite	Doornkamp et al. (1980)
Central Peru	Tertiary shales, siltstones and sandstones	McCauley et al. (1977b)
Borkou, Sahara	Palaeozoic and lower Mesozoic sandstones	Mainguet (1972)
Saudi Arabia	Cambrian sandstones	
Lut Desert, Iran	Pleistocene clays, silts, etc.	Gabriel (1938)
Central Andes	Ignimbrites	de Silva et al. (2010)
Qaidam Basin, China	Neogene sandstones	Rohrmann et al. (2010)

3.8 Inverted Relief

Deflation of desert surfaces can cause relief inversion to occur (Figure 3.19). This is the case, for example, with old gravel channels, that, being resistant to deflation, may be left upstanding as sinuous ridges (Oviatt et al., 2003), which has sometimes been termed suspendritic drainage (Miller, 1937). Raised channels of this type are widespread in Arabia (Beydoun, 1980; Maizels, 1987) but are also known from Australia (Pain et al., 2007). Equally, valley calcretes, because of their ability to

Figure 3.18 Yardangs, mud lions, developed in lacustrine beds in the Bahariya Oasis, Egypt. (ASG)

Figure 3.19 A possible inverted channel composed of river gravels of an ancient course of the Kuiseb River, near Gobabeb, Namibia. (ASG)

Table 3.13 *Landform-based deflation estimates*

Place	Feature	Rate of deflation (mm yr^{-1})
Algeria	Biskra alluvium	0.5–2.0
Algeria	Sebkha Mellala	0.41
Argentina	Yardangs in ignimbrites	0.07–0.3
Central Asia	Lop Nor yardangs	20
Chad	Bodélé Depression	2.6–3.1
China	Yardangs	0.01–0.40
Kharga, Egypt	Yardangs	1.0
Mali	Araouane Basin	0.09
Oman	Palaeochannels	2
Saharan lakes	Yardangs	0.4–4.0

Source: Based on data in Washington et al. (2006b, table 1), Beydoun (1980) and Wang et al. (2011).

create resistant low-lying deposits, may be left upstanding during periods of increased aeolian activity (McLaren, 2004). Other inverted relief may be associated with interdunal lake deposits, as has been found in the Wahiba Sands of Oman (Radies et al., 2005).

3.9 Rates of Deflation and Abrasion

It is possible to establish estimates of rates of deflational lowering by examining remnants of landform features of known thickness and age. This has been done for Holocene lakebeds and alluvium (Table 3.13), but it needs to be stressed that these rates are for susceptible materials and would not relate to bedrock outcrops. Rohrmann et al. (2010) have used cosmogenic radionuclide studies to estimate Neogene sandstone erosion rates in the Qaidam Basin of China and found that rates were between 0.04 and 0.34 mm per year, whereas Kapp et al. (2011) suggest that since the late Pliocene, strata have been removed at rates of >0.12–1.1 mm per year. They assert (p. 7) that wind erosion rates 'in the sandblasted portion of the Qaidam basin are comparable to rates of fluvial and glacial erosion in tectonically active mountain ranges'. They also remark (p. 9) that '[h]undreds to thousands of meters of vertical strata have been removed from above Qaidam basin folds since 2.8 Ma', and suggest that this material has been a major source of loess downwind.

In general, however, there is limited data on the rate at which wind can abrade rocks and related materials. Sharp (1949) described ventifacts of granite, gneiss and quartzitic sandstone from the Big Horn Mountains of Wyoming and inferred a rate of about 1 mm per year, which is similar to rates of abrasion of a dacite boulder field at Mono Craters, California, obtained by Williams (1981). Goudie (1995, p. 161) estimated that brick buildings at Kolmanskop in Namibia had been abraded at rates of 2.2–5 mm per year, whereas Sharp (1964) measured abrasion rates for bricks and other materials at Garnet Hill, California, and found rates of 50 mm per year. At the other end of the scale, McCauley et al. (1979) discovered Upper Palaeolithic sandstone hearthstones that had abraded at rates of only 1–2 mm per thousand years.

3.10 Ventifacts

'Ventifacts are rocks abraded by windblown particles, characterized by their distinctive morphology and texture' (Laity and Bridges, 2009, p. 202). Ventifacts are by no means restricted to warm deserts, however, for they occur widely in polar and periglacial regions. The term itself was coined by Evans (1911, p. 335), and seminal studies were conducted by Sharp (1949, 1964); a good review is provided by Knight (2008). Ventifacts are found on both Earth and Mars (Greeley et al., 2008) and were probably first discussed by Blake (1855) in the context of the south-west United States. Ventifacts occur widely in the deserts of the United States (see, for example, Needham, 1937). Spectacular images of features from Egypt, both large and small, are given in Hume (1925). Indeed, huge expanses of the chalks and limestones of the Western Desert are moulded into razor-sharp ridges. Ventifact forms include facets – which often abut at sharp keels – polished surfaces, differential erosion forms caused by inherent inhomogeneities in the rock and features with essentially universal shapes (pits, grooves, flutes, scallops, etc.) (Várkonyi and Laity, 2012). Rock properties, such as hardness, texture, bedding, and so forth, play a role in determining ventifact morphology. Hard, fine-grained rocks, such as marbles, become facetted, with relatively little roughness, whereas heterogeneous, coarse-grained rocks are more likely to develop pits, flutes or grooves. Small ventifacts, consisting of wind-facetted pebbles, are normally called dreikanters.

There has been some controversy in the literature about the prime abrasive agent involved, with dust being seen as a possible candidate, but Laity and Bridges (2009) made a strong case for saltating sand being the most important abrasive. Abrasion occurs dominantly on windward faces, and their orientation is that of the highest velocity winds to which they are exposed. Early simulation work by Schoewe (1932) demonstrated that most wind-facetted pebbles are made under conditions of directionally rather constant winds, with the Brazil-nut-shaped variety representing the end stage in their development. Simulation of pitting has been undertaken by Bridges et al. (2010).

3.11 Coastal Sabkhas

The coastlines of many arid regions are bounded by flat, salty, marshlands called sabkhas, and aeolian processes contribute to their formation. Offshore sand movement is especially important in the sedimentation process in which dunes move seaward and

Figure 3.20 Satellite image of the Great Sabkha of Abu Dhabi. Scale bar is 10 km. (©Google Earth 2010, ©GeoEye 2010)

interdigitate with lagoonal sediments, as on the coasts of Mauritania, Kuwait, eastern Saudi Arabia, central Namibia and northwest Australia. Dust fallout also plays a role, and wind erosion causes some planation.

The Abu Dhabi sabkha (Figure 3.20) is the best developed example in the world, but other sea-margin sabkhas occur in the Middle East on the Gulf of Suez, the Red Sea (Bahafzullah et al., 1993), the Sinai (Gavish, 1980), the Nile Delta (Wali, 1991) as well as elsewhere in the Gulf itself: Kuwait (Robinson and Gunatilaka, 1991), Qatar (Abu-Zeid et al., 1999) and eastern Saudi Arabia (Barth, 1998). Coastal sabkhas also occur in Baja California (Vogel et al., 2010) and Tunisia (Lakhdar et al., 2006).

The best developed examples of this landscape type are found on the eastern coast of Arabia (Evans et al., 1969; Kirkham, 1997; Alsharhan and Kendall, 2003; Gunatilaka, 2011). The biggest and best of these sabkhas spreads along the coastline of the Abu Dhabi emirate between Jabal Dhanna and Ras Ganada, a distance in excess of 300 km; it is up to 24 km wide.

Sabkhas display billiard-table flatness and lie between land and sea. They slope at about 0.4 m per km. They are characterised by a thin crust of white salt (halite) and a rubbery mat of almost-black stromatolitic algae, underlain by sand, silt or clay, with a locally cemented hard layer of gypsum ($CaSO_4.2H_2O$). Marine shells are common

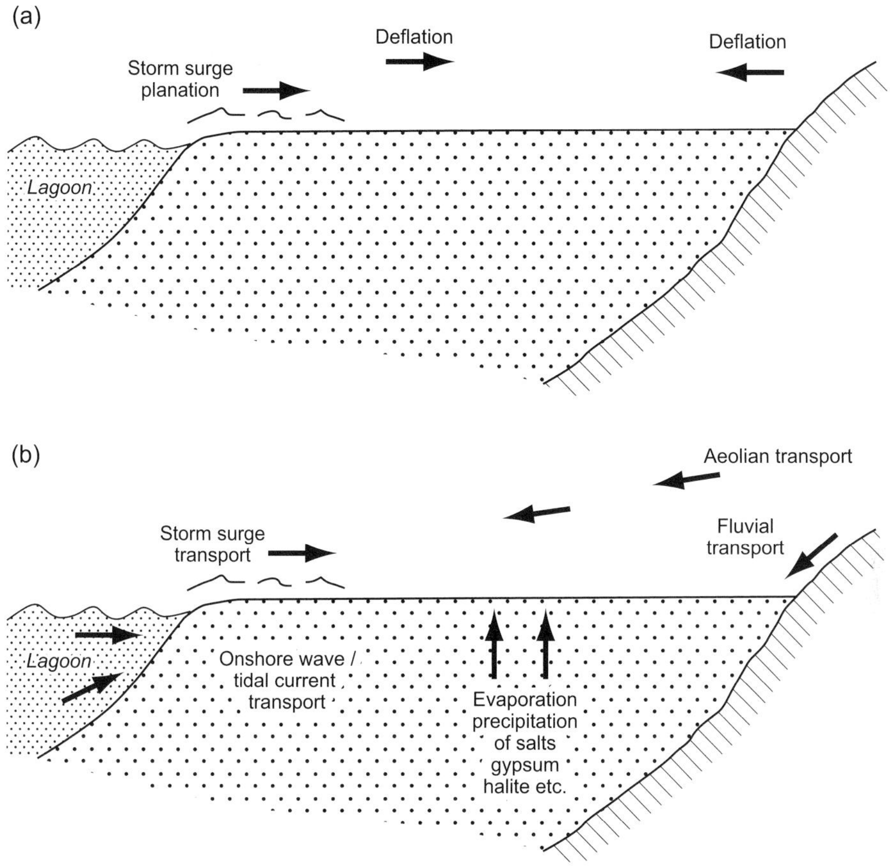

Figure 3.21 Processes of erosion (a) and deposition (b) on coastal sabkhas. (Modified from Evans, 1995, fig. 9, in Goudie, 2002, fig. 8.8)

and the gypsum is often altered to anhydrite under the prevailing high-temperature conditions. Dolomite formation occurs in association with microbial mats (Bontognali et al., 2010). The sabkha surface is also noted for its large polygonal features (up to 166 m in diameter) and tepee structures (Lokier and Steuber, 2009).

The sabkha surface (Figure 3.21) is generated by erosional processes, including planation by surges produced by shamal winds from the north-west and deflation to the water table. It is also generated by depositional processes, including evaporite precipitation from seawater and continental runoff (Wood et al. 2002), aeolian sand and dust deposition and organic processes (e.g. faecal pellet production). It is the balance between erosional and depositional processes in this arid, highly salty, low-energy environment that leads to the special development of the sabkhas. The main Arabian sabkhas lie to the landward of protective reefs, banks and shoals. Of special note is the Great Pearl Bank, which is particularly well developed between Qatar and the Straits of Hormuz. The sabkhas have for the most part formed in the last 6,000 years, when a rapid transgression infilled inter-dunal Pleistocene lagoons, depressions and dunes. Since about 4,000 years ago, sea level has been relatively stable, and the sabkhas on average have prograded at a rate of about 0.7–1.03 km every 1,000 years (Lokier and Steuber, 2008).

Although the Abu Dhabi sabkha is a 'world-class' landform assemblage, deserving of World Heritage status (see Section 6.15), and has been crucial to understanding many ancient hydrocarbon-bearing carbonate-evaporite formations, it has been irrevocably damaged in recent decades because of the relentless growth of economic activity and engineering excesses. Such growth is not without its problems, however, for sabkhas are potentially hazardous environments for humans. This is partly because they are prone to flooding by shamal-induced storm surges, and partly because they are very aggressive and corrosive environments in terms of salt attack on engineering structures (Goudie and Viles, 1997).

3.12 Soil Erosion by Wind

Wind erosion of soils is an important facet of land degradation or desertification, causing dust storms, leading to a reduction in soil depth, causing loss of fines and nutrients and causing abrasion of plants (J. Li et al., 2009; Ravi et al., 2011) (Table 6.2).

Important controls of the susceptibility of soils to erosion include inherent properties of the soils themselves: their grain-size characteristics, surface roughness and their aggregate stability. The first includes clay content, which promotes cohesion, while the last is greatly affected by soil organic content. Soils that are dry, have a low clay content and little binding humus are highly susceptible to wind erosion. The relative susceptibilities of different surfaces to wind erosion can be determined by a number of techniques, including portable wind tunnels (Maurer et al., 2010).

Erosion occurs when the shear stress exerted by the wind on the surface exceeds the ability of the surface material to resist detachment and transport. The wind-shear velocity needs to exceed a certain minimum value, the 'threshold shear velocity', for erosion to occur. It has long been recognized (Bagnold, 1941; Chepil, 1945) that the threshold velocity for particle movement increases as grain size increases, due to the effects of gravity, but that it also increases for the smallest particles due to particle cohesion. The balance of these two effects produces an optimum particle size ($\sim$60–80 μm) for which the threshold friction velocity is at a minimum. Land-surface roughness is also a key factor. On the one hand, the threshold velocity required to initiate dust emission is increased in areas with higher surface roughness. On the other hand, the drag coefficient is also increased, leading to higher wind friction and thus possibly to higher dust emissions (Prigent et al., 2005).

Other important controls on a soil's erodibility include the degree of cover by non-erodible elements such as rocks and vegetation (e.g. Merrill et al., 1999); the moisture content, which affects the adhesive properties of the soil (Ravi et al., 2004); the

Table 3.14 *Wind threshold values for surface types in the U.S. Southwest*

Surface type	Threshold speed (m/s)
Mine tailings	5.1
River channel	6.7
Abandoned land	7.8
Desert pavement, partly formed	8.0
Disturbed desert	8.1
Alluvial fan, loose	9.0
Dry wash	10.0
Desert flat, partly vegetated	11.0
Scrub desert	11.3
Playa (dry lake), undisturbed	15.0
Agriculture	15.6
Alluvial fan, crusted	16.0
Desert pavement, mature	16.0

Source: After Clements et al. (1963), Nickling and Gillies (1986) and Brazel (1991).

entrainment of saltating sand grains (Wiggs et al., 2004); and the amount of disturbance to which the soils are exposed by human activities (Maurer et al., 2010), including cattle trampling of crusted surfaces (Baddock et al., 2011). Snow cover (Kurosaki and Mikami, 2004) will reduce wind erosion during winter months, although blowing snow can also break down soil aggregates. Seasonal freeze-thaw action is another way in which aggregate stability can be reduced (Bullock et al., 2001). Any surface crusts will also control rates of soil erosion (McKenna-Neuman et al., 1996; Singer and Shainberg, 2004; Langston et al., 2005; Ravi et al., 2011), although these may be abraded and broken down by saltating sand grains (Houser and Nickling, 2001). Table 3.14 illustrates the nature and direction of the effects on wind erosion of a range of soils, vegetation and landform conditions while Table 3.15 summarises the various controls on rates of wind erosion.

The erosivity of soils is controlled by a range of wind variables, including velocity, frequency, duration, magnitude, shear and turbulence. Such wind characteristics vary over a whole range of timescales, from seconds to millennia. For example, Bullard et al. (1996) have shown how dune activity varies in the south-west Kalahari in response to decadal scale variability in wind velocity, whereas over a larger timescale (see Section 3.3), there is evidence that trade-wind velocities may have been elevated during the Pleistocene glacials (Rea, 1994).

The controls on wind erosion on agricultural land are often expressed as a Wind Erosion Equation:

$$E = f(C, I, L, K, V)$$

Table 3.15 *Some key physical factors influencing wind erosion*

Climate	Soil	Vegetation	Landform
Wind speed (−)	Soil type	Type	Surface roughness
Wind direction	Particle composition	Coverage (+)	Slope (+)
Turbulence (−)	Soil structure		Ridge
Precipitation (+)	Organic matter (+)		
Evaporation (−)	Calcium carbonate (+)		
Air temperature (+)	Bulk density		
Air pressure (−)	Soil aggregation (+)		
Freeze-thaw action (+)	Soil water (+)		

Note: (+) means wind erosion becomes weaker and (−) erosion becomes greater as the factor increases.
Source: Modified from Shi et al. (2004).

where E is the potential erosion loss, C is a local climatic index, I is a soil erodibility index, L is a factor relating to field shape in the prevailing wind direction, K is a ridge roughness factor for ploughed ground and V is a vegetation cover index. The equation was developed initially for the U.S. Midwest (see Woodruff and Siddoway, 1965) and drew attention to the factors which could be manipulated by farmers (namely, I, L, K and V).

The climatic factor (C) was a simple combination of two key climatic variables: (1) annual wind speed and (2) a moisture index. Plainly, dry and windy areas are likely to be most susceptible to wind erosion. The soil erodibility factor (I) is more complex, and needs to be seen in terms of both individual grain size and aggregate characteristics. Fine sands and silts are likely to be most susceptible, partly because of the relatively low velocities required for their entrainment but also because the presence of clay tends to produce wind-stable clods. The presence of large clods reduces the risk of wind erosion. The fetch distance over which the wind acts (L) is related to field size and the presence or absence of shelter belts of differing heights, spacing and permeability. The ridge roughness factor (K) is based on the experimental observation that the rougher the surface – up to about 6 cm – the lower the wind speed at the surface. Thus furrows at right angles to the wind will tend to dampen down rates of wind erosion. The vegetation factor (V) is absolutely fundamental, for a dense vegetation cover – especially if, like grass, it has short stalks and narrow leaves – does more than anything else to reduce erosion rates. The amount of material that is eroded depends on the size of unvegetated gaps on which the wind can act and the height and density of the vegetation (Field et al., 2009). Grasslands that have been burnt may be subject to accelerated wind erosion (Stout, 2012).

The Wind Erosion Assessment Model (WEAM) is a predictive physical model which aims to account for the combined effect of climate, soil, vegetation and land use. Its fundamental physical viewpoint (Shao, 2000) is that wind erosion is a result

Table 3.16 *Wind erosion models*

Field scale	
Wind Erosion Equation (WEQ)	Woodruff and Siddoway (1965)
Revised Wind Erosion Equation (RWEQ)	Fryrear et al. (1998)
Wind Erosion Prediction System (WEPS)	Hagen (1991)
Wind Erosion Stochastic Simulator (WESS)	van Pelt et al. (2004)
Local to regional scale	
Wind Erosion on European Light Soils (WEELS)	Böhner et al. (2003)
Australian Land Erodibility Model (AUSLEM)	Webb et al. (2006)
Wind Erosion Assessment Model (WEAM)	Shao et al. (1994)
Integrated Wind Erosion Modelling System (IWEMS)	Lu and Shao (2001)
Continental to global scale	
Dust Production Model (DPM)	Marticorena and Bergametti (1995)
Dust Entrainment and Deposition Model (DEAD)	Zender et al. (2003a)

of two opposing forces: (1) the capacity of the wind to start and maintain erosion, and (2) the ability of the soil to resist it. The wind's capacity to start and maintain erosion is the friction velocity u^* (the wind shear or drag on the soil), while the opposing quantity offered by the soil is the threshold friction velocity u^*_t (the minimum friction velocity that is required for erosion to occur). The former is determined by wind-flow conditions and the surface roughness, whereas the latter is determined by such surface factors as soil texture, aggregate structure and moisture content.

There have been many other wind erosion models that have been produced at scales ranging from the individual field, through the local, regional and continental, to the global (Table 3.16). These have been expertly reviewed by Shao (2000; 2008) and Webb and McGowan (2009). Modelling the response of wind erosion to climatic variables on agricultural land is vastly complex, however – not least because of the variability of soil characteristics, topographic variation, the state of plant growth and residue decomposition and the existence of wind breaks. To this needs to be added the temporal variability of aeolian processes and moisture conditions and the effects of different land management practices and disturbance regimes (Leys, 1999; Maurer et al., 2010). Techniques for dealing with soil erosion by wind are discussed in Section 6.3.

4

Dunes

4.1 Introduction

Although the romantic view of deserts envisages landscapes dominated by ever-changing sand dunes, with oases, camels and men in flowing robes, only about one-third to one-quarter of the world's deserts are covered by aeolian sand, so its role in deserts should not be exaggerated. Indeed, in the U.S. deserts, sand dunes occupy less than 1 per cent of the surface area. In contrast, the proportion of Australia covered by dunefields may approach 40 per cent (Hesse, 2010). Nonetheless, great ergs, or seas of sand (Figure 4.1), are found nowhere else on Earth, and they form some of the most beautiful, repetitious and regular landforms that our planet (and, indeed, Mars) has to offer. Particularly large ergs occur in Arabia and the Sahara. The fascination of dunes was well expressed by Bagnold (1941, p. xxi):

> Here, instead of finding chaos and disorder, the observer never fails to be amazed by a simplicity of form, and exactitude of repetition and a geometric order unknown in nature on a scale larger than that of crystalline structure. In places vast accumulations of sand weighing millions of tons move inexorably, in regular formation, over the surface of the country, growing, retaining their shape, even breeding, in a manner which, by its grotesque imitation of life, is vaguely disturbing to an imaginative mind.
>
> Elsewhere the dunes are cut to another pattern – lined up in parallel ranges, peak following peak in regular succession like the teeth of a monstrous saw for scores, even hundreds of miles, without a break and without a change in direction, over a landscape so flat that their formation cannot be influenced by any local geographical features. Or again we find smaller forms, rare among the coastal sand hills, consisting of rows of coarse-grained ridges even more regular than dunes. Over large areas of accumulated sand the loose, dry, uncemented grains are so firmly packed that a loaded lorry driven across the surface makes tracks less than an inch in depth. Then, without the slightest visual indication of change, the substance only a few inches ahead is found to be dry quicksand through which no vehicle can force its way.

Bagnold was the twentieth century's greatest figure in sand dune studies, but there is now a huge literature on dunes, and the history of dune research has been summarised

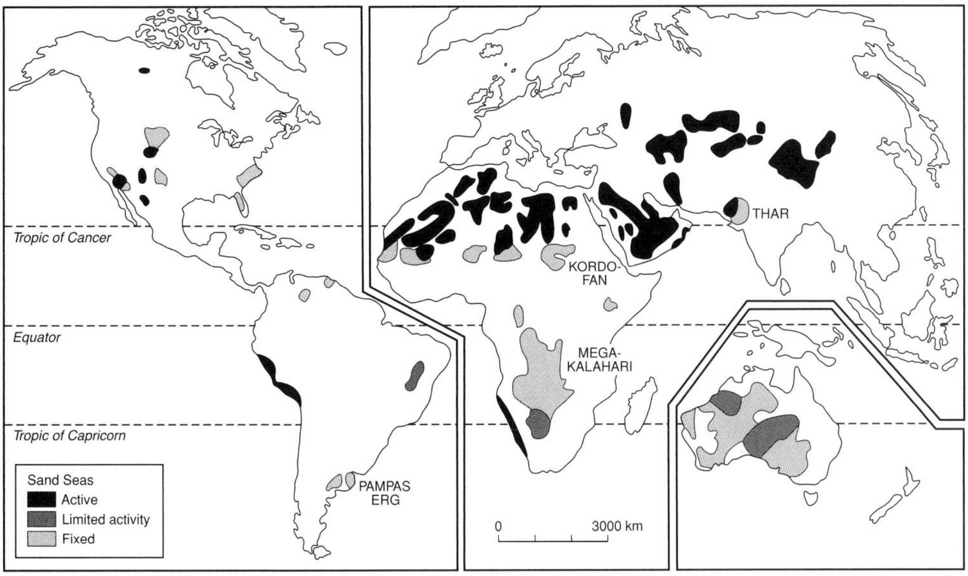

Figure 4.1 Location of the major sand seas (ergs) of the world. (From Goudie, 2002, fig. 1.7)

by Goudie (1999c). There are also various excellent recent texts, including those by Pye and Tsoar (1990) and Lancaster (1995a), which provide a synthesis of information and ideas.

The sand that makes up dunefields is a result of wind winnowing grains of an easily transported size from alluvial plains, lake shores, seashores and weathered rocks such as sandstone and granite. When wind velocity exceeds the threshold velocity required to initiate sand grain movement (generally this is about 20 km h^{-1}), the grains begin to roll along the ground, but after a short distance this gives rise to a bounding or jumping action called saltation. Grains are taken up a small distance into the airstream (often only a matter of a few centimetres) and then fall back to the ground in a fairly flat trajectory. The descending grains bombard and dislodge further particles and thereby the process of saltation is maintained across country (Durán et al., 2011). The process may be enhanced as saltating grains become electrically charged during transport (Kok and Renno, 2006; 2008). The saltating grains (which generally have a diameter between 0.15 and 0.25 mm) shift larger grains (with diameters generally in the range of 0.25–2.00 mm), which move forward by surface creep close to the ground. The smallest grains (<0.15 mm) may be carried high up into the air in suspension and form the major component of dust storms. There is also a further type of motion which is called reptation. This is the low hopping of grains dislodged by descending high-energy particles. Such grains differ from those in creep because they continually pass between the reptation and saltation modes (Livingstone and Warren, 1996, p. 17).

Sediment entrainment by wind on a stable, non-eroding surface occurs when the wind's shear stress (which is a function of wind speed, turbulent energy and surface roughness) overcomes forces of particle cohesion, packing and weight. This relationship can be simplified to two parameters: (1) critical wind shear (u_{*ct}) and (2) particle diameter (d) (Bagnold 1941, p. 86):

$$u_{*ct} = \sqrt[A]{\frac{(\sigma - \rho)}{\rho} g.d}$$

where: σ = the particle density, g = the acceleration due to gravity and A = a constant dependent on the grain Reynolds number (≈ 0.1).

Two thresholds of entrainment may be identified. One, called the fluid threshold, relates only to the drag and lift forces of the wind, whereas the second, the impact threshold, is lower and combines wind forces with additional forces provided by impacting grains already in transport. Once a sediment surface has begun to be eroded by wind forces at the fluid threshold, sediment transport is maintained at the lower impact threshold because energy is also available from the saltating grains. In general, larger grains require a greater wind shear to dislodge them. This relationship is reversed for silt-sized particles (smaller than about 0.06 mm), however, where increased electrostatic and molecular cohesion require larger erosive forces for entrainment. The grain sizes most susceptible to entrainment are sand-sized particles that have diameters between 0.06 and 0.40 mm. It is this susceptibility of sand-sized material to entrainment that allows the accumulation of extensive dunefields in desert regions.

The amount of sand transported by the wind depends on a range of factors, including wind velocity, the electrification of sand grains as a result of collisions (Merrison, 2012), the size of the sand grains and the nature of the ground surface (moisture content, surface crusting, vegetation cover, etc.). For example, wind tunnel studies have shown that the critical moisture content for sand entrainment is between 4 per cent and 6 per cent (Wiggs et al., 2004). There may be a diurnal pattern of blowing sand, with a peak of activity in the early afternoon, when warming by the sun has reduced sand moisture contents and solar heating promotes thermal instability and turbulence in the atmosphere (Stout, 2012). Good introductions to the physics of sand movement are provided by Nickling (1994), Pye and Tsoar (1990), Lancaster (1995a), and Sherman and Li (2012), but the classic work remains that by Bagnold (1941).

Dunes form because saltating grains tend to accumulate preferentially on sand-covered areas rather than on adjoining sand-free surfaces. This seems to result from the check to a strong wind through intensified sand movement over a sand surface and from the lower rate of sand movement where saltating grains 'splash' into loose sand compared with that over firm ground.

Many classification schemes for dune morphology have been developed over the years, and it has for long been recognized that there are a number of possible controls on the development of different dune types in different areas. These include the grain-size characteristics of dune sediments (Wilson, 1972), the amount of sand that is present in a system, the position of the dunes within a sand sea, whether sand budgets are positive or negative (Mainguet and Chemin, 1990), the directional variability of wind (Fryberger and Dean, 1979), the presence of vegetation and high groundwater levels (both of which may favour the formation of parabolic forms) and wind strength. As these conditions vary from desert to desert, the relative importance of different dune type also varies (Table 4.1), as shown by analysis of Landsat-derived maps of some of Earth's deserts from Breed et al. (1979) by Fryberger and Goudie (1981). They showed that the most common type of aeolian depositional surface is that of sand sheets and streaks (c 38%), followed by linear dunes (c 30%), crescentic dunes of predominantly barchanoid type (c 24%), star dunes (5%) and dome dunes (c 1%). There are, however, major regional differences between ergs. Star dunes form about 24 per cent of the dune area in the north-east Sahara (but are nowhere else above 10%), while the Thar Desert parabolics cover around 29 per cent of the dune area. The Kalahari is notable for the predominance of linear dunes (c 86%), a characteristic it shares with the Australian deserts, whereas the Ala Shan of central Asia has very few. Details of the percentage area of different dune types in the Namib are given by Bullard et al. (2011, table 1). Here, c 19 per cent of the dunes are transverse types formed under unidirectional wind regimes; 57 per cent are linear forms produced under bidirectional wind conditions; c 29 per cent are star, network and dendritic dunes formed under multidirectional wind regimes; and about 1.3 per cent consist of sand sheets.

4.2 Ergs (Sand Seas)

'An erg is an area where wind-laid sand deposits cover at least 20% of the ground and which is large enough to contain draas' (large sand ridges or mega-dunes) (Wilson, 1973). It has a similar meaning to the term 'sand sea'. Active ergs tend to occur in areas where the current mean annual precipitation is less than 150 mm. Ergs range in size from 1–1,000,000 km^2, but Wilson calculated that 85 per cent of windblown sand occurs in those larger than 32,000 km^2. Their mean was calculated to be just over 100,000 km^2. Within ergs there are three different scales of bedform, with the draa (or mega-dune) being the largest, the dune being intermediate and the ripple being the smallest. Many ergs show a history of expansion and contraction, and relict or inactive ergs are widespread in the tropics. Notable examples occur on the south side of the Sahara, in the Kalahari Basin in central Africa, and in northern Australia (see Section 1.8).

Most major dunefields occur in basins rather than on uplands. They are generally absent from highlands, probably because of wind acceleration over highlands and its divergence around them. Basins also contain a large supply of sediment. This is because sediments tend to accumulate in lowland basins so that there is a ready source

Table 4.1 *Relative importance of types of dunes (%)*

	Thar	Taklamakan	Namib	Kalahari	Saudi Arabia	Ala Shan	South Sahara	North Sahara	North-east Sahara	West Sahara	Average
(A) Linear dunes (total)	13.96	22.12	32.55	85.55	49.81	1.44	24.08	22.84	17.01	35.49	30.54
Simple and compound	13.96	18.91	18.50	85.85	26.24	1.44	24.08	5.74	2.41	35.49	23.26
Feathered	–	–	–	–	4.36	–	–	3.56	1.13	–	0.91
With crescentic superimposed	–	3.21	–	–	–	–	–	4.02	7.32	–	1.46
With stars superimposed	–	–	14.34	–	19.21	–	–	9.52	6.15	–	4.92
(B) Crescentic (total)	54.29	36.91	11.80	0.59	14.91	27.01	28.37	33.34	14.53	19.17	24.09
Single barchanoid ridges	8.96	3.21	11.80	–	0.59	8.62	4.08	0.06	–	0.65	3.80
Megabarchans	–	–	–	–	–	–	–	7.18	1.98	–	0.92
Complex barchanoids ridges	16.65	33.70	–	–	14.32	18.39	24.29	26.10	12.55	18.52	16.45
Parabolics	28.68	–	–	0.59	–	–	–	–	–	–	2.93
(C) Star dunes	–	–	9.92	–	5.34	2.87	–	7.92	23.92	–	5.00
(D) Dome dunes	–	7.40	–	–	–	0.86	–	–	0.80	–	0.90
(E) Sheets and streaks	31.75	33.56	45.44	13.56	23.24	67.82	47.54	35.92	39.25	45.34	38.34
(F) Undifferentiated	–	–	–	–	6.71	–	–	–	4.50	–	1.12

Source: Analysis of maps in Breed et al. (1979) by A.S. Goudie.

of sand supply for dune formation. In Australia, for example, the Mallee dunefield occupies the western portion of the Cenozoic Murray Basin, while the Cenozoic Lake Eyre Basin underlies both the Simpson and Strzelecki Dunefields. The Kalahari dunes of central and southern Africa also occur in a great zone of subsidence, as do the Rub 'Al Khali of Arabia and the sand seas of the Tarim Basin in central Asia. The western dunefields of Australia (e.g. the Great Victoria and Great Sandy), however, have developed in a fundamentally different setting – on a subdued ridge and valley topography (Hesse, 2010).

4.3 Anticyclonic Swirls, Whirls and Whorls

Even at the greatest scale of investigation – on a continental scale – dunes show a remarkable regularity of pattern. They tend to occur as a wheel-round or swirl in an anticlockwise direction related to dominant continental wind patterns. This is especially evident in Australia (Wasson et al., 1988; Hesse, 2011), but it is also a feature of the Kalahari and Arabia. However, the correspondence between dune direction and present-day resultant winds may be less exact than appears from a mere brisk inspection, and the orientations of the winds forming the whorl may have shifted through time, as has been demonstrated for the Simpson Desert in Australia (Hollands et al., 2006).

4.4 Ripples

Wind ripples are the smallest of aeolian bedforms and are present on almost all sand surfaces except those undergoing very rapid deposition. They generally trend perpendicular to the sand-transporting winds, although on sloping surfaces where the downwind component of grain movement is supplemented by gravity, their flow may be slightly flow oblique. Typically, they have a wavelength of 13–300 mm and an amplitude of 0.6 to 14 mm. Like dunes, ripples have gentle windward slopes (in general between 8 and 13°) and rather steeper lee slopes (up to 30°). Granule ripples are 'aeolian bedforms comprised of a sandy core that is covered by a surface layer of granules, particles that are typically 1–2 mm in diameter' (Zimbelman et al., 2009); they are also known as gravel ripples or megaripples (Isenberg et al., 2011). They tend to be significantly larger than wind ripples formed in well-sorted fine sand, have bimodal sediment distributions and have greater sinuosity than ordinary sand ripples (Yizhaq et al., 2012). Some of the largest examples, known from a windy, high-altitude environment in Argentina, are composed of large, low-density pumice clasts and have wavelengths up to 43 m and heights up to 2.3 m (Milana, 2009). In the Selima Sand Sheet in northern Sudan there are some giant ripples that have amplitudes of up to 10 m and wavelengths up to 1 km (Maxwell and Haynes, 2001, p. 1624). These dimensions are, however, exceptional, and those in the Negev have a wavelength of 75 cm and a height of 10 cm (Yizhaq, 2008). In the Kumtagh Desert of China, wavelengths range between 0.31 and 26 m, and heights from 0.015 to 1 m (Qian et al., 2012). Although granule ripples are mainly constructional features, they are commonly found in erosional settings, such as interdunal corridors or the windward slopes of dunes, where coarse grains tend to concentrate (Fryberger et al., 1992). It is important to recognize that ripples are neither small dunes nor proto-dunes. They are a distinct class of bedform, whose sizes seldom overlap with those of dunes. They also tend to be short-lived and travel much more rapidly than dunes.

4.5 Dune Types

Dune forms have a vast diversity (Figure 4.2) and vary between and across dunefields. Moreover, one form may be superimposed on another. There are, however, some basic dune types that have widespread expression in many of the world's deserts. Tsoar et al. (2004) proposed a threefold classification: *migrating* dunes (exemplified by transverse forms); *elongating* dunes (exemplified by linear dunes) and *accumulating* dunes (exemplified by star dunes). In addition to these free forms, there are also a series of dunes fixed or anchored in the landscape by topographic feature or vegetation.

4.6 Major Controls on the Nature of Dune Type

Sand Supply and Wind Directional Variability

Dune type is determined by a number of factors, including the amount and size of available sand (Eastwood et al., 2011), the transport capacity and variability of wind, the nature of vegetation cover, the evolutionary stage that the dunes have reached, the presence of bedrock highs and the areal extent of the dunefield. In very arid areas, moist phases may be required to create aggradation of fans and alluvial washes from which sand can be deflated to enable dune accretion to occur (see, for example, Lancaster and Mahan, 2012).

Wasson and Hyde (1983) determined that there were two crucial controls on dune type: (1) sand availability and (2) the directional variability of wind (Figure 4.3). They argued that 'wind strength does not appear to be important, vegetation has an ambiguous role, and the particle size of dune sediments is unimportant'. They demonstrated that sand availability (expressed as equivalent sediment thickness – EST – being defined as the sediment thickness that would exist if the dunes were leveled) was a major control on dune height. More important, they concluded that barchans tend to occur where sand supply is limited and winds are unidirectional, that transverse dunes occur where sand is more abundant and winds moderately variable, that linear or longitudinal dunes occur where winds are more variable but there is little sand and that star dunes occur where there is plenty of sand and multidirectional winds. Accurate determination of EST has always been a problem, but this is now

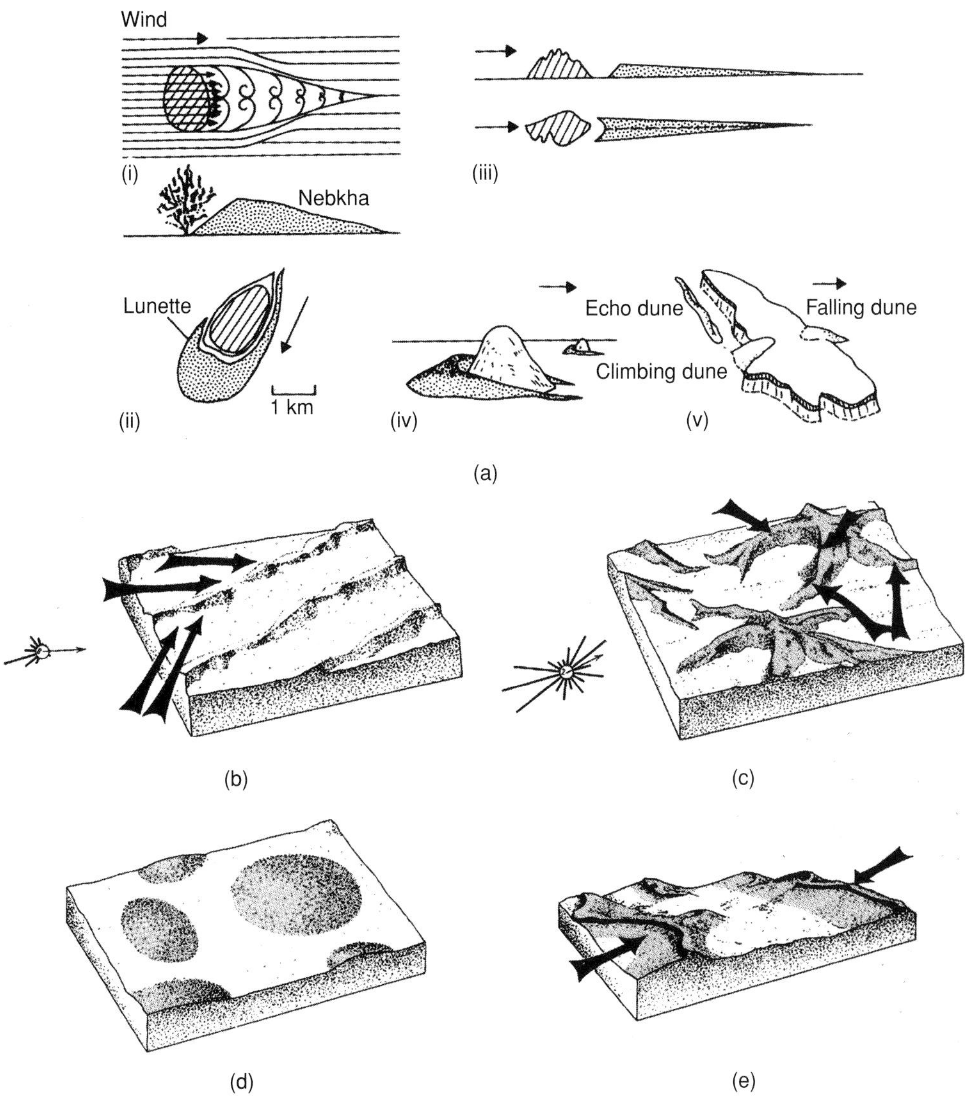

Figure 4.2 Some of the main types of dune form encountered in the world's deserts: (a) obstacle or topographic dunes: (i) a small dune or nebkha in the low-velocity area to the lee of a shrub; (ii) a crescentic lunette formed to the lee of a small desert depression (playa); (iii) wind-shadow dunes formed in the lee of some hills; (iv) a dune formed to the windward of a hill; (v) dune development in the proximity of a plateau; (b) linear dunes, or seifs. The arrows show probable dominant winds; (c) star dunes. The arrows show the effective wind directions; (d) dome dunes; (e) reverse dunes. Arrows show the wind directions; (f) parabolic dunes. Arrow shows prevailing wind direction; (g) barchan dunes. Arrow shows prevailing wind direction; (h) barchanoid ridge. Arrow shows prevailing wind direction; (j) traverse dune. Arrow shows prevailing wind direction. (From Goudie, 1984a, fig. 5.8)

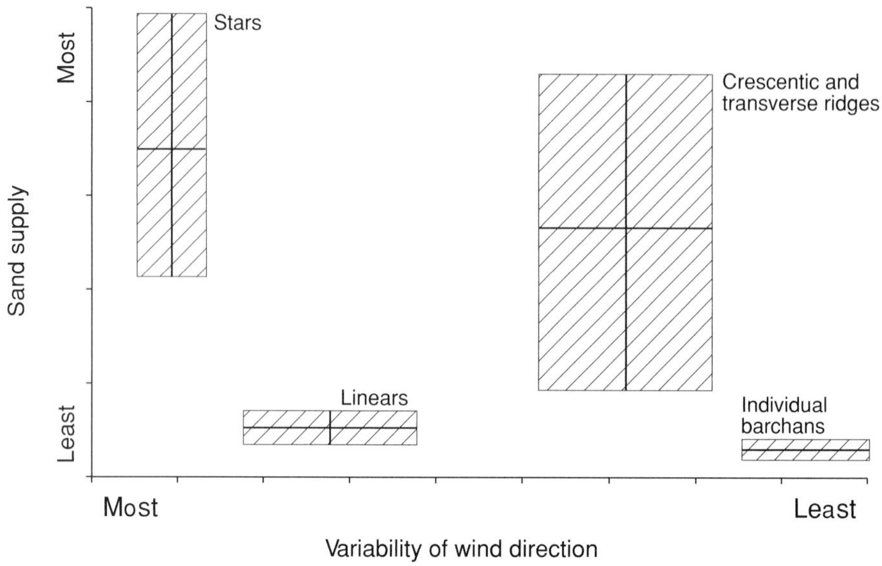

Figure 4.3 A simple model of the relationship between dune form, wind variability and sand supply. (From Wasson and Hyde, 1983, fig. 1, ©Nature Publishing Group)

facilitated by the use of digital elevation models (DEMs) (Hugenholtz and Barchyn, 2010).

The Wasson and Hyde model has proved very durable (Bullard and Livingstone, 2010), and general support for it is provided when one considers the directional index for three main types of dune (Table 4.2). This index (DI) is the ratio between the resultant drift (or sand transport) potential (RDP) and the total drift potential from all directions (DP). High DI values indicate near unimodal wind regimes, whereas low values indicate complex wind regimes. Overall, Fryberger (1979) found that crescentic dunes occur in areas the directional index does not exceed 0.5 (and averages 0.68), that linear dunes occur where the mean directional index is about 0.45 and that star dunes occur where the mean directional index is less than 0.35 and has a mean of 0.19.

Computer simulations using a cellular automaton approach (e.g. Werner, 1995) have demonstrated that each of the major dune types is independent of initial conditions, forming an attractor in a complex system in which wind regime variability is the main determinant of dune type and orientation.

Another major control of dune form proposed for the Sahara by Mainguet and Chemin (1990) is whether the sedimentary budget of a sand sea is negative, with more sand leaving the system than is coming in, or positive. A negative budget produces *dunes d'érosion*, with linear dunes separated by deflational corridors. This is a feature of the central Saharan ergs such as Iguidi, Chech, Ténéré, Rebiana, Calalansho and Bilma. A positive budget is a feature of the northern ergs, such as the Grand Erg Occidental and the Grand Erg Oriental. This is because of the abundance of alluvial sediments deposited by wadis during the Quaternary. A positive budget produces

Table 4.2 *The directional index for three different types of dune*

Location	Source	Directional index
1. Barchans		
Morocco (Tarfaya)	Elbelrhiti et al. (2008)	0.91
Namib	Lancaster (1989a)	0.83
Peru (Alto Ilo)	Londoño et al. (2012)	0.91–1.0
Peru (Chimbote)	Elbelrhiti et al. (2008)	0.91
Saudi Arabia (Dhahran)	Breed et al. (1979)	0.78
USA (Imperial Valley)	Muhs et al. (1995)	0.89–0.90
Western Desert, Egypt	Hereher (2010)	0.78–0.82
2. Star Dunes		
China	Wang et al. (2005)	0.31–0.49
Grand Erg Occidental	Fryberger (1979)	0.07–0.11
Grand Erg Occidental	Fryberger (1979)	0.15
Namib	Lancaster (1989a)	0.23
3. Linear dunes		
Western Desert, Egypt	Hereher (2010)	0.5–0.68
Australia	Fryberger (1979)	0.23–0.46
Kalahari	This book	0.51
Namib	Lancaster (1989a)	0.52
Sharjah, UAE	Breed et al. (1979)	0.58
Taklamakan	Wang et al. (2002)	0.58

dunes d'accumulation. If winds in such areas are multidirectional, pyramidal dunes develop, as in the Grand Erg Oriental and Occidental. If the winds are undirectional, crescentic and transverse chains develop, as in the Kanem, Manga, Haoussa and Mauritanian sand seas (see also El-Baz et al., 2000).

Grain Size

Wilson (1972) believed that grain size was a major control of the size of aeolian bedforms. Observations in the Sahara suggested to him that their wavelength could be related to the grain size of the 20th percentile (i.e. the coarse tail) of the sediment obtained from the crests of the features. Although the wavelengths of the groups overlap, his data plotted in three distinct groupings associated with ripples, dunes and draas. Within each group the wavelength of the bedform tended to increase with grain size, implying that larger features were formed under greater wind speeds. Some subsequent studies have failed to find this relationship, however (e.g. Lancaster, 1989a).

The Role of Vegetation and Groundwater

Although vegetation cover may be minimal in hyper-arid areas, it is of increasing significance in the moister types of desert. As Thomas and Tsoar (1990, p. 473) wrote,

'The geomorphological importance of vegetation lies in its situation at the interface between the atmospheric boundary layer and the surface of potentially deflatable sediments.' Depending on whether the sand deposition rate exceeds the tolerance of vegetation to being buried, it acts as a stabilizer of surfaces (Barchyn and Hugenholtz, 2012), and some dune types, including linear ridges, may be sufficiently stable for vegetation colonization to occur. If the vegetation cover is greatly reduced by fire or drought, however, dune activity may be substantially increased (Wiggs et al., 1994; Hesse and Simpson, 2006; Strong et al., 2010). Vegetation also acts as an accretion focus for coppice dunes and nebkhas. Moreover, parabolic dunes are a feature of areas with a partial vegetation cover (Hack, 1941). Whether dunes occur close to the water table or well above it may also be a major control of dune type and sand-sea state (Kocurek and Havholm, 1993). It also plays a major role in the formation of inter-dune deposits and their bedding structures (Mountney, 2011).

Time

Dune forms and patterns evolve through time, so dune morphology is partly determined by the evolutionary process. This is an area that has received relatively little attention until very recently. Through time, dunes may merge, link up laterally, change size, etc. (Kocurek and Ewing, 2005). Dune patterns should evolve downwind, and this has been found to be the case at White Sands in New Mexico and on the Skeleton Coast in Namibia (Lancaster, 2009, p. 587). The granulometry of dunes may also change through time, with a build-up of their content of silt and clay, and this may make older dunes more resistant to reactivation than younger dunes with less cohesion (Werner et al., 2010).

Dunefield Shape, Size and Topography

Dune systems need space to develop, so large dune areas may have different characteristics than small ones. This partly relates to the role of time that was discussed in the previous section. Ewing and Kocurek (2010) wrote about the importance of the 'geomorphic container' in which the dunes are located and found that dune spacing increased as dunefield area increased. The topography that underlies a dunefield may influence dune patterns, although little work has been undertaken on this (Ewing and Kocurek, 2010; Hugenholtz and Barchyn, 2010).

4.7 Obstacle Dunes

One class of dune is that formed by the interference of an obstacle – such as a hill – with wind and sand movement. Where the wind velocity is checked by the hill, sand will be deposited sometimes to the lee and sometimes to windward. The windward dunes are often referred to as climbing dunes, and the lee ones as falling dunes. When the upwind slope of a topographic obstacle is greater than about 50°, an echo dune is formed, detached from the escarpment or cliff by an upwind distance of about three times its height. The relationship between echo dunes and the slope of the topographic obstacle which they abut has been modelled by Qian et al. (2011). Climbing dunes, which may ascend either gullies or spurs, often cascade over the topography through local low points (passes) to become falling dunes. Falling dunes tend to be preferentially transported down gullies as the path of least resistance. Downwind of narrow obstacles, where flow takes sand around the hill flanks, lee dunes may extend some considerable distance downwind. They may have a linear form and then with distance break down into barchans (see, for example, figure 6.11 in Pye and Tsoar, 1990).

Obstacle dunes, in addition to occurring on Mars (Chojnacki et al., 2010), are prevalent in the Aravalli Hills of north-western India (Figure 4.4) and against the inselbergs of Saurashtra (Allchin et al., 1978), Iran (Thomas et al., 1997), the escarpments of Kuwait (Al-Enezi et al., 2008), Niger (Rendell et al., 2003), in the central Namib (where Goudie [1970, 1972a] called them 'sand glaciers'), in the mountains at the edge of the Rub 'Al Khali in the United Arab Emirates and in parts of the American Southwest (Evans, 1962; Ellwein et al., 2011; Bateman et al., 2012). The obstacle dunes of north-west India can be very large features, extending over many kilometres. They are often deeply gullied, contain old weathering surfaces and preserve large numbers of Stone Age tools (Allchin et al., 1978). The largest dune in North America is the great mass that has accumulated against the Sangre de Cristo Mountains in Colorado, which is some 250 m high. Kádár (1934, p. 475) termed them 'sand-dams' or 'tail-dunes', respectively. He found one against the north-east rampart of the Gilf Kebir in the Libyan Desert that was 126 m high.

These features have also been called sand ramps, multiple generations of which can be identified. They often are contaminated by talus, stone horizons and colluvium derived from the slopes which they abut, and contain palaeosols (Bateman et al., 2012). Sand ramps may have lower slope angles than true obstacle dunes (Thomas et al., 1997). A full discussion of the sand ramps in the Mojave – some of which are as much as 70 m thick, and many of which are inactive today – is provided by Lancaster and Tchakerian (1996).

On top of an escarpment or plateau, just downwind from the crest, a zone of reduced wind velocity has been observed. Consequently, sand often accumulates in such a position as what is called a cliff-top dune.

Although most topographic dunes are associated with positive relief features (cliffs, hills, etc.), there are some that appear to develop in association with negative relief features, and an example of these are the linear dunes that develop along the margins of river valleys in the Kalahari (Bullard and Nash, 2000), mirroring the course of extant and ancient channels. Such a relationship between riverbeds and flanking dunes was

(a)

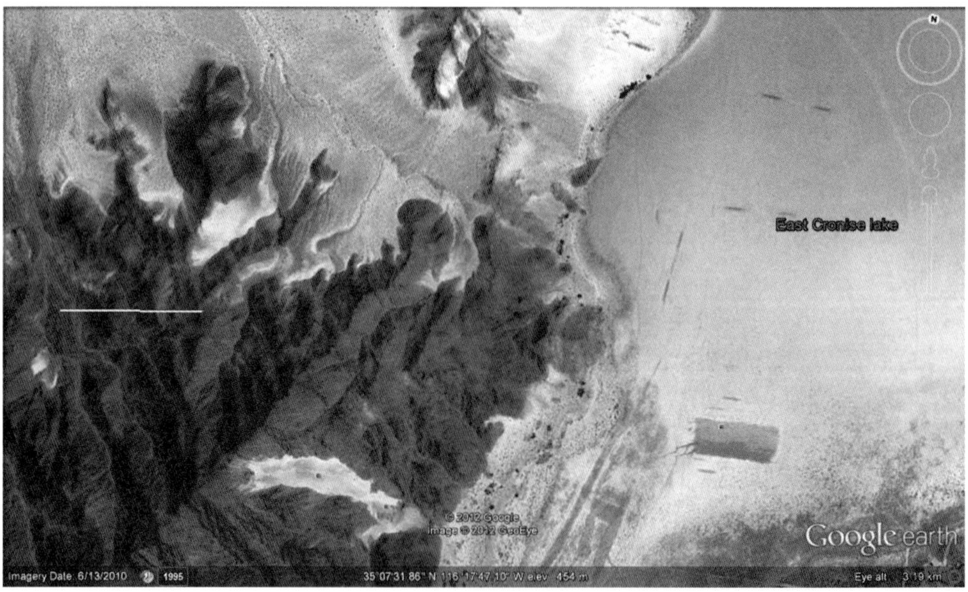

(b)

Figure 4.4 Topographic dunes. (a) A topographic dune developed on a col in the Aravalli Mountains, Thar Desert, India. (ASG). (b) A climbing dune, the Cat, in the Cronese Mountains, Mojave Desert, USA. Scale bar 0.5 km. (©Google Earth 2012, ©GeoEye 2012)

Figure 4.5 Small nebkha dunes developed on the flanks of linear dunes, Gobabeb, Namibia. The plant involved is *Stipagrostis sabulicola*. (ASG)

also noted in the Tarim Basin of China and along the 'lost rivers' of the Thar by Stein (1942).

4.8 Nebkhas

'Nebkha', or 'nabkha', is an Arabic term given to mounds of windborne sediment (sand, silt of pelletized clay) that have accumulated to a height of some metres around shrubs or other types of vegetation (Figure 4.5). Plants that gain their water supply from high groundwater levels (e.g. tamarisk) may often form the core around which accumulation occurs. The largest nebkhas (mega-nebkhas) accumulate around clumps of trees. In the Wahiba Sands of Oman, these can be 10 m high and up to 1 km long (Warren, 1988). Nebkhas are sometimes called shrub-coppice dunes (see Rango et al., 2000). They may occur on bigger dunes, in inter-dune areas, on pan surfaces, near wadis and on or behind beaches and sabkhas. Morphometric data are provided by Tengberg and Chen (1998) and Khalaf et al. (1995). Some nebkhas are more or less circular mounds, whereas others show clear elongation and consist of a long plume to the lee of the anchoring vegetation.

Figure 4.6 One of the world's largest lunettes, developed on the lee of the Sebkha el Kelbia, central Tunisia. The lunette rises 146 m above the lake floor in the background. (ASG)

Nebkhas have been reported from many desert areas, including the Kalahari (Dougill and Thomas, 2002), Sinai (El-Bana et al., 2002), Kuwait (Gunatilaka and Mwango, 1987; Khalaf and Al-Awadhi, 2012), China (X. Wang et al., 2006), Mali (Nickling and Wolfe, 1994), the United States (Langford, 2000; Laity, 2003), Burkina Faso and Tunisia (Tengberg and Chen, 1998). Nebkhas are often enriched in organic matter and nitrogen and so may be involved in a feedback mechanism whereby these nutrients fertilize the vegetation that traps windblown sediment (Dougill and Thomas, 2002). They can build up rapidly, perhaps in response to vegetation degradation, but can also degrade, when, for example, groundwater levels fall and the vegetation that promotes their development dies off (Laity, 2003). They also respond to variation in wind energy conditions (X. Wang et al., 2006), and relict forms, which may have created some of the pimple mounds of the south-central United States, could indicate the former greater extent of drought and aeolian activity (Seifert et al., 2009).

4.9 Lunettes

Some obstacle or topographic dunes develop in the lee of desert depressions, and these crescentic features are called lunettes. They are transverse and roughly crescentic aeolian accumulations that occur on the downwind margins of pans (Figure 4.6). Although they had been described before, they were named as such in Australia by

Hills (1940), although the basis of his etymology is unclear. Good regional descriptions of lunettes are provided for the High Plains of the United States by Holliday (1997), for Tunisia by Perthuisot and Jauzein (1975), for the Kalahari by Lancaster (1978), for Mauritania by Mohamedou et al. (1999), for the Pampas of Argentina by Dangavs (1979), for south-western Australia by Harper and Gilkes (2004) and for Spain by Rebollal and Pérez-González (2008).

They tend to occur in areas where present-day precipitation levels are between about 100 and 700 mm, but their stratigraphy can give a good indication of past changes in climate and hydrological conditions (Lawson and Thomas, 2002). Lunettes may accumulate rapidly, with rates in the Kalahari reaching as much as 10 m in 1,000 years (Telfer and Thomas, 2006).

Some basins may have two or more lunettes on their lee sides (some western Australian pans were reported by Harper and Gilkes [2004, p. 223] as having as many as seven). These may have different grain size and mineralogical characteristics. Lunettes may be some kilometres long and in exceptional circumstances may attain heights in excess of 60 m. The materials that make up lunettes can vary from clay-sized material (which in the case of clay dunes [Bowler, 1973] can make up 30–70% of the total) through to sand-sized material. Equally, some lunettes are carbonate rich (Goudie and Thomas, 1986), whereas others are almost pure quartz. Lunettes may also contain appreciable quantities of evaporite minerals derived from the basins to their windward (Mohamedou et al., 1999).

Various hypotheses have been put forward to explain lunette composition. Hills (1939) believed that the lunettes were built up when the pans contained water, and that they were composed of atmospheric dust captured by spray droplets derived from the water body. Stephens and Crocker (1946) pointed out that his could not account for those lunettes that were not predominantly silty. They also suggested that many of the lunettes were built up of aggregates transported from the pan floors. Campbell (1968) believed that this deflation hypothesis could indeed account for many lunette features. As she remarked (p. 104), '[T]he close similarity between the composition of the lunette and its associated lake bed suggested that the two are causally related, i.e. that the material in the lunette was derived from the lake bed.' She also recognized, however, that some of the material could be derived from wave-generated beaches and so could be analogous to primary coastal foredunes.

This was a view that was developed by Bowler (1973), who saw sandy facies as being associated with a beach provenance (at times of relatively high water levels), whereas clay-rich facies, which also may have a high content of evaporite grains, formed during drier phases when deflation of the desiccated lake floor was possible. Lunettes can therefore provide evidence for understanding past hydrological changes (Page et al., 1994). Some lunette sediment may be derived from river deltas on the margins of playas, as has been proposed for the Etosha Pan in Namibia (Hipondoka et al., 2004).

Figure 4.7 The steep avalanche slope of a barchan, Skeleton Coast, Namibia. (ASG)

4.10 Reversing Dunes

Reversing dunes are dunes that change their shape in response to reversals in wind direction. That said, they have been the subject of only modest study, and they range greatly in size and shape. As Livingstone and Warren (1996, p. 82) remarked, they can be regarded as transverse dunes that reverse or as a special case of dune networks in which the winds of two seasons are diametrically reversed. These dunes tend to have a sharp crestline and a triangular cross section rather than the convex form common to most transverse flow dunes (McKenna-Neuman et al., 1997), and the high rates of sediment flux, erosion and deposition that have been observed in the crestal areas of such dunes are a reflection of their need to adjust to changing wind directions.

4.11 Barchans

Most dunes do not require an obstacle, whether hill, shrub, depression or dead camel. Indeed, the most regular dune forms develop on the most regular surfaces. These are termed free dunes.

Probably the best known and most common basic dune form results from winds which have a single dominant direction and the dune being oriented with its axis at right angles to the wind direction. Such dunes range from small crescent-shaped types (barchans) through parallel rows of barchanoid ridges to essentially straight ridges known as transverse dunes. These dunes (Figure 4.7) are all characterised by slip faces in one direction and represent unidirectional wind movement. Barchan dunes occur where sand supply is limited and transverse dunes where sand is more abundant (Eastwood et al., 2011). They can move at fast rates, up to tens of metres a year, and this is discussed in detail in Section 6.4. Barchans occur in two main situations: (1) on the margins of sand seas and dunefields and (2) in sand transport corridors linking sand-source zones with depositional areas. Barchan dunes are one of the most active dune types, always changing their forms in response to changing environmental conditions but seldom, if ever, attaining a stable equilibrium state (Elbelrhiti and Douady, 2010). Collisions between individual dunes play an important role in the size distribution and structure of barchans dunefields (Durán et al., 2011). As Katsuki and Kikuchi (2011, p. 2) remarked:

> Looking at aerial photographs of a field of barchans, it can be seen that a group of them form a characteristic pattern in which the horns of each barchan point to the center of the barchan(s) on its leeward side(s). As a result, they form a convoy with a triangular pattern similar to that of flying geese or align themselves in a slanted line.... The formation of the triangular pattern should be attributed to some interaction between a given barchan and its leeward barchan neighbors.

Moreover, Hugenholtz and Barchyn (2012) found that collisions between barchans moving at different speeds can cause the ejection of a barchan from the wake of an upwind dune.

Barchans are individual mobile dunes of crescentic shape, the two horns of which face in the direction of dune movement (Figure 4.8). The slopes of their windward sides are generally between 8 and 20° and those of their lee sides around 33–34°. Sand avalanching takes place on these steeper lee sides, although superficial slab or slumping failures may also occur in dunes moistened by fog, dew or hygroscopic salts.

Barchans are generally regarded as occurring in areas of limited sand supply, on planar surfaces, with a low precipitation (usually less than 100 mm per annum) (Table 4.3) and vegetation cover and where winds are narrowly bimodal in direction (with a directional index that is normally around 0.7–0.9). They develop from small patches of sand (Elbelrhiti, 2012) or dome dunes. At a global scale they are quantitatively of limited significance – less than 1 per cent of all dune sand on Earth is contained within them – but in some deserts, notably in southern Morocco, there may be many thousands of them.

Barchans are variable in size, ranging in height from a few metres or less to more than 500 m in the case of megabarchanoids (Kar, 1990; Bishop, 2010). Some data on barchans widths are presented in Table 4.4. Using the formula of Hesp and Hastings (1998), it is possible to calculate slip-face height from horn width:

$$W = 8.82\,h + 7.65 \qquad (1)$$

However, Finkel (1959) indicated the relationship was

$$W = 10.3\,h + 4.0 \qquad (2)$$

Figure 4.8 Migrating barchan dunes at Algeria in the Kharga Oasis, Egypt. (ASG)

In general, the height of a barchan is about one-tenth of its width (Hesp and Hastings, 1998). Alternatively, the barchan height/width relationship can be expressed as

$$\mathrm{H} = \mathrm{W}/2 \tan \Phi \quad (3)$$

Table 4.3 *Precipitation levels for some major barchan areas (mm per year)*

Location	Precipitation
Arequipa, Peru	100
Dharan, Saudi Arabia	108
Doha, Qatar	80
Imperial Valley, California, USA	70
In Salah, Algeria	20
Kharga, Egypt	0
Kuwait	106
Laayoune, Morocco	50
Liwa, UAE	40
Luderitz, Namibia	10
Namibia, central	10
Namibia, Cunene erg	<50
Taklamakan, China	50

Table 4.4 *Selected examples of barchan widths (m)*

Location	Width	Source
Cunene erg, Namibia	range 130–280	Goudie (2007b)
Arequipa, Peru	range 12–70	Hastenrath (1967)
Southern Peru	range 11.4–66	Finkel (1959)
Jafurah, Saudi Arabia	range 230–450	Shehata et al. (1992)
Qatar	range 15–1,000	Embabi and Ashour (1993)
Kuwait	range 12–78	Al-Awadhi et al. (2000)
Walvis Bay, Namibia	range 76–343	Barnes (2001)
Imperial Valley, California, USA	mean 125.6	Long and Sharp (1964)
Kharga, Egypt	mean 155.5	Embabi (1978)
Kharga, Egypt	range 89–485	Stokes et al. (1999)
Dahkla, Egypt,	mean 173.6	Embabi (1986–87)
Southern Morocco	range 30–100	Sauermann et al. (2000)

where H is dune height, W is the width between horns, and Φ is the average angle on one side of the dune. If, following Walmsley and Howard (1985), we take a slope angle of 11° as the typical mean for the side slope of a barchan, then this value can be incorporated into the equation

$$H = W/2 \tan 11^\circ \tag{4}$$

Barchans have generated a very substantial literature over the years (see, for example, Beadnell, 1910 and Bagnold, 1941) and recently have been the subject of renewed attention both by modellers (e.g. Kroy et al., 2005; Schwämmle and Herrmann, 2005) and those seeking analogues for Martian dunes (e.g. Parteli et al., 2005). Much work has been done on establishing relationships between their morphometric characteristics (e.g. height, width, length) and relating these to rates of movement. In general, however, very little work has been done on the morphological variety of barchans, although Long and Sharp (1964), on the basis of the ratio between length of the windward slope and horn-to-horn width, divided them into Fat (≥1), Pudgy (0.75), Normal (0.5) and Slim (0.25). Howard et al. (1978) argued that barchan shape is a function of such factors as grain size, wind-flow velocity and sand saturation and wind variability. They also suggested that smaller grain size or higher wind speed both produce a steeper and blunter stoss side, and that low saturation of the inter-dune sand flow produces open crescent moon–shaped dunes. McKenna-Neumann et al. (2000) remarked that dunes in areas characterised by coarse sediment (with a higher effective wind threshold) and/or low speeds would tend to have lower, longer profiles compared with those in areas with fine sediments and/or strong winds. In some barchans, dune crests and slip-face brinklines may be separate, whereas in others they may be coincident (Hesp and Hastings, 1998). Hastenrath (1967) suggested that the windward profile is very nearly a straight line for tall dunes, but becomes increasingly convexly curved for smaller ones, a finding confirmed by Schwämmle and Herrmann (2005). Equally, Herrmann (2002), Herrmann et al. (2005) and Sauermann et al. (2000) found that not only does the relative position of the slip face within the whole dune vary with dune size but also that the ratio of horn length to total length increases with the height. Shape is, therefore, not necessarily scale invariant, although Andreotti et al. (2002) report cases where barchans of the same height in the same dunefield show both separation and coincidence of the brink and the crest. Bourke (2010) found that some barchans developed asymmetry, partly as a result of non-unimodal wind regimes and partly because of dune collisions.

Bourke and Goudie (2009) compared the shapes of barchans in the Namib and on Mars and developed the following classification of barchan types:

Classic symmetrical barchans – slim. The simplest form of barchan is the classic individual crescentic feature. Some of these are elegantly slim as shown by examples on the rocky plains to the south and east of Luderitz and Elizabeth Bay in Namibia and from the plains of the Bodélé depression in the central Sahara. They also appear to be rather angular in plan. They display a wide range of sizes, with some having widths as great as 500–600 m, and some being only a few tens of metres in width. The slim symmetrical type of barchan is a feature of areas with unidirectional winds and with low sand influx and high values for shear velocity (Parteli et al., 2007).

Classic symmetrical barchans – pudgy and fat. Some simple crescentic forms possess a larger area in relation to their width than the examples given above (Figure 4.9). The horns are relatively small in relation to the total mass of the dune and may be nearly absent, as seen with some examples from the Peruvian Desert. Such dunes have shapes reminiscent of kidneys, broad beans, fortune cookies and pectens. Fat dunes occur in areas where there is a substantial sand influx and lower shear velocities (Parteli et al., 2007). Many of the world's barchans described in the literature appear to be fat rather than slim (see, for example, surveyed outlines of barchans from the Western Desert in Egypt (Stokes et al., 1999). Examination of satellite images of south Morocco, southern Peru and Qatar seems to confirm this.

Classic symmetrical barchans – large, fat and unstable. Some barchans are large features, which may be termed megabarchans. More than 500 m in width, they often have secondary features on their flanks, which may be indicative of instability. They may also shed small barchans onto the desert plains downwind. This appears to be an example of what Elbelrhiti et al. (2005) describe as 'surface-wave-induced instability'. They argue that dune collisions and changes in wind direction destabilize larger dunes and generate surface waves on their lee flanks. The resulting surface waves propagate at a higher speed than the dunes themselves, producing a series of small, newborn barchans by breaking the horns of large dunes. Examples of instability are reported from southern Morocco by Hersen et al. (2004).

Classic symmetrical barchan composed of smaller barchans. In southern Namibia, a single classic barchan form was found that is approximately 400 m across and 700 m long and predominantly made up of a cluster of smaller barchans. It may be an extreme

Figure 4.9 Satellite image of large and small barchans from Djoura, Chad, central Sahara. Scale bar is 100 m. (©Google Earth 2009, ©Digital Globe 2010)

an example of a proto-megabarchan (Cooke et al., 1993, p. 327). Similar features occur in much greater profusion in the central Sahara in the area that lies on the border region between Niger and Chad.

Barchans developing into linear dunes. Following on from the classic model of Bagnold (1941), it is evident that some simple crescentic forms are deformed into linear (seif) features when they move into areas with changing wind regimes. Linear ridges some kilometres long can develop downwind from the original barchans, creating a tadpole shape. Good examples of this can be found in the western Sahara, the Namib and the Lut of Iran.

Barchan dunes developing into transverse ridges. There are many examples of classic individual barchans merging together with their neighbours to form ridges transverse to the formative winds. The original barchanoid and linguoid elements are clearly visible. It is generally believed that sand availability is a crucial control, and that with greater sand supply transverse dune ridges rather than individual isolated barchans will occur (see Section 4.12).

Barchan convoys developing into linear ridges. Some intriguing linear dune ridges appear to be formed by convoys of approximately equally sized barchans. Wang et al. (2004) proposed this style of barchan merging in their model of complex linear dune formation. Another type seems to have formed downwind of major nebkha fields.

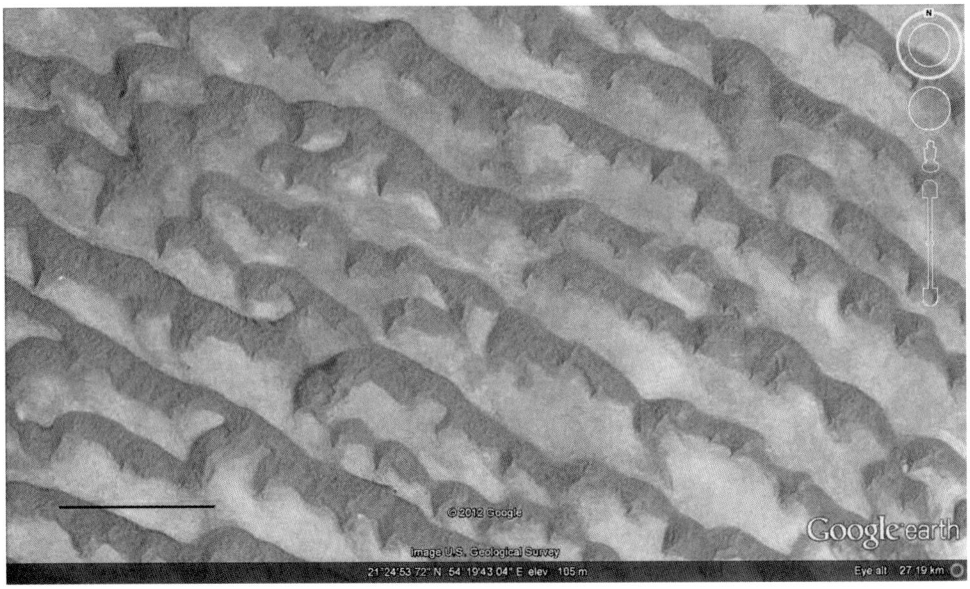

Figure 4.10 Transverse ridges in Arabia. Note the barchanoid and linguoid elements. Scale bar is 5 km. (©Google Earth 2012)

These develop from sand that has accumulated around bushes rather than through the normal style of evolution from a non-anchored sand pile. One issue that needs to be resolved is the extent to which barchan convoys are the origin of linear dunes or the consequence of their decay (Parteli et al., 2007).

4.12 Transverse Ridges, Zibars and Dome Dunes

Transverse dunes (Figure 4.10) form more or less perpendicular to the dominant wind direction, and they are separated by interdune areas for which the airflow patterns have been described by Baddock et al. (2007). As we have just seen, individual transverse ridges are called barchans, but where there is greater sand availability more continuous ridges occur (Reffet et al., 2010; Eastwood et al., 2011). They are a very common type of dune, but like many dune types they show great variability in shape and size. Like barchans, they tend to occur in areas with a more or less unimodal wind regime. Also like barchans, they have steep slip faces on their lee sides and more gentle slopes on their stoss sides. Some transverse dunes can be large features. Those in the Badain Jaran desert of China, for example, are generally 150–350 m high (Z. Dong et al., 2004), and those in the Murzuk Sand Sea in Libya are more than 100 m high. Where barchans collide and climb on top of one another, a network of dunes may develop where the interdunal spaces become increasingly smaller, producing a pattern that French workers call *aklé* (Mainguet, 1984). Transverse dunes may develop in inter-dunal corridors as is seen in the Namib example shown in Figure 4.11.

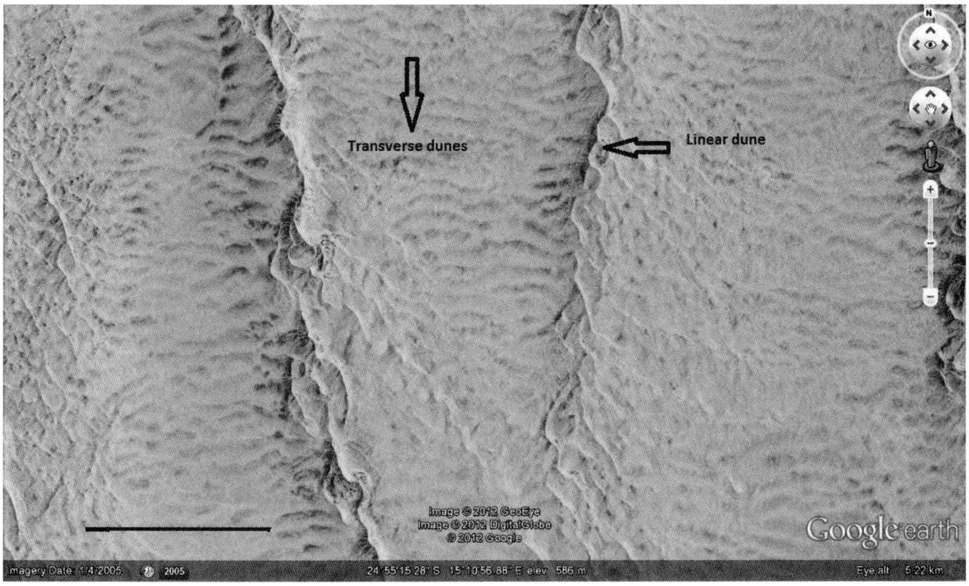

Figure 4.11 Transverse dune ridges developed within corridors between large linear dunes in the Namib Sand Sea. Scale bar is 1 km. (©Google Earth 2012, ©GeoEye 2012)

Zibars are 'coarse-grained, low relief, slipfaceless, eolian bedforms that occur on sand sheets and within interdune corridors of many eolian sand seas' (Nielson and Kocurek, 1986, p. 1). They have fairly regular spacings of up to 400 m, a maximum relief of less than 10 m and form an undulatory surface. Zibar stoss-slope and lee-slope angles are generally less than 5 and 15°, respectively. Examples are reported from the Tenéré Desert (Warren, 1971), Algeria, Saudi Arabia, Sinai, the Namib, Algodones in the United States and the Kumtagh Desert of China (Wang et al., 2009). They tend to be transverse to the modern prevailing wind. The sediments that form them are coarse in comparison to normal dune sand, and this coarseness may be one of the primary factors that suppress slip-face formation.

Dome dunes normally show no obvious slip face and are circular or elliptical in plan view. They may result from strong winds that truncate the top and flatten the lee slopes of barchans.

4.13 Parabolic Dunes

A class of dune in which the form owes much to the presence of a limited vegetation cover or some soil moisture is the parabolic. These are hairpin-shaped with the nose pointing downwind. They were described by Hack (1941, p. 242) as

> long, scoop-shaped hollows, or parabolas, of sand, with points tapering to windward. The windward slope is much gentler than the leeward slope. Such dunes form by the removal of sand from the windward hollows by the wind and the deposition of sand on the leeward slopes.

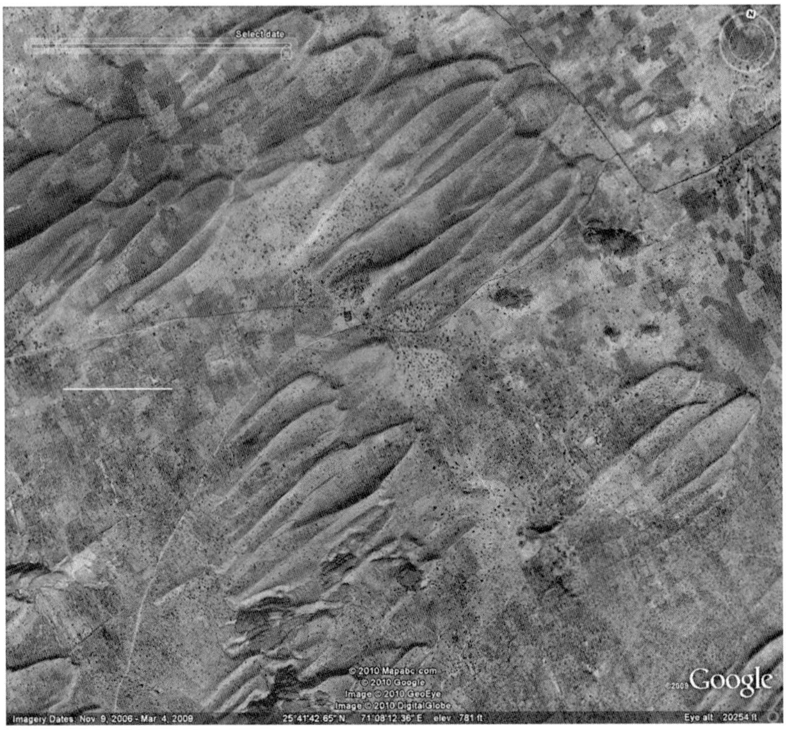

Figure 4.12 Rake-like parabolic dunes in the Thar Desert of India. Scale bar is 1 km. (©Google Earth 2010, ©GeoEye 2010)

Hack drew a distinction between 'parabolic dunes of deflation', caused by deflation of a pre-existing sand layer, and 'parabolic dunes of accumulation', which occur where sand is actively deposited by the wind. Parabolic dunes, which may have a U-shaped or V-shaped form, have three basic features: (1) a depositional lobe on the downwind side, (2) trailing arms or ridges and (3) a deflation basin between the trailing arms.

A classification of parabolic dunes forms is provided by Wolfe and David, 1997). They recognized that some individual parabolics may be open upwind or closed by a back ridge, this having formed by sand blown out of the deflation depression. They also showed that some dunes are more filled in with sediment than others. They suggested that unfilled parabolic dunes have narrow heads and back slopes and have resulted from a relatively limited supply of sand, whereas partially filled and filled dunes with prominent heads and back slopes develop in areas of greater sand supply. Finally, they give examples of merged (compound) and superimposed dunes. They may occur in clusters, creating rakelike forms, most notably in the Thar Desert of India (Allchin et al., 1978) (Figure 4.12).

Although many parabolic dunes are associated with sea and lake coastlines (McKenna, 2007; Luna et al., 2011), some of which may be in humid environments, they also very widespread in the arid and semi-arid dunefields of the plain lands of North America (e.g. Hack, 1941; Márin et al., 2005; Hugenholtz et al., 2008). Particularly fine examples are also evident on the coastline of north-east Libya near Adjabiya. In addition, parabolics occur sporadically in parts of Australia, including the Mallee (Hesse, 2011), in the Jafurah sand sea of eastern Arabia (Anton and Vincent, 1986) and in some formerly more arid parts of South America, such as Roraima (Latrubesse and Nelson, 2001). Examples are also known from the fog belt of southern Peru, where fog moisture plays a significant role, while in the Salar de Uyuni in Bolivia, the parabolic dunes are composed of halite (Svendsen, 2002). The wind regimes associated with parabolic dunes are generally unimodal or acute bimodal.

Parabolic dunes may have partially vegetated parallel arms and the less vegetated nose may be as much as 10–70 m high. Vegetation or dampness in the lower sides of parabolic dunes retard sand motion and so function like an anchor. Vegetation also protects the less mobile arms against aeolian processes, thereby allowing the central section to advance downwind to produce the hairpin form. Blowouts, 'the prototype of the golf bunker' (Cooke et al., 1993, p. 360), may occur in the nose of the dune. If vegetation cover increases or wind velocities slacken, parabolic dunes may replace more active forms such as barchans (Tsoar and Blumberg, 2002; Yizhaq et al., 2007; Ardon et al., 2009; Pelletier et al., 2009; Wolfe and Hugenholtz, 2009; Reitz et al., 2010). Conversely, there are examples in the literature of parabolic dunes losing their vegetation and turning into active transverse dunes. An interesting analysis of the transition between barchanoid ridges and parabolic dunes in the White Sands, New Mexico, has been undertaken by Jerolmack et al. (2012). They argue that dune form is influenced by the internal boundary layer of the dunefield. At the upwind margin of a dunefield, the dunes themselves cause an abrupt increase in surface roughness. This thickens downwind, causing a spatial decrease in the surface wind stress, which in turn leads to a downwind decline in sand flux. At a crucial threshold, the declining sand flux triggers vegetation growth, for vigorous sand movement prevents the establishment of plants (see also Hugenholtz et al., 2008). The presence of this vegetation leads to the growth of parabolic dunes, a process that has been modelled by Duran et al. (2005), Baas and Nield (2007) and Nield and Baas (2008).

The importance of vegetation cover in areas covered with parabolic dunes, such as the Canadian prairies or the U.S. High Plains, suggests that such factors as droughts and human disturbance may cause rapid changes of state to occur (Hugenholtz and Wolfe, 2005a, 2005b; Márin et al., 2005; Hugenholtz et al., 2010). There is certainly clear stratigraphic evidence that parabolic dunes have undergone repeated phases of activity and stability at different points in the Holocene (e.g. Catto and Bachhuber, 2003) in response to phases of drought (Forman and Pierson, 2003; Forman et al., 2006; Forman et al., 2009).

Parabolic dunes also tend to occur in areas with high groundwater levels, although because salinity affects vegetation growth, areas with lower salinity may have a larger vegetation biomass to encourage parabolic dune accumulation (Langford et al., 2009).

Figure 4.13 A small linear dune, c 15 m high, in the United Arab Emirates. (ASG)

4.14 Linear Dunes

Linear dunes, or seifs, are straightish ridges with slip faces on both sides that run more or less parallel to the resultant wind trend (Figure 4.13). Linear dunes are also sometimes called sand ridges or longitudinal dunes, but 'linear dune' is now the preferred term, partly because it has no genetic connotations (Livingstone and Warren, 1996, p. 76). They often develop a sharp crest, which explains why they are called *seif* (a sword) in Arabic. They may also display a meandering tendency (Parteli et al., 2007). The ridges can extend for tens, or even hundreds, of kilometres and link together in tuning-fork-shaped junctions (Goudie, 1969) that almost invariably point downward and may have a similar dendritic pattern to that recognized in stream systems. They occur in loose sand in areas where there is seasonal or diurnal change in wind direction – that is, a bimodal wind regime and where sand supply is relatively high (Parteli et al., 2009). They can also occur in areas with a more unimodal wind regime if the sand is locally stabilised by vegetation, sediment cohesion (due to the presence of salt, moisture or mud) or topographic shelter (Rubin and Hesp, 2009). In the Kumtagh Desert of China, however, the complex ridges have a feathered form caused by the action of three winds intersecting at an acute angle (Qu et al., 2011).

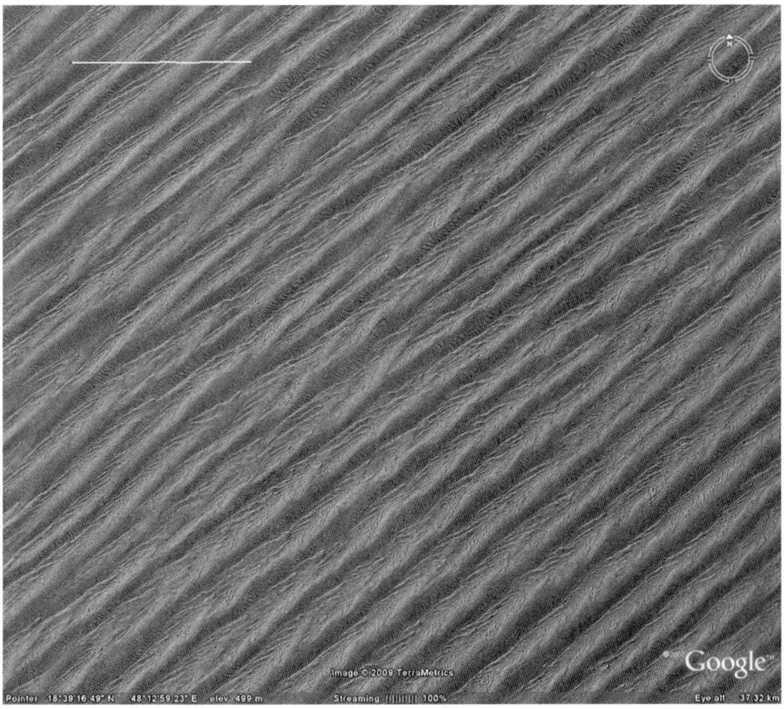

Figure 4.14 Linear dunes in Saudi Arabia. Scale bar is 10 km. (©Google Earth 2009, ©TerraMetrics 2009)

Some linear dunes may be modest in height (10–20 m) and spacing (a few hundreds of metres), but others can be considerably larger, with heights in excess of 150 m, and a spacing of 1 or 2 km. Examples of the former are the dunes of Australia and the Kalahari, whereas examples of the latter are the dunes of the central Namib and those of the Rub 'Al Khali (Figure 4.14). The larger linear dunes are often described as complex or compound forms and may have multiple subdunes superimposed on a large plinth or draa. In general, as linear dunes get higher, they get more widely spaced (Lancaster, 1995a, p. 63). The relationship seems to take a log-log form.

That there is a great range in the size and morphology of linear dunes suggests it would be an error to expect to be able to explain them by any one simple model. Some linear dunefields show a whole range of morphologies, and this is the case, for example, with the south-west Kalahari (Bullard et al., 1995), where there are no less than five different types. Hesse (2011) has also discussed the range of linear dune crests in Australia and points to straight, wavy, network, dendritic and chain forms (Figure 4.15). The presence of river channels may modify the patterns of linear dunes, as is the case with the valley-marginal dunes in that area (Bullard and Nash, 2000). There is also considerable morphological variability in linear dune types in Australia, and substrate type and sand supply are crucial controls (Fitzsimmons, 2007). Dunes

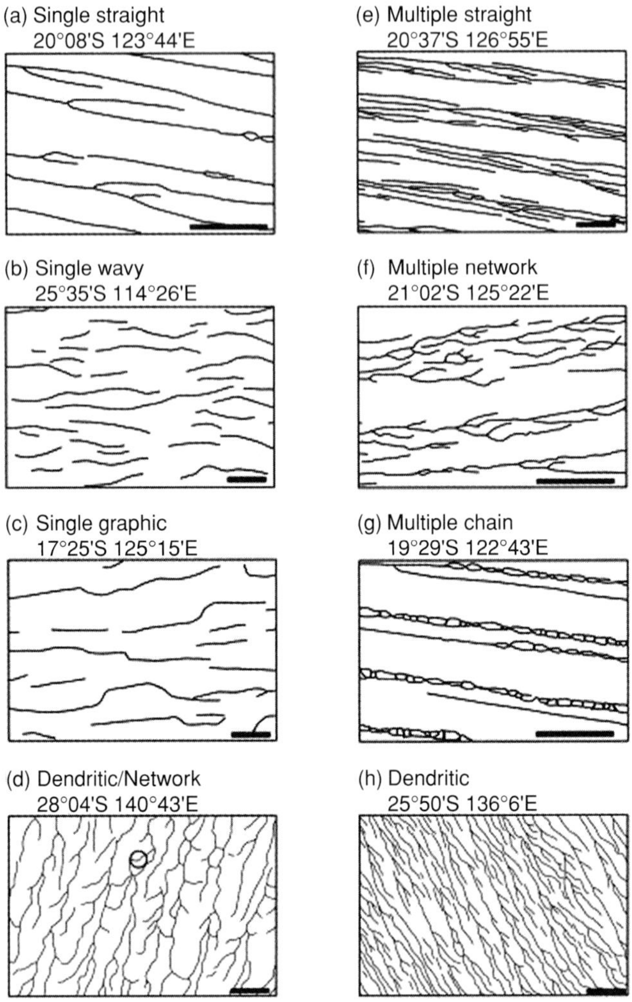

Figure 4.15 Crest types of linear dunes in Australia. (From Hesse, 2011, fig. 3)

on stony substrates are fewer in number and more widely spaced than those located on alluvial surfaces.

One early theory for linear dunes was that they were moulded by thermally generated helical roll vortices (Bagnold, 1953). These vortices, sometimes known as Langmuir circulation, are created by shearing in the boundary layer of the atmosphere. Bagnold suggested, speculatively, that paired, horizontal roll vortices, whose axes are parallel to the dominant wind direction, might sweep sand out of interdune troughs and onto sand ridges where currents would meet and ascend. In this model the wind pattern would create the dune, and the dune spacing would represent the width of a pair of vortices. Roll vortices do exist, but a number of arguments have been developed that suggest this model is not of general applicability (Livingstone, 1988). First, there is little coincidence in the lateral spacing of linear dunes and the

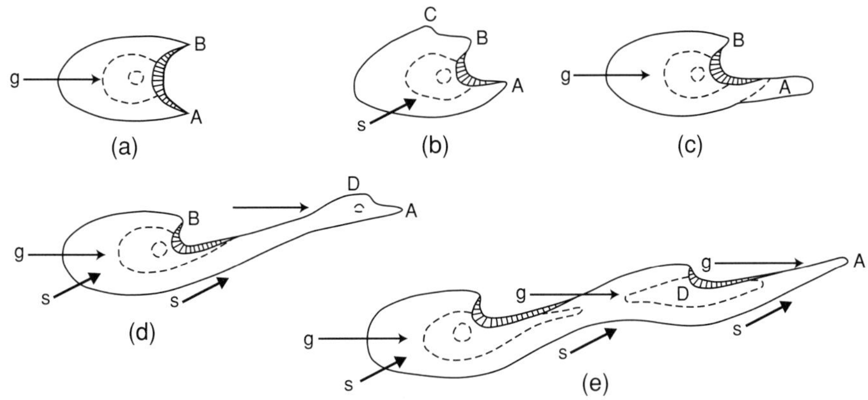

Figure 4.16 Bagnold's model of the development of a seif dune from a barchan due to a bidirectional wind. (Modified from Bagnold, 1941, fig. 78, in Goudie, 2002, fig. 5.11)

measured sizes of roll vortices – the latter are generally much greater. Second, roll vortices display measured transverse velocities well below that required to move sand. Third, the model requires that winds blow parallel to the dune trend, an event which occurs rarely in many linear dunefields.

Bagnold (1941) had also argued that linear dunes could form from barchan dunes (Figure 4.16) that became deformed as they moved into a regime which had less unimodal winds. This may happen in local situations, but it scarcely seems a model that can apply, for example, in Australia, where linear dunes are near ubiquitous, but barchans are almost absent. Verstappen (1968) suggested that in the Thar Desert linear dunes could arise from the progressive elongation of parabolic dunes. This model may again have local applicability, but many linear dunes occur in areas where parabolic dunes are absent.

Modern models, based on field measurement of wind directions and velocities, relate linear dune development to the effects of bimodal wind regimes. Notable here are the studies of Livingstone (1986, 1989, 1993) and of Tsoar (1978, 1983). Both workers showed that the crest of the linear dune migrated laterally in response to seasonally bimodal wind regimes but that net sand transport was along the dune.

Assuming that linear dunes result from the operation of bimodal wind regimes, there are two different ways in which they may develop. On the one hand, there is the downwind extension model which envisages that linear dunes extend longitudinally by progradation along their length. They are fed by sediment from upwind, and the dunes become progressively younger downwind. Telfer (2011) found evidence for this in the south-west Kalahari by dating a linear dune extending into a pan. On the other hand, there is the wind-rift or aeolian scouring model, which envisages that sediment is scoured from inter-dune swales onto the dunes, causing upward accretion of sediment (Hollands et al., 2006). In such dunes the sediment source tends to be local in origin, as has been demonstrated for some of the Australian linear dunefields (Pell et al., 2000, 2001; Fitzsimmons, 2007). They also show no significant downwind decline in ages (Cohen et al., 2010). In reality, many linear dunes may display both modes of accretion (Telfer, 2011).

Because bimodal wind regimes may not be symmetrical – winds from one direction may be more powerful than from the other – linear dunes may have an asymmetric form and also be subject to some lateral movement as well as to elongation (Rubin and Hunter, 1985). Some empirical studies of lateral movement of linear dunes show lateral migration (e.g. Hesp et al., 1989; Bristow et al., 2005), while other studies do not (Livingstone, 2003; Tsoar et al., 2004). In the case of a large linear dune in the Namib, c 300 m of lateral migration took place during 2,000 years (Bristow et al., 2007), whereas in the case of much lower dunes in Sinai, 13 m of lateral migration took place over a 26-year observation period (Rubin et al., 2008).

Another theory of linear dune formation is that they have a wind-rift (erosional) origin, and that they are thus in a sense similar to yardangs and are formed from more extensive former dune sand layers. This theory has been advocated both in Australia (Hollands et al., 2006) and in north-west China (Zhou et al., 2012).

Finally, the recent application of optical dating and ground-penetrating radar has indicated that linear dunes often have a long and complex history, with multiple phases of episodic accumulation and stabilization (Bristow et al., 2007). Buried palaeosols may occur (Fitzsimmons et al., 2009). Optical dating has also enabled rates of dune accumulation to be determined. Those in the Rub 'Al Khali, for example, can reach 2.2–25 m per thousand years (Goudie et al., 2000; Atkinson et al., 2011), whereas those in Mauritania were 2.5 m per thousand years during the Younger Dryas (Lancaster, 2008). In the Negev, dune accretion at rates of '10 m in several hundreds of years' have been established (Roskin et al., 2011).

4.15 Star Dunes

Star dunes (or rhourds) have three or more arms extending radially from a central peak (Figure 4.17a and b). Each arm has a steep-sided, sinuous crest, with avalanche faces. Star dunes are up to 150–300 m high and between 1 and 2 km across. They may be the largest dune type found in many sand seas (Zhang et al., 2012). They have been recorded from numerous deserts, but have until comparatively recently been the subject of relatively little research (Nielson and Kocurek, 1987). They are particularly important in the Grand Erg Oriental of the Sahara, where they comprise about 40 per cent of the dunes (Mainguet and Chemin, 1990). Other Saharan examples include those on the east side of the Murzuk Sand Sea in Libya, the north end of the Erg Chech and the north end of the Erg Iguidi. Good examples also occur on the margins of the Rub 'Al Khali in south Oman, on the eastern margins of the Namib Sand Sea and the Badain Jaran desert of China. They appear to be absent from the Australian, Kalahari

(a)

(b)

Figure 4.17 Star dunes. (a) 150–200 m high star dunes in the Sossus Vlei area of Namibia. These are among the world's largest dunes. (ASG). (b) Star dunes in the Rub 'Al Khali of Oman. Scale bar is 5 km. (©Google Earth 2010, ©Cnes/Spot Image 2010) and Thar Deserts, however, and also from some of the Saharan ergs (Lancaster, 1995a, p. 71). They appear to be preferentially developed in areas with multidirectional or complex wind regimes, and are often located close to topographic features which modify the regional wind regimes and tend to increase wind variability.

4.16 Spatial Superimposition

Although we have so far discussed individual dune types, one often finds that at any one location one type may be superimposed on another (Kocurek and Ewing, 2005; Eastwood et al., 2011). In the Gran Desierto of Mexico, for example, Beveridge et al. (2006) found that there were five separate patterns of different ages. The oldest pattern consisted of linear dunes believed to have formed between 26 and 12 ka. This was followed by the construction of three different phases of crescentic dunes. Finally, during the last few thousands of years, star dunes and a sand sheet were formed. These different patterns can be related to changes in aridity, wind direction and sediment supply in the late Quaternary. Equally, in the Algodones dunefield of California, Derickson et al. (2008) found five dune types with varying degrees of superimposition: linear dunes, zibars, simple crescentic dunes, compound crescentic dunes and parabolic dunes, in addition to a variety of shrub-coppice and shadow dunes.

Another example of spatial superimposition is provided by the Agnetir Sand Sea of Mauritania (Ewing et al., 2006). Here, there are three sets of superimposed linear dunes that differ in orientation, size and age: north-east trending, degraded large linears, up to 50 m high and dated to c 18.5 ka, north-north-east trending linears up to 20 m with active crests and dating to c 8.5 ka, and small, active north-trending dunes dating from c 3.5 ka. In the north-eastern Rub ‘Al Khali in the United Arab Emirates, large linear dunes trend from approximately south-west to north-east, and are overlain by much smaller linear ridges that trend from north-west to south-east (Goudie, Colls et al., 2000; Atkinson et al., 2011). There are also many different dune forms present in the Wahiba Sands of Oman. Warren and Allison (1998) saw these in terms of a hierarchy of features of different longevity. On the one hand, there were the small ephemeral dunes (no more than a few metres high) adjusted to diurnal wind regimes, the meso dunes (no more than a couple of tens of metres high) adjusted to annual wind regimes, and the tall mega dunes adjusted to wind regimes that have persisted for longer periods – perhaps tens of thousands of years. In the Badain Jaran Desert of China, Dong et al. (2009) found that there were three dune generations. The smallest dunes (generally less than 15 m tall) appeared to form in response to seasonal winds over a period of several years. Barchanoid dunes (up to 107 m tall) respond to variations in the average wind regime over periods of years, whereas the largest dunes (up to 480 m tall) respond to changes in wind regime, sand supply and sand availability over century to millennial timescales (X. Yang et al., 2011).

4.17 Sand Sheets and Wind Streaks

Besides occurring as dunes, sand may occur in some deserts as sand sheets. 'Sand sheets are areas of predominantly aeolian sand where dunes with slip faces are generally absent' (Kocurek and Nielson, 1986, p. 795). They may be rippled or unrippled, but their deposits are largely composed of wind-ripple laminae. They are widespread in many of the world's ergs, and there are famous examples in the United States and Mexico (e.g. the Gran Desierto) and in Egypt and the Sudan (the Selima Sand Sheet). It is evident that they form in ergs where conditions are not favourable for dunes or particular factors act to interfere with dune formation. These factors include a high water table, surface cementation or binding, periodic flooding, a significant coarse-grain-size component and vegetation (Kocurek and Nielson, 1986, p. 812). Details of their sedimentary characteristics are given for the Great Sand Dunes of Colorado by Fryberger et al. (1979). Investigations of the Selima Sand Sheet show that it has had a long and complex history, with phases of land planation producing a planar surface, the presence of approximately horizontal sedimentary rocks to contribute to the flatness, alternations of pluvial and non-pluvial conditions, phases when river systems were active and the presence of granule ripples. As a result, the underlying sediments are varied and may be of some antiquity (Maxwell and Haynes, 2001).

Another aeolian form, often created by a combination of deflation and deposition, is the wind streak. These albedo-contrast features (Greeley and Iversen, 1985, p. 209) occur widely on Mars but also occur on Earth (Rodriguez et al., 2010). They often occur in the lee of topographic obstacles such as pans and craters. They are low-relief surfaces, composed of aeolian material, the surface of which may be rippled. They can be quite long, and Zimbelman and Williams (1996) describe 2–4 km long streaks from the Mojave Desert of California, and Greeley et al. (1989) describe 15 km long streaks downwind from hills in the Altiplano of Bolivia.

4.18 Sources of Sand

The sources of the sand that make up dunes and sand sheets are extremely varied. Some of the sand is derived from coastlines, some from river channels or floodplains, some from weathered rock outcrops, some from re-worked volcaniclastic material (Edgett and Lancaster, 1993; Tripaldi et al., 2010) and some from lake basins. Many sand seas contain sand derived from all or many of these sources. The Gran Desierto of Mexico, for example, is the product of the interaction of two main sand systems: the Colorado River and more local materials derived from local Basin and Range Sands (Scheidt et al., 2011). The importance of different sources varies through time. For instance, at times of low sea level, exposure of deltaic materials derived from rivers may have provided a ready source of sand. This seems to be the case with respect to the Late Glacial Maximum (LGM) supply of sand from the exposed Nile Delta to feed the linear dunes of the Negev (Roskin et al., 2011).

Probably rather little dune sand is directly derived from rock weathering (Cooke et al., 1993, p. 317); rather, much of it comes from alluvium. Many dunefields are located downwind from large river channels, with, for example, the Karakum downwind from the Oxus, the Thar and the Thal located downwind from the Indus and its tributaries, the Gran Desierto downwind from the Colorado (Scheidt et al., 2011) and the Grand Erg Occidental in Algeria downwind from the Saoura. Some dunes – source-bordering dunes – plainly develop from flood plain alluvium (G. Han et al., 2007).

The sources for the sand that make up the Namib Sand Sea are probably varied and could include erosion of the older Tsondab Sandstone Formation (Besler and Marker, 1979), weathering of granites and Karoo sandstones, deflation of sand from riverbeds and derivation from the Atlantic shoreline. Lancaster and Ollier (1983) believe that much of the sand may have been supplied by the Orange River to the coastal zone and then been blown inland, and this view has been corroborated by Vermeesch et al. (2010) and Garzanti et al. (2012).

In the Rub 'Al Khali in the United Arab Emirates, there also appear to be diverse sources of sand (White et al., 2001). Some of the sand is derived from the Arabian/Persian Gulf and therefore has many of the characteristics of marine carbonates. Other portions of the sand, with a more quartzitic character – and often a redder colour – are derived from sediments weathered out and washed down from the mountains of Oman (El-Sayed, 1999). In the neighbouring Wahiba Sands of Oman, different parts of the erg derive their sands from a variety sources. In the north, carbonates come from marine limestone outcrops in the Hajr Mountains and have been supplied to the dunefield by the wadi systems. In the south, the carbonate grains are probably derived from underlying aeolianites and coastal sands (Pease and Tchakerian, 2002).

In an attempt to judge the importance of the coastal contribution to sand provenance in India, Goudie and Sperling (1977) attempted to map the extent of the foraminiferal sands in the Thar. This survey showed that foraminifera are found in dune sands at least up to 800 km from the nearest point downwind on the coast. The frequency of tests, moreover, declines in a progressive manner with distance from the sea. This indicates that foraminifera are susceptible to long-distance aeolian transport. The presence of the foraminifera also indicates that the coast has been an important source area for the sands of the Thar and that, during dry phases in the Pleistocene, vegetation conditions must have permitted the saltation of coastal sand grains over some hundreds of kilometres.

Dunes derived from lake basins or coastal sabkhas may be highly gypsiferous, as is the case with some of the dune material derived from the chotts of southern Tunisia and the nebkhas of Kuwait (Khalaf and Al-Awadhi, 2012). The Estancia, White Sands, Guadalupe and Cuatrocienegas dunefields of the Rio Grande Rift of the United States and Mexico comprise what are probably the largest known aeolian gypsum sand dune accumulations on Earth (Szynkiewicz et al., 2009; Jerolmack et al., 2011). There are even examples of small dunes composed of halite (sodium chloride) developed on the Salar de Uyuni, Bolivia (Svendsen, 2002). Sand derived from coastlines may be carbonate rich (as is the case with aeolianites). Sand derived from some coastal lagoons or lake basins may be rich in clay pellets, as is the case with much lunette sand (see Section 4.9).

Most of the world's desert dune sands are composed of quartz, however, and this often forms more than 90 per cent of the mineral content, as in the Namib, the Kalahari, the Jafurah sand sea of Saudi Arabia and the Great Sandy Desert of Australia. Such dunes are said to be 'mineralogically mature' (Muhs, 2004). The reasons for mineralogical maturity include derivation from a quartz-rich source sediment and the chemical weathering or abrasion of less resistant minerals such as feldspars. The mineralogy of sand grains can be studied either through traditional methods of sample collection and analysis or by such methods as thermal infrared remote sensing (Ramsey et al., 1999; Scheidt et al., 2011).

4.19 Aeolianite and Miliolite

Although most desert dunes are dominantly composed of quartz, this is not invariably the case. Indeed, there are many examples of dunes in drylands being dominantly composed of calcium carbonate. These are normally called aeolianites (Brooke, 2001). Such carbonate dunes are not restricted to arid regions or indeed to coastal regions (McLaren, 2007).

One of the largest expanses of carbonate-rich and lithified dunes occurs in the Wahiba Sands of Oman (Goudie et al., 1987). Highly carbonate-rich dunes occur as far as 100–200 km from their coastal source. Calcareous dunes are also common in Bahrain (Doornkamp et al., 1980) and in the United Arab Emirates (Teller et al., 2000); in Egypt; along the coast of Israel and Palestine (Sivan and Porat, 2004), where it is called kurkar or kurkur; at various other places around the Mediterranean; and along massive stretches of the Australian coast (Hearty and O'Leary, 2008). The structures, textures and mineralogy of aeolianites are described by McLaren (2007), as are models for their diagenesis.

In India, there are some controversial deposits which have been called miliolite. These are composed of large quantities of calcium carbonate (up to 99%) made up of the tests of various foraminifera (mainly rotalids and miliolids), small molluscan fragments, broken echinoderm spines, comminuted pieces of coral, lime pellets, micrite and microspar. Included within the dominant carbonate mass are rare detrital grains of quartz and other minerals. Generally the miliolite is highly porous, but, especially on exposure to air, is hard enough to form a widely used building material.

Miliolite occurs widely in Kutch and in Kathiawar (Figure 4.18) but to a minor extent also in mainland Gujarat. The origin of the miliolite is a matter of dispute (see Sperling and Goudie, 1975). It was named miliolite on account of its miliolid foraminiferal content (Carter, 1849), and many workers have regarded its high content

Figure 4.18 Cemented miliolite obstacle dune, now quarried, at Junagadh, Saurashtra, Gujarat, India. (ASG)

of marine carbonate fragments as evidence for a supposed shallow water or littoral marine origin (see, for example, Lele, 1974). However, other workers have postulated that the miliolite is in reality an aeolianite – a calcareous dune sand (see, for example, Biswas, 1971).

Although the content of marine micro-fossils in the miliolite gives strong prima facie support to the marine hypothesis, very few other substantial arguments have been put forward which bear on this point of view. Many arguments can be put forward to support the aeolian hypothesis, however. First, the miliolite is not recorded as containing marine mollusca, although land snails have been reported. The miliolite may also contain obvious colluvial material from adjacent hills, together with some rolled fluvial debris (Biswas, 1971). The foraminifera themselves are well capable of having been blown inland.

The geomorphic evidence provides a second main line of evidence in support of the aeolian hypothesis. This was acutely recognized by Blake (1897), working in Kutch, for he recorded that all the localities where miliolite occurred could 'be described as spots where a wind coming from the west or south would be stopped by an obstacle, or where a shelter-spot exists in a long scarp'. In Kathiawar too, the miliolite occurs where topographic barriers and traps would collect windborne sediments with characteristic geometries (Dasgupta and Bandyopadhyay, 2008). This is directly analogous to the situation presented by the quartzose, topographically induced dunes of the interior and margins of the Rajasthan Desert (Verstappen, 1970; Goudie et al., 1973). Moreover, near Una, Kathiawar, one also finds distinct parabolic forms composed of cemented miliolite, oriented transverse to the south-westerly winds and with steep lee sides. A third characteristic of the miliolite which supports its aeolian origin is the nature of its cross-bedding and laminations. Biswas (1971, p. 155) described foreset beds dipping at 20–30°, and noted the frequent reversals in the direction of dip. Other characteristics of the laminations support the aeolian hypothesis. As Blake (1897) pointed out,

> That the laminations should dip towards the rock on which the concrete (miliolite) rests, on the side nearest to the rock, is what we would expect in a windblown deposit. For when sand is blown against an obstacle it is thrown back again and the wind has to pass away on either side, so that in such places we always find an intervening valley between the mound and the obstacle, the surface of the mound thus sloping toward the obstacle.

Thus, the miliolite at inland situations is as an aeolianite produced by the deflation of marine carbonates from the intertidal and supratidal zone of a warm sea. The deflated materials were transported inland, deposited against topographic barriers and inter-stratified with various subaerial formations.

4.20 Dune Colour

Dunes display many different colours. Some can be dazzlingly white, while others may be a dark shade of orange. In many sand seas, one can see clear spatial trends in dune colour. Important controls of colour include the age of the dune grains, the stability of the dunes, climate and their mineralogical composition.

The reddening of dune sands has been the subject of some interest, and good reviews of mechanisms are provided by Gardner and Pye (1981) and Pye (1983a, 1983b). The red colour results from the presence of iron oxides, which occur either as coatings on individual sand grains or as staining on clay minerals which may form an inter-granular matrix. It is particularly associated with the presence of haematite. This may be derived from the original host sediment, may be caused by in situ weathering of iron-bearing mineral grains contained within the sediment, may be precipitated from percolating iron-bearing groundwaters or may be introduced by the deposition of aeolian dust or by surface water flow from higher ground.

In coastal areas of the Namib, dominated by barchanic dunes, the sand is yellowish brown (10YR 5/4) to light yellowish brown (10YR 6/4), whereas in eastern areas it becomes yellowish red (5YR 5/8). The reasons for this change in colour have been investigated by Walden and White (1997) and Walden et al. (1996). Four main hypotheses have been suggested to account for the reddening of dunes as one progresses inland:

1. The increasing age of the sands inland allows greater time for weathering processes to develop the iron (haematite) coatings around quartz grains.

2. In areas of active sand transport and high energy winds, coating may be lost or fail to develop.
3. Different sand-source materials
4. A regional climatic gradient with warmer and wetter conditions inland providing a control on the rates of weathering processes which generate the haematite coatings.

On the basis of detailed analysis, Walden and White (1997) suggest that different sand-source materials play a major role, but that so also do age and environmental gradients.

Another discussion of dune colour trends is given for the U.S. High Plains by Holliday and Rawling (2006). They found that as one moved downwind across the Muleshoe Dunes, there was an overall decrease in redness from dominantly 5YR Munsell hues in the west, to 7.5YR in much of the central dunes, to 10YR in the eastern end of the dune field. They attributed this to abrasional loss of clay coatings from sand grains with greater distance of travel from source. This finding was confirmed by White and Bullard (2009).

In the Negev, Roskin et al. (2012) looked at the degree of reddening of dunes of different ages and found, contrary to some previous studies and assumptions, that dune redness did not vary greatly with dune age in a transect across dunes dating from the late Pleistocene through to the mid-Holocene and the present. They suggested that the dune sand may have been derived from already-red sources in the Nile Delta.

4.21 Grain Shape and Size

The shape of dune sands has often been said to be well rounded by abrasion. Indeed, some dune sands are rounded – especially the coarser grains and those composed of carbonates or gypsum – but numerous studies over recent years have demonstrated that most quartz dune-sand grains (Table 4.5) are in reality sub-angular and sub-rounded (Folk, 1978; Goudie and Watson, 1981; Fitzsimmons et al., 2009) and that rounding cannot be taken to be diagnostic of aeolian transport. The meanings of numerical values for roundness used in the table are as follows:

Very angular	0.5
Angular	1.5
Sub-angular	2.5
Sub-rounded	3.5
Rounded	4.5
Well-rounded	5.5

There are a number of reasons why dune-sand size characteristics may vary (Livingstone et al., 1999). These include differences in source sediments, the effects of gravity (either by opposing upslope grain transport or by causing falling or sliding

Table 4.5 *Roundness values of dune sand*

Desert	Mean roundness value of 2.50 fraction	Percentage in rounded and well-rounded class of Powers (1953)
Bahrain	3.51	12.25
California	2.89	6.33
Kalahari	3.21	2.00
Namib	3.53	15.30
Thar	2.91	2.68
Tunisia	4.01	50.00
United Arab Emirates	2.77	–
Wahiba	3.35	9.45
Mean	2.85	9.64

Source: Goudie et al. (1987) and El-Sayed (1999).

on a slip face), the changing pattern of shear stress over a dune, wind velocities and post-depositional alteration by weathering and dust accretion.

Coarse sand grains are rare because they need very high velocities to move them. Equally, active dunes normally do not contain large amounts of fines (silt and clay) because during saltation, fine particles are lifted into the air and carried away in suspension. That said, some dunes contain clay aggregates, as is the case with the Strzelecki and Simpson Deserts of Australia (Wasson, 1983; Fitzsimmons et al., 2009). Should they become stabilized, however, dunes may also start to contain fines because of dust additions from above or because of the weathering of their constituent minerals by chemical and physical processes (Pye, 1983a, 1983b), including salt weathering (Goudie et al., 1979). Sands that have been weathered since their deposition may develop quite large fine contents. In the Kimberley of north-west Australia – in an area with c 700 mm of mean annual rainfall – degraded, red, quartzose linear dunes of probable late Pleistocene or early Holocene age have a silt and clay content of 19.9–25.6 per cent. This compared with silt and clay contents of 26–32 per cent for the degraded Qoz dunes of Sudan, and 26.32 per cent for the stabilized dunes of north-west India (Goudie et al., 1993). Werner et al. (2010) have shown that in the High Plains of Kansas, older dunes have a higher fines content than younger dunes, which means that older dunes tend to be less easily reactivated than younger dunes. If dunes with a high content of fines are destabilized by, for example, vegetation removal by fire, then the fines may be liberated and blown away (Strong et al., 2010).

Some data on dune-size characteristics are presented in Table 4.6. Using the Folk and Ward (1957) scale, 1–2ø is medium sand, 2–3ø is fine sand and >3ø is very fine sand. In terms of the sorting index (the graphic standard deviation), values of <0.35 are very well sorted, 0.35–0.50 are well sorted, 0.50–0.70 are moderately well sorted and 0.70–1.00 are moderately sorted. With respect to phi skewness (graphic skewness),

Table 4.6 *Dune grain-size characteristics*

Location	Dune type	Mean (ø)	Mean (mm)	Sorting	Skewness	Kurtosis
Namib[1]	Crescentic/ barchans crest	2.20	0.22	0.55	0.19	0.15
	Slip face	2.32	0.20	0.54	0.07	0.49
	Mid stoss	2.24	0.21	0.73	0.16	0.47
	Base stoss	2.10	0.23	0.84	0.26	0.48
	Linear compound crest	2.25	0.21	0.39	0.19	0.50
	Complex linear crest	2.49	0.18	0.36	0.13	0.51
	Star dune crest	2.29	0.20	0.29	0.13	0.53
Namib[12]	Linear	2.07	0.24	0.73	0.26	0.97
Wahiba Oman[2]	Crest, linear megadune	2.47	0.18	0.27	–	–
	Megabarchans	3.03	0.12	–	–	–
Gran Desierto[3]	Linear dunes	2.72	0.15	–	–	–
	Star dune crest	2.49	0.18	–	–	–
Australia[4]	Linear crests					
Simpson	"	2.46	0.18	0.57	–	–
Strcelecki	"	2.18	0.22	0.68	–	–
Tirari	"	2.22	0.22	0.68	–	–
Australia[5]	Linear crests					
Mallee (Lowan)	"	2.22	0.22	0.59	0.02	–
Woorinen, south	"	2.41	0.19	0.61	0.15	–
Woorinen, north	"	2.04	0.24	0.69	0.80	–
Saudi Arabia[6]	Barchans	2.52	0.17	0.67	0.05	0.69
Saudi Arabia[13]	Barchans	2.02	0.28	0.95	0.35	1.12
Abu Dhabi[7]	Longitudinal	2.53	0.17	0.50	0.06	1.07
	Barchan	2.59	0.17	0.55	0.04	0.99
	Transverse	2.68	0.16	0.54	0.03	0.99
Taklamakan[8]	Crescent	3.08	0.12	0.37	0.07	1.02
	Dome	3.21	0.11	0.36	0.02	1.02
	Linear crests	2.63	0.16	0.94	−0.25	0.67
	"	2.87	0.14	0.52	−0.01	0.91
	"	3.27	0.10	0.33	0.01	1.06
	Stars	2.81	0.14	0.51	0.01	1.04
Kalahari[9]	Linear crest	2.13	0.23	1.01	0.16	1.18
Kalahari[10]	Linear crest	2.21	0.22	–	0.03	1.35
India, Thar[11]	Linear crest	2.72	0.15	–	–	1.02
Nebraska, USA[13]	Barchans	2.66	0.16	0.61	–	1.20
Pakistan, Thar[14]	Linear	2.51	0.18	0.61	–	0.80
Peru[15]	Barchans	2.80	0.14	0.28	–	1.11
Egypt (Kharga)[16]	Barchans	2.25	0.21	0.64	–	0.79
Egypt, Western	Barchans	1.97	0.26	0.61	0.03	0.49
Desert[17]	Longitudinal	1.80	0.29	1.54	0.15	0.52
Australia[18]	Linear crests	2.53	0.17	0.43	0.11	0.52
(Simpson)	Linear flanks	2.75	0.15	0.57	–	–
Mean	–	2.47 (n = 40)	0.18 (n = 40)	0.57 (n = 35)	0.14 (n = 27)	n = 0.82 n = 29

Sources: [1]Lancaster (1989a). [2]Goudie et al. (1987). [3]Lancaster (1995a). [4]Pell et al. (2000). [5]Pell et al. (2001). [6]Binda (1983). [7]Alsharhan et al. (1998). [8]Wang et al. (2003). [9]Thomas (1987). [10]Goudie (1970). [11]Goudie et al. (1973). [12]Watson (1986). [13]Warren (1976). [14]This book. [15]Hastenrath (1967). [16]Embabi (1967). [17]Maxwell (1982). [18]Folk (1971).

values of +0.3–1.0 are very positively skewed, +0.1–0.3 are positively skewed, +0.1–0.1 are symmetrical and –0.1–0.3 are negatively skewed. As regards kurtosis, samples with values of 0.41–0.67 are said to be very platykurtic, 0.67–0.90 platykurtic, 0.90–1.11 mesokurtic and 1.10 to 1.50 leptokurtic. On this basis we can say that most dune sand is in the fine-to-medium sand category, is well-to-moderately well sorted, is symmetrical or slightly positively skewed and is platykurtic or very platykurtic.

There appear to be differences among sand seas, with relatively fine sands (150–180μm) occurring in the Grand Erg Oriental, Thar and Gran Desierto and rather coarse sands in the Kelso and Algodones dunefields of the United States. Those of the Kalahari and Namib are intermediate in size (Lancaster, 1995a, p. 104). In Australia, Fitzsimmons et al. (2009) found that dune sediments in the northern Strzelecki and southern Tirari Deserts had dominant modal peaks in the medium-sand range, whereas in the southern Strzelecki and northern Tirari they had dominant modal peaks in the fine-sand range. The reasons for such differences may include the nature of sand sources, the distance over which sand has been transported and differences in wind energy, with low-energy ergs having finer grain sizes.

Grain size also varies across individual dunes. There may be notable differences between crest and base, flanks and slip face, between rippled and non-rippled surfaces and between individual laminae (Pye and Tsoar, 1990, p. 71). The windward slope and horns of barchans are often coarser than the crest and slip-face sands. In the case of obstacle dunes, windward dune sands tend to be coarser than lee dune sands. Many workers have found that the crests of linear dunes have finer sand than their flanks, but the reverse has also been encountered, and this may in part be due to differences in dune size. Livingstone et al. (1999) looked at the differences between the large linear dunes of the Namib Erg and the much lower and more subdued dunes of the Kalahari and suggested that in the latter case slope effects were less effective in stopping coarser grains from ascending the ridges. Grain size may also change downwind as a result of either sorting and/or abrasion of grains. In the case of the gypsum dunes of White Sands National Monument, United States, elongate large grains get abraded with transport and so become rounder and smaller, and small grains are produced by the breakage of protuberances from larger particles (Jerolmack et al., 2011). Progressive differentiation from coarse to fine sand in the direction of sand flow is also to be expected as a result of coarser material being transported more slowly than finer material and thereby remaining in upwind areas (Wilson, 1973, p. 105). Such a reduction in grain size along a sand-flow path was confirmed for the southern Sahara by Mainguet (1977).

4.22 Internal Structures of Dunes

The internal structures of dunes (Figure 4.19) provide clues to their growth and dynamics (Zhou et al., 2012), and good reviews are provided by Pye and Tsoar (1990, chapter 7), Kocurek (1991) and Mountney (2011).

Figure 4.19 Cross-bedding structures developed in a linear dune near Ras Al Khaimah, United Arab Emirates. (ASG)

There are three primary modes of deposition on dunes (Hunter, 1977): migration of wind ripples, fallout of grains in the lee of the dune crest and avalanching on the steep lee slope of the dune. These three processes form three main types of sedimentary structure: climbing translatent strata (wind ripple laminae), grainfall laminae and grainflow cross-strata. Primary sedimentary structures of these sorts are separated by bounding surfaces (erosional discontinuities). These in turn can be classified into various types. As Lancaster (1995a, p. 88) explained:

Third order or primary bounding surfaces occur within sets of laminae and represent reactivation episodes resulting from short-term changes in wind strength and/or direction; second order or growth surfaces bound sets of strata and form by erosion or non-deposition as dunes grow episodically; first order or stacking surfaces may divide the accumulations of laterally migrating dunes, or in some environments may represent episodes of deflation to the water table. . . . In addition, regional scale or super surfaces form as a result of hiatuses in sand sea accumulation due to sediment supply and/or climatic changes.

Some dunes may display few or no structures in cross section, and this may be because of active post-depositional bioturbation processes (Bateman et al., 2007). Different dune types may display different mixes and arrangements of structures, and our knowledge of these is increasing, partly because of the availability of sections created by large-scale excavation by the construction industry, and partly because structures can be revealed by ground-penetrating radar (Bristow et al., 2000; Hugenholtz et al., 2007; Tatum and Francke, 2012; Vriend et al., 2012).

In linear dunes, McKee (1979, p. 103) suggested that one of their most important and distinctive characteristics is that 'slipfaces develop on both sides of the crest, and foreset dip directions form two clusters of points in early opposite directions, rather than a single grouping'. In transverse dunes, the structures are relatively well known (McKee, 1979, pp. 93–94) and have been described thus by Livingstone and Warren (1996, p. 120):

As an active transverse dune migrates downwind, sand is eroded from the windward slope and deposited in grainfall and sandflow deposits on the lee slope. These high-angle deposits facing downwind are called *foresets*, and this form of stratification is termed *cross-bedding* (or *cross-strata*). The great majority of sand dune deposits consist of this kind of bedding, and indeed these structures are considered a diagnostic feature of dune sands in the sedimentary record.

The nature of cross-strata in barchans and barchanoid ridges have been described by McKee (1979, pp. 89–92). Parabolic dunes have lee-side deposits composed of concave-downward strata that develop from grainfall on a cohesive slip face and are interbedded with avalanche cross-strata. They tend to contain structures that are indicative of moisture (e.g. adhesion laminae) and vegetation (e.g. root tubules) (McKee, 1979, pp. 94–96; Halsey et al., 1990).

5

Rivers and Slopes

In this chapter we will consider the processes and landforms associated with rivers and slopes. As will become evident, slope forms, mass movement processes, runoff generation and river morphology are all very closely linked. We consider first some of the landforms that result from fluvial processes, and then second how it is that runoff and sediment yields are generated by desert surfaces. We then consider the importance of mass movement in moulding desert slopes.

5.1 Fluvial Processes and Forms: Introduction

As we saw in Chapter 1, there have been longstanding debates about the relative importance of fluvial and aeolian processes in moulding desert landscapes. It is, however, undoubtedly true that, notwithstanding their dryness, deserts all show the impact of fluvial processes. Indeed, there are certain characteristics of deserts, including their sparse vegetation cover, which enable water erosion to be a potent force. As Reid and Frostick (1997, p. 225) remarked, 'Rivers play an important role in shaping the Earth's deserts despite the fact that they run for only a vanishingly small fraction of the time as one moves towards the hyper-arid core regions.' In addition, many deserts have been subjected to more humid conditions during their evolution.

Although there have been many debates about the relative power of fluvial and aeolian processes, hard and fast empirical data which compare the operation of the two types of process are sparse. Zhang et al. (2011) provide an exception. Nonetheless, in reality they interact in a whole range of ways (Draut, 2012). Aeolian deposits may be reworked to form the alluvium of desert rivers, alluvial spreads may be a source of dune sand or of dust, valley-marginal sand dunes may have a distinctive morphology (Bullard and Nash, 2000), the presence of aeolian silt affects the infiltration capacity and runoff from desert surfaces, dunes may be degraded by fluvial processes, rivers may be dammed or diverted by dunes and, because of climate change, most deserts have alternated between humid and arid conditions – sediment production is often related to humid periods and sediment then becomes available to the aeolian system during arid phases (Bullard and McTainsh, 2003). Equally, short term climatic variability and fluctuations can lead to both flash floods and severe dust storms occurring in an area as a matter of course (Bullard and Livingstone, 2002).

5.2 Drainage Systems

Desert drainage systems are not always very highly integrated. Much of the drainage – because of high evaporation or high infiltration losses into alluvial fans or aeolian materials – does not reach the sea, and because of the existence of closed basins, much of the drainage may be endoreic (i.e. centripetal, flowing towards the centre). This will end up in closed basins (playas, pans, etc.). In the Altiplano of northern Chile, there is a clear climatic control of drainage, with endoreic patterns being predominant in the hyper-arid areas and exoreic forms (those that flow directly to the Pacific) in the moister northern portion (García et al., 2011).

Some desert rivers, including many of those that are of most use to humans, have their sources outside the desert realm, and such allogenic or exogenous rivers may be through-flowing and perennial. An example of an allogenic stream is the Mojave River in California, which rises in the San Bernardino Mountains and flows north into the Mojave Desert; another is the Oued Saoura, which flows south to the Sahara from the Atlas Mountains in Algeria. Other examples include the Senegal, Niger, Logone, Chari and Nile Rivers (Figure 5.1a), which flow north from humid to arid areas in central Africa; the Euphrates, which flows from the mountains of Turkey to the Arabian/Persian Gulf; and the Indus River which flows from the Himalayas and Karakorams into the Indian Ocean (Figure 5.1b). Geomorphologically, allogenic streams may be associated with landforms and deposits in deserts which are alien to the desert environment and reflect landforms more common in humid areas, such as well-defined terrace sequences.

Runoff of streams rising within deserts is characterised by the same attributes as their rainfall: brief, infrequent, localized, sometimes of high intensity and varying greatly from year to year and season to season. Perennial flow is rare in deserts, except when the drainage is allogenic. Intermittent flow, in which there is an alternation of dry and flowing reaches along a drainage channel, is more common. Most channels in deserts, however, only carry water during storms, and flow in them is therefore ephemeral. As a result, flow in most river systems is only rarely, if ever, integrated, and much river flow occurs in the form of flood events with very steep hydrographs – flash floods (Reid and Frostick, 1997). Moreover, those rivers that flow from a subhumid area into a more arid climate may suffer an attenuation of flow downstream, which is exacerbated by transmission losses into sediments. This is the case with the Kuiseb River in Namibia, which runs from the moist mountains of the interior to the hyper-arid Atlantic coast at Walvis Bay. Only in rare years does it manage to reach the sea.

Figure 5.1a The Blue Nile derives its flow from the relatively wet mountains of Ethiopia. (ASG)

Between 1836 and 2007, only sixteen floods did so. Peak discharges for a ten-year flood event average 90 m^3/s at Gobabehb, but at Rooibank, just at the head of the Kuiseb Delta, this figure has dropped to 0.9 m^3/s (Heidbüchel, 2007).

Knighton and Nanson (1994) found that transmission losses amounted to more than 75 per cent of discharge at certain flows along Cooper's Creek, Australia, whereas in the Kairouan area of Tunisia, Besbes (2006) reported transmission losses of 40–50 per cent of flood volume. In the Nahal Zin watershed of Israel, transmission losses decrease with flood magnitude from c 86–100 per cent in small floods to c 10 per cent for large ones (Greenbaum et al., 2006). Transmission losses occur because of percolation into channel beds and at high flows into channel terraces (Dahan et al., 2008), palaeochannels and the like. They may also occur because of evapotranspiration where floodwaters become disconnected in low-lying areas, such as billabongs (McMahon et al., 2008b). Transmission losses go up with increasing channel wetted perimeter (Mudd, 2006) and also with the permeability of bed and bank materials, which in turn relates to whether they are sand or clay rich. This is discussed in the context of Australia by Dunkerley and Brown (1999) and Dunkerley (2008). Some arid rivers have also suffered reductions in discharge because of a diminution in the contribution of groundwater to their flow as a consequence of aquifer overpumping

Figure 5.1b The Karakoram/Himalayan-fed River Indus forms a meandering green thread through the Thar Desert in Pakistan. Its channel is very unstable and prone to flooding. (ASG)

(Kustu et al., 2010).). Such discharge decreases can cause a concomitant downstream decrease in channel dimensions and sediment size (Kemp, 2010).

The main differences between rivers of the arid zone and those of the humid zone are well summarised in Table 5.1.

5.3 Pediments

Pediments are gently sloping, generally concave, rock-cut surfaces that abut mountain fronts (Applegarth, 2004) and connect eroding slopes to areas of sediment deposition at lower levels. They are often smooth and are only weakly incised by immature drainage networks (Strudley et al., 2006) (Figure 5.2a and b) and are preferentially formed on weathering-resistant rocks (Pelletier, 2010); however, that said, they have been described on a huge range of lithologies (Strudley and Murray, 2006). They tend to have average longitudinal slopes that are between 2° and 4°, rarely exceeding 6°.

Slope profiles in arid areas, being little obscured by vegetation, are visually more dramatic than those of most humid regions, and many workers have commented on their apparent angularity. Their form can be analysed in terms of an idealised slope profile with four components: (1) an upper convexity (the waxing slope), (2) a cliff (free face), (3) a straight segment (constant slope) and (4) a basal concavity

Table 5.1 *A comparison of arid zone and humid zone rivers*

	Arid zone	Humid zone
INPUT	Low and unreliable	Relatively high and dependable (often seasonally so)
	Limited duration but often high-intensity storms	Long duration (often frontal) precipitation of variable intensity
	Extremely variable at the event and annual scales	Temporal variability is much less
	Spatially concentrated events	Large areas generally affected
THROUGHPUT	Horton overland flow dominant	Infiltration, throughflow and groundwater flow more significantly
	Rapid onset of surface runoff	Longer lag between precipitation and runoff
	Relatively high runoff coefficient	Lower runoff coefficient
	Decreasing discharge downstream due to transmission losses	Increasing discharge downstream due to tributary inflows
OUTPUT	Mostly intermittent	Largely perennial
	Extremely flashy regime	Relatively steady regime
	Sharply peaked runoff hydrograph	Runoff hydrographs have lower amplitude
	Considerable interannual variability	Dependable interannual flows
CHANNELS	Drainage densities can be high but networks may not be fully integrated	Well-integrated drainage networks
	Floods as major channel controls	Channels adjust to a range of more frequent discharges
	Long recovery time after disturbance	Channel recovery is quicker
	Transient behaviour dominant	Tendency for channels to equilibriate

Source: Derived from Knighton and Nanson (1997, p. 186).

(pediment). The pediment and the constant slope are often separated by an abrupt break of slope (Figure 5.3). The pediments may coalesce and cover extensive areas (called pediplains), and from them may rise isolated residual outliers called buttes, which possess one or both of the steep-slope components. French workers have often called pediments *glacis d'érosion*, and many such features are known from North Africa (Dresch, 1957; White, 1991). However, some workers have drawn a distinction between a *glacis d'érosion* and a pediment, with the former truncating softer, less resistant rocks (often young sedimentaries) adjacent to a more resistant upland, and the latter being a feature where there is no change in lithology between upland and pediment (White, 2004, p. 469). The North African *glacis* are also distinguished by the fact that they often exhibit multiple levels, which can be traced back into the upland drainage basin where they form river terraces.

(a)

(b)

Figure 5.2 Pediments. (a) A classic pediment developed in granites in Namaqualand, South Africa. (b) A pediment developed in granitoid rocks in the Mojave Desert, California, USA. (ASG)

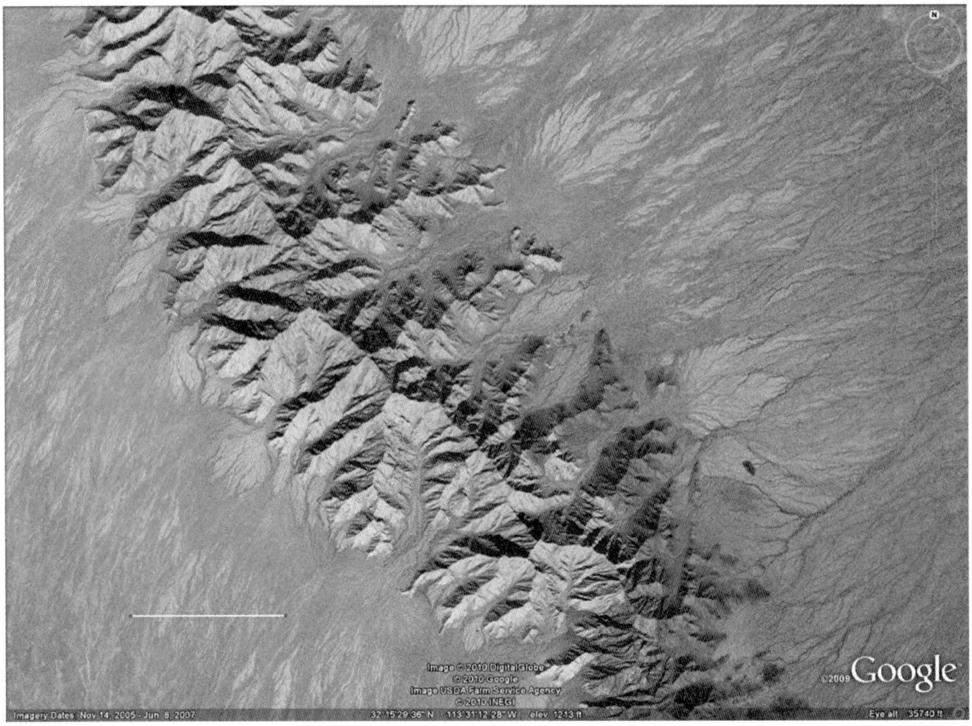

Figure 5.3 Pediment slopes, drained by an intricate network of small stream channels developed around steep mountains in south-west Arizona, USA. Scale bar 2 km. (©Google Earth 2010, ©Digital Globe 2010, ©INEGI 2010)

Whether desert slopes develop by parallel retreat to give pediplains and steep-sided residuals or evolve by gradual slope decline through time depends very much on rock type. This is demonstrated very clearly by a study that Schumm (1956) made in the semi-arid Dakotas in the United States. In that area there are two sedimentary rock types occurring under identical climatic conditions, the Oligocene Brule and the Eocene Chadron. The Brule had steep (44°) straight slopes, and erosion had produced outliers with a similar slope profile. By contrast, the Chadron had rounded slopes with broad interfluves, and residuals had lower slope angles than the scarp from which they had become detached by erosion. The explanation for this can be seen in terms of the processes operating on the two different rock types. Because it had a low infiltration capacity, the Brule was characterised by surface wash, and so the steep slopes were produced by stream incision. The Chadron produced a surface of clay aggregates with an open texture which had a much higher infiltration capacity, and so the creep of the clay aggregates when saturated with moisture was the main slope-forming process.

Pediments are widespread in North American deserts, and it was here that McGee (1897) introduced the term to the scientific literature. They occur on a wide range of lithologies (Dohrenwend, 1987), but within the Mojave and Sonoran Deserts they are preferentially developed on deeply weathered, coarse-crystalline granite rocks, although locally they also cut discordantly across volcanic and metamorphic rocks. In the west-central Great Basin, however, pediments are preferentially developed on sedimentary rocks and on Tertiary volcanics. A second general characteristic of pediments is that they are 'mostly located in stable or quasi-stable environments where erosional and depositional processes have been approximately balanced for relatively long periods of time . . . regional trends suggest a general correspondence between tectonic stability and pediment development' (Dohrenwend, 1987, pp. 322–23). They tend not to occur in areas where, for either climatic or tectonic reasons, incision predominates, or in areas where climatic conditions engender deep-weathering profiles (Strudley and Murray, 2006).

Pediments may be ancient. Dohrenwend (1987, p. 323) notes that 'the ubiquitous presence of deep-weathering profiles beneath extensive pediments on granite rocks suggest at least pre-Quaternary ages for the original surfaces of these pediments'. Moreover, some U.S. pediments appear to have been buried by late Miocene and Pliocene basalt flows. It is therefore likely that they have evolved under a range of climatic conditions. However, dating using cosmogenic nuclides in the tectonically active, hyperarid conditions of southern Peru has indicated the formation of five pediment surfaces over the last million years (Hall et al., 2008). These have formed in response to ongoing Andean uplift.

An abrupt boundary normally occurs between a pediment and the mountain slope behind it. Dohrenwend (1994, p. 326) suggests that the junction will be particularly well defined where: (a) an active fault bounds the mountain front, (b) marginal streams flow along the base of the upland slope, (c) slopes are capped with resistant caprock, (d) slopes are coincident with bedrock structure (dykes, joints, fault line, steep-dipping to vertical beds, etc.) or (e) a pronounced contrast exists between debris size on the hillslope and on the piedmont surface. With respect to point (e), granites often break down discontinuously from boulders or corestones into constituent mineral grains, and this may contribute to the excellence of pediment development on granite rocks in the Mojave. This is a view that has been championed by Oberlander (1997, p. 151), who has argued that 'on non-granite rocks the absence of a well-defined piedmont angle may reflect a continuum of clast sizes in transit from hillslope to pediment as opposed to the common bimodal size distribution of granitic debris'.

For a long time, the origin of pediments has intrigued geomorphologists. Gilbert (1877, p. 125) called them 'of planation' and attributed their origin to lateral planation by streams. McGee (1897) invented the term 'pediment' and gave a vivid account of a sheetflood that he experienced which went at 'race-horse speed'. He believed that these floods of unconcentrated water were the agent that caused the planation of the pediment surfaces (pp. 108–9):

Over a dozen or scores of square miles in carefully examined localities, hard rocks like those of the mountains, and with no sign of decomposition, are planed off almost smooth as the subsoil by the plowshare, with nothing either in configuration or covering to indicate that streams have flowed over them, and extended consideration has yielded no other suggestion as to the eroding agent than that also found in analogy with the observed sheetflood.

Some subsequent workers have accepted that sheetfloods are important (e.g. Blackwelder, 1931; Davis, 1938), and Rahn (1967) provided dramatic photographs of such events in Arizona, although Graf (1988, p. 106) was rather more dismissive of their role in pediment formation.

Another early view was that the whole suite of pediment, playa and inselberg landforms was the result of extreme aeolian planation. This was a view that was most forcibly put forward by Keyes (1912).

One of the leading German climatic geomorphologists, Büdel (1982), made two main points about pediments. First of all, he noted that they were a feature of deserts with cold winters (e.g. in Tibet and Iran) and suggested that frost may have contributed to slope evolution. He believed that frost would continually loosen masses of debris, which would then be intermittently swept away across the unvegetated footslopes by occasional cloudburst. He believed they were less well developed in what he termed the lower-latitude, frost-free, 'trade-wind deserts'. Second, and more important, he felt that pediments were relicts of long-continued Tertiary moist tropical conditions and were ancient etchplains associated with deep weathering and subsequent stripping. He averred (p. 246):

No relief-developing mechanism other than that of the seasonal tropics can accomplish the planation of huge erosional surfaces, and at the same time weather back and sharpen the slopes of small inselberg ranges and large mountain blocks.

Other workers saw pediments as the result of the parallel retreat of mountain fronts as a result of weathering and fluvial erosion (Lawson, 1915; Kesel, 1977) or as interfluves between stream basins (Lustig, 1969), whereas yet others have seen a role for lateral planation by streams leaving the mountain front. Johnson (1932a, 1932b) believed that if this were the mechanism, then piedmont zones should show a series of relatively flat semi-cones called rock fans. Parsons and Abrahams (1984) accepted that lateral planation by rivers turning abruptly after issuing from a mountain mass could contribute to pediment formation, but they also argued that hillslope processes would contribute to mountain-front retreat. Finally, some workers, following the proposal of Mabbutt (1966) – opaquely expressed as it is – believed that pediment surfaces are old weathering fronts exposed by fluvial stripping.

As Cooke et al. (1993, p. 189) observed, however:

The explanations of pediments have, until recently, made gross assumptions about processes. For example, sheet flooding, a common *deus ex machina*, has been observed only rarely on pediment surfaces; backweathering of mountain fronts is more honoured in the deduction than in the observation; and lateral stream planation is not a phenomenon commonly seen on pediments. Coupled with deductions concerning process is the common cause-and-effect

Figure 5.4 A classic alluvial fan in the Karakoram Mountains of Pakistan. Note the debris flow tracks. (ASG)

error arising from relations between a deduced process and the visible landform. Clearly, for instance, sheetflooding cannot produce a planar surface, because a planar surface is necessary for sheetflooding to occur. Again, widespread weathering and occasional removal of weathered debris is unlikely to produce a pediment surface; it is more likely to maintain, probably at a lower level, a pre-existing form. In short, there is frequently confusion between *pediment-forming* and *pediment-modifying* processes. And yet a more justifiable assumption concerning processes is frequently ignored: processes have almost certainly changed in many areas in nature, magnitude, frequency, etc. as a result of climatic changes during the course of pediment evolution.

5.4 Alluvial Fans

Alluvial fans are coned-shaped landforms composed of fluvial sediments (Figure 5.4). Very often they radiate downslope from points which are commonly where a feeder stream leaves a mountain tract. Although Graf (1988, p. 185) has remarked that the '[a]lluvial fan literature is voluminous in comparison to the amount of space that fans occupy in drylands', alluvial fans are one of the most important landforms of the arid realm. They cover about one-fifth of California (Bull, 1963), and they can dominate large swathes of terrain in tectonically active and mountainous deserts, such as those of China, Chile, Mongolia, central Asia, the Oman Peninsula, Iran and the Mediterranean lands (Harvey, 2002; Harvey et al., 2005). The features are highly variable in size, ranging from mega-fans that are tens or hundreds of kilometres across (Blechschmidt et al., 2009), to features that are only a few metres across. Fan surfaces tend to have concave long profiles and to show a progressive decrease in grain size down fan (Stock et al., 2007). The planform morphology of a fan is generally approximately semicircular, whereas the three-dimensional morphology is conic.

Alluvial fans dominate some areas, whereas pediments dominate others. In the context of the U.S. Southwest, Denny (1965, pp. 58–9) attributes this to tectonic history and mountain size.

Denny noted that where the mountains are recently elevated, as in much of the Death Valley region, the piedmonts consist largely of fans. Where the mountains are small compared with the adjacent basin, however, as in parts of the Mojave, the piedmonts include extensive areas of exposed pediment, especially near the highlands. He argued that as the mountain is reduced in size, it will supply less and less debris to the washes. Thus erosion will become the dominant process on a large segment of the piedmont and will cause the area of pediment to increase.

Although alluvial fans occur in many non-desert environments, they are especially well developed in arid lands for four main reasons (Harvey, 1997, p. 234):

> First, as a result of sparse vegetation cover, intense storm rainfall and the dominance of overland flow processes on the hillslopes, desert mountains have high rates of storm sediment production. Second, steep, flashy desert mountain streams have high rates of sediment transport and delivery to mountain-front locations. Third, the episodic nature of sediment transport is accentuated in arid regions as the result of the dominance of rare high-magnitude storm events in sediment transport. Fourth, arid region fluvial systems are characterised by spatial discontinuity, resulting from high evaporation rates and high transmission losses that accentuate down-channel decreases in stream power through alluvial fans.

More than 100 years ago, Gilbert (1882) indicated in general terms the processes by which fans are constructed:

> [W]hen water leaves the margin of the rocky mass [i.e. a desert range] it is always united into a comparatively small number of streams, and it is by these that the entire volume of detritus [from the mountains] is deposited. About the mouth of each gorge a symmetric heap of alluvium is produced – a conical mass of low slope, descending equally in all directions from the point of issue; and the base of each mountain exhibits a series of such alluvial cones, each with its apex at the mouth of a gorge and with its broad base resting upon the adjacent plain or valley. Rarely these cones stand so far apart as to be completely individual or distinct, but usually the parent gorges are so thickly set along the mountain front that thc concs are more or less united and give to the contours of the mountain base a scalloped outline.

Alluvial fan deposition is not caused by an abrupt change of stream channel gradient but because of changes in the hydraulic geometry of flow after the stream leaves the confines of the trunk stream channel (Bull, 1977). It becomes free to diverge and to infiltrate, so that stream power dissipates and deposition ensues. Fan shape results from frequent radial shifts of fan-feeder channels about the nodal point of entry from the drainage catchment to the depositional basin. The deposition is achieved both by streams and by debris flows (Trowbridge, 1911; Hooke, 1967; Dühnforth et al., 2007; Wasklewicz et al., 2008; Hardgrove et al., 2009). These two processes may often be related through time during a single flood. Eyewitness reports of alluvial fan flooding in the White Mountains of California and Nevada (Beaty, 1963, p. 531) provide an illustration:

> Two-and-a-half hours after a heavy thunderstorm in the mountains, masses of debris advanced down channel in a series of waves along a low front of boulders and mud; the lower ends of this debris flow consisted of silts and fine sands with cobbles and small pebbles which constituted a rapidly moving mudflow; after the debris-flows had halted, streamflows continued for up to 48 hours and dissected the newly laid sediments and older deposits.

During periods of aggradation, the ratio of sediment to water flux is high and sediment is deposited across the fan. In contrast, entrenchment occurs during periods when there is a lower ratio of sediment to water flux. On a stream-dominated fan, incision at the apex produces a fan-head trench in which the longitudinal stream channel is at a lower elevation and lower gradient than the surrounding fan surface. This trench is deepest at the apex and becomes shallower down fan (Blainey and Pelletier, 2008).

Lithology also plays a role in determining which process is dominant. Blair (1999) argued that fans derived from andesites and granites are not rich in clay and silt, whereas those derived from sedimentary rocks are, so the former tend to be dominated by streamflows and the latter by debris flows. In the case of Cucomonga fan in California, which has developed from eroded colluvium, as much as 88 per cent of it is built by debris-flow deposits (Blair, 2003). By contrast, in the Black Hills of Arizona, only 5–10 per cent of fan deposits were found to have been deposited by mudflows (Blissenbach, 1954).

The viscosity of debris flows is an important control of surface form. Low viscosity flows (i.e. those with low sediment concentrations) tend to move further down fan where they spread out and thereby smooth out prior topography and fill in channel traces. By contrast, high viscosity flows (i.e. those with high sediment concentrations) are unable to move from the upper fan, where they remain as rough features bounded by boulder levees (Whipple and Dunne, 1992).

The size of fans depends on such factors as the size of the upstream drainage-basin area, but drainage-basin lithology is also important. For example, in the San Joaquin and Deep Springs Valleys of California, the fans derived from erodible mudstones are roughly twice the size of their source areas, whereas fans derived from resistant quartzite source areas are only about one-sixth the size of their source areas (Hooke, 1968). Lecce (1991), however, working in the western White Mountains of California and Nevada, found that it was resistant rather than erodible basins that had the larger alluvial fans. This was attributed to the fact that fan size may be influenced by sediment storage as well as lithologically controlled morphological variables such as valley-side slope. Canyons that developed in resistant rock types, and which are steep and narrow and provide little space to store sediment, contribute more sediment to fans. Conversely, gently sloping basins in erodible rock types store more sediment, decreasing its delivery to the fans.

Tectonics influence fan development in a very significant way, and as Bull (1997, p. 248) has remarked:

> Thick alluvial fans are orogen deposits, not only because uplift creates mountainous areas that provide debris and increased stream competence, but also because the loci of deposition on alluvial fans are controlled by the rate and magnitude of uplift of the adjacent mountains.... Optimal conditions for accumulation of thick sequences of fan deposits occur where the rate of uplift exceeds the rate of downcutting of the trunk stream-channel at the mountain front.

Tectonic activity can also cause segmentation and incision (Bull, 1964; Denny, 1967; Hooke, 1967).

Climatic changes may have been another important factor in determining fan characteristics (Ritter et al., 2000). Traditionally, two models have developed, both of which suggest that episodes of alluvial fan aggradation are due either to periods of increased magnitude or intensity of precipitation, or to decreased vegetation density and thus increased sediment transport during transitions from wet to dry intervals (Eppes and McFadden, 2008). Variable rates and types of weathering may also have been important, however. In the case of Iran, Walker and Fattahi (2011) identified a major phase of fan aggradation from c 30–10 ka, followed by deep incision. They interpret the aggradation as being a possible consequence of a relatively high rate of sediment production through the action of freeze-thaw weathering under cold and arid conditions during the Last Glacial Maximum (LGM). In Israel, by contrast, Enzel et al. (2012) argued that fan deposition in the Nahal Yael watershed at 35–20 ka was a result of accelerated sediment production on slopes associated with frequent extreme storms. In the case of the Cuyama River fan in California, Delong et al. (2011) argued that Holocene fan aggradation occurred in relatively dry periods, and incision occurred as a result of flood events.

Recent availability of optical and cosmogenic dating techniques (e.g. Dühnforth et al., 2007) has made it possible to establish the ages of multiple phases of fan aggradation and to relate these to climatic changes. One finding of this work is that some fans may be old features. This is the case, for example, of alluvial mega-fans, exceeding 100 km in length, which have developed in the interior of Oman (Blechschmidt et al., 2009). These may be of Miocene-Pliocene age. Later fans in the same area seem to have aggraded during pluvial phases when there was enhanced erosion of hillslopes and higher transport rates during strong monsoon phases. In Peru, Steffen et al. (2009, 2010) also related fan aggradation to more humid phases of the late Pleistocene. Multiple phases of fan aggradation and fluvial incision in Morocco have been related to changing base levels of a master stream, the Draa River, but superimposed on that trend were multiple phases related to climatic changes over at least the past four glacial cycles (Arboleya et al., 2008). In the Chinese Tian Shan mountains, at least four major phases of alluviation have also been identified over the last 550 kyr, with aggradation in the late phases of glacial cycles and incision during glacial to interglacial transitions (Lu, Burbank and Li, 2010). In Baja California, Mexico, eight regional scale-fan surfaces have been identified, and it has been suggested that these relate to Milankovitch scale climate changes (Spelz et al., 2008).

In the western United States, Holocene fans tend to be coarser than Pleistocene ones. Wet late Pleistocene environments, which created large lakes where now there are salty playas, produced thick soils and extensive vegetation which limited hillslope runoff and stream power (Dohrenwend, 1987). Moreover, most Holocene fan surfaces have a distinctive and ubiquitous bar and swale micro-topography, but this is absent or extremely subdued on most late Pleistocene surfaces. This may be because of greater degradation of the older fans or because the features develop better in the coarser Holocene sediments. Responses may have varied according to the climatic conditions in the different deserts of the south west, however, and Harvey et al. (1999) have noted differences in fan histories between the Mojave and the Great Basin. In the Mojave, fan aggradation phases occurred at 14–9 ka and 6–3 ka, and this can be related to phases of enhanced summer monsoon rainfall (Miller et al., 2010). In the Sonoran Desert of Arizona, intensification of El Niño-Southern Oscillation (ENSO) conditions led to fan aggradation from 3,200–2,300 years ago (Bacon et al., 2010). Harvey and Wells (1994) reported that in the Mojave, the late Pleistocene to Holocene transition saw a switch away from widespread hillslope mass movement processes (e.g. debris flows) towards fluvial processes involving episodic fan-head dissection and distal progradation.

Alluvial fans are often unstable, and this has implications for human occupancy (see also Section 6.7). The debris flows that are so important in moulding their forms can themselves be hazardous, the channels on fans are prone to avulsion (Field, 2001) and fans are also susceptible to recurring phases of cut and fill (Dühnforth et al., 2007).

5.5 Drainage Density

Drainage density (Dd), the total length of stream channels per unit area of a drainage basin, is an important property of a river network and defines the extent to which streams dissect an area. It depends on a wide range of factors, including the rainfall characteristics of a region, the vegetation cover, the infiltration capacity of the surface, the amount of available relief and also the age of a surface. Given the diversity of factors involved, it is likely that there is no simple relationship between aridity and Dd. Indeed, some desert surfaces, called badlands (see Section 5.6), may show very high levels of dissection (Figure 5.5a and b), whereas some flat, plateau surfaces may show almost none.

(a)

(b)

Figure 5.5 Intricate dissection producing high drainage densities in (a) weak mudstones in Death Valley and (b) in Tertiary sediments in Red Rock Canyon, California, USA. (ASG)

Nonetheless, various studies have examined the relationship between various measures of rainfall or effective moisture and Dd. For example, Melton (1957), working in the south-west United States, found an inverse relationship between Thornthwaite's P-E index of effective precipitation and Dd. Abrahams (1972) found that in eastern Australia the highest values of Dd occurred where mean annual rainfall (MARF) was 28 cm. Gregory and Gardiner (1975) found a more complex relationship between MARF and Dd: (1) for a MARF of <50 cm, Dd increases with precipitation, (2) for MARF between 50 and 100 cm, Dd decreases with precipitation and (3) for MARF >100 cm, Dd seems to show a slight positive trend with precipitation (Figure 5.6b). Moglen et al. (1998) found that Dd increases steadily as effective annual precipitation increased up to 35 cm, and then declines thereafter. Similarly, in New Mexico, Newman et al. (2006) found an increase in Dd up to a MARF value of 30 cm, and near constant values between 30 and 50 cm. Dd values in areas with rainfall less than 300 were below c 0.5–0.6 km/km^2, whereas in moister areas Dd values were around 1.0 km/km^2.

Carlston (1966, p. 62), recognized the differences between arid and semi-arid regions and stressed the importance of surface material types:

> A progressive increase in aridity results in a decrease in soil and vegetal cover which greatly magnifies the range of drainage densities characteristic of semi-arid regions. In such regions, where the land surface has a good infiltration capacity rainfall sinks rapidly into the dry soils . . . and runoff is virtually zero, as is drainage density. Impermeable terranes devoid of vegetal and soil cover reject the rain, runoff is briefly total and drainage density may be greatly magnified, as in the South Dakota badlands, where drainage density runs into the hundreds.

5.6 Badlands and Gully Erosion

Arid regions are notable for the development of a whole range of gully systems caused by water erosion of susceptible materials (Poesen et al., 2003). Such gullies are of intrinsic interest because of their sometimes striking morphology, but they are also major sources of sediment for dryland rivers (Poesen et al., 2002) and can be a major manifestation of desertification (Avni, 2005). Some of the most striking gully landscapes, with high drainage densities, are called badlands (Figure 5.7).

This term is probably derived from the French *mauvais terres*. As Geikie (1882, p. 223) remarked, 'This expressive name has been given to some of the strangest, and, in many respects, most repulsive scenery in the world. They are tracts of irreclaimable barrenness, blasted and left for ever lifeless and hideous.' Similar forms occur in industrial spoil, on volcanic ash, on areas devegetated by smelter fumes and in areas of anthropogenically accelerated soil erosion in non-arid areas. However, they are undoubtedly common in arid areas on appropriate lithologies, especially smectite-rich shales and mudstones, loess and volcanic ash but also on sandy conglomerates with clay matrices and weakly cemented sandstones (Bryan and Yair, 1982). Examples

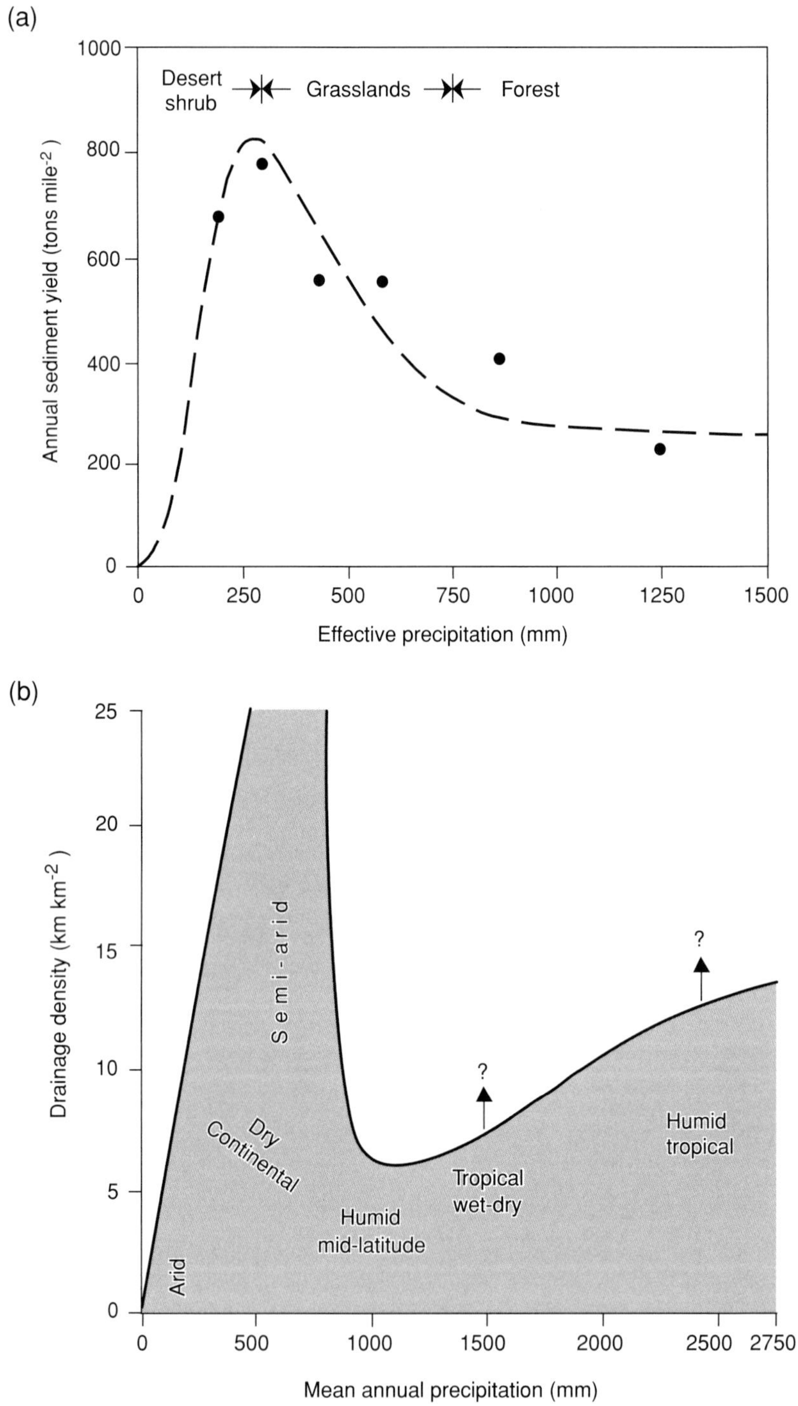

Figure 5.6 Relationships between climate and geomorphology (Modified after Langbein and Schumm, 1958 and Gregory, 1976 in Goudie, 2002, fig. 1.3). (a) Variation of sediment yield with climate as based on data from small catchments in the United States. (b) Relationship between drainage density and mean annual precipitation.

Figure 5.7 The classic badlands of South Dakota, USA. (ASG)

are recorded from the drier parts of Alberta, Canada (Kasanin-Grubin and Bryan, 2007), the western United States (Godfrey et al., 2008), the Negev (Kuhn et al., 2004; Yair et al., 2012), the loess lands of China, south-east Spain (Lázaro et al. 2008) and elsewhere in the Mediterranean Basin (Gallart et al., 2002), and the Karoo in South Africa (Boardman et al., 2003). As Howard (2009, p. 265) remarked, 'Badlands appear to offer in miniature spatial scale and a shortened spatial scale many of the processes and landforms exhibited by more normal fluvial landscapes, including a variety of slope forms, bedrock or alluvium-floored rills and washes, and flat alluvial expanses similar to large-scale pediments'. In areas with steep slopes and exposed bedrock of suitable lithology, they may have well-developed pinnacle forms, as is well displayed in Bryce Canyon, Utah.

The forms that make up badlands overlap with some of the forms encountered in arroyos and have been summarised thus by Campbell (1997, pp. 262–3).

Regardless of their origin badlands are barren and usually intricately dissected by rills and gullies. Steep-sided residuals rise above gently sloping alluvial or pediment surfaces with interviewing slope angles often reflecting the erosional resistance of different lithologies. Microrelief is complex; deep desiccation cracks, pipes, rills, knife-edge divides and vertical faces alternate abruptly with rounded forms.

Likewise, in the Henry Mountains of Utah a contrast occurs between badlands on the Mancos Shale and those in the Morrison Formation. The former tends to have nearly linear profiles with narrow, rounded divides, whereas the latter tends to have rounded, creep-dominated slopes.

Aspect appears to be an important control of slope form in some badlands (Yair et al., 2012). Churchill (1981), working in South Dakota, found that south-facing slopes were generally significantly shorter, steeper and straighter in profile than their north-facing counterparts. More intense fluvial erosion occurs on the latter, as is made evident by significantly greater densities of rills. Because they receive far less direct solar radiation than south-facing slopes, north-facing slopes retain higher moisture levels, have higher antecedent moisture levels and have a greater susceptibility to fluvial erosion and rilling. Moreover, saturation of the regolith, which increases internal pore pressures, enhances susceptibility to slope failure. Aspect may also partially control the development of biological crusts, which in turn can affect erosional processes and badland stability (Lázaro et al., 2008).

Badlands are sites of high rates or erosion, and many contribute large amounts of sediment to river basins of which they are, areally speaking, only a small part. Indeed, erosion may be so rapid that a regolith or soil cannot develop. This does not favour vegetation establishment, and in a positive feedback, the lack of vegetation enhances erosion. Gonzalez (2010, p. 37) has averred that 'badlands are landscapes where the erosive power of slopewash reaches its maximum expression on Earth'. The rates at which the inter-rill portions of badlands erode have been discussed by Clarke and Rendell (2010), whose monitoring studies in the Mediterranean lands suggest that erosion rates show a strong positive correlation with cumulative rainfall amounts. There is, however, a great range in published rates, which vary by 3–4 orders of magnitude.

It is possible that wind action plays some role in moulding badlands on susceptible materials (Geikie, 1882), and wind fluting can be detected on many shale outcrops in the badlands of South Dakota. However, badlands in general are the result of weathering-induced production of regolith, various mass movement processes (creep, slab failure and mudflows) together with fluvial rilling (Godfrey et al., 2008; Kasanin-Grubin and Bryan, 2007), sub-surface piping (Gutierrez et al., 1997; Bull and Kirkby, 2002; Kuhn et al., 2004) and splash and wash (Howard, 2009).

5.7 Arroyos

Related to badlands are various other types of gully system. In the south-western United States, many broad valleys and plains became deeply incised with valley-bottom gullies (arroyos) over a short period between 1865 and 1915, with the 1880s being especially important (Cooke and Reeves, 1976). This incision had a rapid and detrimental effect on the flat, fertile and easily irrigated valley floors, which are the most desirable sites for settlement and economic activity in a harsh environment. Some arid areas of the south-west United States were, in pre-Columbian times, 'bonanzas of human enterprise' that were terminated by arroyo incision (Bryan, 1941). The arroyos can be cut as much as 20 m into the valley floor, be more than 50 m wide and tens or even hundreds of kilometres long.

There has been a long history of debate as to the causes of incision (Bull, 1997; Elliott et al., 1999; DeLong et al., 2011; Harvey and Pederson, 2011) and an increasing appreciation of the scale and frequency of climatic changes during the Holocene (McFadden and McAuliffe, 1997) – changes that could have led to changes in channel and slope behaviour. For example, Waters and Haynes (2001) have argued that arroyos first appeared in the American Southwest after c 8,000 years ago, and that a dramatic increase in cutting and filling episodes occurred after c 4,000 years ago. They believe that this intensification could be related to a change in the frequency and strength of El Niño events.

Human actions (timber felling, overgrazing, cutting grass for hay in valley bottoms, compaction along well-travelled routes, channelling of runoff from trails and railways, disruption of valley-bottom sods by animals' feet and the invasion of grasslands by scrub) could also have caused the entrenchment. The apparent coincidence of settlement and arroyo development in the late nineteenth century tended to give credence to this viewpoint. On the other hand, study of the long-term history of the valley fills shows that there have been repeated phases of aggradation and incision, and that some of these took place before the influence of humans could have been significant. Elliott et al. (1999), for example, recognized various Holocene phases of channel incision at 700–1200 BP, 1700–2300 BP and 6500–7400 BP. Huckleberry and Duff (2008), working in New Mexico, identified two phases of incision: at AD 900–1050 and AD 1300–1400, while French et al. (2009) recognized multiple phases of channel entrenchment in New Mexico: at 4100–3700 BC, 2900–2400 BC, 2200–400 BC, before and after AD 900–1200 and in the late 1800s.

Huntington (1914) argued that valley filling could be a consequence of a climatic shift to more arid conditions. This, he believed, would cause a reduction in vegetation cover, which in turn would promote rapid removal of soil from devegetated mountain slopes during storms and would overload streams with sediment. With a return to humid conditions vegetation would be re-established, sediment yields would be reduced and entrenchment of valley fills would take place. Bryan (1928a) put forward a contradictory explanation. He argued that a slight move towards drier conditions, by depleting vegetation cover and reducing soil infiltration capacity, would produce significant increases in storm runoff which would erode valleys. Another climatic interpretation was advanced by Leopold (1951), involving a change in rainfall intensity rather than quantity. He indicated that a reduced frequency of low-intensity rains would weaken the vegetation cover, while an increased frequency of heavy rains at the same time would increase the incidence of erosion. Support for this contention comes from the work of Balling and Wells (1990) in New Mexico. They attributed early-twentieth-century arroyo trenching to a run of years with intense and erosive rainfall that succeeded a phase of drought conditions in which the protective ability of the vegetation had declined. Mann and Meltzer (2007) also recognized that in New Mexico incision occurred in the Medieval Warm Period and aggradation in the Little Ice Age, and they argued that when the North American monsoon system is strong, more frequent summer thunderstorms cause increased flooding. Wetter summers over periods of decades to centuries cause the vegetation cover to increase, which reduces sediment input from hillslopes at the same time that floods are eroding the valley fills; incision therefore occurs. That large floods have been important was put forward by Hereford (1986). He argued that erosion and entrenchment result from larger flood regimes, with streams having a large sediment transport capacity. With lower flood regimes, however, a reduction in channel width and sediment storage occurs. If there are no floods, then no alluviation of floodplains is possible. Conversely, periods of prolonged summer drought reduce forest cover, increase slope erosion and thus cause sediment to accumulate in the valleys, promoting aggradation (Graf et al., 1991).

It is also possible, as Schumm et al. (1984) pointed out, that arroyo incision may not be the result of climatic change or human influence but rather of some intrinsic natural geomorphological threshold (such as stream gradient) being crossed. Under this autogenic argument, conditions of valley-floor stability decrease slowly over time until some triggering event initiates incision of the previously ‘stable’ reach. It is also likely that different reaches of river valleys may have responded differently to environmental changes. For instance, upstream incision may have created sediment pulses that led to downstream aggradation (DeLong et al., 2011; Harvey et al., 2011; Gellis et al., 2012).

In conclusion, it is possible that arroyo incision and alluviation result from a whole range of causes (Gonzalez, 2001), both natural and anthropogenic, and that the timing of incision or aggradation will have varied from area to area and that individual arroyos will have had unique histories. As Cooke and Reeves (1976, p. 189) concluded:

> Apparently similar arroyos can be formed in different areas as a result of different combinations of initial conditions and environmental changes.

5.8 Arid Zone Floodplains

The floodplains and channel types in dryland rivers may have certain characteristics that can make them different from humid rivers. Graf (1988, p. 232) stressed the importance of their highly variable discharges:

> Because arid-region rivers are subject to wide fluctuations in discharges, their channels change configuration to accommodate the variations in input of mass and energy. They may adopt different channel configurations from one place to another or the same reach may change configuration from one time to another. These spatial and temporal changes in form occur abruptly and are not accommodated by general mathematical models of channel behavior.

Graf also makes it clear that many arid-region rivers have both meandering and braided characteristics, with a meandering channel that carries low flows nesting within a much larger braided channel that is only occupied during infrequent high discharges. In addition, the high mobility of bed materials and the general lack of cohesive bank materials result in substantial locational instability. In general, stream channels are much wider than one would normally expect to find in more humid environments. They also are notable for having remarkably subdued bed topography (Reid and Frostick, 1997, p. 215).

So then, dryland rivers have a diversity of channel types, but anabranching forms are perhaps unusually common. For example, many of the rivers of the Australian interior, from the Fitzroy in the northwest to the rivers of the Lake Eyre Basin in the centre (McMahon et al. 2008) and to the Murray-Darling system in the south, are characterized by anabranching (Tooth and Nanson, 1999) or anastomosing forms (Gibling et al., 1998). The archetypal example is the Channel Country of the north-eastern part of the Lake Eyre Basin, which comprises rivers such as the Cooper, Diamantina and Georgina. Billabongs (waterholes) are a widespread feature of the anabranching systems (Knighton and Nanson, 2000). The rivers occur in a semi-arid environment, and they have very low gradients (Bourke and Pickup, 1999). They display extreme flow variability, with tropical air from the north producing intense but erratic rainfall that brings floods. During floods, the river expands to exceptional widths, resulting in streets of water 70 km wide on the Cooper below Windorah and up to 500 km wide on the Diamantina and adjacent channels above Birdsville (Queensland) (Gibling et al., 1998). Normally, they consist of multiple channels separated by large islands and channel-train ridges. In-channel trees, such as *Eucalyptus camaldulensis*, act as substantial obstacles and influence local patterns of streamflow and sediment transport (Dunkerley, 2010), but islands and ridges can be both erosional and depositional forms.

Some of the anabranching systems may be relicts of different hydrological/climatic regimes, and this is the case with the so-called prior streams and dead rivers of the Murrumbidgee valley. Langford-Smith (1960), arguing on the basis of the statistical relationship between meander wavelength and stream discharge, believed that the large prior stream channels were associated with larger discharges under pluvial conditions. Butler (1961) disputed this, however, and suggested that the presence of coarse sediment in the area of the prior streams was indicative of a copious supply of large-calibre material from a catchment that had a limited vegetation cover, caused by aridity. Thermoluminescence dating of the prior stream deposits (Page et al., 1996) suggests that major palaeochannels developed between about 105 and 10 ka years ago (i.e. during the last full glacial cycle). There were four phases of palaeochannel activity over this period. Some appear to have been associated with moist conditions, whereas others appear to have been associated with seasonal snow melt and increased peak flows in periods flanking the LGM. The dating of the transition from the palaeochannel regime to that of modern fluvial activity varied from system to system and may have taken place at 7–10 ka in the Upper Murray and before 11–13 ka on the Goulburn River (Ogden et al., 2001). Study of the Lachlan River in south-eastern Australia has suggested that, for much of the last glacial cycle, the rivers carried between three and eleven times the discharge of present-day rivers (Kemp and Rhodes, 2010). Shifts from meandering to braided patterns are also evident in the history of some rivers of north-west India, where rivers meandered during enhanced monsoonal rainfall conditions during the period 130–120 ka (Juyal et al., 2006).

The nature of arid zone channels may be markedly affected by riparian vegetation. It is a mistake to assume that because this is generally less dense in deserts than elsewhere, it is of no consequence. In fact, phreatophytes such as tamarisk, cottonwood and willow, which draw their sustenance from groundwater at depth, can significantly influence channel geometry by increasing bank resistance to erosion, inducing deposition and increasing roughness and by taking up so much water that discharge is reduced (Dunkerley, 2010). Such vegetation can lead to significantly reduced channel width and, as a result, to increased overbank discharge (flooding) (Birken and Cooper, 2006). Thus, Graf (1978) showed that major channels of the Colorado River and the plateau country in the United States had an average width reduction of 27 per cent when tamarisk, a highly effective invasive species (Tickner et al., 2001), was established after 1930. Hadley (1961) demonstrated a similar effect on an arroyo in northern Arizona, and Graf (1981) demonstrated that in the ephemeral Gila River channel, sinuosity was increased from 1.13 to 1.23 as phreatophyte density increased in the 1950s. Tamarisk, because of its root system, is capable of resisting the hydraulic stresses of flash floods. It is also drought tolerant and has an ability to produce denser stands than native species, which gives it a competitive advantage. However, although a temporal link between channel narrowing and the spread of plants such as tamarisk, Russian olive and cottonwood has been established, care has to be taken in proposing a causal relationship. In Arizona, for example, Cadol et al. (2010) found that the channel narrowing in the Canyon de Chelly National Monument may also have occurred as a result of another factor that occurred at the same time – a diminution in sediment load resulting from a great reduction in grazing by sheep in the Navajo Reservation as a result of government decree.

Allogenic rivers that debouch from highlands onto low-lying plains – such as the Himalayan and Karakorum Rivers onto the north Indian plains – may be prone to major shifts in their courses. The Indus has shown great variability both in its character and in its route to the sea (Meadows and Meadows, 1999). Its low gradient, its uncohesive banks, its presence in a tectonically active area, its large sediment load and its propensity to flood all contribute to the instability of its wide floodplain. In late Pleistocene times it seems to have incised itself to a low glacial base level, but during the Holocene it has shifted its course repeatedly, and many ancient courses (dhoros) are evident (Flam, 1993; Shroder, 1993). Throughout the Punjab, north-west Rajasthan and Sind, there are traditions of rivers that suddenly ceased to flow, or changed their courses (Wilhelmy, 1969). Many old channels have been traced and mapped, and some serve as a basis for irrigation canals.

It is probably wise not to exaggerate the distinctiveness of arid zone river channels. As Nanson et al. (2002, pp. 32–33) wrote:

> It is debatable whether on average the channel geometries of dryland rivers are distinctly different from rivers in humid environments. Despite major variations in climate and hydrology from region to region, the hydraulic properties of running water and its influence on sediment transport hardly change, so rivers in different settings with similar sediment types still exhibit many similar geomorphic characteristics in response to changes in discharge and bank strength.

5.9 Floodouts

Associated with the anabranching and anastomosing channels of some dryland river systems is the phenomenon of floodouts. As Tooth (1999) explained (p. 220):

> In many arid and semi-arid regions of the world, headwater channels debouch into lowland plains where few tributary contributions are received. As a result of a combination of diminishing downvalley flows, an over-supply of sediment relative to the capacity for onward transport, declining gradients and sometimes aeolian, structural or hydrological obstructions to flow, many channels fail to reach the lowest point in the drainage basin and channelized flow largely, or completely, disappears.

This phenomenon, whereby channels increasingly lose definition downstream, is termed a floodout in Australia. Floodouts may also produce what are termed terminal splays – sediment lobes at their downstream end (Fisher et al., 2008); fans of fine sediment may acculuate (Nichols, 2009). The common occurrence of floodouts can be largely related to the increasing aridity of the Australian continent in late Tertiary and Quaternary times, resulting in the retraction and disintegration of formerly better integrated drainage networks. At a more local scale, they can result from such factors as aeolian and structural barriers.

Aeolian barriers, which include late Pleistocene dunefields, are a feature not only of Australia but also of the courses of the Niger River in its inland delta and of some of the rivers of the ancient Kalahari sandveld in northern Botswana and Angola (Shaw and Goudie, 2002). The rivers of the Skeleton Coast of northern Namibia also demonstrate the effects of aeolian ponding, with evidence for periods of alluvial deposition behind and within the sand sea followed by periods of flood breakouts through the dune cordon, sometimes as catastrophically large flows (Krapf et al., 2003; Svendsen et al., 2003).

5.10 Groundwater Sapping Forms

Normally, when rivers incise themselves into an upland area, they form a dendritic pattern of V-shaped valleys and intervening ridges. Some valley systems do not have this form, however. Rather, they have deep, amphitheatrical heads, flat floors, a generally 'stubby' network and a lack of well-developed tributaries. For more than 100 years, it has been recognized that this particular type of valley form and network requires a different explanation from normal fluvial networks. Generally it has been argued that they have been moulded by emerging spring water, a process known as seepage erosion or groundwater sapping. In the last thirty years, however, this supposition has taken on greater significance because of the presence of networks with a similar distinctive morphology being found on Mars (Higgins, 1982) and Titan (Tomasko et al., 2005). River networks that could perhaps have been caused by groundwater sapping seem to be particularly common in drylands and include those of the Gilf Kebir in Egypt (Peel, 1941), the Kharga Oasis in Egypt (Luo et al., 1997), the Colorado Plateau in the United States (Laity and Malin, 1985; Howard and Kochel, 1988), the *quebradas* of northern Chile (Hoke et al., 2004) and the valleys of the central Kalahari (Nash et al., 1994).

Sapping can be defined as 'the process leading to the undermining and collapse of valley head and side walls by weakening or removal of basal support as a result of enhanced weathering and erosion by concentrated flow at a site of seepage' (Laity and Malin, 1985, p. 203). It is likely to be a feature of moderately fine-grained rock types that are permeable and can transmit groundwater in substantial quantities, including sandstones, basalts, chalks and volcanic tuffs. Sapping may also be directed by the presence of joints and faults. Evidence for spring activity may be provided by the presence of freshwater carbonate accumulations (tufas) (see Section 2.25). In Table 5.2 is a list of supposedly diagnostic features of valley networks developed by groundwater seepage.

However, the sapping hypothesis has been the subject of some doubts in recent years. Some stubby networks may be an artifice of low-resolution satellite images, but more importantly, some examples that have been cited as classic manifestations of groundwater sapping have been found to show evidence of flash floods and plunge pools (Lamb et al., 2006). It is possible that palaeofloods have played a role (Lamb et al., 2008). Moreover, because seepage erosion and weathering is a slow process, it is difficult to actually observe it in operation. In addition, some of the morphological characteristics of the valleys may essentially be controlled by the horizontal disposition of cliff-forming sandstones. What is more, coarse sediment has to be evacuated

Table 5.2 *Diagnostic features of valley networks developed by groundwater sapping*

Abrupt channel initiation, possibly with amphitheatre valley headwalls
Alcove development with springs or seepage zones in headwater regions
Flat or stepped longitudinal profile
Long main valley with constant valley width
Low drainage density
Possible parallelism of tributaries
Short first-order tributaries with a paucity of downstream tributaries
Small basin area to canyon area ratio
Steep valley walls with an abrupt angle to a flat valley floor
Structurally controlled tributary asymmetry

Source: After Nash (1997, table 15.2).

from the valley for headwall retreat to occur, and this may require the sorts of velocity associated with surface runoff. Lamb et al. (2006, p. 15) concluded:

> While we know of no unambiguous case of seepage eroding an amphitheatre–headed valley in resistant rock, several examples exist of valley formation by runoff and mass wasting processes in the absence of seepage erosion. Instead of a particular hydraulic process, amphitheatre heads might instead be indicative of a substrate that, because of rock strength and fracture orientation, is relatively unstable to headwall retreat, but resistant to incision at the rim of the headwall. Amphitheatre valley heads should not be used as a diagnostic indicator of seepage erosion on Earth.

5.11 Long Profiles

One of the most intriguing features of some desert rivers is that in contrast to most 'normal' rivers, they display convex long profiles. This is true of the westward flowing ephemeral Namib rivers (such as the Kuiseb, Swakop, Omaruru and Ugab), and those of neighbouring Angola (Roberts and White, 2010). The same is also true of some of the rivers of the Indian arid zone in Gujarat, including the Mahi. Convex tracts have also been identified in the courses of some of Atacama's rivers in northern Chile (Riquelme et al., 2003). Whether these examples of convexity are because of the nature of the uplift on these tectonic margins or the fact that river discharges diminish downstream (or a combination of both) is a matter of debate. The traditional explanation is that in dryland rivers there is a diminution in flow downstream because of transmission losses (see Section 5.2), and so the ratio of sediment to flow often increases downstream, leading to aggradation and the development of a convex profile. It is also possible that there is a less clear diminution in grain size of sediment downstream in comparison with humid climate rivers. Certainly, studies of drainage basins in the High Plains of the United States (Zaprowski et al., 2005) indicate that, in tectonically stable settings, areas with higher intensity rainfall and greater mean annual precipitation have increasingly concave long profiles. A strong case has been made, however, that uplift rate histories explain many of the main characteristics of the long profiles of those rivers that drain the tectonic swells of Africa (Roberts and White, 2010).

5.12 Processes: Runoff Generation

Desert surfaces have characteristics that enable them to generate considerable runoff from quite low rainfall intensities. First, the limited vegetation cover provides little organic litter on the surface to absorb water. Second, the sparseness of vegetation means that humus levels in the soil are low, and this – combined with the minimal disturbance by plant roots and a greatly reduced soil fauna – makes the soil dense and compact in texture. Third, as there is virtually no plant cover to intercept the rainfall; rain is able to beat down on the soil surface with maximum force, and fine particles, unbound by vegetation, are redistributed by splash to lodge in pore spaces and to create a puddled soil surface of reduced permeability – a physical or non-organic crust (Ben-Hur and Lado, 2008) (see Section 2.31). There are also biological crusts which may play a similar role. Studies in various areas have shown that, where such crusted and impermeable soils exist, the infiltration rate is only a few millimetres per hour, so that a rainfall rate in excess of this is likely to produce overland flow. In semi-arid Arizona, Polyakov et al. (2010) found that 10 mm of rainfall within thirty minutes was required to initiate runoff.

Sodic soils, those with an accumulation of excess sodium, are a feature of arid and semi-arid climates (Qadir and Schubert, 2002) and have various characteristics which affect runoff and erosion. Clay-rich sediments with a high Exchangeable Sodium Percentage (ESP) tend to lose their structure when wetted and so may be eroded to form intricately dissected gully systems (Gutiérrez et al., 1997). They also tend to lose their permeability, so that their infiltration capacities are reduced.

Runoff on desert slopes is commonly believed to be characteristically Hortonian flow (Horton, 1945; Yair and Lavee, 1985); that is, runoff is ultimately produced from circumstances in which the rate of supply of rainfall is greater than the infiltration capacity of the soil or surface debris. As we have already seen, infiltration capacity will be controlled by a range of soil characteristics such as soil structure, texture, vegetation cover, biological activity, moisture content and surface condition (such as stoniness). Initially, rainfall will infiltrate into the surface. As the supply rate begins to exceed infiltration capacity, puddle or detention storage will commence. Eventually this will be exceeded, and overland flow will begin, usually as unconcentrated flow (i.e. inter-rill or sheet flow). While the Hortonian model appears to be more appropriate in arid and semi-arid areas than in many well-vegetated humid areas (Descroix et al., 2007), it may not be the only acceptable explanation of slope runoff. The Hewlettian saturation excess overland flow model, however, may have only limited applicability in arid regions.

Desert surface infiltration capacities are highly variable, and one cause of such variability is the nature of vegetation cover. This can be important at the local scale (as with the banded vegetation patterns discussed in Section 2.32) and more generally in terms of the difference between bare, grassland and shrub surfaces. The differences between grass and shrub surfaces have been the subject of an extensive literature (Ravi et al., 2009), and Parsons et al. (1996), for example, found that as a result of rainfall simulation experiments, compared to grasslands, the inter-rill portions of shrubland hillslopes had higher runoff rates, higher overland flow velocities and greater rates of erosion. In general, it can be argued that under a shrub canopy, infiltration capacities will be relatively high because the addition of organic matter and the activity of roots increase soil porosity. On the other hand, the decreased infiltration capacity in intercanopy areas may more than offset higher infiltration capacities under the shrub canopies, with the net result that runoff from shrub-dominated hillslopes may be many times greater than those dominated by grassland (Zhang et al., 2011). Furthermore, as Turnbull et al. (2010, p. 410) explain on the basis of their erosion and runoff plot studies in New Mexico:

> Over the grassland to shrubland transition, the connectivity of bare areas where runoff tends to be preferentially generated increases. Therefore, from the grass, grass-shrub, shrub-grass to shrub plots, flow lines become increasingly well connected, which increases the capacity for flow to entrain and transport sediment, leading to the greater sediment yields monitored over shrubland.

Another very important control of water behaviour on desert slopes is the nature and extent of the surface stone cover. Stones may intercept some rainfall and absorb droplet impact energy, cause surface detention of water by roughening the surface, provide a non-absorbing surface that reduces infiltration per unit area and increase the tortuosity of overland flow paths (Dunkerley, 1995). Stony soils, however, with a great deal of pore space, will have high infiltration capacities and limited runoff (Verbist et al., 2009). Foody et al. (2004) found that infiltration capacities of stone pavements in the Eastern Desert of Egypt were a mere 0.7 mm per hour, and Meadows et al. (2008) indicated that infiltration capacities were low on desert pavements, particularly as they grow older and become plugged by dust accumulation (Young et al., 2004).

Infiltration capacities may be modified by various human activities that cause soil compaction, including livestock grazing and trampling, off-road vehicular movements (Webb, 1982) and the replacement of grasslands by shrublands (Bhark and Small, 2003). The relationships between grazing pressures and soil infiltration capacities are complex. On the one hand, moderate stocking levels may increase infiltration capacities by breaking down surface biological or rainbeat crusts, while on the other hand, high stocking levels may remove all vegetation cover, cause breakdown of soil aggregates and produce severe trampling and soil compaction, thereby decreasing soil infiltration rates (du Toit et al., 2009).

5.13 Hydrophobicity

Water repellency (hydrophobicity) is a characteristic that has been reported from many arid zone soils (Doerr et al., 2000), although it is by no means restricted to such environments. It tends to be more characteristic of shrub than grass areas (Glenn and Finley, 2010). This phenomenon reduces the affinity of soils to water such that they resist wetting for periods ranging from a few seconds to hours, days or weeks. Implications include the reduced infiltration capacity of many soils, enhanced overland flow and accelerated soil erosion (Glenn and Finley, 2010). Watershed experiments in the chaparral scrub of Arizona, involving denudation by a destructive fire, indicated that whereas erosion losses before the fire were only 43 tonnes per square kilometre per year, after the fire they were between 50,000 and 150,000 tonnes per square kilometre per year. Equally, streamflow in fire-impacted years is higher than in non-fire-impacted years (Loáciga et al., 2001). Post-wildfire flooding and increased sediment yields have also been noted in the chaparral of coastal California (Warrick et al., 2012).

The causes of the marked erosion associated with chaparral burning are particularly interesting. There is normally a distinctive 'non-wettable' layer in the soils supporting chaparral. This layer, composed of soil particles coated by hydrophobic substances leached from the shrubs or their litter, is normally associated with the upper part of the soil profile (Mooney and Parsons, 1973) and builds up through time in the unburned chaparral. The high temperatures which accompany chaparral fires cause these hydrophobic substances to be distilled so that they condense on lower soil layers. This process results in a shallow layer of wettable soil overlying a non-wettable layer. Such a condition, especially on steep slopes, can result in severe surface erosion (Debano, 2000; Shakesby et al., 2000; Letey, 2001; Ravi et al., 2009). In chaparral terrain it is possible to envisage a fire-induced sediment cycle (Graf, 1988, p. 243). It starts with a fire that destroys the scrub and the root net, and changes surface-soil properties in the way already discussed. After the fire, a precipitation event of low magnitude (with a return interval of around one or two years) is sufficient to induce extensive sheet and rill erosion, which removes enough soil to retard vegetation recovery. Eventually, a larger precipitation event occurs (with a return interval of around five to ten years) and, because of limited vegetation cover, produces severe debris slides. Slowly the vegetation cover re-establishes itself, and erosion rates diminish. In due course, however, enough vegetation grows to create a new fire hazard.

5.14 Sediment Yield and Rates of Denudation

A useful measure of geomorphological activity is sediment yield per unit area over a period of time. This can be determined by such means as recording the quantity of sediment accumulation in reservoirs (Nichols, 2006) or measuring the amount of material carried by rivers. Sediment yields vary greatly in both space and time, and although runoff volume is a major controlling factor, other variables are also important, including basin area, the ratio of basin area to main channel length, the nature of the drainage net, drainage density, channel slope, vegetation cover and soil and sediment characteristics and availability.

Studies suggest that in drylands, as rainfall rises from zero, yields of sediment increase rapidly because more and more runoff becomes available to move sediment, and yet there is still plenty of bare ground susceptible to rainwash erosion. as the amount of vegetation cover increases, however, sediment yields start to decline. Thus, while in extreme deserts sediment yields are very small, in semi-arid areas they may be among the highest in the world. Moreover, many streams are so full of sediment that there are frequent occurrences of mudflows, in which solid matter may account for between 25 and 75 per cent of the flow. Such large sediment concentrations are important in the formation of alluvial fans and can lead to rapid sedimentation behind engineering structures such as dams.

Gilbert's classic studies in the Colorado Plateau on rates of denudation in arid regions (Gilbert, 1876) foresaw the subsequent conclusions of Langbein and Schumm (1958) and are worth quoting at length. He wrote (pp. 92–3):

Vegetation is intimately related to water supply. There is little or none where the annual precipitation is small, and it is profuse where the latter is great and especially where the temperature is at the same time high. In proportion as vegetation is profuse the solvent power of percolating water is increased, and, on the other hand, the ground is sheltered from the mechanical action of rains and rills. The removal of disintegrated rock is greatly impeded by the conservative power of roots and fallen leaves, and a soil is invariably preserved.

Hence the general effect of vegetation is to retard erosion and since the direct effect of great rainfall is the acceleration of erosion, it results that its direct and indirect tendencies are in opposite directions.

In arid regions of which the declivities are sufficient to give thorough drainage, the absence of vegetation is accompanied by absence of soil. When a shower falls, nearly all the water runs off from the bare rock, and the little that is absorbed is rapidly reduced by evaporation. Solution becomes a slow process for lack of a continuous supply of water, and frost accomplishes its work only when it closely follows the infrequent rain. Thus weathering is retarded, and transportation has its work so concentrated by the quick gathering of showers into floods, as to compensate, in part at least, for the smallness of the total rainfall from which they derive their power.

Hence in regions of small rainfall, surface degradation is usually limited by the slow rate of disintegration; while in regions of great rainfall it is limited by the rate of transportation. There is probably an intermediate condition, with moderate rainfall, in which a rate of disintegration greater than that of an arid climate is balanced by a more rapid transportation than consists with a very moist climate, and in which the rate of degradation attains its maximum.

A highly influential study that related sediment yields to climatic influences was that of Langbein and Schumm (1958) (Figure 5.6a). They used gauging data relating to 94 catchments (mean area 3,885 km^2) and reservoir sedimentation data from 163 catchments (mean area 78 km^2) in the United States. They found that sediment yields reached a peak of 198 mm 1,000 a^{-1} at the semi-arid/grassland precipitation boundary (mean annual precipitation of c 300 mm at a mean annual temperature of 10°C). Sediment yield minima occurred in very dry regions and in more humid ones. The explanation given for this pattern is that at effective levels of less than about 300 mm a^{-1}, there appears to be insufficient runoff to produce maximum erosion, whereas above it the erosive effects of increased runoff are more than counteracted by the presence of a more or less continuous vegetation cover. The precise position of the peak varies with mean annual temperature because rainfall effectiveness becomes less as temperatures rise.

Some more recent studies have tended to confirm the general trend of Langbein and Schumm's curve. For example, Inbar (1992) analysed sediment yield data for four areas with a Mediterranean climatic regime: Israel, Spain, California and Chile. He inferred from these data (Figure 5.8) that there is a decrease of sediment yield with increasing precipitation, with a sediment yield that peaks with 300 mm of annual rainfall in the case of Israel and Spain and with 400 mm in the case of Chile and California. Similarly, Kosmas et al. (1997), on the basis of the analysis of eight sites in southern Europe, found that erosion rates peaked at around 280–300 mm of mean annual rainfall. In the Loess Plateau of China, Xiu et al. (2011) found that sediment yields increased up to 460 mm of mean annual rainfall and then declined after that.

Yair and Enzel (1987), however, working in the Negev, were critical of Langbein and Schumm's curve as it relates to the very driest conditions (p. 133):

> The prevailing idea that the denudation rate increases from zero to about 300 mm average annual rainfall is based on the assumption that surface properties are on the whole uniform within this wide range of climatic conditions and that erosivity is mainly controlled by the amount of annual rainfall. This assumption encounters serious difficulties when applied to the northern Negev, where a converse relationship seems to fit the reality better. . . . The detailed study of the present-day processes clearly indicates that surface properties, namely, the ratio of bare bedrock outcrop to soil cover, play a predominant role in runoff generation and in erosion processes. Runoff and erosion are positively related to this ratio. Since it is higher than in adjoining semi-arid areas, sediment yields are higher in the former than in the latter areas.

Rates of sediment yield are increasingly influenced by miscellaneous human activities, including vegetation removal, trampling by domestic stock, replacement of grassland with shrubland and increases in fire frequency. Techniques for controlling such accelerated erosion are discussed in Section 6.7. Kosmas et al. (1997) list erosion rates for a range of land-use types in southern Europe and find that rates are especially high on slopes planted with vines. Although the spread or intensification of agriculture is often seen as the cause of accelerated erosion, there are examples known of erosion rates increasing as a consequence of abandonment of agricultural land. In semi-arid south-east Spain, soils that are no longer ploughed and are only slowly colonized by vegetation develop crusts that may increase runoff and erosion, something that

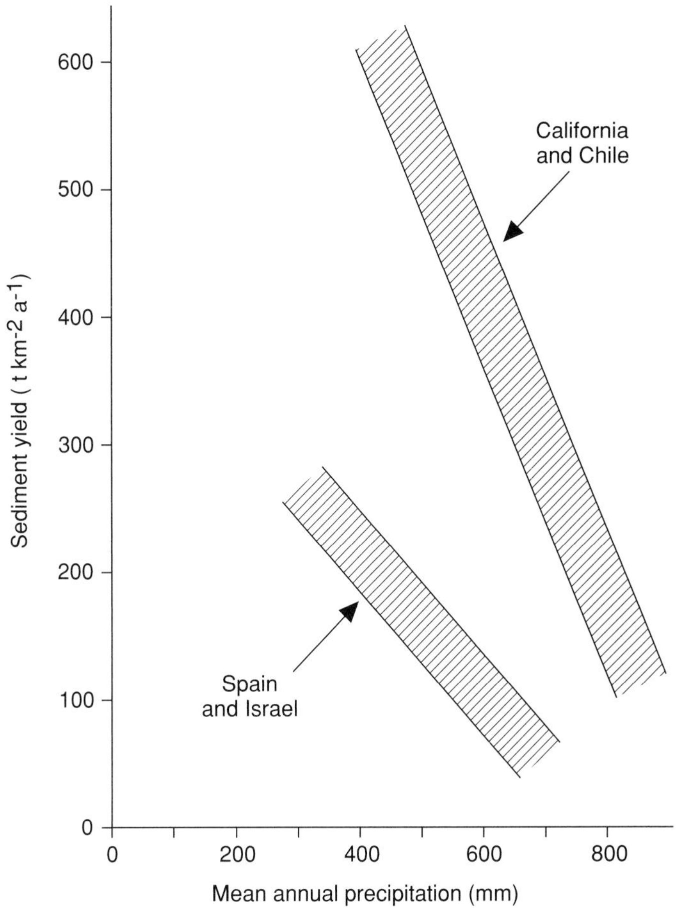

Figure 5.8 The decreasing rate of sediment yield with increasing precipitation in Spain-Israel and California-Chile. (From Inbar, 1992, fig. 6)

will be exacerbated if long-established terrace systems are not maintained (Lesschen et al., 2008). Also in south-east Spain, Sougnez et al. (2011) found on the basis of estimation of sedimentation rates behind check dams that erosion rates in areas with degraded vegetation are not invariably high, and in catchments developed in areas with thin soils formed on Betic metamorphic rocks, they were generally low – very often below 2 t ha^{-1} y^{-1}. They contrasted this with the much higher rates that have been found for catchments developed on poorly consolidated and erodible Neogene sediments.

5.15 Runoff and Erosion in the Negev and Judean Deserts

In recent decades, process studies in the Negev and Judean Deserts have greatly influenced our views on runoff generation, flooding and sediment transport. These studies have indicated how important fluvial processes are in such a desert environment. One reason for this is that rainfall intensities can be high (Schick, 1988; Greenbaum et al., 2010). In the Nahel Yael catchment, over a seventeen-year period, intensities exceeding 14 mm per hour accounted for nearly one-half of the total rain (223 mm out of 449). Of this intense rain, 37 per cent fell in intensities exceeding 2 mm per minute. Extreme flooding can follow major rainfall events, as was demonstrated by the storms that afflicted southern Israel and Jordan in 1966 (Schick, 1971).

However, a major contributing factor in runoff generation is the nature of some of the surfaces. For example, with dry conditions and a limited vegetation cover, silty soils, associated with loess deposition – which has been widespread in the Negev – rapidly become sealed under the influence of raindrop impact and so have diminishing infiltration capacities. Even on moderate slopes, silty soils generate substantial runoff (Evenari et al., 1982). Biological crusts are another runoff-generating surface type. These contain cyanobacteria which partially plug soil pore space, particularly when they swell after they are moistened by rain (Verrecchia et al., 1995). Their importance has been demonstrated by Yair et al. (2011), who found that, on a similar sandy substrate in a transect between 86 and 160 mm of mean annual rainfall, the frequency and magnitude of ruff declined with increasing rainfall. The explanation for this somewhat counter-intuitive finding was that under wetter conditions a thicker biological crust develops, which absorbs and retains a great deal of rainfall.

Bare rocks are another type of surface that generates runoff. Yair (1990) indicated that the threshold level of daily rainfall that generated runoff in rocky areas is a mere 1–3 mm. This compares with 3–5 mm for stony colluvial soils and more than 10 mm for stone-free loess soils. Because arid areas have a greater exposure of bare rock than semi-arid areas, they may generate more runoff and sediment (Yair and Enzel, 1987). Studies of experimental catchments over several decades (Schick and Lekach, 1993) indicated high sediment yields. Ephemeral streams in the region move vast quantities of sediment during each flood event. The Nahal Yael (mean annual rainfall only 31 mm) had a mean annual sediment yield of 170 tonnes per square kilometre.

Great amounts of bedload transport have been recorded in the Nahal Yatir and neighbouring areas (H. Cohen et al., 2010). Reid et al. (1998) showed that, although channels are only hydrologically active for about 2 per cent of the time (c seven days per year) and only have overbank flow for about 0.03 per cent of the time (three hours per year), the bedload flux is remarkably high. Indeed, the Nahal Yatir is about 400 times more effective at transporting coarse material than its perennial counterparts in humid zones (Laronne and Reid, 1993). Even higher bedload discharges have been reported from the Nahal Rahaf (Cohen and Laronne, 2005). The explanation for these high values (Reid and Laronne, 1995) is that the stream beds are not armoured with coarse material. The unvegetated nature of the catchment provides ample supplies of sediment of all sizes, and this, together with the rapid recession of the flash-flood hydrographs and the extended periods of no flow, discourages the development of an armour layer.

Suspended sediment concentrations may also be high in comparison with humid environments, reaching 10–40 per cent, whereas in humid rivers they tend to be lower than 1 per cent (Cohen and Laronne, 2005). In some localities, sediment yields may be increased as a result of the burrowing activities of porcupines and the excretion of soil cubes by isopods. They produce material that is easily removable by shallow flows (Yair, 1974).

Studies in the small, semi-arid Eshtemoa catchment over a fifteen-year period have enabled an assessment to be made of the relative importance of the three main modes of sediment transport: suspended, dissolved and bedload (Alexandrov et al., 2009). The average total sediment and solute load was 291 Mg km^{-2} of, which 275 Mg km^{-2} (94.5%) was suspended sediment, 15.3 Mg km^{-2} (5.3%) was bedload, and only 0.6 Mg km^{-2} (0.2%) was dissolved load. The majority of the sediment was transported by as few as eight of the seventy-four flow events of the fifteen-year record.

It needs to be remembered, however, that it would be premature to extend the remarkable data assembled in the small, steep catchments of Israel to a wide range of dryland rivers (Nanson et al., 2002, p. 33). The amounts and types of sediment carried in lower-gradient and larger catchments, such as those of the Cooper Creek system in Australia, are very different.

5.16 Sheetfloods

A sheetflood can be defined (Hogg, 1982, p. 59) as 'a sheet of unconfined flood water moving down a slope'. Sheetfloods seem to occur on alluvial fans and on pediments. Among the factors that contribute to their development are intense rainstorms, insufficient relief to confine flow, a lack of vegetation and low permeability of the surface. Other workers have preferred use of the term 'sheetflow' (e.g. Jutson, 1919).

Early recognition of the nature and possible importance of sheetfloods came from McGee (1897, p. 100), working in the Sonoran Desert of the American West. He witnessed a short-lived, impetuous flood that came roaring out of a canyon in the mountains. This advanced at 'race-horse speed' in the form of a lobate wall of water generally 20–30 cm in depth, with well-developed transverse wave and 'foaming breakers', a large sediment load and was more than a mile wide. This paper had a great impact on Davis (1938), who championed their role in removing sediment from granite domes ('sheet-flood robbing') and in leading to deposition in the bajada zone. Another detailed description of sheetfloods was made in Arizona by Rahn (1967). Some workers argued that sheetfloods might be used to explain the erosion of the gently sloping rock-cut surfaces that make up pediments. Graf (1988, p. 106), however, was rather dismissive of their role in pediment formation:

> There are at least four major considerations in determining the importance of sheetfloods in dryland fluvial processes. . . . First, sheetflow and unconfined flows of water occur across desert surfaces, but they are minor in depth and extent if pure sheets are required in the definition. Second, early descriptions of sheetfloods may be unintentionally exaggerated (Lustig, 1969), a particularly important point given the lack of photographic evidence despite the recent expansion of dryland populations and the increased opportunities for observation. Third, even if sheetfloods exist, they are not likely to form surfaces, since the sheetflood depends for its own existence on a pre-defined planar surface. Sheetfloods could only modify surfaces. . . . Finally, because observations and measurements of sheetflows are generally lacking, no conclusions are possible about their physical impact.

5.17 Debris Flows

Debris flows, which are a mass movement type that typically consists of churning, water-saturated mixtures of poorly sorted sediment and miscellaneous detritus, can be major processes of sediment transport in arid regions of high relief (Magirl et al., 2010). They play a major role in sediment transport across alluvial fans, can create rapids in the trunk valleys into which they debouch their coarse load (Griffiths et al., 2004) and cause severe channel constriction or blocking (Cerling et al., 1999). They result from high-intensity rainfall events and in the Negev seem to be associated with convective rainfall cells that produce rainfall intensities of more than 30 mm per hour for durations of at least one hour (David-Novak et al., 2004). Broadly similar rainfall intensities have also been determined for the Grand Canyon region in the western United States (Griffiths et al., 1997). In hyper-arid areas, however, this means that more than 60 per cent of the annual rainfall must be concentrated into a single hour or less to produce a debris flow, so that they are by no means common a phenomenon. The hazards posed by desert mass movements are discussed in Section 6.6.

5.18 Palaeofloods

Many arid river systems have been impacted by past flood events of considerable magnitude. During the last thirty years, geomorphologists have succeeding in dating what are termed slackwater deposits laid down by past flood events and preserved in suitable topographic situations (caves, tributaries to large gorges, etc.) (Harvey and Pederson, 2011). Using information on the height of the deposits and the slope of the channel, they have also succeeded in reconstructing past discharges and so have been able to build up a flood record for the Holocene and late Pleistocene. Such work has now been undertaken by V.R. Baker and his associates in many parts of the world, including Arizona (Partridge and Baker, 1987; O'Connor et al., 1994), Utah (Ely et al.,

1993), Australia (Baker et al., 1983; Wohl et al., 1994) and Israel (Greenbaum et al., 2000; Greenbaum et al., 2006). It enables flood flows to be provided back beyond the period of gauged flows and for rivers for which there are no gauging data. For example, O'Connor et al. (1994), working on the Grand Canyon, recorded 15 large floods over the last 4,500 years, including a flood, which occurred 1,600–1,200 years ago, that had a discharge of more than 14,000 m^3 sec^{-1}, which is a flow rate more than twice that of the largest gauged flood. Past floods have also been estimated by relating the size of transported boulders to velocity through various theoretical and empirical methods (Kehew et al., 2010). The classic methodology, developed for small, steep bedrock channels in the Colorado Front Range, has been described by Costa (1983).

A very comprehensive study of flood histories in the south-western United States is provided by Harden et al. (2010). They assembled a large number of dates and estimates of past flooding discharges for bedrock and alluvial streams. They found that there were seven episodes of increased flooding at 11,250–10,400, 8800–8350, 8230–7600, 6700–5700, 5600–4820, 4550–3320 and 2000–0 years BP. They found clear differences between the Holocene flood records of the two types of river system, however. Alluvial rivers tend not to preserve records of the very largest floods because these are usually lost through channel erosion and enlargement during high-magnitude events. Conversely, evidence for large flood events is more commonly preserved in bedrock reaches because of their more stable geometries.

The interpretation of slackwater deposits may not always be straightforward, as has been demonstrated by debates relating to the accumulation of laminated, silty valley fills such as occur along the Wadi Feiran in Sinai (Figure 5.9a) (Issar and Eckstein, 1969; Smykatz-Kloss et al. 2003), the Kuiseb and other rivers in Namibia (Smith et al. 1993; Eitel et al. 2001; Srivastava et al. 2006) (Figure 5.9b) and the Brachina Gorge in South Australia (Haberlah et al. 2010). Hypotheses for their development include flood deposition as slackwater deposits, accumulation in a lake or swamp as a result of a drainage line being dammed by dune accumulation, river end-point deposits and accumulation as a result of large aeolian silt deposition in the catchment and its translocation from slopes into the channel. Each of these mechanisms has very different environmental/climatic significance (Leopold et al. 2006).

5.19 Some Slope Forms: Hillslopes in Massive Rocks

In a review of desert hillslopes, Mabbutt (1977a, p. 39), wrote:

> On a broad scale, there is a striking lack of slopes of intermediate angle. The hills rise abruptly, island-like, from plains of gentle declivity, a characteristic which suggested to early investigators that the deserts were abandoned seabeds, or which led to erroneous views of the hills as projections through depositional surfaces in landscapes drowned in their own detritus.

(a)

(b)

Figure 5.9 Alluvial silts on desert rivers. (a) The horizontal silt deposits of the Wadi Feiran in Sinai. (b) The Homeb Silts on the Kuiseb River, Namibia. Their origin has been the subject of debate. (ASG)

As Davis pointed out in his cycle of arid erosion (see Section 1.1), isolated domes and hills – often called bornhardts or inselbergs – are common in many arid landscapes, rising up dramatically above the surrounding pediments and pediplains. Such features are by no means restricted to arid lands, but features such as Uluru (Ayers Rock) developed in arkose in central Australia, or Spitzkoppe, formed in granite in central Namibia, are iconic examples of these forms. There have been major debates as to whether these forms have been produced under more humid conditions with deep weathering and etchplanation (Büdel, 1982), have formed as a result of slope retreat or are simply the result of erosion of rocks with variable degrees of susceptibility to weathering and massiveness of jointing. With respect to lithological control, studies in Kenya and Zimbabwe have shown that granitic inselbergs develop preferentially in granites that are potassium rich and which therefore weather relatively slowly (Pye et al., 1984, 1986).

5.20 Scarp and Cuesta Forms

In areas like the Colorado Plateau in the western United States, the erosion of sequences of sedimentary rocks dipping at gradual angles creates a landscape of scarps and cuestas, often capped by a resistant layer. Scarp retreat proceeds as the weaker underlying beds (shales, soft sandstones, etc.) are eroded, undermining the more resistant caprock and leading to cliff development, rockfalls and backwasting of the scarp. Groundwater sapping processes may be important. Slickrock slopes occur on the tops and crests of sandstone cuestas. These are areas of low, generally rolling relief with typically convex to convexo-concave forms (Howard and Kochel, 1988), although more substantial relief may develop where sandstone units are thick. Slickrock slopes are said to be weathering limited in that transport processes are potentially more rapid than weathering processes. Thus loose debris tends to be removed from the slopes as fast as it is produced by weathering, so that little or no loose residuum covers the bedrock surfaces. The bedding is often emphasized by the grain-by-grain loosening of the sandstone, and exfoliation of thin sheets of sandstone is prevalent. Weathering pits produced by calcite solution may also occur.

The mesas and escarpments developed in the gently dipping sedimentary rocks of the Colorado Plateau intrigued J.W. Powell (1895, p. 32):

> After the canyons, the most remarkable features of the country are the long lines of cliffs. These are bold escarpments scores or hundreds of miles in length, – great geographic steps, often hundreds or thousands of feet in altitude, presenting steep faces of rock, often vertical. Having climbed one of these steps, you may descend by a gentle, sometimes imperceptible, slope to the foot of another. They thus present a series of terraces, the steps of which are well-defined escarpments of rock. The lateral extension of such a line of cliffs is usually very

Table 5.3 *Rates of scarp retreat in the Colorado Plateau*

Cuesta scarp and caprock	Rate of retreat (km per million years)
Chocolate Cliffs, Sinarump Conglomerate	6.7
Black Mesa, Dakota Sandstone	4.5
Black Mesa, Salt Wash Sandstone	4.5
Mesa Verde, Mesaverde Sandstone	3.2
Black Mesa, Mesaverde Sandstone	3
Cedar Mesa, Cedar Mesa Sandstone	3
Pink Cliffs, Wasatch Formation	2
Red House Cliffs, Kayenta-Wingate Sandstone	1
Grand Canyon, Kaibab Limestone	0.5

Source: Adapted from Schmidt (1989, table 1).

irregular; sharp salients are projected on the plains above, and deep recesses are cut into the terraces above.

Such early investigators were stuck by the amount of denudation that had occurred to produce the cliffs, escarpments and mesas of the region. Schmidt (1989, p. 93) asked why this may have been and, building on the work of Hunt (1956), gave an indication of the quantity of denudation involved:

The Precambrian basement of the Colorado Plateau was covered by a sequence of terrestrial and marine deposits, several kilometres in thickness, during Palaeozoic and Mesozoic times. Laramide deformations resulted in the development of anticlinal uplifts with interspersed basins (e.g. Uinta Basin, San Juan Basin) which were filled with thick Palaeogene erosional debris from the surrounding uplifts.... Today, 60 per cent of the surface of the Colorado Plateau has already been denuded of strata younger than the base of the Cretaceous, and the complete Mesozoic sequence has already been eroded from 25 per cent of the surface.... Taken overall, the mean thickness of the sedimentary cover removed lies between 2500 and 5000 m. What erosional mechanism has been capable of removing such an amount of material since the period of denudation began in a geologically brief timespan, i.e. since the beginning of the Tertiary in the anticlinal uplifts and since the end of the Eocene in the basins?

Schmidt calculated the efficiency of scarp retreat as a denudational agent, and by a variety of means (Table 5.3) estimated that the rates of scarp retreat varied between 0.5 and 6.7 km per million years. He believed that these rates were sufficient for scarp retreat, operating simultaneously and independently at different levels, to remove great proportions of the sedimentary cover of the Colorado Plateau during the Cenozoic. Another study of cliff retreat rates was undertaken along the margins of the Dead Sea by means of cosmogenic dating of rockfalls in Cambrian sandstones (Matmon et al.,

2005). Rates were between 0.4 and 0.7 m ky^{-1}, and these are of a similar order to those from the Colorado Plateau.

There are various reasons why the rates of scarp retreat may have been rapid. First, the area has a semi-arid climate, conducive to denudation. Second, the sedimentary sequence favours slope retreat because of the alternations of sandstones and shales. Third, the sandstones are relatively weak, breaking down to produce easily removed fine-grained debris and producing limited amounts of talus. They are also prone to rockfalls and landslides (Schumm and Chorley, 1966). Fourth, groundwater seepage and salt weathering seem to be highly effective in undermining slopes and causing alcoves, natural arches and box canyons to develop (Gregory, 1917; Bryan, 1928b; Laity and Malin, 1985) (see Sections 2.15 and 2.16). However, some seemingly very delicate sandstone pillars appear to be more persistent than one might think. The Navajo Twins of Utah appear to be very similar in photographs going back to 1875 (Bryan, 1927) to how they appear in modern ones.

Some scarps, such as those of the south-central Sahara, may have series of closed depressions at their feet, created by lower Tertiary weathering and subsequent aeolian denudation (Baumhauer, 2010), but one of the remarkable things about the slopes of some dryland sedimentary rock environments, such as the Al-Qawarah/Rum area of southern Jordan, is the absence of talus at the cliff bases. This is analogous to the situation in the Colorado Plateau, where Schumm and Chorley (1966, p. 22) remarked, 'In view of the evidence of the occurrence of rockfalls, the scarcity of talus is all the more striking.' They argued that one explanation for this is that the talus blocks are rapidly weathered and removed, leaving the foot of the scarp relatively free of rock accumulations.

In the Jordanian situation the same explanation is probably valid. The Disi and Ishrin Sandstones are mechanically weak, so that many blocks disintegrate when they fall from any great height. Rockfall debris appears to suffer from rapid weathering on the desert floor, and the massiveness of the rocks means that intermediate-sized material controlled by the existence of frequent bedding planes is not present. The sandstones are composed of ill-cemented grains of medium quartz sand that are non-cohesive (because of their small clay content) and of a size that is easily removed by wind or water in this sparsely vegetated environment (Goudie et al., 2002).

Talus, however, is not always absent, and many dryland slopes have talus accumulations that have been dissected to form what are called talus flatirons (Gutiérrez-Elorza et al., 2012). These relict slope accumulations have a triangular shape with their apex towards the scarp, although some may have a trapezoidal form (Gutiérrez and Sesé-Martínez, 2001). The facets are separated from the scarp. They may result from environmental changes, with erosional phases resulting from a decrease in vegetation cover, but induration of the slope materials may also be a significant factor. Suites of flatirons may occur, representing multiple cycles of accumulation and erosion. The dating of different ages of flatiron may give an indication of rates of scarp retreat (Gutiérrez et al., 1998), with rates in the Ebro Basin of northeast Spain being about 0.9 to 1.0 mm per year. Three generations of talus flatirons have also been described from the hyperarid Negev by Boroda et al. (2011).

In semi-arid regions, slope aspect affects moisture availability, and because of that it also affects vegetation cover (Sternberg and Shoshany, 2001; Badano et al., 2005; Bochet et al., 2009), soil properties (Kutiel and Lavee, 1999) and weathering (Burnett et al., 2008). These factors in turn affect slope forms and processes and can cause valley asymmetry. For example, Istanbulluoglu et al. (2008), working on sedimentary catchments in New Mexico, United States, found significantly steeper slopes in north-facing aspects and shallower slopes in south-facing ones. The explanation given was that south-facing slopes, being more xeric than those facing north, have less vegetation, and thus have different erosion rates than north-facing ones. However, working in a Morrison sandstone area in northeast Arizona, Burnett et al. (2008) found the reverse: that steeper slopes occurred on the warmer and drier south-facing slopes. The explanation proffered was that these suffered from lower rates of weathering by clay hydration than did the moister north-facing slopes and so maintained cliffs rather than more degraded slopes.

5.21 Long-Term Rates of Overall Denudation from Cosmogenic Nuclides

In recent years, cosmogenic nuclides have been used to determine long term (i.e. millions of years) rates of surface denudation (Fujioka and Chappell, 2011). This technique has now been used in a number of deserts and, as one would expect, given the different lithological, climatic and tectonic characteristics of different areas, there is a considerable range in values. Different rates have also been found according to what materials in an area are selected for analysis (Codilean et al., 2012).

Some of the rates (expressed as m of lowering per million years) are very low. Matmon et al. (2009), working on desert pavement surfaces in the Negev, found rates of only 0.25–0.3. Comparably low rates of 0.3–5.7 were found on the inselbergs of the Eyre Peninsula in Australia (Bierman and Caffee, 2002), on the pediments of southern Peru where the rates were less than 0.5 m (Hall et al., 2008), on the pediments of the Namib, where rates were 0.11–0.15 (van der Wateren and Dunai, 2001), and on the granite inselbergs and other bedrock outcrops of the same desert, where the rates were c 3.2–5 (Cockburn et al., 1999; Bierman and Caffee, 2001). The rates in some other areas, however – especially those with active tectonics and weaker rocks – are much higher, with, for example, values in Tibet of 4.0–24.0 (Kong et al., 2007), in the Nahal Yael watershed of Israel of 26–29 (Clapp et al., 2000), in Yuma Wash, Arizona of 30 m (Clapp et al., 2002), in the drier parts of the San Bernadino Mountains in California of 52 (Binnie et al., 2010) and in the Rio Puerco basin of New Mexico of 100 (Bierman et al., 2005). In northern Chile, Owen et al. (2011) suggested that bedrock erosion rates increased with precipitation following a power law, from c 1 m in the hyperarid regions, to c 40 m in the semi-arid regions. To a certain extent this is corroborated by Karátson et al. (2012), who, using morphometric techniques on Quaternary stratovolcanoes in the Central Andes, found that rates of erosion were lower in the hyperarid Puna than in the relatively moister cordillera of southern Peru.

6

Applied Geomorphology in Deserts: Hazards, Resources and the Future

6.1 Introduction

The development of arid lands requires a knowledge of geomorphological conditions, and this has become increasingly true with respect to the exploitation or rehabilitation of water resources (see, for example, Ghayoumian et al., 2007), the selection of routes and sites for new engineering structures such as pipelines (Fookes et al., 2001) and the spread of urbanisation (Cooke et al., 1982). A general review of the engineering problems of hot deserts is provided by Walker (2012). It is inevitable that such work will become increasingly important in the future if large geoengineering schemes to modify global climate and generate new types of energy are instituted. An example of geomorphology's application to a major engineering project in an arid area is provided by work that has been done on site conditions for the proposed nuclear repository at Yucca Mountain in Nevada (Table 6.1).

Drylands are in many ways very different from most other environments, and have a range of geomorphological hazards (Table 6.2) that impact on the welfare of their inhabitants (Cooke, 1984) and the success – or otherwise – of engineering schemes. Some of these hazards are components of land degradation in drylands – a process often referred to as 'desertification' (Ravi et al., 2010). Included here are the damage to desert surfaces caused by the construction of seismic lines and oil and water pipelines (Kröpelin, 2002). Some hazards may be largely apocryphal, as is the case with the supposed capability of dune sand to engulf the unwary.

In this chapter we examine the various major hazard types, look at some of the techniques that are available for mitigation and also consider ways in which the incidence of hazards may change under conditions of global warming. Geomorphological surveys and maps are also vital in locating appropriate materials for the construction industry, and this was, for example, the motive behind the ground-breaking Surface Materials Resources Survey of Bahrain (Doornkamp et al. 1980). Consideration is also given to the role of geomorphologists in assessing the role of

Table 6.1 *Selected geomorphological studies at Yucca Mountain, Nevada, USA*

Study	Topic
Bryan et al. (2009)	Rates of feldspar dissolution
Coe et al. (1997)	Debris flows
Keefer et al. (2004)	Quaternary faults and palaeoseismology using cosmogenic nuclides
Pelletier et al. (2008)	Channel dispersal of contaminants
Quade et al. (1995)	Spring deposits as groundwater level indicators
Sharpe (2006)	Past climates and infiltration rates
Stirling et al. (2010)	Unstable rock surfaces using varnish micro-lamination technique
Stüwe et al. (2009)	Rates of denudation
Valentine et al. (2007)	Nature of volcanic activity
Wells et al. (1990)	Dating of volcanic activity
Whitney and Harrington (1993)	Stability of slope deposits (colluvium)

Table 6.2 *Examples of geomorphological hazards in deserts*

Aeolian
Dune movement
Dust and sand storms
Soil erosion
Wind abrasion
Fluvial
Arroyo trenching
Avulsion of channels
Clear water erosion below dams
Floods
Piping
Sheetfloods
Siltation behind dams
Soil erosion
Weathering and Mass Movements
Debris flows
Landslides
Salt attack
Salt heave
Surface
Collapsible soils
Ground fissuring
Hydrocompaction
Salt crusts over saturated ground giving 'quicksand' conditions
Sinkholes in evaporites and carbonates
Subsidence
Miscellaneous
Coast erosion promoted by sediment starvation
Lake expansion and shrinkage
Storm surges across coastal sabkhas

Figure 6.1 A dust storm on the Emirates Highway near Dubai, United Arab Emirates. (ASG)

geomorphological changes in prehistory, and contributing to landscape conservation and military planning. Finally, geomorphologists help us to understand the development of arid landforms and processes elsewhere in the planetary system.

6.2 Hazards: Dust Storms

As we saw in Section 3.4, dust storms (Figure 6.1) are of frequent occurrence in many drylands, and globally their occurrence may have doubled in the twentieth century (Mahowald et al., 2010). They have a range of impacts both on the environment and on humans (Table 6.3). Much of the current interest in dust storms relates to their possible role in the Earth System (see Goudie and Middleton, 2006, table 1:1). Dust loadings may affect air temperatures through the absorption and scattering of solar radiation (Durant et al., 2009); modify cloud formation (Toon, 2003) and convectional activity; influence sulphur dioxide levels in the atmosphere, either by physical absorption or by heterogeneous reactions (Adams et al., 2005); and influence marine primary productivity and thus atmospheric carbon dioxide levels (Ridgwell, 2003). Wong et al. (2008) has linked the strength of African dust outbreaks in the Saharan air layer to changes in hurricane intensity in the North Atlantic region (see also Evan et al., 2006). There are still huge uncertainties about the relationship between atmospheric mineral dust aerosol levels and radiative forcing, and an increasing number of observational and modelling studies are investigating this issue (see, for example, Zhu et al., 2007).

Table 6.3 *Some environmental consequences and hazards to human populations caused by dust storms*

Environmental
Algal blooms
Butterfly transport
Calcrete development
Case hardening of rock
Climatic change
Clouds
Coral reef deterioration
Desert varnish formation
Easterly wave intensification
Glacier mass budget alteration
Loess formation
Ocean productivity
Ocean sedimentation
Plant nutrient gain
Playa (pan) formation and relief inversion
Radiative forcing
Rainfall acidity/alkalinity
Rock polish
Salt deposition and groundwater salinisation
Sediment input to streams
Silcrete development
Soil erosion
Soil nutrient gain
Stone pavement formation
Terra rossa formation
Tropospheric ozone
Ventifact sculpture
Human related
Air pollution
Animal madness
Animal suffocation
Asthma incidence
Car-ignition failure
Closing of business
DDT transport
Disease transmission (human)
Disease transmission (plants)
Drinking-water contamination
Electrical-insulator failure
Machinery problems
Microwave propagation
Radioactive dust transport
Radio communication problems
Rainfall acid neutralisation
Reduction of property values
Reduction of solar power potential
Respiratory problems and eye infections
Transport disruption
Warfare disruption

It remains a major research priority, not least because of the possibility that dust aerosols could be an accelerant of aridity trends.

The possible role that dust may have played in Pleistocene climatic changes is another area of active research. It is possible that the observed high dust loadings during glacial maxima may, through their effect on radiation budgets, have accentuated the degree of cooling (Calov et al., 2005). It is possible, however, that high quantities of dust deposition on ice surfaces would have had a negative feedback effect by lowering snow albedo and amplifying snowmelt. Indeed, Krinner et al. (2006) have used simulations which indicate that strong Asian dust emissions may have prevented the establishment of permanent snow cover in northern Asia and may even have played a role in determining the position and extent of the last great ice sheets. Bar-Or et al. (2008) believe that, through the albedo effect, dust may have had a significant impact on the speed of ice-sheet retreat. A study of how dust causes snow cover to decrease is provided by Painter et al. (2007).

Another important area of research has been to identify the role of dust in biogeochemical cycling and soil formation (Yang et al., 2008). The global dust cycle plays a major role in the delivery of iron to the oceans (Mahowald et al., 2005), together with nutrients such as phosphorus (Mahowald et al., 2008; Pulido-Villena et al., 2008) and nitrates (Chen and Chen, 2008). Dust fertilisation could be an important control of the productivity of marine phytoplankton Moreover, there are numerous studies showing that dust transport can affect geochemical conditions at long distances from dust sources. For example, Saharan dust has influenced the nature of soils in the Canary Islands (Menéndez et al., 2007; Muhs et al., 2010b Suchodoletz et al., 2011), Mallorca (Muhs et al., 2010a), the mountains of Cameroon (Dia et al., 2005) and more remarkably in Barbados, the Bahamas and Florida (Muhs et al., 2007), Jamaica (Muhs and Budahn, 2009) and the Andes (Boy and Wilcke, 2008; Fabian et al., 2009). Dust storms also remove materials from the soil, and wind erosion can be seen in terms of a whole range of on-site and off-site effects, and these are summarized in Table 6.4. Downwind from playa source areas, salt levels may build up with deleterious consequences for soil quality (Quick and Chadwick, 2011).

Dust storms affect humans in a variety of ways. One of these is health (De Longueville et al., 2010). Dust storms cause transport accidents for both civilians and the military and cause the closure of airports. Perhaps more important, dust emissions from dried lake basins (e.g. the Aral) introduce fine particles, salts and chemicals (including herbicides) into the atmosphere, with a suite of health impacts, including not only respiratory complaints but also other serious illnesses (Small et al., 2001). Dust may also contain dangerous arsenic concentrations (Soukup et al., 2012). Dust storms can lead to particulate levels that exceed internationally recommended norms (Ozer et al., 2006; Chu et al., 2008) and transport allergens, including bacteria and fungi (Kellogg and Griffin, 2006). The annual meningococcal meningitis

Table 6.4 *Some on-site and off-site effects of wind erosion*

On-site effects	Off-site effects
Soil degradation	**Short-term effects**
1. Fine material may be removed by sorting, leaving a coarse lag.	1. Reduced visibility, affecting traffic safety
2. Evacuation of organic matter	2. Deposition of sediment on roads in ditches, hedges, etc.
3. Evacuation of soil nutrients	3. Deposition of dust in houses, on cars, washing, etc.
4. Degrading water economy in the topsoil	4. Penetration of dust in machinery
5. Degrading soil structure	5. Deposition of dust on agricultural and industrial crops, ruining their quality
6. Stimulated acidification of the topsoil	
Abrasion damage	**Long-term effects**
1. Direct abrasion of crop tissue, resulting in lower yields and lower quality	1. Penetration of dust and its constituents in the lungs, causing lung diseases and other respiratory problems
2. Infection of crops due to the penetration of pathogens	2. Absorption of airborne particulates by plants and animals, leading to a general poisoning of the food chain
3. Stimulated dust emission due to sandblasting of the surface layer	3. Deposition of heavy metals and other eroded chemical substances infecting the soil
Other damage	4. Contamination of surface and groundwater via deposition of airborne particles
1. Infection, with pathogens or soil constituents, of adjacent uncontaminated fields and crops	5. Increased eutrophication of surface and groundwater
2. Accumulation of low-quality windblown deposits on fields	6. Infection of remote uncontaminated areas, transforming these into new potential sources
3. Building of sand accumulations at field borders, covering of drainage ditches	
4. Burial of plants	
5. Loss of seeds and seedlings	

Source: From Goossens (2003, table 1).

outbreak in the Sahel of Africa (Sultan et al., 2005) and coccidioidomycosis (valley fever) outbreaks in the south-west United States (Zender and Talamantes, 2006) have been related to dust storm activity. In China and Taiwan, dust events have been seen as a risk factor for respiratory and cardiovascular complaints and for conjunctivitis (Yang, 2006). Dust storms may also transport radionuclides produced by the testing of nuclear bombs (Masson et al., 2010).

Dust storms can also be a menace in terms of other human activities, particularly with regard to communications. For example, the great Sydney dust storm of 23 September, 2009 – ‘Red Dawn’ – largely derived from the Lake Eyre region, caused chaos over large parts of eastern Australia, led to the closure of the main international airport (Li et al., 2010; Box et al., 2010) and caused air quality standards to be massively exceeded (Leys et al., 2011).

6.3 Dust Control

Introduction

An array of techniques have been used for wind erosion and dust storm control, most of them developed to protect cultivated fields from soil loss (Bennett, 1938; Middleton, 1990; Riksen et al., 2003a; Sterk, 2003; Nordstrom and Hotta, 2004; Ravi et al., 2011). These techniques are normally classified into three categories: (1) crop management practices, (2) mechanical tillage operations and (3) vegetative barriers. These all aim to decrease wind speed at the soil surface by increasing surface roughness and/or increasing the threshold velocity that is required to initiate particle movement. The numerous crop management practices, also commonly referred to as agronomic measures, can influence both the detachment and transport phases of soil particle movement, particularly when combined with good soil management. Mechanical methods, by contrast, effectively do little to prevent soil detachment but tend to be more effective in preventing soil transport (Morgan, 1995).

Agronomic Measures

Agronomic measures use living vegetation or the residues from harvested crops to protect the soil and to absorb the wind's shear stress. The maintenance of a sufficient vegetative cover is the 'cardinal rule' for controlling wind erosion (Skidmore, 1986). Planting strips of vegetation perpendicular to the erosive wind direction is one option. Tree planting can be rendered more successful by means of irrigation, composting and inoculation of trees with plant growth–promoting bacteria and arbuscular mycorrhizal fungi (Bashan et al., 2012).

The maintenance of a crop residue or mulch on cropland is recognized as an efficient method for reducing wind erosion losses. The wise management of crop residues is especially valuable in poor countries; in the Sahel of West Africa, millet mulches of around 2 tonnes per hectare have proved to be highly effective (Bielders et al., 2001). In an experiment to determine the loss of topsoil prevented by a millet mulch in Niger, Michels et al. (1995) found a relative difference in surface elevation of 33 mm after just one year between bare millet plots and those spread with 2,000 kg ha^{-1} of mulch as a result of wind erosion and sediment deposition. Also working in the Sahel, Toure et al. (2011) found that a crop residue cover of about 2 per cent was the minimum required to reduce wind erosion. Dung, which is widely used in subsistence agriculture because of its fertilising properties, also provides effective protection to the soil against particle creep and saltation – initiators of suspension – even at a very low level of cover (de Rouw and Rajot, 2004). Rock fragments are another widely accepted stabiliser. Pebble and gravel mulches have been used by farmers in northwest China for more than 300 years to dampen down soil erosion and to trap dust carried by the wind (Li et al., 2001). The accumulation of dust may supply valuable additional nutrients to gravel-mulched fields (Li and Liu, 2003). In some countries, sandy soils can be stabilised by the addition of clay. This process is often called marling, and it reduces erosion risk by increasing aggregate stability. A range of synthetic materials have also been evaluated for their applicability to wind erosion control (e.g. Armbrust and Dickerson, 1971; Armbrust and Lyles, 1975).

Soil Management

Soil management techniques involve preparing the soil to promote good vegetative growth and to improve soil structure in order to increase resistance to erosion. The application of organic matter is a form of soil management that can decrease soil erodibility as well as enhance its fertility, but most soil management methods for erosion control are concerned with different forms of tillage.

Excessive tillage, particularly of light-textured soils, breaks soil clods, reduces surface roughness and exposes soil to wind action. To overcome this destruction of structure, tillage operations must be restricted. The practice of no-tillage agriculture, in which drilling is carried out directly into the stubble of the previous crop (e.g. Phillips et al., 1980) reduces soil and moisture losses, enhances soil organic matter content and maintains a good soil structure. The effects of various forms of conservation tillage on erosion rates, soil conditions, and crop yields has been the subject of many studies (see, for example, López et al., 1998; Merrill et al., 1999), and the results show the success of the system to be highly soil specific and also to depend on how well weeds, pests, and diseases are controlled (Morgan, 1995). Reduced tillage produces a smaller wind-erodible fraction at the soil surface and a greater percentage of soil cover with crop residues and clods, resulting in lower values of vertical dust flux and of small particles (PM10s) in the air (Sharratt et al., 2010).

Mechanical Methods

Mechanical approaches involve the manipulation of the ground surface to control wind flow, including through the creation of barriers such as fences, windbreaks and shelterbelts. The planting of shelterbelts has a long history, and in the United States, for example, the government established a tree nursery in the Nebraska Sand Hills in 1902 to pioneer the mass production of conifer seedlings for farm protection on the Great Plains, and the Forest Service conducted experiments on the most effective form of planting (Gardner, 2009).

Barriers to wind flow decrease surface shear stress in their lee and act as traps to moving particles, although barriers also create turbulence in their lee which can reduce their effective protection. The most efficient barrier is semi-permeable because, although its velocity reduction is less than for an impermeable fence, the amount of eddies and turbulence in its lee are reduced (Cooke et al., 1982). Windbreaks and shelterbelts need to be designed to optimize the interaction between height, density, porosity, shape and width of the plant barrier (Cornelis and Gabriels, 2005). A barrier oriented perpendicular to winds predominantly from a single direction will decrease wind erosion forces by more than 50 per cent from the barrier leeward to twenty times its height, the decrease being greater at shorter distances from the barrier (Skidmore, 1986). In situations where erosive winds come from several directions, grid or herringbone layouts provide better all-round protection. Most barriers reduce the amount of space that would otherwise be used for crops, however, and perennial barriers grow slowly, can be difficult to establish and compete with crops for water and plant nutrients (Dickerson et al., 1976; Frank et al., 1977; Lyles et al., 1983). In China there was a very ambitious shelterbelt programme, but its overall effectiveness in controlling dust emissions is not at all clear (Wang et al., 2010).

The ploughing of ridges is a common anti-erosion measure that roughens the soil surface and thus reduces the average wind velocity for some distance above the ground. Ridges also trap entrained particles on their leeward sides (Chepil and Milne, 1939). Tillage to produce ridges across the path of the erosive wind is usually carried out by chisel and is successfully used temporarily to control wind erosion in an emergency (Woodruff et al., 1957). In poorer farming regions such as the Sahel, however, where mechanical measures depend on animal traction, the technique is not so widely used and, because of sandy soils, ridges and furrows are short-lived, it is often broken down during rainstorms (Bielders et al., 2000).

Miscellaneous Methods to Reduce Dust Emissions

Fugitive dust emissions require the use of suppression techniques, which include the application of water by means of trucks, hoses and/or sprinklers prior to conducting any activities that might disturb the soil surface. Such short-term control techniques may be complemented by the cessation of activities at times of high wind velocity. Surfaces can be stabilised for longer periods by paving dirt tracks and roads or by applying dust-suppressant chemicals. The stabilisation of desiccated lakebeds is a particularly important issue with respect to locations such as Owens Lake in California or the Aral Sea in central Asia. In an ideal world, streamflows that are currently being diverted would be returned to the basin. Given that this is unlikely to be possible, other techniques have been experimented with in the Owens Valley (Gill and Cahill, 1992), including sand fences to catch coarse particles, chemical surfactants, the spreading of gravel, mechanical compaction, sprinkler irrigation and re-vegetation. It is also important to recognize surface materials that may be especially prone to wind erosion if they are disturbed, as for instance by off-road driving, and to restrict access to them.

Figure 6.2 The main railway line between Walvis Bay and Swakopmund in Namibia had to be realigned because of sand encroachment. (ASG)

For example, to this end, Goossens and Buck (2009) examined soil properties in Nevada using a Portable In Situ Wind Erosion Laboratory. Indeed, geomorphologists have now made extensive use of mobile wind tunnels.

6.4 Dune Migration and Encroachment

Introduction

As Cooke et al. (1993, p. 339) remarked, 'Mobility is a very striking property of dunes, matched only by their sound-production. Sound and movement ostensibly bring dunes closer to life than anything else in the inorganic world. Movement is inexorable and can be exasperating'. The migration of dunes can lead to abandonment of settlements, the overwhelming of agricultural land, the infilling of canals and the blocking of railway lines (Figures 6.2 and 6.3), runways and roads (see, for example, Al-Harthi, 2002; Han et al., 2003; Dong et al., 2004b; Mainguet et al., 2008). The Tarim Desert Highway has suffered severe damage along more than 64 per cent of its length (Lei et al., 2008), and extensive shelterbelts have been planted. The same is true of the 'Road of Hope' linking Nouakchott and eastern Mauritania. Begun in the early 1970s, by 1991 the Nouakchott-Boutilimit section was sanded over for more than 60 per cent of its length (Jensen and Hajej, 2001). Although movement is a natural and normal part of dune development, human pressures, such as trampling, burning and deforestation,

Figure 6.3 Sand encroaching on mining buildings at Kolmanskop, southern Namibia. (ASG)

can make dunes less stable. For example, the lowering of groundwater produced by water abstraction can lead to the death of phreatophytes, which can in turn lead to aeolian destabilisation (Laity, 2003). Equally, the removal of such anthropogenic pressures can enable recovery to occur (Seifan, 2009).

Methods of Study

Many methods have been developed to establish dunefield activity and rates of dune migration. For example, Chinese workers have successfully used ancient archival data to identify phases of dune encroachment. Topographic maps and air photographs have also been employed to establish dune movement histories during the last ninety or so years (Bailey and Bristow, 2004; Forman et al., 2008). GPS enables rapid and accurate remapping of dunes that have been identified on old maps and air photographs (see, for example, Stokes et al., 1999). Increasing use is being made of remotely sensed imagery to monitor dune migration (Yao et al., 2007) and dune activity (Seifan, 2009). Dune movement can also be monitored on the ground using stakes, markers and pins (Bristow and Lancaster, 2004; Hugenholtz et al., 2008), as has been done for some decades in the Namib by Livingstone (1989). Radiocarbon dates on buried trees can also be used (Wiles et al., 2003). In addition, optical dating can be employed to establish rates of dune accumulation and movement (Stokes and

Bray, 2004), and this can be used in tandem with ground-penetrating radar (Bristow et al., 2005; Bristow et al., 2007). Data from such sources can be incorporated into Geographical Information System (GIS) databases (Marín et al., 2005; Mitasova et al., 2005).

Rates of Dune Movement

Barchans, which are described in detail in Section 4.11, are one of the most dangerous of dune types because of their ability – particularly if small – to move fast. Many travel at rates of some tens of metres per year, and small barchans move more quickly than large ones. Cooke et al. (1993, table 23.24) show a large number of plots of dune height against rates of movement. As Thomas (1992) explained, 'Size dependency is not surprising as the whole dune migrates forward by a "rolling" process whereby sediment exposed at the foot of the stoss slope is transported up to the dune crest to be deposited on the slipface.... The larger the dune, therefore, the more sediment that has to be moved per unit of forward movement.' There is some debate as to whether the form of the relationship between dune size and rate of movement is linear or exponential. Rates of movement will also depend on wind energy conditions, and Lancaster (1989a, p. 94) has commented on the very high rates of barchan movement in the southern Namib and attributed this to the extremely vigorous wind energy conditions that exist there. Equally, in his long-term study of rates of barchan movement in the Western Desert in Egypt, Embabi (1986–87) found that dunes at Kharga between 1930 and 1961 had moved on average some 280 m, while those at Dakhla had averaged some 170 m. The explanation he offered was in terms of differences in sand drift potential between the two areas, with the Kharga figure being 245 vector units compared with 73 at Dakhla. Indeed, in the Libyan Desert we have some of the longest records of barchan movement rates that are known from any arid area. Haynes (1989) re-surveyed one of Bagnold's barchans and calculated a mean rate of advance for a 16.6 m high dune of 7.5 m per year. Using Differential (Kinematic) Global Positioning System surveying techniques, Stokes et al. (1999) obtained rates of movement from Kharga for the period 1930–97, which varied between 3.6 and 20.3 m per year (Figure 6.4). The overall mean for the period 1930–61 was 5.8 m per year, and for 1961–97 it was 9.3 m per year. Studies such as these built on the pioneer studies of Beadnell at Kharga (Beadnell, 1910). Over a two-year period, he found that the annual rates of movement of five barchans varied between 10.2 and 20.6 m (with an average rate of 15–6 m per year).

Data on the movement of linear dunes are sparser and less easy to interpret. In general, however, they are less threatening than barchans and appear to undergo relatively little movement of the dune form as a whole (Thomas, 1992, p. 35). Linear dunes, may, however, undergo some lateral movement, as GPS and optical dating studies in the Namib have shown (Bristow et al., 2007). Furthermore, the amount of

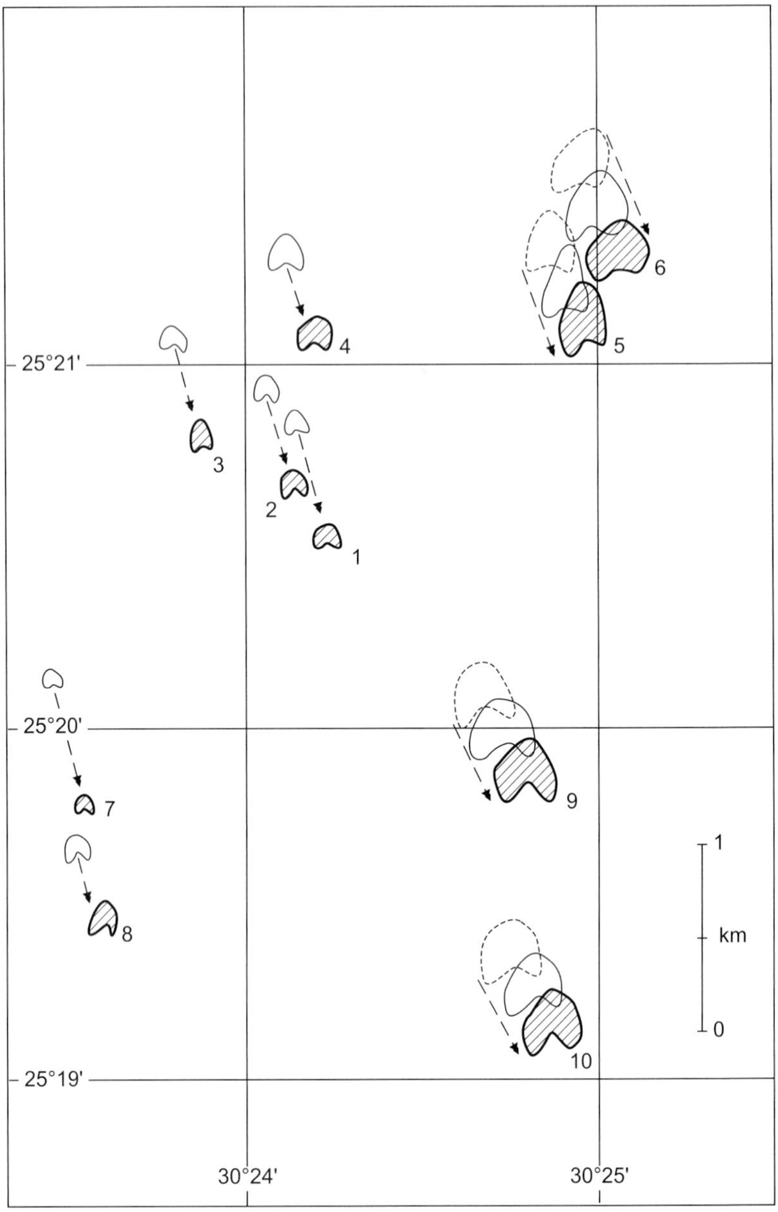

(a)

Figure 6.4 Dunes in the Kharga Depression, Egypt. (a) Barchan dune positions over time. Dotted outline is 1930, single unfilled outline is 1961 and filled solid outline is 1997. (Modified from Stokes et al., 1999). (b) Road realignments caused by barchan encroachment from the north. Scale bar 0.5 km. (©Google Earth 2012, ©Cnes/Spot Image 2012)

(b)

Figure 6.4 (*continued*)

movement that occurs on linear dunes varies from decade to decade in response to changes in vegetation cover and wind strength (Bullard et al., 1996).

With regard to parabolic dunes, in Saskatchewan, Canada, Hugenholtz et al. (2008) found that over a sixty-year period, parabolic dunes migrated at average rates of around 3.3–3.5 m per year. The migration of some parabolic dunes in Colorado has shown temporal variability, with rates in dry years (30 m per year) being about six times those in wet years (Márin et al., 2005). Indeed, the occurrence of mega-droughts is very important in determining the degree of dune activity in the western United States (Hanson et al., 2009). Data on rates of movement for transgressive transverse ridges (sometimes called precipitation ridges) are provided by Hesp and Thom (1990, p. 261), who suggest that the rates of advance of their slip faces may range from almost stationary to as much as 10–20 m per year.

Control

Techniques used to control drifting sand include promotion of deposition of drifting sand (upwind of the problem area) by such devices as ditches, barriers and fences and vegetation belts; enhancement of the transportation of sand by means of aerodynamic streamlining and surface treatments; reduction of the sand supply by surface treatments (e.g. water spraying, chemical stabilisers, mulches), fences and vegetation strips (Pye and Tsoar, 1990, pp. 303–306; Watson, 1990; Jensen and Hajej, 2001); and deflection of moving sand by fences, barriers and tree belts.

Figure 6.5 Checkerboard palm frond dune-control structures at Erfoud, Morocco. (ASG)

For the control of moving dunes, the main techniques that are available are dune removal by mechanical excavation and transportation to a new location; the dissipation of a mobile dune by disrupting its aerodynamic profile by means of reshaping, trenching or surface treatment; and dune immobilization by surface strips, fences and so forth (Figure 6.5).

Experience suggests that often these techniques are not particularly or entirely successful. Frequently, the best solution is to site and design engineering structures to allow free movement of sand across them. Alternatively, by mapping different dune types and knowing their direction and rate of movement, structures can be located out of harm's way. A good axiom is that avoidance may be better than defence.

Different stabilisation techniques have been evaluated in recent years. For example, Zhang et al. (2004) found that the best means of stabilising moving dunes in Inner Mongolia, China, were wheat-straw checkerboards and the planting of *Artemisia halodendron*. This finding was confirmed by a study in the Kerqin Sandy Land of northern China (Y. Li et al., 2009). Along a highway in the Taklamakan, checkerboards, reed fences and nylon nets were found to be effective (Dong et al., 2004). In northwest Nigeria, Raji et al. (2004) found that shelterbelts were the most effective technique and were superior to mechanical fencing. Success has also been claimed for chemical stabilisers (Z. Han et al., 2007) and geotextiles (Escalente and Pimentel, 2008), but

Table 6.5 *Examples of archaeological and architectural heritage and modern cities in arid areas being damaged by salt attack*

Location	Source
Ancient	
Temples of Karnak, Egypt	Bromblet (1993)
Temples of Luxor and Thebes, Egypt	Smith (1986)
Giza Sphinx, Egypt	Hawass (1993)
Temples of Kharga Oasis, Egypt	Salman et al. (2010)
Mohenjo Daro, Pakistan	Goudie (1977); Fodde (2007)
Islamic buildings, Uzbekistan	Cooke (1994)
Petra, Jordan	Albouy et al. (1993); Fitzner and Heinrichs (1991, 1994)
Pueblo buildings, USA	Brown et al. (1979)
Jiaohe ruins, northwest China	Shao et al. (2012)
Modern	
Bahrain	Doornkamp et al. (1980)
Suez City, Egypt	Cooke et al. (1982)
Saharan roads, runways, etc.	Horta (1985)
Los Angeles, USA	Robinson (1995)

some devices can be hugely expensive (e.g. chemical fixers) (Dong et al., 2004), while others, such as checkerboards, are much cheaper.

6.5 The Salt Weathering Hazard

As we saw in Section 2.6, salt weathering is a potent process in drylands. Some of the world's great cultural treasures are afflicted by salt weathering (as extensively reviewed by Goudie and Viles, 1997) (Table 6.5). Many important World Heritage Sites are recorded to have problems of salt weathering. Mohenjo-Daro in Pakistan (Figure 6.6), for example, is known to be affected by increasing groundwater levels and salinity, with sodium sulphate producing particularly aggressive deterioration (Lohuizen de Leeuw, 1973; Fodde, 2007). Similarly, in Uzbekistan the ancient towns of Kiva, Bukhara and Samarkand (all World Heritage Sites) have suffered from irrigation-induced groundwater rise and salinisation, causing deterioration to the lower courses of the buildings (Akiner et al., 1992). The same applies to some of the new cities of the Middle East, including those of Bahrain, Egypt and the United Arab Emirates. The problem is particularly serious where groundwater levels are high and the upward movement of salts takes place into buildings and their foundations through capillary rise – a process termed the wick effect. It is not only buildings that are affected; bridges, roads and runways have also been recorded as suffering serious salt problems in Australia, southern Africa, Algerian Sahara, India and the United States (see Januszke and Booth, 1984; Horta, 1985).

Figure 6.6 The Harappan City of Mohenjo-Daro, Sind, Pakistan. The bricks are suffering from severe attack by sodium sulphate and other salts. (ASG)

The Sphinx and Other Ancient Egyptian Archaeological Sites

There are many threats to Egypt's ancient cultural heritage sites, with salt damage and other problems relating to rising groundwater being perhaps the most serious, as identified by Keatings et al. (2007) at the mudbrick ruins of Hawara Pyramid. Studies by Smith (1986) and Wüst and Schlüchter (2000) illustrate the widespread threat of groundwater rise and salt crystallization processes around Thebes (Luxor), with damage caused by sodium chloride the major problem. Weathering problems are also notable and alarming in Cairo, and Fitzner and Heinrichs (2002) note the multiple threats posed by rising groundwater over the last few decades and associated salt weathering (affecting the lower parts of monuments) as well as increasing air pollution linked to gypsum crust damage to the upper parts. Kamh et al. (2008), in a study of Islamic archaeological sites in the city, note the synergy between salt weathering and the 1992 earthquake. Some of the sites most badly damaged by the earthquake were those previously weakened by intense salt attack.

The complex nature of deterioration and conservation of the Giza Sphinx illustrates the general problems of salt weathering of ancient Egyptian sites. This huge monument is carved into limestone layers of variable durability and has palpably deteriorated since it was first photographed in 1850. By one estimate, loss of stone is occurring at the rate of about 30 cm per century (Selwitz, 1990, p. 854). Some of the limestones are very prone to deterioration by sodium chloride and calcium sulphate, especially in the lower parts where water can build up (Livingston, 1989; Gauri et al., 1995). However, aeolian abrasion has been hypothesized to cause damage to the windward, upper parts of the Sphinx (Camuffo, 1993). Pharaonic and Coptic wall paintings near Luxor have also been damaged by sodium chloride and gypsum attack (Moussa et al., 2009). In the Kharga Oasis, there are various temples and cemeteries that are being damaged by salt weathering, and, as elsewhere in Egypt, the problem is being exacerbated by changes in groundwater levels brought about by irrigation (Salman et al., 2010).

Petra, Jordan

At Petra in Jordan the lower portions of many of the Nabatean monuments show substantial decay (Figure 6.7). According to Wedekind and Ruedrich (2006), more than 50 per cent of their surfaces are now damaged by weathering phenomena, with almost 12 per cent totally destroyed by salt-induced cavernous weathering (Heinrichs, 2008). There is also abundant evidence in the sandstone cliffs and buildings of Petra of the development of large numbers of cavernous weathering forms, ranging in size from small honeycombs to huge tafoni.

The serious, multiple hazards posed by weathering at Petra are summarised by Wedekind and Ruedrich (2006, p. 261) who note, 'Today, the existence of the unique rock architecture of these monuments is in danger due to decomposition, poor maintenance and lack of conservation.' Salt weathering, especially by sodium chloride and calcium sulphate, plays a key role in the deterioration as discussed by Albouy et al. (1993), Fitzner and Heinrichs (1991, 1994) and Paradise (2005). Links between salt weathering and other hazards have also been noted, with Bani-Hani and Barakat (2006) illustrating the role of salt weathering in reducing compressive strength and increasing the susceptibility of monuments to earthquake shaking.

Mohenjo-Daro, Pakistan

The largest and most important of the Harappan sites in the Indus Valley is Mohenjo-Daro. There it is clear that the disintegration of the burnt bricks of which the site is built has accelerated since the mid-1920s, before which they were in a state of relatively good preservation. The development of modern irrigation systems has caused groundwater levels to rise, leading to waterlogging and consequent salinity. At ground level and on the lower parts of exposed walls, disintegration of natural and artificial materials takes place with great efficiency, a timespan of two to twelve years often being sufficient for complete breakdown to occur (Goudie, 1977). The disintegration is associated with the development of a white efflorescence on brick and stone surfaces. The disintegration is caused by the hydration and crystallization of various salts (Fodde, 2007). Even though the efflorescences are composed of more than one salt,

Figure 6.7 Salt decay to the base of the Royal Tombs of Nabattean age at Petra, Jordan. (ASG)

the predominant component is aggressive sodium sulphate, with some other sulphates (including calcium sulphate, magnesium sulphate and potassium sulphate).

Corrosion Effects on Reinforcements

Many engineering structures, including pipelines, are made of concrete that contains iron reinforcements (Figure 6.8). The formation of the corrosion products of iron causes a volume expansion to occur, thereby exerting pressure on the surrounding concrete. This may cause the concrete cover over the reinforcements to crack, which in turn permits the ingress of oxygen and moisture, which then aggravate the corrosion process. In due course, spalling of concrete takes place, the reinforcements become progressively less strong and the whole structure may deteriorate severely. Rates of corrosion are accelerated by chloride ions which may occur in a concrete because of the

Figure 6.8 Salt weathering has destroyed these water pipes in the central Namib near Rössing. (ASG)

use of contaminated aggregates or because of penetration from a saline environment (Soroka, 1993, section 10.5). Corrosion of metals can also be produced by sulphates (Hong Naifeng, 1994, p. 33), for there are often sulphate-reducing bacteria in saline soils containing sulphates, which can cause strong corrosion of metals.

Sulphate Attack on Concrete

Sulphates can cause severe deterioration of concrete (Mehta, 1983; Bijen, 1989). They react with the alumina-bearing phases of the hydrated cement to give a high sulphate form of calcium aluminate known as ettringite. Its formation involves an increase in the volume of the reacting solids, a pressure build-up, expansion and, in the most severe cases, cracking and deterioration (Soroka, 1993). The volume change on ettringite formation is very large; it is even greater than that produced by the hydration of sodium sulphate. Magnesium sulphate is also particularly aggressive (Addleson and Rice, 1991, p. 407). Another mineral formed by sulphates coming into contact with cement is thaumasite. This causes both expansion and softening of cement (Crammond, 1985) and has been seen as a cause of disintegration of rendered brickwork and of concrete lining in tunnels (Lukas, 1975).

Accelerated or Enhanced Salinity

Human activities cause enhanced or secondary salinisation in drylands in a variety of ways (Goudie and Viles, 1997). In Table 6.6 these mechanisms are grouped into

Table 6.6 *Causes of enhanced salinisation*

1. **Irrigation salinity**
a. Rise in groundwater
b. Evaporation of water from fields
c. Evaporation of water from canals and reservoirs
d. Waterlogging produced by seepage losses
2. **Dryland salinity**
a. Vegetation clearance
3. **Urban salinity**
a. Water importation and irrigation
b. Faulty drains and sewers
4. **Inter-basin water transfers**
a. Mineralisation of lake waters
b. Deflation of salts from desiccating lakes
5. **Coastal zone salinity**
a. Over-pumping
b. Reduced freshwater recharge
c. Sea-level rise
d. Ground subsidence

five main classes: (1) irrigation salinity, (2) dryland salinity, (3) urban salinity, (4) salinity brought about by inter-basin water transfers and (5) coastal zone salinity. Human-induced salinisation affects about 77 million ha on a global basis, of which 48 million ha are in susceptible drylands (Middleton and Thomas, 1997).

In recent decades, there has been a rapid and substantial spread of irrigation across the world, which has brought about a great deal of waterlogging and salinisation (Rhoades, 1990). The irrigated area in 1900 amounted to less than 50 million hectares. By 2000, the total area amounted to five times that figure.

A second prime cause of dryland salinity extension is vegetation clearance. By reducing losses through interception and evapotranspiration, this allows a greater penetration of rainfall into deeper soil layers, which causes groundwater levels to rise, thereby creating conditions for the seepage of sometimes saline water into low-lying areas. Groundwater levels have increased some tens of metres since clearance of the natural vegetation began. This is a particularly serious problem in the wheatbelt of Western Australia. Here, the clearance of eucalyptus forest has led to an increased rate of groundwater recharge and to the spreading salinity of streams and bottomlands. Salt ‘scalds’ have developed. Dryland salinity is also a major problem on the Canadian prairies and in south-west Niger (Leduc et al., 2000) because of the replacement of natural woodland savannah with millet fields and associated fallows.

Recent decades have seen a great growth of cities in drylands (see Section 1.16). Urbanisation can cause a rise in groundwater levels by affecting the amount of moisture lost by evapotranspiration. The spread of impermeable surfaces (roads, buildings, car parks, etc.) interrupts the soil evaporation process so that groundwater levels rise (Shehata and Lotfi, 1993). Urbanisation can lead to other changes in groundwater conditions that can aggravate salinisation. In some large desert cities, such as Cairo (Smith, 1986; Hawass, 1993), the importation of water and its usage, wastage and leakage can produce the ingredients to feed this phenomenon.

Increases in levels of salinity are brought about by the changing state of lakes caused by inter-basin water transfers. The most famous example of this is the shrinkage of the Aral Sea, the increase in its mineralisation and the deflation of saline materials from its surface and their subsequent deposition downwind (Saiko and Zonn, 2000; Kravtsova and Tarasenko, 2010). Its mineral content has increased more than threefold since 1960. Another illustration of the effects of inter-basin water transfers is the desiccation of Owens Lake in California. Diversion of water to feed the insatiable demands of Los Angeles has caused the lake to dry out, so that saline dust storms have become an increasingly serious issue (Gill, 1996; Tyler et al., 1997). Future climate changes may also contribute to increasing levels of salinity in lake basins (Sereda et al., 2010).

Another prime cause of the spread of saline conditions is the incursion of seawater brought about by the over-pumping of groundwater. Salt water displaces less-saline groundwater through a mechanism called the Ghyben-Herzberg principle. The problem presents itself on the coastal plain of Israel, parts of North America (Barlow and Reichard, 2010), Bahrain, the coastal aquifers of the United Arab Emirates and Oman and in the Nile Delta.

Avoidance and Zoning

An effective way to cope with salt attack is not to build in aggressive areas, particularly those where groundwater level and salinity are the crucial controls, as in the Middle East's low-lying coastal cities. Here, the potential intensity of the salt weathering hazard is basically a function of the elevational relationship between the ground surface and the limit of capillary rise as well as the salinity of the rising water. The capillary fringe limit is a particularly important boundary and can often be identified on air photographs because of the presence of a well-defined tonal boundary. Hazardous zones can be identified (Jones, 1980) by gathering and analysing data on ground-surface elevation, the depth to water table, the nature and distribution of different types of surface material (to facilitate determination of the potential height of capillary rise) and spatial variation in the salinity of groundwater (Figure 6.9).

Reduction in Groundwater Level

Given that groundwater level is such an important control of salt attack, efforts need to be made either to keep groundwater levels from rising (e.g. by control of irrigation developments) or, if they have already reached critical levels, to make them fall. The latter of these two strategies is probably best approached by pumping groundwater via tube wells and its evacuation in drainage canals (disposal channels). This was the

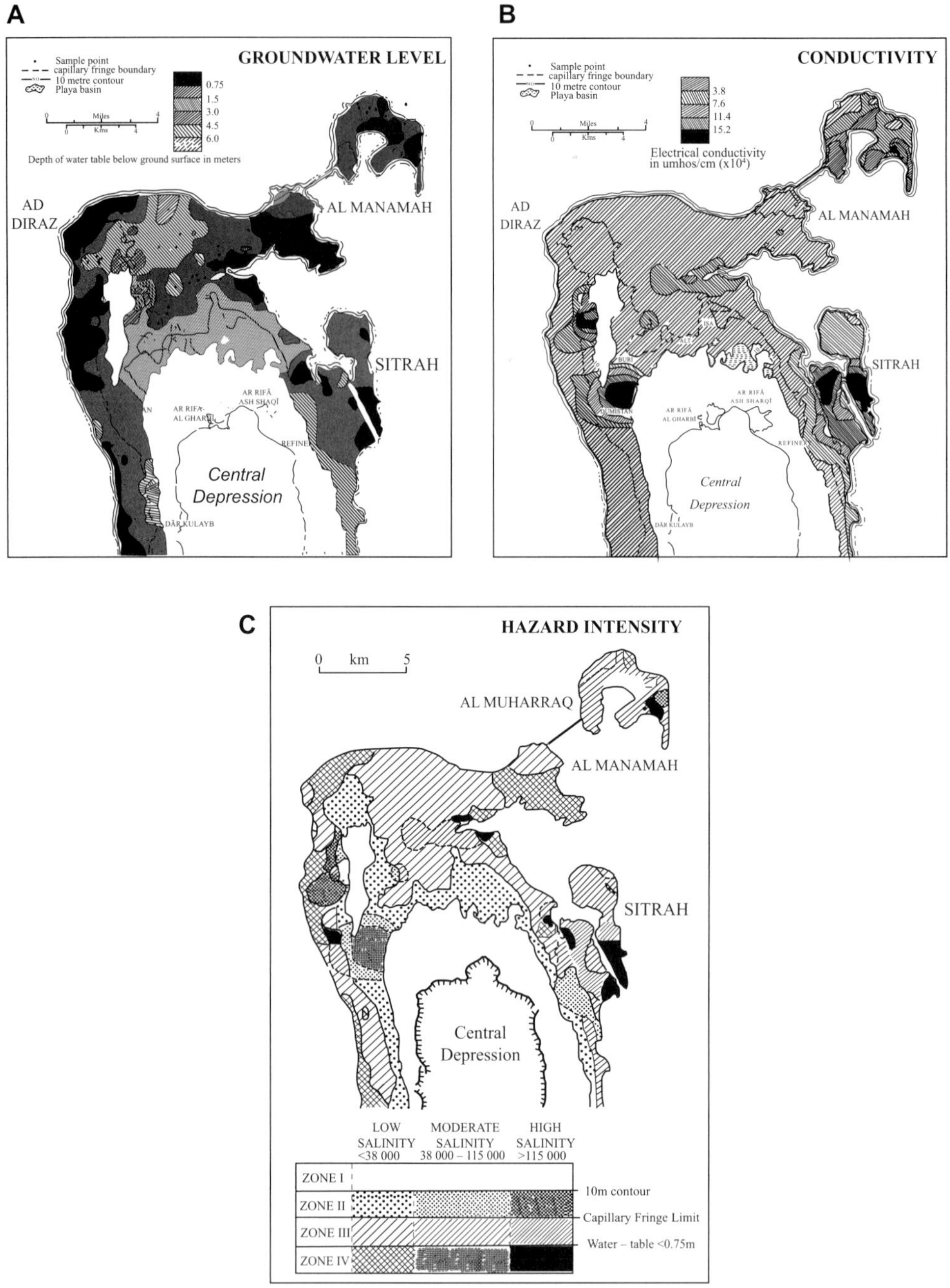

Figure 6.9 Salt weathering conditions in Bahrain. (a) Groundwater level. (b) Water conductivity (salinity) with values in μmhos cm^{-1} (c) Hazard intensity. (Modified after the work of the Bahrain Surface Material Resources Survey in Cooke et al., 1982, pp. 177, 178, 180)

Figure 6.10 A rockfall in sandstone in the Wadi Rum area of Jordan. (ASG)

approach adopted in the Master Plan for Mohenjo-Daro (Lohuizen-de Leeuw, 1973), where rings of tube wells were constructed (Jansen, 1996). Afforestation can also cause groundwater levels to fall, thereby reducing topsoil salinity (Bari and Schofield, 1992; Hbirkou et al., 2011).

6.6 Slope Instability

Recent studies of slopes in arid areas, particularly those developed in sedimentary rocks, have shown how extensive landslide phenomena can be (Figure 6.10). This is the case for much of the central Sahara (Busche, 2001), the sandstone terrains of south Jordan (Goudie et al., 2002) the cliffs of the Dakhla Oasis in the Libyan Desert (Brookes, 1993) and the loess terrains of Afghanistan. Mass movements are, however, by no means restricted to sedimentary rocks, and in California's Anza-Borrego Desert State Park, Hart (2008) has described large translational slides and long run-out avalanches developed in tonalite and gneiss. Slope failure is a critical problem in the Los Angeles region (Cooke, 1984), where severe instability has been caused by El Niño events.

While slopes in hyperarid areas may display the effects of former wetter conditions, there are various current factors that lead to slope instability – the paucity of vegetation cover, the steepness of some slopes, the role of bush fires in destabilising both vegetation cover and soils, the effect of salts on the slaking behaviour of surface materials and the role of occasional high-intensity rainfall events.

In mountainous deserts such as those of the Andes or Karakorams, the high available relief, the deep incision of drainage and the occurrence of seismic activity mean that large landslides are a major hazard. In Pakistan, for example, the Karakoram Highway is frequently blocked by landslides and rockfalls, and from time to time the major rivers, such as the Hunza, may be impounded behind landslide dams (Jones et al., 1983). The lakes thus formed may subsequently collapse and cause severe downstream flooding (Moreiras, 2005). One of the most impressive landslides occurs in the high Andes. Socompa, on the border between Argentina and Chile, is a large stratovolcano that rises to 6,051 m above sea level and forms part of the Andes range (Wadge et al., 1995). Sometime around 6,400 years ago, the northwestern section of the volcano collapsed, sending a vast debris avalanche nearly 40 km to the northwest over a relatively flat, arid plain (the Monturaqui Basin). The amphitheatrical scar from which the collapse took place is marked by a great triangular embayment. Because of the area's aridity, this is arguably the best preserved and most pristine example of a major volcanic debris avalanche on Earth. At the mouth of the embayment or amphitheatre is a series of ridges several hundred metres high. These are gigantic blocks of the volcano which slid into their present location, rotated, but maintained some degree of coherence and some of their original structure. Such blocks are called toreva blocks. The largest of these is 2.5 km long, 1 km broad and is bounded by scarps some 400 m high. The avalanche, which covers an area of about 490 km^2, flowed down over a vertical interval of c 3,000 m, and may be as much as 90 m thick. Its limits are steep and well defined, commonly with a distinct levee. The debris was reflected back of the western margin of the Monturaqui Basin oblique to the primary flow direction. Some of the debris consists of 10 m mega-blocks. The avalanche was probably a very rapid event. Such phenomena are often associated with velocities of no less than 20–100 metres per second.

Elsewhere in the mountains of South America there are records of other landslides and rock avalanches (sturzstroms) (Hermanns and Strecker, 1999), some of them of great size (e.g. the Lluta Collapse of northern Chile) (Strasser and Schlunegger, 2005), the paleolandslides of the Tafi del Valle in Tucuman Province, Argentina (Fernández, 2005) and the Sierra Laguna Blanca in the Puna (Hermanns et al., 2001).

As long-term monitoring by repeat photography and other techniques has shown (Griffiths et al., 2004), debris flows are a common phenomenon in desert valleys and on alluvial fans (see Section 5.4) and can cause loss of life. For example, in Antofagasta, Chile, a debris flow event in 1991 killed 103 people and destroyed 500 houses (Sepúlveda et al., 2006). Debris flows are also a threat to settlements on alluvial fans in northwest Argentina (Marcato et al., 2012). In the western United States, debris flows are especially common after wildfires have occurred, for these destroy the vegetation cover, leave the soil bare and help to create changes in the hydrophobicity of soils that makes them more likely to generate runoff and sediment. They occur in areas of chaparral scrub. Numerous examples of fire effects on debris flows are known from the western United States (Gartner et al., 2008). Debris flows have also threatened properties in the Tucson metropolitan area of Arizona (Dorn, 2012), and models have been developed to produce debris flow hazard delineation zones (Magirl et al., 2010). A classic study of a mudflow was that undertaken by Sharp and Nobles (1953) at Wrightwood in the San Gabriel Mountains of southern California. It was precipitated by rapid snowmelt and ran in a series of surges for approximately 24 km from its source. Surge-front velocities were c 3 m per second. Roads and other infrastructure were buried.

6.7 Fluvial Hazards

Someone once remarked that more people were killed by drowning in deserts than by thirst. While this may be something of an exaggeration, there is no doubt that occasional extreme rainfall events – and the general propensity of some desert surfaces to generate runoff – has caused flood events that have led to great loss of life and damage to properties and infrastructure. Burckhardt (1829), for example, provided an early description and history of flooding that had inundated the Holy City of Mecca. Rare floods can be exceptionally large, and in arid regions the ratio between the magnitudes of the annual flood and the 100-year flood are markedly greater than in humid regions or in basins fed by snowmelt (Pitlick, 1994; Molnar, 2001). Large floods can carry huge boulders that can cause culverts (Figures 6.11a and b) and bridges (Figure 6.12) to be blocked or damaged. They may also carry large amounts of tree debris that can act as battering rams on engineering structures (Figure 6.13).

Alluvial fans (see Section 5.4) are especially hazardous environments because of the shifting nature of their stream courses, their propensity to cut and fill, their occasionally high flood discharges and velocities (Chawner, 1935; Woolley, 1946; House, 2005) and the prevalence of debris flows across their surfaces. Thus when urban centres expand across them, they may encounter problems (Robins et al., 2009). This is, for example, the case with Eilat in southern Israel (Schick et al., 1999). The Negev Desert wadis are dangerous because they are sporadically active systems that can carry large sediment loads. One reason for this is the nature of rainfall events in the region; rainfall intensities can be high (Schick, 1988). In the Nahel Yael catchment, close to Eilat, over a seventeen-year period, intensities exceeding 14 mm per hour accounted for nearly one-half of the total rain (223 mm out of 449). Of this intense rain, 37 per cent fell in intensities exceeding 2 mm per minute. The October 2004 flood in the Sedom region of Negev was produced by a rainfall intensity that reached 175 mm h^{-1} (Greenbaum et al., 2010). Extreme flooding in the wadis can follow

(a)

Figure 6.11 (a) and (b). Desert stream channels often carry large clasts, and in these cases the engineers of the Karakoram Highway, Pakistan, have built culverts of inadequate dimensions. (ASG)

major rainfall events, as was demonstrated by the storms that afflicted southern Israel and Jordan in 1966 (Schick, 1971).

Severe flooding occurred in January 2010 in the Sinai Desert of Egypt, creating havoc in the resort town of Sharm El Sheik; many roads in Sinai are subject to regular disruption by flash floods (Abdel-Lattif and Sherief, 2010; Youssef et al., 2010; Cools et al., 2012). Equally, Salalah in Oman, where there has been considerable encroachment into wadis, has suffered frequent and expensive flooding damage that has required the construction of interceptor dams, raised banks along wadis and the installation of drains (Chakraborty, 2009). Various methods are now available to

(b)

Figure 6.11 (*continued*)

Figure 6.12 A road bridge in southern Tunisia that has been partially blocked by a large boulder brought down in a flood. (ASG)

Figure 6.13 Flood debris along the Kuiseb River, Namibia, August 2011. (ASG)

assess and predict flash-flood discharges, and some recent examples of such work are given in Table 6.7.

Urbanisation may exacerbate flooding problems. In one of the few studies of this phenomenon in drylands, Chin and Gregory (2001), working in Fountain Hills, Arizona, found that streets were often built to serve as storm drainages, this being less costly than constructing infrequently needed storm sewers. In addition, the urban process generates large expanses of impermeable surfaces. Fire, through its effect on hydrophobicity and runoff, can also increase flood risk (Figure 6.14).

Table 6.7 *Examples of studies to estimate and predict flash floods in arid catchments in Egypt and Arabia*

Location	Study
Makkah, Saudi Arabia	Dawod et al. (2011a and b); Subyani (2011)
Western Saudi Arabia	Subyani et al. (2012)
Southern Sinai, Egypt	Masoud (2011); Youssef et al. (2010); Cools et al. (2012)
Wadi Hudain, Egypt	El-Bastawey et al. (2009)
Wadi Rahbaa, Egypt	Soussa et al. (2012)
Oman	Al-Qurashi et al. (2008); Al-Rawas and Valeo (2010)

Figure 6.14 A road sign in Yucca Valley, California, indicating the risk of flash floods following fire. (ASG)

An important contribution of geomorphologists is to record flood events and their effects. Their aim is to try to establish – in the general absence of gauging stations – the history, magnitude and frequency of past events through palaeoflood analysis (see Section 5.18), including dendrochronological work (Jacoby et al., 2008); which landform elements are safe in terms of events of particular magnitudes; and to produce maps of flood hazard risk (Foody et al., 2004; Pelletier et al., 2005; Robins et al., 2009), which may also contribute to the establishment of flood insurance rate zones (House, 2005).

Another type of fluvial hazard is gully formation or arroyo trenching (see Section 5.7). This can lead to changes in the agricultural suitability of bottomlands, modify

Figure 6.15 Soil and water conservation structures built across an ephemeral channel in loess at Matmata, Tunisia. (ASG)

local aquifers, modify sediment inputs into reservoirs and cut into engineering structures. Arroyo incision can also lead to the draining of riverbed marshes (*cienegas*). It may even have produced settlement abandonment (Hereford et al., 1995). Monitoring and various types of analysis of the archaeological record can be used to show the speed at which incision and headward erosion occur (e.g. Avni, 2005). Among the techniques that have been developed in an attempt to control gully development is the check dam. Check dams have a long history of use, as for example, in the American Southwest (Doolittle, 1985), but they have also been used and evaluated in the Loess Plateau of China (Xu et al., 2004) and in south-east Spain (Castillo et al. 2007).

Water erosion on slopes is an active process, particularly in semi-arid regions, and numerous techniques have been applied to try to reduce the amount of sediment that is moved (Figure 6.15). These are well reviewed by Morgan (1995). Techniques – the appropriateness and effectiveness of which need to be the subject of geomorphological study – include the establishment of a good vegetation cover (e.g. by afforestation, fertilisation and the replacement of bush with grass), careful land management (e.g. suppression of fires, control of overgrazing, crop residue preservation, mulching and minimum tillage), modification of soil structures by the addition of gypsum and/or polymers (Graber et al., 2006) and control of slope runoff (e.g. by terracing, transverse hillside ditches, contour ploughing and vegetation strips). It is important to stress that such techniques are not always successful, however. For example, in semi-arid Tunisia, where approximately 1 million hectares of agricultural land have been installed with anti-erosive contour benches, the life of the benches has proved to be limited, with those on gypsum clays being especially prone to failure (Baccari et al., 2008). It was also found that soil conservation was in part achieved at the cost of reduced runoff and thus a reduction in available water resources for irrigation. Indeed, soil erosion countermeasures can be counterproductive, as has been made evident by reviews of large-scale afforestation schemes in China (Cao et al., 2010) and Spain (Romero-Diaz et al., 2010). The latter study showed that aggressive land sculpting and the subsequent planting of inappropriate trees on scrublands actually accelerated rates of erosion on marl slopes by between 1 and 2 orders of magnitude. Bulldozing can be a curse in Mediterranean environments. One also needs to be aware that some species of exotic plant that have been introduced for control of water erosion can prove to be highly invasive, as with mesquite (*Prosopsis juliflora*) in Kenya (Muturi et al., 2009). Table 6.8 lists some studies that have evaluated the various techniques for controlling water erosion on slopes and in channels in drylands.

One particular area of concern with regard to water erosion and transport in drylands is the dispersal of mine waste from exposed mine tailings, especially after storm events (Sims et al., 2012). This has been described from many deserts, including those in Mexico (Razo et al., 2004), the Atacama (Oyarzún et al., 2003), Spain (Navarro et al., 2008), the south-west United States (Sims and Francis, 2008), Namibia (Taylor and Kesterton, 2002) and Australia (Taylor and Hudson-Edwards, 2008). This waste can contain harmful substances, including mercury, arsenic, cyanide and toxic metals such as lead (Hayes et al., 2009; Kim et al., 2012), so that stabilisation of tailings with appropriate plants is a major priority (Mendez and Maier, 2008).

6.8 Subsidence

Ground subsidence is a major geomorphological hazard in some arid areas because of the removal of subterranean water and hydrocarbon resources. Changes in level, which can be appreciable, can cause problems for engineering structures and also lead to waterlogging, flooding and salinisation. Groundwater abstraction in the Central Valley of California has caused around 9 m of subsidence, while in the Los Angeles area, 9.3 m of subsidence occurred as a result of the exploitation of the Wilmington oilfield between 1928 and 1971. The Inglewood Oilfield displayed 2.9 metres of subsidence between 1917 and 1963, and some coastal flooding problems occurred at Long Beach because of this process. Central Las Vegas has suffered about 2 m of subsidence since about 1935, although there is some evidence that subsidence rates have declined with a reduction in groundwater exploitation (Amelung et al., 1999). Parts of Arizona, such as the Eloy area, have seen 4.6 m of subsidence (Gelt, 1992). The extent of subsidence

Table 6.8 *Examples of the evaluation of methods to control water erosion on slopes in drylands*

Techniques	Source
Addition of fertiliser	Lasanta et al. (2000)
Afforestation	Cao et al. (2010); Romero-Diaz et al. (2010)
Aleppo pine plantations	Chirino et al. (2006)
Bench terracing	Ternan et al. (1996)
Blade ploughing and exclosure	Eldridge and Robson (1997)
Control of trampling by sheep	Eldridge (1998)
Crop residue management	Unger et al. (1991)
Engineering trenches	Marston and Dolan (1999)
Fallow cropping	Valentin et al. (2004)
Geotextiles	Rickson (2006)
Gypsum addition to improve structure	Graber et al. (2006)
Hedges	Kiepe (1996); Smolikowski et al. (2001); Poesen et al. (2003)
Juniper control	Belski (1996)
Matorral species	Bochet et al. (1998)
Mulching	Smith et al. (1992); Bautista et al. (1996)
No tillage	Kabakci et al. (1993); Hansen et al. (2012)
Plant strips	Martinéz Raya et al. (2006)
Rock mulches	Poesen et al. (1994)
Soil compaction	Poesen et al. (2003)
Synthetic polymers and biopolymers	Graber et al. (2006); Orts et al. (2007)
Trash and stone lines	Quinton et al. (1997); Wakindiki and Ben-Hur (2002)
Vegetation cover	Rogers and Schumm (1991); Snelder and Bryan (1995); Durán Zuazo et al. (2004); Sandercock and Hooke (2011)

that has taken place in the United States as a result of groundwater abstraction has been assessed by Chi and Reilinger (1984). Further data on amounts and rates of subsidence is provided in Goudie (2006, table 6.6).

In areas where thick halite deposits have accumulated in saline lake basins, solution may cause sinkholes to develop. An example of this is provided by the Dead Sea basin, where the recent decline in level caused by water abstraction has promoted their formation. More than a thousand potentially dangerous sinkholes have developed along its shorelines since the early 1980s as a result of the flow of undersaturated groundwater dissolving the evaporites. As Yechieli et al. (2006, p. 1075) explain it in the context of the Israeli side of the basin:

> The abrupt appearance of the sinkholes, and their accelerated expansion thereafter, reflects a change in the groundwater regime around the shrinking lake and the extreme solubility of halite in water. The eastward retreat of the shoreline and the declining sea level cause an eastward migration of the fresh-saline water interface. As a result the salt layer, which

originally was saturated with Dead Sea water over its entire spread, is gradually being invaded by freshwater at its western boundary, which mixes and displaces the original Dead Sea brine.

Similar problems have been encountered on the Jordanian side of the Dead Sea, with karstic collapse creating major problems for the chemical plants on the Lisan Peninsula (Closson et al., 2007). Salt dissolution has also created severe problems for a new highway across the Great Kavir of Iran (Ghazifard and Khorashadizadeh, 2010). Gypsum deposits may also be subject to dissolution, although they can also lead to ground heave if they form as a result of the hydration of anhydrite in the presence of groundwater (Yilmaz, 2001).

Karstic collapse in carbonate rocks in deserts is probably rendered unusual because of the low precipitation levels, causing solution rates to be low. There are, however, examples of sudden sinkhole development in urban areas caused by a combination of declines in groundwater level and downward infiltration of excess irrigation water (see, for example, the studies made in Kuwait [Shaqour, 1994], Saudi Arabia [Amin and Bankher, 1997] and Iran [Atapour and Aftabi, 2002]).

In certain circumstances, ground-surface subsidence is accompanied by ground cracking (Lee et al., 1996; Hoffmann et al., 1998; Al-Harthi and Bankher, 1999) that can lead to the formation of giant ground or earth fissures – some of them more than 10 km long, 3 m deep and 4 m wide. They are major hazards to agricultural and urban areas in drylands, such as central California, the Las Vegas area of Nevada, Arizona and Australia. Irregularities in the ground surface can seriously disrupt drainage, dams, aqueducts and sewerage and irrigation systems. Wells may fail, roads and linear services can be broken and damage can be caused to buildings, bridges and other structures. In the case of Bicycle Playa in the Fort Irwin National Training Centre of the Mojave, subsidence-induced fissuring has posed problems for aircraft runways (Densmore et al., 2010). In addition, fissures may act as conduits which can lead to groundwater pollution, and litigation may result (Corwin et al., 1991).

6.9 Hydrocompaction and Collapsible Soils

Hydrocompaction, also known as 'collapse compression', 'hydrocompression', 'hydroconsolidation' and 'saturation shrinkage' (Charles, 1994) is the compaction and reduction in volume of soils and sediments that occurs when their moisture content is increased. The process causes ground subsidence when unconsolidated sediments (collapsible soils) of low density are wetted, as for example, by the application of irrigation water, the disposal of wastewater or runoff from urban surfaces. It is a feature of arid and semi-arid lands (Houston et al., 2001) where materials such as windblown loess, debris flow deposits or certain alluvial sediments above the water table are not normally wetted below the root zone and have high void ratios. When dry, such materials may have sufficient strength to support considerable effective stresses without compacting. When they are wetted, however, their inter-granular strength is weakened because of the rearrangement of their particles. The associated subsidence may create fissures in the ground and is a process that needs to be considered during the construction of canals, pipelines, dams and irrigation schemes (Al-Harthi and Bankher, 1999). It can cause severe structural deterioration of buildings. Collapsible soils can be identified by such techniques as plate load tests.

The propensity of clay-rich materials to swell and contract on wetting and drying is a related problem for buildings and roads, as has been discussed in the case of Tabuk, Saudi Arabia (Sabtan, 2005).

6.10 Lake Shrinkage and Expansion

Closed depressions are widespread in arid lands, and their water levels and salinity respond rapidly and profoundly to climatic changes (Grimm et al., 1997; Scuderi et al., 2010) (see Sections 1.9 and 1.12). This generalisation applies to both large and small lakes. In the twentieth century, for example, some of the largest arid zone lakes (e.g. Chad, the Aral Sea, the Caspian and the Great Salt Lake of Utah) have shown large variations in their extents, partly in some cases because of human activities, but also because of climatic fluctuations within their catchments.

For example, from the early 1950s to the mid-1980s, the total area of lakes in China with an individual area of more than 1 km^2 declined from 2,800 to 2,300 km^2, and the whole area of China's lakes has been reduced from 80,600 km^2 to 70,988 km^2 (Liu and Fu, 1996). An increasingly warm and dry climate was the principal cause of the reduced lake area on the Qingzang Plateau, north-west China, the Inner Mongolian Plateau and the North China Plain.

Likewise, in the early 1960s, prior to the development of the Sahel drought, Lake Chad had an area of 23,500 km^2 but by the 1980s had split into two separate basins and had an area of only 1,500 km^2. The Caspian Sea was –29.10 m in 1977 but in 1995 had risen to –26.65 m, an increase of 2.45 m in just seventeen years. Similarly impressive changes have occurred in recent decades in the level of the Great Salt Lake in Utah, with a particularly rapid rise taking place between 1964 and 1985 of nearly 6 m, although the lake receded and reached another low stand in 2005.

Such changes in lake level have an impact on a diverse range of human activities, ranging from fisheries and irrigation to recreation and transport infrastructure. Moreover, the drying up of lakebeds can have adverse effects on air quality and human health through the liberation of dust, as has been found as a result of the human-induced desiccation of the Aral Sea and Owens Lake (Reheis, 1997). Military installations located on salt playas may also be affected (French et al., 2005).

Even the world's largest lake, the Caspian, has been modified by human activities. The most important change was the fall of 3 metres in its level between 1929 and the late 1970s. This decline was undoubtedly partly the product of climatic change

Figure 6.16 Shorelines showing the recent recession in the level of the Dead Sea, Jordan. This recession has promoted sinkhole development. (ASG)

(Micklin, 1972). Nonetheless, human actions have contributed to this fall, particularly since the 1950s, because of reservoir formation, irrigation, municipal and industrial withdrawals and agricultural practices. In addition to the fall in level, salinity in the northern Caspian increased by 30 per cent since the early 1930s. An amelioration of climate since the late 1970s has caused some recovery in the lake's level.

Between 1960 and 1990, the Aral Sea – largely because of diversions of river flow (Saiko and Zonn, 2000) – lost more than 40 per cent of its area and about 60 per cent of its volume, and its level fell by more than 14 m (Kotlyakov, 1991). By 2002, its level had fallen another 6 metres. By 2008, the area of the Aral Sea was just 15.7 per cent of that in 1961 (Kravstova and Tarasenko, 2010). This has lowered the artesian water table over a band 80–170 km in width, has exposed extensive expanses of former lakebed to desiccation and has created salty surfaces from which salts are deflated to be transported in dust storms, to the detriment of soil quality. The mineral content of what remains has increased almost threefold over the same period.

Water abstraction from the Jordan River has caused a decline in the level of the Dead Sea (Figure 6.16). During the last four decades, increasing amounts of water have been diverted from surface and groundwater sources in its catchment. The water level has dropped about 28 m over that period, causing subsidence to occur, karstic collapse to take place and landslides to develop (Closson et al., 2010).

The development of various remote sensing devices now make it easier to obtain a record of lake fluctuations and inundation expanses in ephemeral lake basins (see, for example, the work of Bryant and Rainey [2002] on the North African chotts and Verdin [1996] in Niger using the advanced very high-resolution radiometer [AVHRR]). One role of the geomorphologist is to try to establish the potential areas that may be subject to inundation in response to particular climatic events – for example, to establish such parameters as 100-year-flood hazard zones (French et al., 2005). Light Detection and Ranging (LiDAR) is an important tool for obtaining precise height data for what are very gentle surfaces, while various types of surface features, such as mud crack polygons, may be indicative of flooding history (Lichvar et al., 2008).

6.11 Piping

Piping is a term used to describe subterranean channels formed by water moving through and eroding incoherent and insoluble sediments (Bryan and Jones, 1997). It is associated with subsurface ephemeral discharge in many drylands, including those in the United States (e.g. Parker, 1963; Parker and Jenne, 1967); northeast Victoria, Australia (Downes, 1946); the Hoggar area of the Sahara; Israel; southeast Spain (Gutiérrez et al., 1997); Saskatchewan (Bryan and Yair, 1982); and the drier loess areas of Asian deserts (Zhu, 1997).

Piping poses a subsidence/collapse hazard in drylands, threatening roads and railroads and other engineering structures such as dams (Richards and Reddy, 2007), bridges, culvert facings, and retaining walls. A major survey was by Bryan and Yair (1982), who showed that piping is often associated with intensely dissected, vegetation-free badlands (see Section 5.6), such as the dongas in southern Africa (Rienks et al., 2000). Piping, and consequent surface collapse and gully development, can cause damage to human activities, such as disruption of banks and fields (e.g. Baillie et al., 1986). It can be a particular hazard in irrigated areas leading to accelerated erosion, sinkhole development, small landslides and fan accumulation (García-Ruiz et al., 1997).

Pipes are commonly formed adjacent to steep free faces, such as alluvial channel banks, and in alluvial slopes as a result of rapid water movement down steep hydraulic gradients through permeable, easily dispersed clay, silt or other sediment. Piping often occurs where high infiltration capacities are found in material of low intrinsic permeability, which allows concentrated infiltration (Bryan and Yair, 1982). Initially, water seeping through the alluvium carries with it dispersed and disaggregated clay and silt particles, and a small hole develops. The hole is enlarged by this process, but as it becomes larger it acquires more water. The increased flow leads to more rapid development of the pipe by corrasion and wall caving. Eventually, the roof of the pipe collapses, creating a sinkhole down which surface water may be funnelled, augmenting the flow and the erosion of the pipe still further. Ultimately, the sinkholes

Figure 6.17 A series of pipes developed in sediments in southern Tunisia. Such pipes can coalesce to form discontinuous gully systems. (ASG)

merge and a ragged gully pattern is created (Figure 6.17). Hoffmann et al. (1998) provide a full discussion of pipe-related subsidence from near Tucson in Arizona, and Zhu (2012) discusses the role of tunnel erosion in gully formation in the Loess Plateau of China.

Many badland areas have the basic conditions necessary for piping to occur (Campbell, 1997, p. 274):

> These include, not necessarily in priority: strongly alternating wet and dry climatic patterns (preferably with at least some intense rainstorms to generate large volumes of surface runoff and prolonged periods of dry weather); deep desiccation cracking (or other fractures for entry of moisture); steep slopes (providing sufficient hydraulic head to assist subsurface flow);

presence of swelling clay minerals such as montmorillonite or illite, and large amounts of exchangeable sodium (to develop strong expansion and contraction stresses and to cause deflocculation and dispersion of the clay aggregates along moisture routes); alternating layers of permeable and impermeable strata (to facilitate lateral flow of water in some rock units); and an outlet for the pipe system – such as a gully floor or a porous layer.

Pipes also occur in materials which may not appear, at first sight, to be very permeable, such as clays, silts, loess and volcanic ash. All of these materials may contain swelling clays, usually montmorillonite, but often illite or bentonite (Parker, 1963). Swelling clays have two important qualities in this connection: When they are dried they shrink and create cracks, thus rendering impermeable materials more permeable, and when wet, they become highly dispersed, non-cohesive and, therefore, easily removed in suspension by water – even slow-moving water. For both of these effects to be realized, wetting and drying are essential, and this is common in many arid and semi-arid areas.

Piping is often a precursor of gully development. For example, Harvey (1982) found in south-east Spain that some gully development was promoted by piping in three circumstances: (1) where soils have a light surface crusting over a less consolidated layer in thick, poorly consolidated alluvium; (2) where similar soils occur over relatively weak bedrock with differential strength or porosity; and (3) where deep trenching has produced near-vertical walls that lead to tension cracking.

6.12 Dam Problems

A particularly important consequence of impounding a reservoir behind a dam is the reduction in the flow and in the sediment load of the river downstream. This not only affects the river itself but may have an impact on aeolian landscapes as well (Draut, 2012).

In the United States, Fitzhugh and Vogel (2011) have shown that in the more arid portions of the country the reductions in flow resulting from dam construction have been higher than in more humid regions. A clear demonstration of the effect of sediment retention behind dams is provided by the data for the Colorado River in the United States. Prior to 1930, it carried around 125–150 million tonnes of suspended sediment per year to its delta at the head of the Gulf of California. Following a series of dams, the Colorado now discharges neither sediment nor water to the sea (Schwarz et al., 1991). Equally, various human interventions mean that the flow of the Murray-Darling system into the ocean is about one-quarter to one-third of its natural level. The Indus River has also undergone a marked reduction in flow and sediment transport, especially since the completion of the Tarbela Dam in the early 1970s. Between 1931 and 1947, the river at Kotri had a flow of about 70–150 cubic km per year and a sediment discharge of 190–330 million tonnes per year. For the period from 1962–86, the respective figures were 10–100 cubic km per year and 10–130 million tonnes per year (Meadows and Meadows, 1999). Another dryland river that has suffered a marked diminution in sediment load in recent decades is the Huanghe (Yellow River) in China. It now only carries about 14 per cent of what it did previously (H. Wang et al., 2007).

Such interventions also mean that channel morphology has often been transformed with, for example, channel capacity downstream being reduced, as was the case for the Medjerda River downstream of the Sidi Salem Dam in Tunisia (Zahar et al., 2008), the Rio Grande in Texas (Everitt, 1993) and the Green River in Utah (Allred and Schmidt, 1999).

Sediment retention is also well illustrated by the Nile. Since the completion of the Aswan High Dam, the silt load has been lower throughout the year, and the seasonal peak has been removed. Petts (1985) indicated that the Nile now only transports 8 per cent of its natural load below the Aswan High Dam. The Nile sediments, on reaching the sea, used to move eastward with the general anticlockwise direction of water movements in the eastern Mediterranean, generating sandbars and dunes which contributed to delta accretion. About a century ago, an inverse process was initiated and the delta began to retreat. For example, the Rosetta mouth of the Nile lost about 1.6 km of its length from 1898 to 1954. In addition, large amounts of sediment are retained in an extremely dense network of irrigation channels and drains that has been developed in the Nile Delta itself (Stanley, 1996). Much of this coast is no longer nourished with sediment, and, as a result of this overall erosion of the shoreline, the sandbars bordering Lake Manzala and Lake Burullus on the seaward side are eroded and vulnerable to collapse.

Sediment removal can cause accelerated riverbed erosion as less sediment is available to cause bed aggradation. The last process is often called clear-water erosion (see Beckinsale, 1972), and in the case of the Hoover Dam it affected the river channel of the Colorado for 150 kilometres downstream by causing incision. Incision has also occurred downstream from the Glen Canyon Dam, and this and other geomorphological effects have been monitored by Grams et al. (2007). In turn, such channel incision may initiate headward erosion in tributaries and may cause the lowering of groundwater tables and the undermining of bridge piers and other structures downstream of the dam. On the other hand, in regions such as northern China, where modern dams trap silt, the incision of the river channel downstream may alleviate the strain on levees and lessen the expense of levee strengthening or heightening.

Clear-water erosion does not always follow from silt retention in reservoirs, however. There are examples of rivers for which, before impoundment, floods carried away the sediment brought into the main stream by steep tributaries. Reduction of the peak discharge after the completion of the dam leaves some rivers unable to scour away the sediment that accumulates as large fans of sand or gravel below each tributary mouth (Dunne and Leopold, 1978). The bed of the main stream is raised, and if water intakes or other structures lie alongside the river they can be threatened again by flooding or channel shifting across the accumulating wedge of sediment. Rates of aggradation of a metre a year have been observed, and tens of kilometres of channel have been affected by sedimentation (Petts, 1985). In the Rio Grande, vertical channel accretion of 2.75–3.0 m occurred between 1991 and 2008 (Dean and Schmidt, 2011). The reduction in peak floods below dams can also permit the expansion of riparian vegetation, such as tamarisk, which may trap sediment (Allred and Schmidt, 1999; Birken and Cooper, 2006) and so add to the amount of accretion that occurs. Rapid aggradation has also been noted in the channel of the Yellow River in China, partly because inputs of aeolian sands are no longer so effectively flushed out (Ta et al., 2008).

The life of dams and their associated reservoirs can be greatly shortened because of rapid sedimentation. The total annual loss of storage due to reservoir siltation is 0.5 per cent per year for Morocco, 0.7 per cent for Algeria and 1.2 per cent for Tunisia (Lahlou, 1996). Some small reservoirs in semi-arid Tunisia only have a lifespan of about fourteen years (Jebari et al., 2010). Conversely, some large reservoirs, such as Lake Nasser, may have a lifespan of some hundreds of years. Geomorphologists, through sediment budget and erosion-rate studies need to assess the probable speed at which siltation is likely to occur, to identify the hotspots for sediment generation in a catchment and to identify effective erosion control measures to reduce sediment inputs to reservoirs (see, for example, Haregeweyn et al., 2006).

6.13 Future Climates

Introduction

If global warming occurs, the Earth's whole climatic system will be transformed. Higher temperatures will in themselves cause rates of soil moisture loss to be greater through their effect on rates of evapotranspiration. The International Panel on Climate Change (IPCC, 2007) suggested that in drylands temperatures could increase by between 1 and 7°C by 2017–2100 compared to 1961–90, and that precipitation levels could decrease by as much as 10–20 per cent in the case of the Sahara but increase by as much as 10–15 per cent in the Chinese deserts. In addition, it is likely that many areas that are currently dry, such as the Sahel (Sylla et al., 2010), may see enhanced aridity because of reductions in precipitation, although even within an area such as the Middle East, some parts may become wetter and some may become drier. The Saudi Arabian deserts may, for instance, become wetter because of a more northerly intrusion by the Intertropical Convergence Zone (Evans, 2010). There is also an emerging consensus that the south-western United States will become more arid (Seager et al., 2007; Seager and Vecchi, 2010). A review of global climate models for Australia suggests that over many areas, droughts will become more intense and more frequent by 2070 (Kirono et al., 2011). Zeng and Yoon (2009) have suggested that as conditions become drier and vegetation cover is reduced, there may be vegetation-albedo feedbacks which will serve to enhance any aridity trend. By 2099, their model suggests that, globally, the warm desert area may expand by 8.5 million km^2, or 34 per cent. Some areas may experience more frequent hurricane activity, however, and there may also be changes in El Niño-Southern Oscillation (ENSO) frequency and intensity, although this latter aspect of climate change is still characterised by highly divergent model results (Latif and Keenlyside, 2009).

Wind Erosivity and Erodibility

Changes in climate could affect wind erosion either through their impact on erosivity or through their effect on erodibility. The former is controlled by a range of wind variables, including velocity, frequency, duration, magnitude, shear and turbulence. Unfortunately, General Circulation Models (GCMs) as yet give little indication of how these characteristics might be modified in a warmer world, so that prediction of future changes in wind erosivity is problematic. Erodibility is largely controlled by vegetation cover and surface type, both of which can be influenced markedly by climate. In general, vegetation cover, which protects the ground surface and modifies the wind regime, decreases as conditions become more arid. Likewise, climate affects surface materials and their erodibility by controlling their moisture content, the nature and amount of clay mineral content (cohesiveness) and organic levels. Modelling the response of wind erosion to climatic variables on farmland, however, is vastly complex, not least because of the variability of soil characteristics, topographic variations, the state of plant growth and residue decomposition and the existence of windbreaks. To this needs to be added the temporal variability of aeolian processes and moisture conditions and the effects of different land-management practices (Leys, 1999), which may themselves change with climate change.

Future Dust Storm Activity

Wind erosion leads to dust storms. Future dust activity will depend on three main factors: (1) anthropogenic modification of desert surfaces (Mahowald and Luo, 2003; Mahowald et al., 2006), (2) natural climatic variability (e.g. in the ENSO or the North Atlantic Oscillation) and (3) changes in climate brought about by global warming. If soil moisture declines as a result of changes in precipitation and/or temperature, there is the possibility that dust storm activity could increase in a warmer world (Wheaton, 1990). If dust storm activity were to increase as a response to global warming, it is possible that this could have a feedback effect on precipitation that would lead to further decreases in soil moisture (Miller and Tegen, 1998). A comparison between the U.S. Dust Bowl of the 1930s and model predictions of precipitation and temperature for the Great Plains of Kansas and Nebraska indicates that mean conditions could be similar to or worse than those of the 1930s under enhanced greenhouse conditions (Rosenzweig and Hillel, 1993). Munson et al. (2011) have also argued that with increased drought brought about by reduced precipitation and higher temperatures, there will be a reduction in perennial vegetation cover in the Colorado Plateau and thus an increase in aeolian activity. In contrast, however, in northern China, there is some evidence that dust storm activity has decreased in recent warming decades, partially in response to changes in the atmospheric circulation and associated wind conditions (Jiang et al., 2009), and so it might decrease still further in a warming world (Zhu et al., 2008).

The impact and occurrence of dust storms will depend a great deal on land-management practices, and recent decreases in dust storm activity in North Dakota and the High Plains have resulted from conservation measures (Todhunter and Cihacek, 1999).

Sand Dunes

Dunes, because of the crucial relationships between vegetation cover and sand movement, are highly susceptible to changes of climate – although there are huge challenges in predicting just how much dune systems may change (Knight et al., 2004; Thomas and Wiggs, 2008). One problem is that active and stabilised dunes often coexist under the same climatic conditions (Yizhaq et al., 2007). There may also be lags in terms of response, and these may be related to physical-biological interactions. So, for example, a dune may become vegetated when the wind power is sufficiently low, and once vegetated a much higher wind stress is needed to destroy the vegetation and reactivate the dunes. Vegetation dynamics are a major consideration (Hugenholtz and Wolfe, 2005a, 2005b). In addition, there are multiple environmental factors involved in determining dune activity: wind energy, rainfall, snow cover, soil moisture, groundwater conditions, vegetation cover, fire frequency, and so forth.

Some areas, such as the south-west Kalahari (Stokes et al., 1997) or portions of the High Plains of the United States (Gaylord, 1990), may have been especially prone to changes in precipitation and/or wind velocity because they are located in zones close to a climatic threshold between dune stability and activity. Explorers' accounts (Muhs and Holliday, 1995) and optical dating of dunes have led to the realisation that such marginal dunefields have undergone episodic and repeated phases of change at decadal and century timescales in response to extended drought events during the Holocene (Forman et al., 2006, 2008). In the Canadian Prairies, changes in precipitation amounts and decreases in wind velocity have led to increasing dune stability since the mid-1900s (Hugenholtz and Wolfe, 2005a, 2005b), while over the last few hundreds of years, relatively stable parabolic dunes have tended to replace active barchans (Wolfe and Hugenholtz, 2009).

The mobility of desert dunes (M) is directly proportional to the sand-moving power of the wind but indirectly proportional to their vegetation cover (Lancaster, 1995a,

p. 238). An index of the wind's sand-moving power is given by the percentage of the time (W) the wind blows above the threshold velocity (4.5 m/second) for sand transport. Vegetation cover is a function of the ratio between annual rainfall (P) and potential evapotranspiration (PE). Thus, $M = W/(P/PE)$. Empirical observations in the United States and southern Africa indicate that dunes are completely stabilised by vegetation when M is <50 and are fully active when M is $=200$. Muhs and Maat (1993) used the output from GCMs combined with this dune mobility index to show that dunes and sand sheets on the Great Plains are likely to become reactivated over a significant part of the region, particularly if the frequencies of wind speeds above the threshold velocity were to increase by even a moderate amount. Stetler and Gaylord (1996) suggested that with a 4°C warming, vegetation would be greatly reduced and that as a consequence sand dune mobility in Washington State would increase by more than 400 per cent.

Scenarios for dune remobilization have been developed for the mega-Kalahari (D.S.G. Thomas et al., 2005). Much of this vast region is currently vegetated and stable, but GCMs suggest that by the end of the century all dunefields, from South Africa and Botswana in the south to Zambia and Angola in the north, will be reactivated. Reactivation of these dunes could release large amounts of material for dust storms (Battachan et al., 2012). The methods used to estimate future dunefield mobility are still problematic, however, and much more research is needed before we can have confidence in them (Knight et al., 2004), and so, for example, Ashkenazy et al. (2012) have argued that the Kalahari dunes are unlikely to be subjected to dry enough or windy enough conditions for them to become greatly mobilized by the end of this century.

Once again, however, it is important to stress that the activity of sand dunes and dust storms may be dependent on drought interludes, but that the effects of such interludes – both in the past (see, for example, B.I. Cook et al., 2009; Seifan, 2009) and in the future – may be greatly affected by miscellaneous anthropogenic activities (Yizhaq et al., 2009): deforestation, overgrazing, dune removal, extension of irrigation, surface disturbance, mining, fire, and so forth.

Recent research in China has demonstrated that during the last five decades or so there have been changes in wind activity that have been related to warming trends. The general message is that as warming has occurred, wind velocities have fallen (e.g. Wang et al., 2007). It has also been suggested that as a consequence, dunes in northern China have become more stable (X. Wang et al., 2007), although – partly because of human pressures – this is not invariably true. Thus it is possible that as warming continues and wind velocities decrease, dunes may become still more stable in the region, but a great deal will also depend on future soil moisture conditions, which in turn will depend on precipitation amounts and levels of evapotranspiration. Various modelling studies have suggested that, more generally, in low latitudes extreme wind events will become less frequent with global warming, and this has been confirmed for the United States (Breslow and Sailor, 2002). Atmospheric stilling has been a widespread feature of recent warming decades in Australia and elsewhere (Vautard et al., 2010). Another possible influence on future dune activity, as it has been in the past, may be changes in the frequency and severity of wildfires (Sankey et al., 2009).

Coastal Flooding and Sea-Level Rise

Desert coasts may show major responses to future sea-level changes. There has been a considerable diversity of views about how much sea-level rise is likely to occur by 2100. In general, however, best estimates are of just under 50 cm by 2100. This implies rates of sea-level rise of around 5 mm per year, which compares with a rate of about 1.5 to 2.0 mm during the twentieth century (Miller and Douglas, 2004). Should Greenland melt at a faster rate than is currently predicted, however, then the amount of rise will be greater. Pfeffer et al. (2008), for example, suggest that melting of Greenland ice could plausibly lead to 0.8 m of sea-level rise by 2100.

Coastal flooding and sea-level rise could be a serious matter for low-lying sabkhas (see Section 3.11). These result from the interaction of various depositional and erosional processes that create a low-angle surface in the zone of tidal influence. This means that they are subject to periodic inundation and might be vulnerable to modest sea-level rise and to any increase in storm-surge events. Given the degree of infrastructural development that has taken place in their proximity, this is a serious issue for cities such as Abu Dhabi (Garland, 2010). However, it is likely that many sabkhas will be able to cope with modestly rising sea levels, for a range of processes contribute to their accretion. These include algal growth, faecal pellet deposition, aeolian inputs and evaporite precipitation. Some of these can cause markedly rapid accretion, even in the absence of a very well-developed plant cover. Moreover, as sea level (and groundwater) rises, surface lowering by deflational processes will be reduced.

Deltas subject to subsidence and sediment starvation (e.g. the Nile), and areas where ground subsidence is occurring as a result of fluid abstraction (e.g. California), will be susceptible to sea-level rise. Whereas the IPCC prediction of sea-level rise is 30–100 cm per century, rates of deltaic subsidence in the Nile Valley are 35–50 cm per century, and in other parts of the world rates of land subsidence produced by oil, gas or groundwater abstraction can be up to 500 cm per century.

Rising sea levels can be expected to cause increased flooding, accelerated erosion and accelerated incursion of saline water up estuaries and into aquifers. Coastal lagoon, spit and barrier systems (such as those of Ras Al Khaimah) may be especially sensitive (Goudie et al., 2000), as will coastlines that have been deprived of sediment nourishment by dam construction across rivers.

Sea-level rise is especially dangerous where there are coastal freshwater aquifers, as along the Batina coast of Oman or on the seaward margins of the Nile Delta. The balance between fresh aquifers and seawater incursion is a delicate one. Modelling, using the Ghyben-Herzberg relationship, suggests that in the Nile Delta – where subsidence is occurring at a rate of 4.7 mm per year, over-pumping is taking place and less freshwater flushing is being achieved by the dammed Nile – a 50 cm rise in Mediterranean level will cause an additional intrusion of salt water by 9.0 km into the Nile Delta aquifer (Sherif and Singh, 1999). In all, a metre rise in sea level would lead to one-third of the Nile Delta being lost (Bohannon, 2010).

Rivers and Runoff

Some dry regions will suffer large diminutions in precipitation (Seager and Vecchi, 2010) and soil moisture levels (Wetherald and Manabe, 2002). Droughts may become more prevalent over wide areas (Dai, 2011). Annual runoff may decline by as much as 60 per cent or more, and drylands appear to be more vulnerable than humid regions in this respect (Guo et al., 2002). The sensitivity of runoff to changes in precipitation is complex, but in some environments quite small changes in rainfall can cause proportionally larger changes in runoff. As rainfall amounts decrease, the proportion that is lost to streamflow through evapotranspiration increases.

Highly significant runoff changes and rates of aquifer recharge (Rosenberg et al., 1999) may also be anticipated for the semi-arid environments of the south-west United States (Thomson et al., 2005). The early model of Revelle and Waggoner (1983) suggested that the effects of increased evapotranspiration losses as a result of a 2°C rise in temperature would be particularly serious in those regions where the mean annual precipitation is less than about 400 millimeters. Projected summer dryness in such areas may be accentuated by a positive feedback process involving decreases in cloud cover and associated increases in radiation absorption on the ground consequent on a reduction in soil moisture levels (Manabe and Wetherald, 1986). Shiklomanov (1999) suggested that in arid and semi–arid areas an increase in mean annual temperature by 1–2° and a 10 per cent decrease in precipitation could reduce annual river runoff by up to 40–70 per cent.

One factor that makes estimates of rainfall-runoff relationships complicated is the possible effect of higher CO_2 concentrations on plant physiology and transpiration capacity. At higher CO_2 concentrations, transpiration rates are lower, and this could lead to increases in runoff (e.g. Idso and Brazel, 1984). It is also important to remember that future runoff will be conditioned by non-climatic factors, such as land-use and land-cover change, the construction of reservoirs, groundwater storage and water demand (Conway et al., 1996).

Channels in arid regions are particularly sensitive to changes in precipitation characteristics and runoff (Nanson and Tooth, 1999). They can display rapid changes between incision and aggradation over short time periods in response to quite modest changes in climate (see Section 5.7). This is particularly true in the case of the arroyos of the American Southwest (Balling and Wells, 1990), which have undergone major changes in form since the 1880s. There has been considerable debate as to the causes of phases of trenching, and it is far from easy to disentangle anthropogenic from climatic causes, but in many cases it is fluctuations in either rainfall amount or intensity that have been the controlling factor (Hereford, 1984; Graf et al., 1991). Sediment delivery by rivers may also be impacted by climate change. It has been suggested, for example, that fire activity in semi-arid regions may increase in a warmer and drier environment, and that this could cause an increase in suspended loads as slopes are subjected to greater erosion and debris-flow generation (Goode et al., 2012).

6.14 Geomorphology and Archaeology

A great disciple of W.M. Davis, E. Huntington (1907, p. 14), working in central Asia, used geomorphological evidence such as shrunken river systems and lakes to reconstruct climate change. He recognised the great influence that such changes could have for human societies in drylands:

> In relatively dry regions increasing aridity is a dire calamity, giving rise to famine and distress. These, in turn, are fruitful causes of wars and migrations, which engender the fall of dynasties and empires, the rise of new nations, and the growth of new civilizations. If, on the contrary, a country becomes steadily less arid, and the conditions of life improve, prosperity and contentment are the rule.

Since Huntington wrote, studies of the geomorphology and Quaternary history of deserts has led to some very valuable collaboration between geomorphologists and archaeologists, as was demonstrated by Butzer's work in Nubia (e.g. Butzer and Hansen, 1968). Also notable is a recent study of the Namib and the Sahara (Bubenzer et al., 2007). Relict landforms (e.g. old lake basins, river courses or spring mounds) can prove to be the sites of former human settlements (Allchin et al., 1978); past climatic changes which have been revealed, for example by changes in lake levels, can help to explain periods of occupation and abandonment of settlements (Table 6.9); and fluctuations in sea level and humidity can help to control human migrations, as for example from Africa to Asia (Osborne et al., 2008). Tectonic history may also have been important, as was the case with the destruction of Sodom and Gomorrah (Neev and Emery, 1995). Sudden shifts in river courses, whatever their cause, may help to explain the changes in the fortunes of the Harappan peoples in northern India (Madella and Fuller, 2006) and in Mesopotamia (Morozova, 2005). Furthermore, geomorphological conditions are highly important in controlling the preservation, exposure and visibility of archaeological remains (Fanning et al., 2009).

There are many examples of the effects of climate change on human societies evident during the Holocene, not least in west Asia (Staubwasser and Weiss, 2006). For example, the dry 8200 cal yr climate event may have forced abandonment of agricultural settlements in northern Mesopotamia and the Levant (Anderson et al.,

Table 6.9 *Examples of some recent studies of climatic influences on human history in drylands in the late Pleistocene and Holocene*

Location	Approximate date	Source
Expansion of cattle keepers into the Sahara	Early- to mid-Holocene	Petit-Maire et al. (1997); Bubenzer and Reimer (2007)
Emergence of Egyptian dynastic state and of Pharaonic civilization	c 6000 BP	Brooks (2006); Kuper and Kröpelin (2006)
Neolithic and Bronze Age in Arabia	c 8000–5000 BP	Parker and Goudie (2007, 2008); Preston et al. (2012)
Anasazi collapse, SW USA	End of thirteenth century AD	DeMenocal (2001); Grove (2004)
Decline of Middle Eastern empires (e.g. the Akkadian)	4200 BP	Weiss et al. (1993); Cullen et al. (2000)
End of Harappan (Indus) civilization	4200 BP	Staubwasser et al. (2003); Madella and Fuller (2006)
Culture change in the Atacama Desert, Chile and Peru	Mid-Holocene	Grosjean et al. (1997); Goldstein and Magilligan (2011)
Cultural change in coastal Peru	Mid-Holocene	Wells and Noller (1999)
Cultural change in Chinese Loess Plateau	C 4000 BP	An et al. (2005)
Occupation of southern Arabia	c 10,000–8,000 BP	Lézine et al. (2010)
Clovis and other early sites in North America	Younger Dryas	Adams et al. (2008) Ballenger et al. (2011)
Bronze Age in Iran	Mid-Holocene	Walker and Fattahi (2011) Schmidt et al. (2011)

2007). Conversely, the 'Greening of the Sahara' in the moist early- to mid-Holocene, may have led to an explosion of activity by Neolithic peoples (Petit-Maire et al., 1999; Drake et al., 2011). From around 6000 cal yr BP, a reduction of rainfall and of monsoonal strength in North Africa, the Near East and Arabia could have forced people out of the deserts into more favourable environments. Around 5200 cal yr BP (Parker et al., 2006a; Staubwasser and Weiss, 2006), a rapid drying and cooling event in the Middle East may have led to the collapse of the Uruk Culture in southern Mesopotamia. Wright (2001, pp. 145–6) sees this as a period of 'differential growth, accelerated inter-regional conflict, the emergence of large polities and their collapse'. Around 4200–4100 BP, another sharp climatic deterioration may also have caused severe problems for many urban centres (Cullen et al., 2000). In general, agricultural intensification and domestication may have been stimulated by episodes of increased aridity (Sherratt, 1997), and there was an association in the mid-Holocene between desiccation and increasing social complexity in the central Sahara and Egypt (Brooks, 2006). Enhanced aridity, Brooks argued, caused population agglomeration in environmental refugia characterised by the presence of surface water (e.g. the Nile Valley). Climate changes may also have impacted on human societies before the Holocene, with pluvials in the Pleistocene (as, for example, during MIS 5), allowing human migrations across the Sahara (Castaneda, 2009; Drake et al., 2011).

6.15 Desert Landforms and World Heritage

There is a considerable opportunity for geomorphologists to become involved in the conservation of geomorphological sites in deserts as part of the World Heritage scheme. In 2011, it was good to see the inclusion of Wadi Rum (Jordan) in the list of newly designated World Heritage Sites. It is clear, however, that the most distinctive landforms and land-forming processes of deserts – aeolian features – are very inadequately represented on the current list of World Heritage Sites. This is the case for dunes, yardangs, pans, dust sources and coastal sabkhas. In addition, weathering forms and various types of crust, rind and varnish are not well represented. The same is true of desert karst features, tufas, various Quaternary landforms (e.g. ancient river systems and pluvial lakes) and some highly important fluvial phenomena, including alluvial fans, pediments and debris flows.

It is therefore important to make nominations for World Heritage status for exceptional features (Goudie and Seely, 2011). A property nominated for inclusion in the Natural World Heritage list will be considered to be of outstanding universal value if the World Heritage Committee finds that it meets one or more of the following criteria, providing it also meets the conditions of good management and of integrity (UNESCO, 2008, clauses 77 and 78):

(vii) Contain superlative natural phenomena or areas of exceptional natural beauty and aesthetic importance;
(viii) Be outstanding examples representing major stages of the Earth's history, including the record of life, significant on-going geological processes in the development and landforms, or significant geomorphic or physiographic features.

The following locations are exemplars of sites that may perhaps deserve inclusion in the World Heritage List in the future. Parts of some of them are on the tentative lists of some countries (see below).

Western Desert, Egypt. The Western Desert (see Section 7.3) is notable as being the site of some of the most formative work that has even been undertaken in aeolian geomorphology, most notably by Ball, Cornish, King, Beadnell and, above all, R.A. Bagnold. It is characterised by classic barchans and linear dunes that have probably been the subject of more serious observation than any other dunes on Earth. However, it also has a full range of other desert features that reflect the area's profound aridity: spring mounds, tufa spreads, groundwater sapping features, closed depressions, yardangs, relict karst, the Selima Sand Sheet and the sandstone topography of the Gilf Kebir (Embabi, 2004).

United Arab Emirates – Sabkha. The sabkha, consisting of marine salt flats on the western side of the Arabian Gulf, is the best developed and most studied example of this landscape type to be found anywhere on Earth (see Section 3.11). It extends along the coastline of Abu Dhabi Emirate between Jabal Dhanna and Ras Ganada, a distance in excess of 300 km. It is also a highly important model for hydrocarbon generation (Evans, 1995).

The Chotts, Tunisia. The closed basins of Tunisia, the subject of much classic French geomorphological research, consist of a series of large basins that were formerly more extensive in pluvial times. They are notable as being examples of saline basins, but they also have within them some of the best world examples of gypsum crusts and gypsum dunes. They are bounded in part by extensive rock ramps, called glacis. The Chott Djerid is probably the most important of these features (Swezey, 1997).

Badain Jaran, China. This interior desert of China is already a Global Geopark and has been the subject of intense study in recent years by Chinese, Japanese, German and UK scientists. In addition to being aesthetically very impressive, it contains the world's tallest dunes (up to 450 m) and a great variety of forms, including star dunes. Within the dunes there are many intriguing inter-dunal lake basins. The area also possesses spectacular weathering features, including tafonis and alveoles.

Death Valley, California, United States. Death Valley is the lowest point in the United States and is bounded on either side by great actively uplifting mountains. It is a classic example of basin and range topography, of a salt lake and of pluvial lake expansion. It is also a classic area to study desert varnish of different ages. One of its most important landform types, however, is the alluvial fan, which develops on the interface between the mountains and the basin (Hunt, 1975).

The Namib-Naukluft Park, Namibia. The ancient, coastal Namib Desert of southern Africa is a well-protected area with an extended period of detailed desert research based on Gobabeb. The modern Sand Sea (see Section 7.5) is underlain by a fossil desert of Tertiary age, represented by the lithified Tsondab Sandstone. In addition to this important example of desert evolutionary history, the Namib also exemplifies the impact of sea-floor spreading since the Cretaceous, with the emplacement of many sub-volcanic complexes and the development of an upwarped marginal escarpment. It also contains the full range of dune types, excellent examples of the ways in which river courses can be blocked by dunes (as at Sossus Vlei), great spreads of calcretes and tufas and many examples of granite weathering and inselberg and pediment formation (Lancaster, 1989a).

Lake Bonneville, Utah, United States. Lake Bonneville was a giant pluvial lake which occupied the basin in which the current Great Salt Lake lies. Made important by the classic work of G.K. Gilbert (1882, 1890), it possesses beautifully developed strandlines (Figure 1.8) and ancient lake coast features (deltas, etc.). Ideas developed about the basin, which displays the interface between mountain glaciation and pluvial lake development, have been very important in understanding the evolution of ideas on climate change in mid-latitude locations and also of hydroisostasy. The history of the basin included episodes of catastrophic flood pulses, which have played a role in the development of neo-catastrophist thinking (Oviatt et al., 1992).

Lut Desert, Iran. This desert contains some of the largest and best developed yardangs found anywhere on Earth. The kaluts form parallel ridges and depressions over an area of 120 × 50 km. Some of the ridges exceed 60 m in height and run parallel – with superbly developed aeolian streamlining – to the formative shamal winds. They are located near Bam, and occur in association with some impressive barchan dunes (Gabriel, 1938).

Bodélé Depression, Chad. This Saharan depression is very easily the largest dust source on Earth (see Section 3.4). It is therefore the best location to study the action of dust storms and to witness the effects of deflation on desiccated Holocene and Pleistocene lake sediments (including diatomites) (Warren et al., 2007). It is also believed to contain the fastest moving barchan dunes on Earth (Vermeesch and Drake, 2008).

Arches National Park, Utah, United States. This area possesses a suite of sedimentary rocks of which various sandstones are the most important. The rocks have been eroded by fluvial and groundwater action to produce the largest collection of natural arches on Earth and some of the largest and most aesthetically pleasing specimens that are known. In addition to the intrinsic value of the arches themselves, the area is one that demonstrates the many forms of weathering features that develop in an arid climate and the importance of groundwater sapping processes for stream development in desert regions. This has some interest in that it can be used as a potential analogue of Martian features (Barnes, 1987).

Kimberley Limestone Ranges, Western Australia. This site, consisting of the Oscar and Napier Ranges, deserves to be designated on the basis of its karst features. It has become the type site for semi-arid karst and displays many karren features, tufas, gorges, box valleys, pediments, tunnel valleys, and so forth. It is also a superb demonstration of the importance of rock lithology on the development of landforms, for the exposed facies of the ancient Devonian reef are crucial in understanding the array of different slope forms that have formed (Goudie et al., 1990).

Hunza Valley, Pakistan. In the Karakoram Mountains of Pakistan, the valley bottoms have an arid climate (less than 100 mm of rainfall). The degree of incision that has taken place is such that between the base of the Hunza River and Mount Rakaposhi, there is the first- or second-greatest relative relief on the Earth's land surface. One thus has a magnificent example of a mountain desert, with the interplay between active tectonics, present and past glaciation, river floods, massive debris flows, landslide damming of lakes and scree formation on an unimaginable scale (Miller, 1984).

In addition, some additional sites that are on the tentative lists of miscellaneous state parties have very considerable merit as exemplars of particular geomorphological processes or landforms. These include:

1. **Afghanistan. Band-E-Amir**. These lakes, located in an arid portion of the country, have been the subject of detailed research by French and other workers, and are of very considerable aesthetic appeal. They are a suite of tufa-dammed lakes which are of comparable importance to the Plitvice Lakes of Croatia.
2. **Algeria. Les oasis a foggaras et les ksours du Grand Erg Occidental**. This site includes some of the great dunes and wadi systems of the Sahara.
3. **Argentina. Las Parinas**. This is an arid area (mean annual rainfall less than 200 mm) at a high altitude in the Puna. There are many volcanic landforms, but there are also major salt flats.
4. **Botswana. Mkgadikgadi Pan Landscape and the Central Kalahari Game Reserve**. These sites include one of the world's largest salt pans and the ancient river systems and palaeolakes of the Kalahari.
5. **Chad. Les Lacs d'Ounianga and gravures et peintures rupestres de l'Ennedi et du Tibesti**. The former site consists of lakes, which occur in the lee of topographic obstructions, are good examples of deflational basins and also occur in an area of sand dunes, some of which may be relict features of former more arid conditions. The latter is a classic mountain area of the central Sahara with excellent examples of many desert features.
6. **Chile. San Pedro de Atacama**. This site (mean annual rainfall less than 100 mm) is located in the Atacama Desert in close proximity to the Andes. It is an important example of a desert landscape in an area with complex block faulting and an enormous salt lake.
7. **China. Taklamakan Desert – *Populus euphratica* forests**. This area includes a whole range of desert landforms and is a major area for dust generation and wind erosion.
8. **India. Desert National Park**. This includes a large area of the Thar (Rajasthan) Desert.
9. **Iran. Hamoun Lake and Lut Desert (the vicinity of Shahdad)**. A large ephemeral lake system forming part of the Seistan Basin and one of the world's most impressive yardang areas.
10. **Israel. Makhteshim Country**. A fine display of the relationship between rock structure and landforms.
11. **Kazakhstan. Steppes and lakes of North Kazakhstan**. The area has one of the largest arrays of closed depressions and associated lakes known. There are many thousands of examples in the area, which makes it comparable in significance to the High Plains of the United States and parts of the interior of southern Africa. However, they have been very little studied.
12. **Mexico. El Pinacate et le Grand Desert d'Altar**. This is a large area in the north-west of Mexico and has a range of dunes, in addition to volcanic landforms. It forms part of the Sonoran Desert.
13. **Mongolia. Great Gobi Desert**. The Gobi in south-west Mongolia is a major desert and is one of the best examples of a relatively high-altitude, cold desert (especially in winter).

It undoubtedly has a very large range of desert landforms, including gravel deserts and so forth.

14. **Namibia. Brandberg National Monument, Fishriver Canyon, Southern Namib Erg and Welwitschia Plains**. These four sites include major inselbergs, gorges, dunes and gravel plains.
15. **Qatar. Khor Al-Adaid Natural Reserve**. This area contains both dunes and sabkhas.
16. **Sudan. Wadi Howar**. This is a major former tributary of the Nile and is one of the most perfect examples of a river system that has ceased to flow through its length because of climate change. It has been the subject of important palaeoclimatic research by German scientists.
17. **Tunisia. Chott El Jerid**. One of the major salt lake areas of the chott zone of North Africa.
18. **Turkmenistan. Repetek Biosphere State Reserve**. Contains a large portion of the south-east Karakum Desert.
19. **United States, Utah. Bryce Canyon, Canyonlands, Capital Reef, Rainbow Bridge, Zion**. There are many truly world-class sites in Utah that contains superb examples of a range of desert geomorphological phenomena.

Some existing World Heritage Sites, which have been included largely as cultural sites, also have geomorphological significance, and there may be scope for the case being made for their designation as mixed properties. These include the Aflaj Irrigation System of Oman (human-made landforms associated with irrigation); the Al-Hijr Archaeological Site, Saudi Arabia (sandstone weathering); ancient Thebes with its Necropolis, Egypt (accelerated weathering); Archaeological Ruins at Mohenjo-Daro, Pakistan (accelerated salt weathering); the Archaeological Site at Volubilis, Morocco (calcrete development); the Champaner-Pavagadh Archaeological Park, India (fossil topographic dunes); Humberstone and Santa Laura Saltpeter Works, Chile (Caliche); Incense Route Desert Cities in the Negev (desert runoff processes at Avdat); Rock art sites at Tadrart Acacus, Libya (rock rinds, desert varnish, sandstone weathering and natural arches); the St Catherine area, Egypt (granite weathering); and Tsodilo Hills, Botswana (ancient linear dunes and pluvial lake deposits).

Finally, there is also the UNESCO Global Geoparks Network. As yet, with the exception of the Alxa Desert Geopark of Inner Mongolia, China, there are no desert sites incorporated in this list. Finally, desert areas with spectacular landforms can be major tourist attractions. For instance, in the United States each year the Grand Canyon attracts c 4.4 million recreational visitors, Joshua Tree 1.4 million and Badlands, Death Valley and Arches around 1.0 million each.

6.16 Desert Landforms and Military Activity

It is an unfortunate fact that many wars have been waged in arid areas over recent decades, including those in Iraq, Sinai, Kuwait, Afghanistan, Palestine and Sinai. Geomorphologists have been employed to advise government agencies on such issues

Figure 6.18 The Coastal Defence Regiment on patrol in the Wahiba Sands of Oman. (ASG)

as the trafficability of terrain, the destruction of ecosystems, the generation of aeolian dust by armoured vehicles and by helicopters (Gillies et al., 2010) and the increase in fluvial sediment yields caused by the operation of tracked vehicles (Fuchs et al., 2003). A recent survey of this type of endeavour, which relates to the Yuma Proving Ground in Arizona and the desert terrain of Afghanistan, is provided by Bacon et al. (2008). A full analysis of the trafficability problems (Figure 6.18) of a very wide range of desert landform types is provided by the Desert Processes Working Group (1991). Advice is provided on which terrains to avoid. Equally, some work is undertaken on the potential environmental consequences of military activity over desert surfaces (e.g. Gilewitch, 2004; Peterson et al., 2008), for the movement of vehicles over, for example, stone pavements can leave very long-lived scars (Belnap and Warren, 2002).

6.17 The Search for Planetary Analogues

One of the consequences of planetary exploration has been an attempt to find analogues on Earth for phenomena that have been identified elsewhere in our solar system, most notably on Mars and Titan (Clarke, 2011). This has stimulated research on a number of landforms and processes (M. Thomas et al. 2005; Tooth, 2009) and has led to investigations in a number of the world's deserts, including Namibia (Bourke and Goudie, 2009), Egypt (El-Baz and Maxwell, 1982) and Australia (Mann et al., 2004).

Table 6.10 *Possible desert features on Mars*

Aeolian scouring (Bishop, 2011)
Alluvial fans (Williams et al., 2006)
Desiccation polygons (El Maarry et al., 2010)
Dunes (Bourke and Goudie, 2009; Chojnacki et al., 2010)
Dust devils and dust storms (Balme et al. 2003)
Gully systems (Morgan et al., 2010)
Inverted relief (Pain et al., 2007; Williams et al., 2009; West et al., 2010)
Mound springs (West et al., 2010)
Mudflow lobes (Heldmann et al., 2010)
Physical weathering (Viles et al., 2010)
Polygonal weathering (Chan et al., 2008; Levy et al., 2010)
Ripples (Balme et al., 2008; Zimbelman et al., 2009; Zimbelman, 2010)
Rock coatings and varnish (Bao et al., 2001; Krinsley et al., 2009)
Sabkha (Sadooni et al., 2010)
Salt weathering (Jagoutz, 2006)
Sapping features (Lamb et al., 2006, 2008)
Split rocks (Bourke and Viles, 2007)
Stone pavements (gibber plains) (West et al. 2010)
Ventifacts (Laity and Bridges, 2009)
Wind erosion basins (Howard et al., 2012)
Wind streaks (Rodriguez et al., 2009)
Yardangs (de Silva et al., 2010; Zimbelman and Griffin, 2010)

Excellent high-resolution images are now available, and for Mars we have information from ground-based rovers. The resolution and quality of the data produced in recent years is very impressive – often comparable to or better than that produced by satellites orbiting Earth. The *Mars Reconnaissance Orbiter*, launched in 2005, produces imagery at resolutions of around 0.25 -1 m from the HiRISE camera. It can pick up individual ripples. There have been successful lander vehicle missions on Mars, starting with the *Pathfinder* lander in 1997 and more recently with the *Spirit* and *Opportunity* rover vehicles. The vehicles which land on the surface and move around have provided remarkable data on their surroundings from on-board cameras. They have also used tools such as the rock abrasion tool (RAT) and the microscopic imager (MI) to view the surface and subsurface of boulders and rock outcrops to a resolution of c 100 microns.

Mars is a predominantly cold, basaltic desert, but it differs in many ways from Earth in terms of its lower surface gravitation, lower atmospheric pressures, lower average temperatures, limited plate tectonic activity and limited water vapour. There are many Martian phenomena for which Earth analogues have been sought (Table 6.10). The Martian dunes include a suite of linear and barchan features (Figure 6.19) which can be ten times the size of similar features on Earth. Globally, dunefields on Mars cover an area in excess of 900,000 km^2. They are found in a series of sand seas that surround the north polar ice cap, on the floors of impact craters in the inter-basin

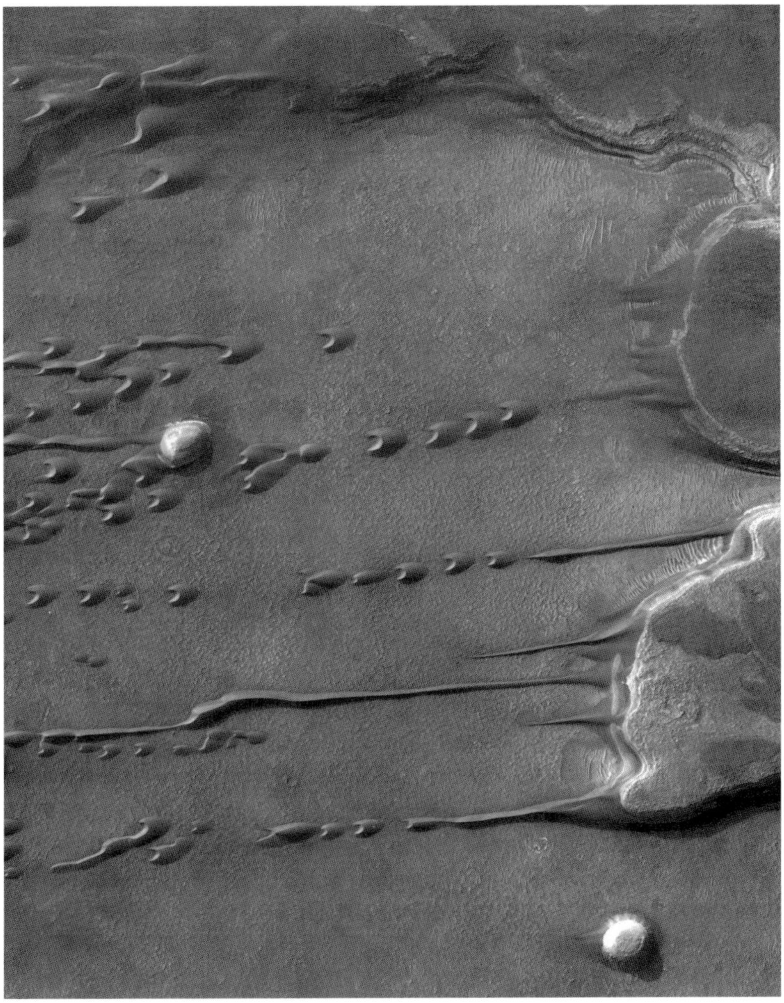

Figure 6.19 Linear dunes and barchans on Mars. (Courtesy of NASA)

plains of the southern hemisphere and in various low-latitude topographic traps. Most of the dunes are formed of basaltic or andesitic material, although some contain a significant proportion of hydrated minerals (Bourke et al., 2010). HiRise images show that substantial sediment transport occurs today on Martian north polar dune faces as a result of grain flow triggered by seasonal CO_2 sublimation (Hansen et al., 2011) and in response to gusty winds (Bridges et al., 2012a). Indeed, Bridges et al. (2012b, p. 341) have argued that over much of Mars, conditions 'are sufficient to move large dunes and transport fluxes of sand equivalent to those on Earth'.

Martian dust events have been observed at scales ranging from local dust devils to storms that envelop the entire planet ('global dust storms'). These planet-encircling events occur approximately one year in three, usually in late southern spring when the planet is closest to the Sun. Regional dust storms may be produced whenever the poleward temperature gradient is sufficiently large to generate intense zonal circulation across the mid-latitudes in the form of baroclinic waves. Other regional dust storms are produced by katabatic outflow from receding frost outliers of the polar caps, with winds descending from areas of high relief. Regional dust storms affect the radiation budget, and this can lead to feedback effects that may cause the development of dust storms of global dimensions. During the Martian summer, in the lower boundary layer of its clear, thin, cold atmosphere, the large temperature gradient that exists above the relatively warm surface may support intense free convection and the formation of dust devils, which are often greater than those found on Earth, reaching several hundred metres across and 8 km high.

Titan is the largest of Saturn's moons, and its geomorphology recently become visible as a result of the *Cassini* mission which first visited Titan in 2004, dropped the Huygens probe onto the surface in 2005 and has flown by several times since then. Titan is comparable in size to our Moon but is atmospherically more like Earth. The images of the surface of Titan sent back by the Cassini mission illustrate a landscape quite similar to that on Earth – except that the very cold temperatures mean that the surface is made up of water ice, not rock, and is sculpted by liquid methane, not water. Titan has some interesting geomorphological features, including thousands of linear dunes (Lancaster, 2006; Rubin and Hesp, 2009), which cover as much as 20 per cent of its surface (Radebaugh et al., 2010). The 12–18 million km^2 of dunes is the largest cover of dunefields in our solar system (Bourke et al., 2010). The main dunefields occupy the lowest elevation areas in equatorial regions (Le Gall et al., 2012). The dunes appear to be composed of sand-sized organic particles and are of a scale similar to large linear dunes on Earth in places such as Namibia. The dunes on Titan are 1–2 km wide, are spaced 1–4 km apart and can be hundreds of kilometres long. There are also some stubby drainage systems that may result from methane-spring sapping (Soderblom et al., 2007) and also some playa lakes that may be comparable in some respects to Race Track Playa in California (Lorenz et al., 2010) or to Etosha Pan in Namibia (Bourgeois et al., 2008).

It is also possible that there are aeolian processes on Venus, for 'the enormous atmospheric pressure at the surface of Venus should allow even large particles to be transported with a slight breeze' (Craddock, 2011, p. 111). However, the evidence for aeolian features there is limited, although wind streaks and dunefields have been postulated.

7

Regional Variety

7.1 Introduction

The purpose of this final chapter is to describe some of the more important factors that make particular warm deserts distinctive. Greater detail is given in Goudie (2002).

7.2 Sahara

The Sahara, the world's largest desert, covers an area of around 7–9 million km^2 (Tucker et al., 1991), which is similar to that of the United States (including Alaska). Occupying approximately one-third of the African continent, it stretches from the Atlantic to the Red Sea and from the Mediterranean to the Sahel zone of West Africa. Its heart is dry, with mean annual rainfall dropping to less than 10 mm. Sometimes it is subdivided into two or more parts, with the eastern portion being called the Libyan Desert and the western part being regarded as the Sahara proper. This division is followed here. The general morphology of the Sahara has been discussed by Mainguet (1983), who suggests that its most distinctive characteristic – save only the relief provided by the Hoggar, Tibesti and other massifs – is its flatness. This is associated with great sandstone plateaux – a series of broad, closed basins (of which Chad/Bodélé is the most notable), and a series of wind-moulded landscapes which include deflational surfaces, fields of yardangs and areas of sand deposition (ergs).

In the Ordovician extensive glaciation of the Sahara and Libyan Deserts occurred the erosional evidence which is still well preserved around the Hoggar massif (Schwarz and Germann, 1999) and elsewhere (Le Heron et al., 2010). In the Mesozoic, marine and non-marine sedimentary rocks were laid down, including the Nubian Sandstone formation of the eastern Sahara and the Continental Intercalaire of the central Sahara. Many of these beds, which are important aquifers, have gentle dips and form extensive plateaux or hamada. Some of the sandstone massifs of the Sahara – such as those of the Ennedi Plateau in Chad, the Tassili n'Ajjer of eastern Algeria and the Acacus Mountains of western Libya – have spectacular landform assemblages that rival those of Wadi Rum in Jordan. They are also notable in some cases for the development of sandstone karst (Busche, 1998), which probably developed during the early Miocene (Busche and Erbe, 1987).

The next major phase in the Sahara's geological evolution, following the retreat of the Cretaceous seas, was Cenozoic weathering and erosion (Swezey, 2009), including the mobilization and deposition of iron and silica, which now protect many of the Mesozoic and Tertiary plateau summits from erosion and produce resistant caprocks (Michel, 1973). Conditions in the early Tertiary were affected by the latitudinal position of the area, which was about 15° further south than at present. This resulted in hot, humid climates that favoured substantial chemical weathering and crust formation. From the middle Eocene onwards, the Atlas Mountains rose up (Gomez et al., 2000; de Lamotte et al., 2009), with the highest relief developing in the western part of the Maghreb. In the late Cenozoic, the uplift of the central Saharan uplands – including Jebel Marra, Tibesti, Hoggar and Aïr – was accompanied by massive outpouring of lava and the creation of striking volcanic craters and pinnacles. Indeed, Tibesti 'is one of the world's major volcanic provinces and a key example of continental hot spot volcanism' (Permenter and Oppenheimer, 2007, p. 609). These uplands form a very important component of the Sahara. They rise to considerable altitudes: Tibesti to more than 3,400 m, Jebel Marra to more than 3,000 m, Hoggar to more than 2,900 m, and Aïr to more than 2,000 m.

Another major event in the Sahara's history was the Miocene Messinian Salinity Crisis, which occurred around 6 Ma. It was a time when the Mediterranean Basin became isolated from the Atlantic at what are now the Straits of Gibraltar (Blanc, 2000). The level of the water in the Mediterranean plummeted, and it became a zone of major salt accumulation. Rivers cut down to low base levels, creating enormous gorges (Clauzon et al., 1996). A large lake was present in the Chad Basin at this time, and sporadically it overflowed to produce river systems – the Sahabi rivers, which have been traced in Chad and Libya (Griffin, 2006, 2011).

The Sahara is the world's largest source of dust in the atmosphere, with an annual production of 400–700 Tg per year (Swap et al., 1996), perhaps contributing as much as 50 per cent of the total. The Bodélé Depression is a true hotspot for global dust emissions. The Sahara displays many other features that result, at least in part, from aeolian action, including some enormous ridge and swale systems produced by wind erosion of bedrock. They are particularly well developed in a zone bordering the Tibesti Mountains, especially on their east, south and west sides. The features sweep in a broad arc around the mountains and are easily identified on satellite imagery. Busche (1998) provides detailed descriptions of wind erosion features, large and small.

The Sahara has many active ergs, whereas on the south side of the Sahara, in the Sahel belt, there are at least three ages of ancient ergs, now fixed, that formed under drier and windier conditions. Dunes cover just over a quarter of the total area. As Lancaster (1996, p. 221) has remarked, 'The Sahara contains a significant proportion of the global inventory of wind-blown sand as well as some of the world's largest sand seas.' The major sand accumulations occur on either side of the uplands such as Tibesti and Hoggar, which themselves have little sand, being zones of transport rather than deposition. There are recognizable sand-transport pathways in which the trade winds tend to move sand from both the sand-depleted eastern and central parts of the Sahara towards thick sand accumulations in the Sahel zone and from the piedmont of the Atlas Mountains and the Mediterranean coast to the northern and western ergs.

The location of the major Saharan ergs is controlled in part by topography. The Grand Erg Occidental and the Marzuq and Ubari Sand Seas have accumulated against the northern slopes of the central Saharan uplands. Sand seas in the Akchar of Mauritania and the Fachi-Bilma region of Chad exist where winds converge in the lee of uplands. Many of the dunes occur near and in topographic depressions, into which sand has been transported by rivers, including numerous palaeorivers revealed on radar images (El-Baz et al., 2000).

On the north side of the Sahara, in a belt that runs through the centre of Tunisia westwards towards the high plateaux of Algeria, is a zone of depressions that are normally called chotts. Swezey (1997) suggested that they lie in a structurally controlled sedimentary basin within which there are certain depocentres (e.g. Chott Djerid) that were created by Miocene to early-Pleistocene compression associated with the Atlas orogeny. Another great hydrological system on the south side of the Sahara is the Chad Basin. Covering a total area of 2.5 million km^2, it stretches for more than 1,500 km from the low watershed of the Congo into the heart of the Sahara, where it receives runoff from Tibesti and Hoggar. About 90 per cent of its water comes from the Chari/Logone river systems, however. The lowest point of the basin is the Bodélé Depression, the altitude of which is only 150 m above sea level. This and other basins contained large freshwater bodies both in the early Holocene and during the last interglacial ($\sim$125 ka).

7.3 The Libyan Desert

The Libyan Desert (which is called the Western Desert in Egypt) forms part of the eastern Sahara and is the largest expanse of profound aridity on Earth. For the most part it is rather flat, and only limited areas reach altitudes more than a few hundred metres above sea level. Much of it is underlain by relatively gently dipping limestones, shales and sandstones that create low escarpments and gently sloping plateaux. Higher land only tends to occur in the south-west of the region, where the Gilf Kebir forms a flat plateau of sandstone attaining heights of more than 1,000 m above sea level and where the granitic Gebel Uweinat rises to more than 1,900 m. The abrupt rise of the Gilf Kebir is made all the more remarkable because of the fact that it climbs above a desert surface that is immensely flat – the so-called Selima Sand Sheet. This covers c 120,000 km^2 and is a largely featureless surface of lag gravels and fine sand broken only by widely separated dunefields and giant ripples. It was thus described by Peel (1941, p. 6):

> This region appears to the eye absolutely flat and featureless save for an occasional line of sand dunes. In reality it slopes very gently eastwards towards the Nile, but for scores of miles no feature larger than a tiny pebble breaks the uniformity of the surface. The solid rock is everywhere covered with a uniform sheet of wind-blown sand, which is probably nowhere more than a few feet thick: the sandstone beneath it would appear to have been worn down almost to a true plain.

The sand sheet is the product of a very long history of erosion and aggradation by river and wind processes (see, for example, Maxwell and Haynes, 2001). Another intriguing feature of the Gilf Kebir region is the occurrence of many crater-shaped structures, the origin of which is still obscure, although it has been hypothesised that they could be the result of meteorite impact or of hydrothermal vent formation (Paillou et al., 2006).

The erodible sedimentary rocks that characterise most of the region, however, have been excavated to produce some great closed depressions – Fayum, Qattara, Farafra, Bahariya, Dakhla, Kurkur, Kharga and Siwa – places where the underground aquifers approach to or attain the surface, so producing oases. The Qattara has been excavated to –133 m below sea level. These depressions may owe some of their form to excavation by Eocene karstic processes or to incision by now-defunct river systems, but wind action has certainly played a highly significant role, aided and abetted by salt attack (Aref et al., 2002). Indeed, because large areas only have a few mm of precipitation per year, and because they are subjected to the persistent northerly trade winds, aeolian processes have produced dunefields and wind-fluted terrain (Embabi, 2004). It has been a classic area for dune research, (Bagnold, 1941). It contains a large erg, which is appropriately called The Great Sand Sea (Besler, 2008).

The closed depressions of the Libyan Desert have been much affected by humid climates in the mid-Holocene and portions of the Pleistocene. Large freshwater lakes existed, as did active rivers such as Wadi Howar (Pachur and Kröpelin, 1987). Hoelzmann et al. (2001) have established the existence of what they term the West Nubian Palaeolake, which covered up to 7,000 km^2 between 9500 and 4000 years BP. The moist phases are also represented by widespread spring deposits, tufas, large mass movements in shales and groundwater-sapped cliffs.

Some of the distinctiveness of the Libyan Desert is created by the existence of the Nile's present and former courses. It is a young river in geological terms. Its course has been affected by the retreat of the Tethys Ocean, the desiccation of the Mediterranean

Basin around 6 Ma and the plate splitting that led to the uplift of the Red Sea Hills and the mountains of Ethiopia (Issawi and McCauley, 1992; Goudie, 2005). For example, at the end of the Oligocene (c 24 Ma), a river (the Gilf system) flowed westward from the newly uplifting Red Sea Hills through Aswan and Dakhla to Siwa, whereas in the middle Miocene (c 16 Ma), drainage in the area (the Qena system) was essentially south-westwards from the Red Sea Hills towards the Chad Basin. Around 6 Ma, at a time of very low sea level in the Mediterranean (the Messinian Salinity Crisis), a precursor of the present Nile (the Eonile) cut back southwards along a great canyon to capture the Qena system. Another great river system of possible Miocene age linked Tibesti to the Mediterranean via Kufra (Paillou et al., 2009; Ghoneim et al., 2012).

7.4 Eastern Africa

There is a large area of aridity in eastern Africa. It includes the Eastern Desert of Egypt, the eastern Nubian Desert in Sudan, the Danakil Depression, the Chalbi Desert of northern Kenya, the Ogaden of Ethiopia and the drylands of Somalia. East Africa is remarkably dry given its latitudinal position. In Kenya and Somalia, notwithstanding their proximity to the equator, much of the region below c 1,000 m is semi-arid or even arid. There are those who argue that if orographic effects were removed, almost the whole of eastern Africa would be dry or subhumid. The reasons why tropical East Africa features anomalously low precipitation relative to its latitudinal position include wind patterns that typically run parallel to the coast, the rapid passage of the Intertropical Convergence Zone (ITCZ), the rain-shadow effect of the mountains, the presence of cold upwelling water off the Somali coast and the loss of moisture in the south-east trades over the mountains of Madagascar before reaching East Africa. The Chalbi Desert lies east of Lake Turkana in the arid rain-shadow corridor between the Kenyan and Ethiopian Highlands. The mean annual rainfall at North Horr, which is at the desert's centre, is 157 mm. Chalbi is an enclosed intermontane basin that contains a playa – Lake Chalbi – that has periodically been wet (Nyamweru and Bowman, 1989), including at a time of enhanced precipitation in the latest Pleistocene and earliest Holocene (Bruhn et al., 2011). The area has been much affected by Pliocene and Quaternary tectonic and volcanic activity.

Central to an understanding of much of this region is the development of the Red Sea and of the highlands along its margin. The Red Sea appears to have experienced its first major continental rifting and concomitant igneous activity in the late Oligocene, about 30 Ma. During the late Oligocene to early Miocene (15–30 Ma), the Afar Depression formed, and extension began to occur in the Red Sea Depression. Uplift of the Red Sea margins began in the middle Miocene (13.8 Ma), and the Gulf of Aden opened in the late Miocene (10 Ma). In the Pliocene, mid-ocean ridge basalts developed in the axial trough that runs down the Red Sea.

Details of the rifting progression on the western side of the Red Sea are given in Kenea et al. (2001) and Wolfenden et al. (2005). The Eastern Desert of Egypt largely consists of mountainous terrain, produced by the uplift of the Red Sea Hills. It is characterised by heavily dissected, typical crystalline rock terrain underlain by the pre-Cambrian basement complex but also by outcrops of Nubian Sandstone and later carbonates (Abdel Moneim, 2005). The eastern portion of the Nubian Desert in Sudan is also mountainous, and here the Tokar Delta, which is located at a gap in the Red Sea Hills, is a major source of dust storms (Hickey and Goudie, 2007).

The Danakil Desert (an area which probably has the world's highest mean annual temperature) occupies the northern end of the East African Rift Valley at its junction with the Red Sea Rift. It is in an area that has active volcanic activity and tectonism. It contains two areas that extend below sea level. One lies to the west of Djibouti and has a sump, Lake Assal Hayk, whose surface lies at 150–155 m and is the lowest point in Africa. It is also reputed to be the saltiest body of water in the world. The other area, the Dallol salt flats, lies in Tigray and Eritrea and extends to 116 m below sea level. The area's closed depressions have experienced fluctuations in their levels in response to climatic changes (see, for example, Gasse, 1978; Street, 1980) and had high stands in the early to mid-Holocene.

The Ogaden, an area disputed between Ethiopia and Somalia, is perhaps the least known of the African deserts. Parts of it are extremely dry, with the mean annual precipitation at Bender Cassim (on the Somali coast) being only 18 mm. A detailed study of the geomorphology of the Eritrean desert is provided by Abul-Haggag (1961), while the long-term evolution of the Red Sea margin in Eritrea is discussed by Balestrieri et al. (2005). The drylands of Somalia are, inter alia, notable for the extraordinary development of banded vegetation stripes (*brousse tigrée*) (see Section 2.32).

7.5 The Namib

The Namib extends for 2,000 km along the South Atlantic coastline of South Africa, Namibia and Angola (Figure 7.1). It is narrow (only 120–200 km wide), being bounded to its east by the Great Escarpment. It is hyperarid (annual rainfall at the coast is often only 10–20 mm), but is characterised by frequent, wetting fogs (Olivier, 1995).

Its landscape demonstrates the importance of its setting on a passive margin. The Great Escarpment (Kempf, 2010), the sloping plains of the Namib itself and its major inselbergs can be explained by the opening of the South Atlantic in the early Cretaceous, the separation of southern Africa from South America and the development of a major hotspot track associated offshore with the Walvis Ridge and the Tristan and Gough Islands (Goudie and Eckardt, 1999). Igneous extrusive and

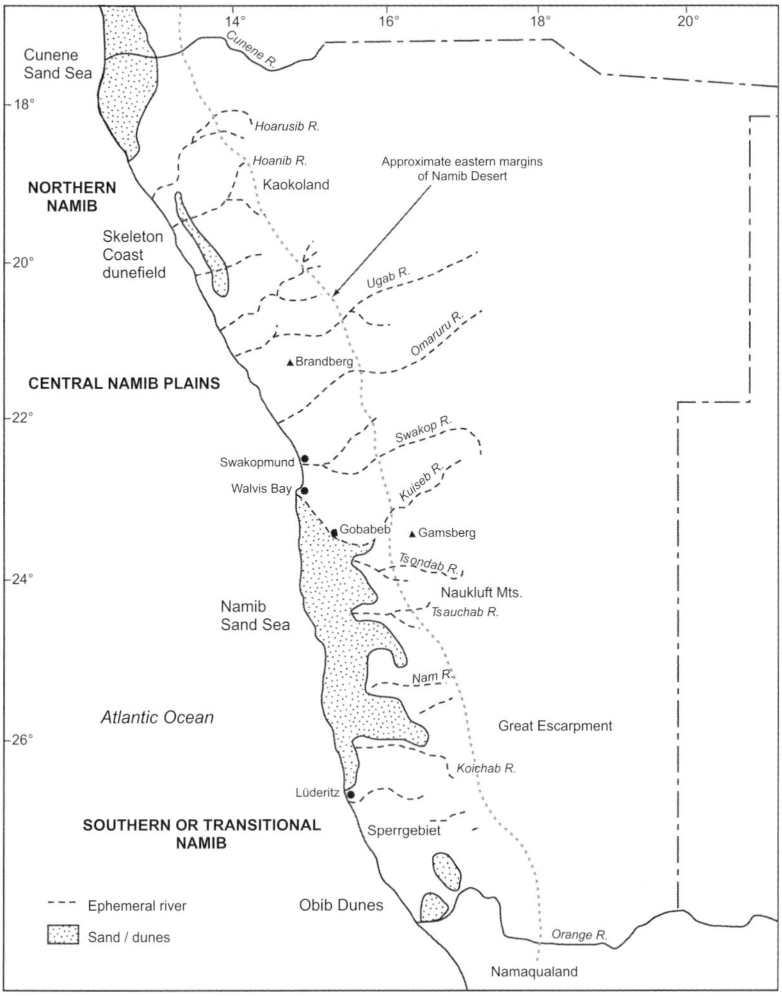

Figure 7.1 The major physiographic features of the Namib. (From Goudie, 2002, fig. 6.2)

intrusive activity produced large spreads of lava (Figure 7.2) (the Etendeka lavas) and caused the development of some large plutons and associated inselbergs (e.g. Brandberg, Erongo and Spitzkoppje). The Great Escarpment formed as a result of uplift and incision following the breakup of Gondwanaland. Deeply incised into it is the Fish River Canyon.

The Namib is one of the world's most ancient deserts, and this also must have been controlled to a considerable extent by its plate tectonic history, which influenced the opening of the seaways of the Southern Ocean, the location of Antarctica with respect to the South Pole and the subsequent initiation of the cold, offshore Benguela Current. The onset of aridity could date back to the early Cretaceous, for dune beds are found

Figure 7.2 The Etendeka lavas of northern Namibia were extruded in the early Cretaceous as a result of the opening of the South Atlantic. (ASG)

interdigitated with Etendeka lavas (Jerram et al., 2000). Ward et al. (1983) believe that the Namib has not experienced climates significantly more humid than semi-arid at any time during the last 80 million years. The present Namib Sand Sea, for which an impressive digital database is now being developed (Livingstone et al., 2010), and which contains a wide range of dune types (Bullard et al., 2011), is underlain by a lithified erg composed of the Tsondab Sandstone, and this dates back to at least the lower Miocene (Senut et al., 1994.) By the late Miocene, offshore dust inputs were increasing, and river inputs were decreasing (Kastanja et al., 2006).

The Namib has a diversity of landforms that includes wind-fluted terrain (yardangs), especially in northern Namibia (just to the south of the Cunene River) (Goudie, 2007b) and in the southern Namib near Luderitz. There are also four major ergs or sand seas (which from north to south are the Baia dos Tigres erg in Angola, the Cunene Erg, the Skeleton Coast Erg and the Namib Sand Sea [Figure 7.3]). High velocity winds blowing out from the interior plateau (berg winds) (Eckardt et al., 2001) have created pans, wind streaks and dust storms.

The coastal portions of the Namib Desert, because of the prevalence of fog and large quantities of salts – including gypsum (Eckardt and Spiro, 1999; Eckardt et al., 2001) – are sites of very rapid salt weathering (Viles and Goudie, 2007, in press). As in the Atacama, this may explain the presence of extensive, featureless plains (Goudie et al., 1997).

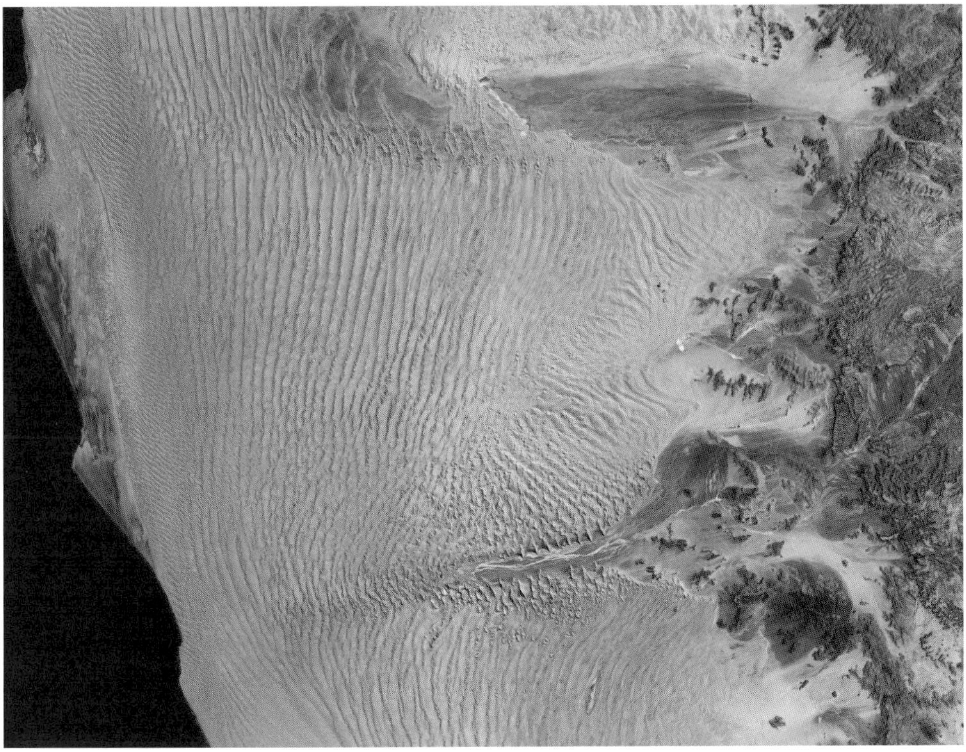

Figure 7.3 Landsat image of the Namib Sand Sea. (Courtesy of NASA)

7.6 The Kalahari and the Karoo

In the interior of southern Africa – much of it in Botswana – lies the Kalahari (Thomas and Shaw, 1991). The Kalahari contrasts to the Namib because of its relatively high rainfall and because of its basinal form. Most of it is not a true desert but rather an extensively wooded 'thirstland'. Over enormous distances the relief is subdued and the landscape monotonous. In the south-west, on the borders of Botswana, Namibia and South Africa, the annual rainfall (< 200 mm) is just sufficient to allow present-day dune movement, but to the north the Kalahari is largely a relict sand desert, which extends into Angola, Zambia, Zimbabwe and the Congo (Shaw and Goudie, 2002) and has a mean annual rainfall in excess of 800 mm.

The Kalahari owes its overall form and subdued relief to the fact that, following the break-up of Gondwanaland, it became an area of downwarping that was bounded on the west by the highlands of Namibia and Angola and on the east by the Drakensberg and Lubombo Mountains (Haddon and McCarthy, 2005). It became a basin of sedimentation, and this largely accounts for its flatness. The mainly unconsolidated Kalahari Beds that fill this basin are often more than 100 m thick, and in parts of the Etosha Basin of northern Namibia they are more than 300 m thick (Figure 7.4). They

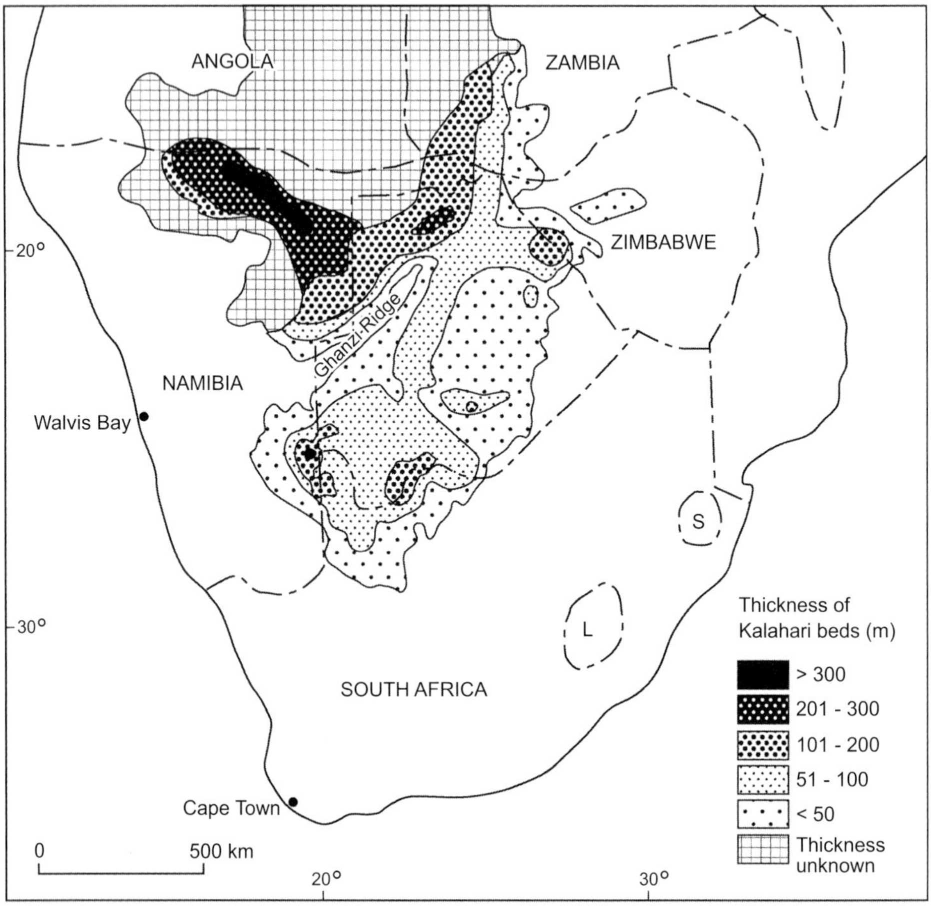

Figure 7.4 The thickness of Kalahari Beds. (Modified after Thomas and Shaw, 1991, fig. 3.1, in Goudie, 2002, fig. 7.3)

consist of aeolian sands, alluvium, calcretes, silcretes and marls. Faulting within the basin has led to the formation of the great Okavango Delta (McCarthy et al., 1988).

Apart from its relict dunes (Thomas, 1984), the Kalahari contains large numbers of pans and their associated lunettes (Goudie and Thomas, 1985), together with two large closed depressions – Etosha and the Mkgadikgadi. Today these are major sources of dust plumes (Washington et al., 2003). In the mid- to late Tertiary, palaeolake Etosha received water via the Cubango, Kunene and Cuvelai drainage systems. It largely dried up at about 4 Ma under conditions of progressively increasing aridity, although it still occasionally floods (Miller et al., 2010). During Pleistocene pluvials – and perhaps because of tectonically controlled water inputs form the Zambezi and other rivers – the Mkgadikgadi was occupied by a huge lake (Grove, 1969), which covered more than 120,000 km^2. Waters may also have been contributed in the past by a series of fossil valleys – mekgacha (Shaw et al., 1992).

The Kalahari shows excellent development of calcrete, silcrete and combinations of the two (Watts, 1980; Nash, Shaw and Thomas, 1994; Shaw and Goudie, 2004). The reason why the calcretes are so well developed probably relates to the area's long history of gentle sedimentation, although the carbonates that make them are largely derived from the great expanses of ancient limestones and dolomites that occur on its margins. Ancient river systems, such as the Molopo and Okwa, are incised into the calcrete.

The third desert of southern Africa is sometimes called the Great Karoo semi-desert. It occurs as a plateau at an altitude of 600–1,000m, tends to be underlain by horizontally bedded Palaeozoic sediments of the Beaufort Series, is bounded on the north and south by mountain ranges and has a primarily winter rainfall regime that produces 130–400 mm per annum.

7.7 Arabia and the Middle East

The Middle East has remarkable landscape diversity. It possesses ancient, stable shields, large rift structures, the lowest point on the Earth's surface, outpourings of recent volcanic lavas, high mountains and escarpments, magnificent anticlinal and synclinal structures, large numbers of salt domes, the world's finest coastal sabkhas, salty lakes and seas, mega-fans and enormous sand seas. Much of this diversity and distinctiveness can be attributed to the interaction of a number of plates, for the tectonic setting of this area is one both of complexity and of great activity.

The Middle East is on, or close to, a group of plate boundaries. In the south-west is the African plate, and to the east of that is the Arabian plate. The two are moving apart along the Red Sea and the Gulf of Aden. The Arabian plate is pushing northwards, colliding with the Turkish and Eurasian plates, and producing crustal deformation in the shape of multiple ripples of synclines and anticlines in the Zagros Mountains of Iran and elsewhere. The Red Sea, the Gulf of Suez, the Gulf of Aqaba and the Afar Depression form one of Earth's most remarkable groupings of tectonic phenomena and give a valuable insight into the process by which the break-up of continental crust can lead to the generation of new ocean crust material. Indeed, the majority of the most spectacular relief in the Middle East and north-eastern Africa is related to the tectonic movements resulting from the opening of the Red Sea (Ghebreab, 1998).

Another factor contributing to the diversity of the Middle East is its range of climates. The southern part comes under the direct influence of the great Asian monsoon and may therefore receive summer rain and winds from the south and south-west. By contrast, the northern part comes under the influence of winter cyclones that track from the west and bring winter precipitation. It tends to have greater annual precipitation totals, as do some of the higher mountain ranges in the south, such as those of Yemen and Oman. There is, however, a great tract of country in eastern Iran, southern Jordan, Syria, Iraq and in the Arabian Peninsula where annual precipitation totals are less than 100 mm; extensive portions of Arabia have rainfall that is less than a third of that figure.

The Dead Sea is the most famous and lowest of the world's great depressions. Its water level stands at around 400 m below sea level. This compares with Lake Assal in the Afar Triangle of East Africa at –156 m, the Qattara Depression in Egypt at –133 m and Death Valley in the western United States at –86 m. The Dead Sea itself is part of a large basin – the Dead Sea Basin – which is one of a group of depressions that lies along a lineament that stretches for more than 1,000 km from a zone of plate divergence along the Red Sea in the south to the Taurus-Zagros zone of plate convergence in the north. It is a rift, shear or transform fault that is located at a plate boundary as a result of the break-up of the formerly continuous Arabo-African continent in the late Cenozoic, and it displays an offset along its sides of around 105 km (Courtillot et al., 1987).

A major geomorphological consequence of the presence of ancient evaporite layers in this area of tectonic activity has been the formation of large numbers of salt diapirs and domes in Arabia, Iran and the Gulf itself. On the western side of the Arabian Gulf are a series of coastal salt flats – sabkhas – that are the best developed example of this landscape type to be found anywhere on the Earth (see Section 3.11).

Arabia's sand seas (Holm, 1960) are enormous, are linked and stretch almost down the whole latitudinal extent of the peninsula from Kuwait to Oman (Edgell, 2006). They account for nearly a third of its land area, covering as they do close to 800,000 km^2. The Rub 'Al Khali itself, with an area of between 500,000 and 600,000 km^2, is often called the Empty Quarter. It is the biggest active sand sea in the world and is approximately the size of France. Some if its linear ridges can be up to 300 km long, although in the Liwa area there are also imposing mega-barchans (Bishop, 2010). Wind erosion has produced, in the Lut Desert of Iran, some of the largest yardangs in the world. As elsewhere in the arid realm, however, even the dry heart of Arabia has experienced major pluvial phases (see, for example, Lézine et al., 2010; Petit-Maire et al., 2010; Fleitmann et al., 2011), with the development of extensive lakes and large tracts of huge, alluvial fans (Blechschmidt et al., 2009).

7.8 The Thar

The Thar of India and Pakistan (Allchin et al., 1978) is not profoundly arid, and very little of it has less than 100 mm of mean annual rainfall. It shows great diversity, however, for there is a striking contrast in relief between the arid foothills and valleys of the Karakorams and Ladakh in the north, the enormous alluvial plain and delta of the Indus River in the west, the salty sabkha of the Rann of Kutch in the south and the ancient Aravalli hills in the east.

The flood-prone Indus, which derives its waters from the high mountains to the north (Shroder, 1993), is a dominant influence on the desert, but in the past the mountains also provided the discharge of a whole series of 'lost rivers' (Wilhelmy, 1969) that are a feature of the Punjab. Other rivers, such as the Luni, flow from the Aravallis, one of the oldest mountains systems – still maintaining some relief – in the world (Spate, 1957). They were moulded by four orogenic events between 3,000 Ma and 750 Ma (Mishra et al., 2000).

Although the Thar is relatively moist and is a low velocity wind environment, it has extensive tracts of dunes, the sand for which comes from a range of possible sources: the coastline of the Arabian Sea, the large alluvial plains and the weathering of extensive areas of sandstones and granites. Uniquely in the world, many of the dunes are rakelike parabolics (Kar, 1993), which have formed transverse to the dominant early-summer south-westerly monsoon winds. Dunes were, however, much more extensive under past, more arid conditions (Goudie et al., 1973; Allchin et al., 1978), and this includes the highly lithified aeolianites (miliolites) of Gujarat (Sperling and Goudie, 1975; Goudie and Sperling, 1977) (see Section 4.19). The Thar has some lake basins, created in part by aeolian disruption of drainage lines, and these provide evidence for former more humid conditions (Wasson et al., 1984: Singh et al., 1990), not least in early to mid-Holocene. Likewise, there have been repeated alternations of fluvial and aeolian accumulation in the Thar during the late Pleistocene (Juyal et al., 2006; Singhvi et al., 2010; Dhir and Singhvi, 2012), together with phases of dune activity and of calcrete formation (Dhir et al., 2010). These reflect fluctuations in the nature of the summer south-west monsoon.

7.9 Central Asia

The Turkestan Desert lies between 36° N and 48° N and between 50° E and 83° E. It is bounded on the west by the Caspian Sea, on the south by the mountains bordering Iran and Afghanistan, on the east by the mountains bordering Sinkiang and on the north by the Kirghiz Steppe. To the north lay an extensive Pleistocene ice sheet, the melting of which brought meltwater into the area via the Volga and other rivers. At the present time the region is not one of extreme aridity, although in a small area in the vicinity of the Aral Sea the mean annual precipitation falls to about 100 mm. In contrast to the deserts of China, Tibet and Mongolia, large tracts lie at low altitudes above sea level.

The Caspian is the largest inland lake of internal drainage on our planet, and is fed by Europe's largest river, the Volga. It has an area of about 400,000 km^2, a water volume of 80,000 km^3, a perimeter of about 2,000 km, a length of 1,200 km and a maximum depth of 1,025 m (Golubev, 1998). It lies around 27 m below sea level. Its history as a closed basin dates from the late Pliocene, as the Paratethys Sea was severed by the rapid uplift of the Caucasus Mountains (Kroonenberg et al., 1997; Mitchell and Westaway, 1999). However, the Quaternary history of the Caspian has been one of great variability, and at some points in the past it has been even more monstrously large than it is today. The Aral Sea, which until its recent shrinkage was the fourth largest inland water body in the world, dates back to the Pliocene. It is bounded on the west by a number of major faults, but long-term deflation has also contributed to its development. Its history has been reviewed by Boomer et al. (2000), who note that a marked regression occurred under cold, dry climatic conditions in the late Pleistocene to early Holocene.

Satellite images show a major concentration of characteristically shaped aeolian pans on the semi-arid Western Siberian Steppes. These occur in the lee of the Urals between latitude 58° N and 48° N and between 60° E and 80° E. Other groups occur to the north of the Caspian Sea in the Zapadno Kazakhstanskaya Oblast. They appear to be developed primarily on Palaeogene, Neogene and Quaternary strata in an area where the rainfall is around 450 mm or less. Those in the vicinity of Chelyabinsk have a preferred orientation with their long axes trending approximately north/north-east–south/south-west (i.e. transverse to the dominant winds) and with bulbous eastern slides.

One of the most striking features of the area – and one it shares with China (Bronger et al., 1998) – is the development of thick (up to 200 m), complex loess deposits dating back to the Pliocene. They are well displayed in the Tajik and Uzbek Republics, where deposition rates were very high in the Pleistocene (Lazarenko, 1984; Frechen and Dodonov, 1998).

The Turkestan desert has extensive areas of sand dunes, which cover about 17 per cent of the area. These include large tracts to the north-west of the Caspian on either side of the Volga River: the Karakum ('black sands') to the south-west of the Amu Darya River and the Kyzylkum ('red sands') between the Amu Darya and the Syr Darya. These two systems are linked. In the east of the region there is the Muyunkum, and to the south of Lake Balkhash there are the Sary-Ishikotrau and Taukum dunefields. The nature of these dunefields is imperfectly known, but some preliminary remote sensing data are provided by Maman et al. (2011). For the most part, the dunes occur in a relatively moist, low-energy wind regime, so that many are stabilized.

7.10 Taklamakan, Tarim and the Other Chinese Deserts

The deserts of China and its neighbours cover a wide range of geomorphological and tectonic settings, from the Turfan (Turpan) Depression at –150 m to the high mountains of the Kunlun and Karakorams. They are all characterised by very low winter temperatures and very high summer ones. The deserts contain many sand seas, details of which are provided by X. Yang et al. (2011).

The Taklamakan is China's largest desert, and its mean annual precipitation drops as low as 10 mm. It occurs within the Tarim Basin, which, with an area of 530,000 km^2, is one of the largest closed basins on Earth. It was initiated by subsidence in the Oligocene and is underlain by huge thicknesses (up to 3,300 m) of Pliocene and Pleistocene sediments (Zheng et al., 2010). The lowest point in the basin is 'the wandering lake' of Lop Nor, at only 780 m above sea level (Dong et al., 2011).

The Tarim Basin is bounded on the south by the Kunlun Mountains and on the north by the Tian Shan. These produce alluvial fans and gravel aprons and feed the basin with sediment. It is for this reason the Taklamakan can lay claim to have the most positive sediment budget of any sand sea in the world (Mainguet and Chemin, 1986). At 337,600 km^2 it is indeed a huge sand sea, with an area of more than 300,000 km^2 and a diverse range of dune types, many of which are 80–200 m high (Zhu, 1984).

Winnowing of fine sediment from this basin has been a major source of material for dust storms and for the areas of loess that reach their ultimate development in the Loess Plateau downwind to the east. Indeed, the Taklamakan is one of the dustiest places on earth (Zhang et al., 1998) because of its aridity, its plentiful supply of mountain-derived sediment and its topographically funneled winds (Washington et al., 2003). In addition, the area is the classic location for yardang formation (Hedin, 1903; Halimov and Fezer, 1989).

Aridity in the area may be of some antiquity (see also Section 1.6). The rapid uplift of Tibet in the Miocene (Molnar et al., 1993) caused a major shift in climate and a transformation in the nature of the monsoonal system (Fluteau et al., 1999; Wang et al., 1999). The Pliocene Red Clay Formation of China, which is in part a product of aeolian dust accumulation and has loessic characteristics, has been dated to around 7.2 to 8.35 Ma (Ding and Yang, 2000; Qiang et al., 2001), although dust derived from the Tibetan Plateau and the Gobi is evident in Pacific cores going back to at least 11 million years (Pettke et al., 2000). Sand dune sediments have been recognized from Neogene sequences in the Tarim Basin and may be as much as 7 million years old (Sun et al., 2009).

In addition to Taklamakan and the Tarim Basin, there are some other Chinese deserts (Figure 7.5). The Junggar (Dzungarian) Basin lies between the Tian Shan and Altai Mountains. It is a temperate desert with extensive tracts of rocky desert (gobi), although in its centre there is a mobile 70,000 km^2 sand desert called Gurbantunggut. The Turpan-Hami Basin, which covers well over 50,000 km^2, is separated from the Tarim Basin by the Kuruktag Mountains. It contains the Ayding salt lake, which at 155 m below sea level is the lowest place in China. It is extremely arid, partly because of its exceptionally high summer temperatures (mean July temperature 34°C). The Qaidam Basin, 850 km long from east to west and 250 km wide from north to south, lies at an altitude of 2,600–3,000 m above sea level and is surrounded by the Kunlun, Altun and Qilian Mountains. As in the Tarim Basin, fans and river deposits occur on the margins, but sand dunes are more scattered and less extensive, and saline playas occupy the basin centre. Wind erosion has produced fields of yardangs (Halimov and Fezer, 1989) and caused the wholesale removal of hundreds of metres of strata (Kapp

Figure 7.5 Complex network dunes in the Tengger Desert of Alashan, China. (ASG)

et al., 2011). Streams draining snow-capped mountains flow into it and in the past formed more extensive water bodies (Wang et al., 1999). The Hexi-Gansu Corridor runs from the eastern end of the Tarim Basin for about 1,000 km to the Yellow River north of Lanzhou. It is bounded on the north by the Mongolian Plateau and the Gobi steppes and to the south by the Qilian Mountains. It is a Cenozoic foreland basin system, the structure of which is characterised by reverse faults that have led to the deformation of terraces and the offset of stream reaches (Li et al., 1999). Piedmont fans and gravels dominate the surface, most of which lies at between 1,000 and 1,500 m above sea level. The Badain Jaran Desert of Inner Mongolia is the second largest sand sea in China. and it contains enormous mega-dunes that may be as much as 460 m high (Yang, 2001; X. Yang et al., 2011), making them among the world's tallest. Within this sand sea there are also some alkaline salt lakes with tall tufa pinnacles or spring mounds (Arp et al., 1998). Towards the east of the Chinese arid zone, and lying within the big bend of the Yellow River, is the Ordos Plateau. It northern portion, the Hobq or Kubuqi Desert, contains extensive areas of mobile dunes, largely composed of sand derived from the Yellow River. Its more southern portion, the Mu Us Desert, which is moister, contains large areas of fixed and semi-fixed dunes. It grades south-eastwards into the Loess Plateau (see Section 3.5).

7.11 Helmand and the Seistan Basin

One of the world's least studied deserts is the Seistan Depression. It is located on the border between south-eastern Iran and western Afghanistan. It is a huge basin of internal drainage, which receives drainage from the Helmand River, the only major perennial river in western Asia between the Tigris-Euphrates and Indus Rivers. In addition, the area is affected by the high velocity 'wind of 120 days' (Middleton, 1986a), particularly from May to July. Early travelers (e.g. Huntington, 1907; McMahon, 1906) have described the ferocity of these dust-generating winds, which flow through mountain passes to the Hamoun lowlands (Hickey and Goudie, 2007). Yardangs are present, not least in proximity to the Gaud-i-Zirreh Lake. As Tate (1910, p. 114) graphically remarked, 'Where such mounds are numerous, the tract looks exactly like an enormous cemetery filled to overflowing with the graves of giants.'

The area is hyper-arid. The mean annual precipitation at Zahedan is only 82 mm. There is great inter-annual variability, however, and the area has been subjected to many alternations of droughts and floods which lead to major changes in the outlines of the various lakes that occur on its floor. The shallow Hamoun lakes complex – fed by the Helmand River – have from time to time been more extensive than today, covering an area ranging from 2,000–4,000 km^2, but equally they have a history of almost complete desiccation. The inland delta of the Helmand has also experienced frequent channel changes (Whitney, 2006).

To the north, Seistan is confined by the southern Hindu Kush, on the west by the East Iranian ranges and to the south by the mountains of Baluchistan. The Helmand, which rises just to the west of Kabul, flows for 1,300 km before ending as a large, marshy inland delta system. To the south of it is a sandy desert – that of Registan.

Although this brief section has concentrated on the Seistan Basin, Afghanistan as a whole is dominated by mountains, alluvial fans and pediments (Bacon et al., 2008).

7.12 North American Deserts

Two main physiographical provinces – the Basin and Range and the Colorado Plateau – contain the most important of the North American deserts. Within the former lies the Sonoran, Chihuahuan, Mojave and Great Basin Deserts (Tchakerian, 1997). These are characterised by block-faulted, more or less north-to-sound trending mountain ranges and basins (Peterson, 1981; Morrison, 1991), which started to develop in the late Oligocene as a response to crustal extension. The juxtaposition of topographic highs and lows provides a situation where there are many alluvial fans (Figure 7.6), extensive pediments, active runoff and the formation of many closed basins, which in pluvial times contained large lakes (Tchakerian and Lancaster, 2001; Kurth et al., 2011). Notable among these were Lakes Bonneville and Lahontan.

Figure 7.6 Death Valley in the Mojave Desert of California, USA, showing a large fan and the mountain front. (ASG)

The Colorado Plateau is morphologically very different. It exhibits gently dipping sedimentary strata, which give rise to extensive mesa and scarp landscapes, sandstone canyons and intricately dissected fluvial landscapes. In the early Tertiary, the Colorado Plateau – made famous by the pioneer research of Powell and Dutton – was a low-lying basin. Uplift started in the mid-Eocene and lasted into the late Miocene, causing the plateau to become a high region into which drainage became incised. Tectonic activity also created some major igneous landforms, including the classic intrusions of the Henry Mountains and elsewhere (Gilbert, 1877). The Colorado Canyon's initiation is probably relatively recent, having started around 5.5 Ma, although some authors suggest that the incision began as early as 17 Ma in the western part of the canyon. K.L. Cook et al. (2009), and Pelletier (2010) review this matter. In some parts of the plateau, incision has taken place at up to c 800 meters per million years (Patton et al., 1991; K.L. Cook et al., 2009), although the rates vary considerably in different tributaries. This rapid denudation has created the great cliffs of the plateau, with their magnificent escarpments, mesas, natural arches, box canyons and ground water–sapped alcoves (Laity and Malin, 1985).

Although these two very different provinces contain some dunefields (Algodones, Kelso, White Sands, etc.) and generate some dust from desiccated lakes, they are not

Figure 7.7 The Atacama Desert near Putre, Chile. (ASG)

areas where aeolian processes and phenomena are generally dominant. Fluvial and slope processes – driven by water and gravity – give these North American deserts their distinctiveness.

In addition, North America has the semi-arid drylands of the Great Plains, a land of aligned drainage, multitudes of pans and lunettes, extensive sheets of calcrete and dunefields (of which the Nebraska Sandhills are the most extensive) that have shown alternating phases of activity and relative stability repeatedly during the Quaternary (Rich and Stokes, 2011).

7.13 Atacama, Altiplano, Monte, Patagonia and Caatinga

To the west of the Andes (Figure 7.7), between latitudes 5° and 30° S, lies Earth's largest west-coast desert (Bowman, 1924) – the Atacama. In Peru, the desert strip is called the Peruvian Desert (or the Sechura Desert). It is also the world's driest desert, and Quillagua in Chile (mean average rainfall 0.05 mm) may be the driest place on Earth (Middleton, 2001). There are wetting fogs and occasional high rainfall years associated with El Niño conditions that cause great floods (Magilligan and Goldstein, 2001), but for the most part aridity is intense. Like the Namib, the Atacama has an extended history that goes back to at least the late Eocene and possibly to the Triassic (Alpers and Brimhall, 1998; Clarke, 2006), although there may have been modestly wetter conditions during, for example, the last interglacial (Contreras et al., 2010).

One consequence of long-continued intense aridity is that the Atacama contains the most famous and important caliche (sodium nitrate) deposits in the world (see Section 2.20). Nitrate, being highly soluble, can only accumulate under very dry conditions. Precipitation seems to have plummeted between 19 and 13 Ma (from $>$ 200 mm per annum to $<$ 20 mm) as the uplift of the Andes blocked the ingress of the South American summer monsoon into the Atacama. Nitrate accumulation may have begun at that time (i.e. in the middle Miocene) (Rech et al., 2010). The combination of fogs and salt at altitudes below c 1,100 m create an aggressive environment for salt weathering (Goudie et al., 2002).

A major influence on the geomorphology of the Atacama has been the growth and presence of the Andes, for the Andes are on an active plate margin. Tectonic uplift and the eastward migration of the Andes volcanic arc have created some of the greatest altitudinal contrasts to be found on Earth. Over a horizontal distance of no more than 300 km, one moves from the Peru-Chile Trench (at some 7,600 m below sea level) to Andean peaks that rise up to over 6,000 m above sea level. There is much evidence of volcanic activity, folding and faulting and, in the Altiplano, basin and range topography containing large closed depressions (Lamb et al., 1997). The main relief features run approximately north to south with a very narrow or non-existent coastal plain, a coastal range (Cordillera de la Costa), a longitudinal Central Valley and then, to the east, the higher-level Andes and Altiplano. Long-term erosion rates as determined by cosmogenic nuclides suggest that rates of erosion are very low in the hyperarid Cordillera de la Costa but become greater under the semi-arid conditions that occur at higher altitudes (Kober et al., 2007). Fluvial impacts on the landscape became greatly reduced in the late Pliocene to early Pleistocene, but fluvial features, such as fans, produced from the late Miocene onwards, are still preserved (Amundson et al., 2012). The Altiplano, a high plateau composed of the sedimentary infill of a series of intermontane tectonic trenches, is characterised by some large basins (salars) which have in the past contained large bodies of water (Rouchy et al., 1996; Placzek et al., 2001). One of these, the dazzling Salar de Uyuni in Bolivia, is now the major source of dust in South America (Washington et al., 2003). Mass movements are of a great size (see Section 6.6), and large fields of yardangs have developed in ignimbrites.

The Monte and Patagonian Deserts both lie essentially in the lee of the Andes. The Monte Desert (Labraga and Villalba, 2008), which is more or less continuous with the deserts to the west, is composed of basin and range topography, including mountain blocks, extensive piedmont surfaces and largely internal drainage. Alluvial fans are widespread (Sancho et al., 2008); volcanic features are also to be found. The main types of landform are listed in Table 1.7d, and a complete description of the physical geography of the full Monte Desert is provided by Abraham et al. (2009).

To the south of the Monte, the Patagonian Desert stretches for more than 500 km between the Andes and the Atlantic. It owes its aridity to the mountains, which block the rain-bearing winds from the west, and to the cold Falkland Current off the coast. The region is dominated by piedmont plains that slope eastwards towards the Atlantic, where they are terminated by marine surfaces, by ephemeral rivers that are entrenched into them and by enclosed drainage basins. Volcanic, glacial and fluvial deposits occur extensively in the region. Several glacial episodes during the Quaternary in the Patagonian Andes certainly influenced the evolution of this arid area strongly, especially in feeding fluvio-glacial gravels into the desert (Mercer, 1976). Aeolian features include pans and dunes (Del Valle, 2008; Del Pilar Alvarez, 2010). Similarly, in the Pampas of Argentina, and also in the Pantanal, there are remains of many aeolian phenomena, including linear dunes (Tripaldi and Forman, 2007; Tripaldi et al., 2010; Zárate and Tripaldi, 2012) and pans, which indicate that arid conditions were once far more widespread (Klammer, 1982), and loess deposition has been taking place for at least 1.9 million years (Heil et al., 2010).

In north-east Brazil, there is an area of semi-aridity that covers about 925,000 km^2 and with places where the mean annual rainfall is as low as about 200 mm. It is covered by dry woodland, called caatinga, and has, in the São Francisco valley, extensive dunefields. Calcretes and tufas are widely reported.

7.14 Australia

Australia is, with the exception of Antarctica, the world's driest continent, but aridity is not especially intense, and nowhere does mean annual precipitation fall below 100–125 mm. This relatively modest degree of aridity reflects the absence of very high-relief barriers against the inland penetration of moist air (particularly of tropical air from the north) and the lack of a definite cold inshore oceanic current along the west coast (contrast this with the hyperarid west coast deserts of the Namib and Atacama). The rainfall in the heart of Australia appears to be more variable than is the norm for other desert areas (McMahon et al., 2008a). Much drainage flows towards Lake Eyre, the fifth largest terminal lake in the world (McMahon et al., 2008b).

Australia has about 40 per cent of its area standing less than about 200 m above sea level. It is dominated by large plain lands, associated with such typically Australian phenomena as stone mantles (gibbers) and duricrusts. Australia is also an ancient continent with extensive venerable shields and surfaces that have been exposed to subaerial processes for hundreds of millions for years (Twidale, 2000). As Oberlander (1994, p. xx) observed, 'The erosional flattening of Australia is so thorough that any sharp protuberance constitutes a major landmark'. It is geomorphologically comatose and a museum of relict features, with some of the lowest denudation rates of any land surface in the world (Gale, 1992).

Australia, a fragment of Gondwanaland, has landscapes and climates that have been affected by northward continental drift that has been ongoing since the middle Jurassic. During the Tertiary, it moved from a high-latitude near-polar climatic zone, through a mid-latitude humid zone, into a zone of tropical and subtropical climates (including desert). This is well reviewed by Fujioka and Chappell (2010). Ancient, broad, infilled valley systems, now dismembered and containing strings of salt lakes – especially in Western Australia – may have been beheaded about 75 Ma by the rifting that initiated separation of Australia from Antarctica. Deep weathering profiles and etchplains, often associated with a range of duricrust types (particularly ferricretes and silcretes), are among the geomorphological phenomena that date back to the early Tertiary and before. Neogene phenomena include some of the karst features of the Nullarbor Plain, the largest karst area on Earth (Miller et al., 2012).

Other phenomena have resulted from more recent environmental changes, however, including a massive anticlockwise whirl of sand deserts (Wasson et al., 1988; Hesse, 2010, 2011), composed very largely of linear dunes, great networks of anastomosing and anabranching rivers created by intense tropical storms (Bourke and Pickup, 1999; Tooth and Nanson, 1999), numerous salt lakes of either a structural and/or deflational origin (Bourne and Twidale, 2010) that were filled by large water bodies at various times in the Pleistocene (Harrison and Dodson, 1993), and clay and sandy lunettes developed on the lee sides of many ephemeral basins.

References

Abdalla, O. and Al-Abri, R.Y. (2010). Groundwater recharge in arid areas induced by tropical cyclones: Lessons learned from Gonu 2007 in Sultanate of Oman. *Environmental Earth Science*, doi:10.1007s12665–010-0688-y.

Abdel-Lattif, A. and Sherief, Y. (2010). Morphometric analysis and flash floods of Wadi Sudr and Wadi Wardan, Gulf of Suez, Egypt: Using digital elevation model. *Arabian Journal of Geosciences*, doi:10.1007/s12517–010-0156–8.

Abdel Moneim, A.A. (2005). Overview of the geomorphological and hydrogeological characteristics of the Eastern Desert of Egypt. *Hydrogeology Journal*, 13, 416–25.

Abele, G. (1983). Flachen hafte Hanggestaltung und Hangzzerschniedung im Chilenisch – Peruanischen Trockengebiet. *Zeitschrift für Geomorphologie*, Supplementband, 48, 197–201.

Abraham, E., del Valle, H.F., Roig, F., Torres, L., Ares, J.O., Coronato, F. and Godagnone, R. (2009). Overview of the geography of the Monte Desert biome (Argentina). *Journal of Arid Environments*, 73, 144–53.

Abrahams, A.D. (1972). Drainage densities and sediment yields in eastern Australia. *Australian Geographical Studies*, 10, 19–41.

Abrahams, A.D, Parsons, A.J. and Wainwright, J. (1995). Effects of vegetation change on interrill runoff and erosion, Walnut Gulch, southern Arizona. *Geomorphology*, 13, 37–48.

Abu Ghazleh, S. and Kempe, S. (2009). Geomorphology of Lake Lisan terraces along the eastern coast of the Dead Sea, Jordan. *Geomorphology*, 108, 246–63.

Abul-Haggag, Y. (1961). *A Contribution to the Physiography of Northern Ethiopia*. London: University of London, Athlone Press.

Abu-Zeid, M.M. (1999). Petrology, mineralogy and sedimentation of the coastal Quaternary carbonates in Qatar, Arabian Gulf. *Carbonates and Evaporites*, 14, 209–24.

Achyuthan, H., Kar, A. and Eastoe, C. (2007). Late Quaternary-Holocene lake-level changes in the eastern margin of the Thar Desert, India. *Journal of Paleolimnology*, 38, 493–507.

Adams, J.W., Rodriguez, D. and Cox, R.A. (2005). The uptake of SO_2 on Saharan dust: A flow tube study. *Atmospheric Chemistry and Physics*, 5, 2679–89.

Adams, K.D., Goebel, T., Graf, K., Smith, G.M., Camp, A.J., Briggs, R.W. and Rhode, D. (2008). Late Pleistocene and Early Holocene lake-level fluctuations in the Lahontan Basin, Nevada: Implications for the distribution of archaeological sites. *Geoarchaeology*, 23, 608–43.

Addleson, L. and Rice, C. (1991). *Performance of Materials in Buildings*. Oxford: Butterworth-Heinemann.

Adelsberger, K.A. and Smith, J.R. (2009). Desert pavement development and landscape stability on the eastern Libyan Plateau, Egypt. *Geomorphology*, 107, 178–94.

Adelsberger, K.A. and Smith, J.R. (2010). Paleolandscape and paleoenvironmental interpretation of spring-deposited sediments in Dakhleh Oasis, Western Desert of Egypt. *Catena*, 83, 7–22.

Agam, N. and Berliner, P.R. (2006). Dew formation and water vapour adsorption in semi-arid environments – A review. *Journal of Arid Environments*, 65, 572–90.

Aires-Barros, L. (1977). Experiments on thermal fatigue of non-igneous rocks. *Engineering Geology*, 11, 227–38.

Akiner, S., Cooke, R.U. and French, R.A. (1992). Salt damage to Islamic monuments in Uzbekistan. *Geographical Journal*, 158, 257–72.

Al-Awadhi, J.M., Al-Dousari, A. and Al-Enezi, A. (2000). Barchan dunes in northern Kuwait. *Arab Journal of Scientific Research*, 18, 32–40.

Al-Awadhi, J.M., Al-Helal, A. and Al-Enezi, A. (2005). Sand drift potential in the desert of Kuwait. *Journal of Arid Environments*, 63, 425–38.

Albouy, M., Deletie, P., Haguenauer, B., Nion, S., Schreiner, E., Rewerski, J. and Seigne, J. (1993). Petra, la cité rose des sables. La pathologie des grès et leur traitement dans la perspective d'une préservation et d'une restauration des monuments. In *Conservation of Stone and Other Materials*, ed. M.J. Thiel. London: E and F.N Spon, 1, pp. 376–85.

Albrecht, C.F., Joubert, J.J. and De Rycke, P.H. (2001). Origin of the enigmatic, circular, barren patches ('Fairy Rings') of the pro-Namib. *South African Journal of Science*, 97, 23–7.

Al-Dousari, A.M., Al-Elaj, M., Al-Enezi, E. and Al-Shareeda, A. (2009). Origin and characteristics of yardangs in the Um Al-Riman depressions (N Kuwait). *Geomorphology*, 104, 93–104.

Aleinikoff, J.N., Muhs, D.R., Bettis, E.A., Johnson, W.C., Fanning, C.M. and Benton, R. (2008). Isotopic evidence for the diversity of late Quaternary loess in Nebraska: Glaciogenic and nonglaciogenic sources. *Bulletin of the Geological Society of America*, 120, 1363–77.

Al-Enezi, A., Pye, K., Misak, R. and Al-Hajraf, S. (2008). Morphologic characteristics and development of falling dunes, northeast Kuwait. *Journal of Arid Environments*, 72, 423–39.

Alexandre, A., Meunier, J.-D., Llorens, E., Hill, S.M. and Savin, S.M. (2004). Methodological improvements for investigating silcrete formation: petrography, FT-IR and oxygen isotope ratio of silcrete quartz cement, Lake Eyre Basin (Australia). *Chemical Geology*, 211, 261–74.

Alexandrov, Y., Cohen, H., Laronne, J.B. and Reid, I. (2009). Suspended sediment load, bed load, and dissolved load yields from a semiarid drainage basin: A 15-year study. *Water Resources Research*, 45, W08408, doi:10.1029/2008WR007314.

Al-Farraj, A. (2008). Desert pavement development on the lake shorelines of Lake Eyre (South), South Australia. *Geomorphology*, 100, 154–63.

Al-Farraj, A. and Harvey, A. (2000). Desert pavement characteristics on wadi terrace and alluvial fan surfaces, Wadi Al-Bih, U.A.E. and Oman. *Geomorphology*, 35, 279–97.

Al-Harthi, A.A. (2002). Geohazard assessment of sand dunes between Jeddah and Al-Lith, western Saudi Arabia. *Environmental Geology*, 42, 360–9.

Al-Harthi, A.A. and Bankher, K.A. (1999). Collapsing loess-like soil in western Saudi Arabia. *Journal of Arid Environments*, 41, 383–9.

Allchin, B. and Goudie, A.S. (1971). Dunes, aridity and early man in Gujarat, western India. *Man*, 6, 248–65.

Allchin, B., Goudie, A.S. and Hegde, K.T.M. (1978). *The Prehistory and Palaeogeography of the Great Indian Desert*. London: Academic Press.

Alley, R.E., Mayewski, P.A., Sowers, T., Stuiver, M., Taylor, K.C. and Clark, P.U. (1997). Holocene climate instability: A prominent, widespread event 8200 yr ago. *Geology*, 25, 483–6.

Allison, R.J. and Goudie, A.S. (1994). The effects of fire on rock weathering: An experimental study. In *Rock Weathering and Landform Evolution*, ed. D.A. Robinson and R.B.G. Williams. Chichester: Wiley, pp. 41–56.

Allred, T.M. and Schmidt, J.C. (1999). Channel narrowing by vertical accretion along the Green River near Green River, Utah. *Bulletin of the Geological Society of America*, 111, 1757–72.

Almeida, M.P., Parteli, E.J.R., Andrade, J.S. and Herrmann, H.J. (2008). Giant saltation on Mars. *Proceedings of the National Academy of Sciences*, 105, 6222–6.

Alonso, R.N., Jordan, T.E., Tubbutt, K.T. and Vandervoort, D.S. (1999). Giant evaporite belts of the Neogene central Andes. *Geology*, 19, 401–4.

Alonso-Zarza, A.M. (1999). Initial stages of laminar calcrete formation by roots: Examples from the Neogene of central Spain. *Sedimentary Geology*, 126, 177–91.

Alonso-Zarza, A.M. and Wright, V.P. (2010). Calcretes. *Developments in Sedimentology*, 61, 225–67.

Alonso-Zarza, A.M., Silva, P.G., Goy, J.L. and Zazo, C. (1998). Fan-surface dynamics and biogenic calcrete development: Interactions during ultimate phases of fan evolution. *Geomorphology*, 24, 147–67.

Alpers, C.N. and Brimhall, G.H. (1998). Middle Miocene climatic change in the Atacama Desert, northern Chile: Evidence from supergene mineralization at La Escondida. *Bulletin of the Geological Society of America*, 100, 1640–56.

Al-Qurashi, A., McIntyre, N., Wheater, H. and Unkrich, C. (2008). Application of the Kineros2 rainfall-runoff model to an arid catchment in Oman. *Journal of Hydrology*, 355, 91–105.

Al-Rawas, G.A. and Valeo, C. (2010). Relationship between wadi drainage characteristics and peak-flood flows in arid northern Oman. *Hydrological Sciences Journal*, 55, 377–93.

Alsharhan, A.S. and Kendall, C.G.St.C. (2003). Holocene coastal carbonates and evaporites of the southern Arabian Gulf and their ancient analogues. *Earth-Science Reviews*, 61, 191–243.

Alsharhan, A.S., Ahmed, E.A. and Tamer, S. (1998). Textural characteristics of Quaternary sand dunes in the Eastern Province of Abu Dhabi, United Arab Emirates. In *Quaternary Deserts and Climate Change*, ed. A.S. Alasharhan, K.W. Glennie, C.L. Whittle, and C.G.StC. Kendall. Rotterdam: Balkema, pp. 91–108.

Ambrose, G.J. and Flint, R.B. (1981). A regressive Pliocene lake system and silicified strandlines in northern South Australia: Implications for regional stratigraphy and silcrete genesis. *Journal of the Geological Society of South Australia*, 28, 81–94.

Amelung, F., Galloway, D.L., Bell, J.W., Zebker, H.A. and Laczniak, R.J. (1999). Sensing the ups and downs of Las Vegas: InSAR reveals structural control of land subsidence and aquifer-system deformation. *Geology*, 27, 483–6.

Amin, A.A. and Bankher, K.A. (1997). Karst hazard assessment of eastern Saudi Arabia. *Natural Hazards*, 15, 21–30.

Amit, R. and Gerson, R. (1986). The evolution of Holocene reg (gravelly) soils in deserts – An example from the Dead Sea region. *Catena*, 13, 59–79.

Amit, R., Enzel, Y., Crouvi, O., Simhai, O., Matmon, A., Porat, N., McDonald, E. and Gillespie, A.R. (2011a). The role of the Nile in initiating a massive dust influx to the Negev late in the middle Pleistocene. *Bulletin of the Geological Society of America*, 123, 873–89.

Amit, R., Simhai, O., Ayalon, A., Enzel, Y., Matmon, A., Crouvi, O., Porat, N. and McDonald, E. (2011b). Transition from arid to hyper-arid environment in the southern Levant deserts as recorded by early Pleistocene cummulic Aridisols. *Quaternary Science Reviews*, 30, 312–23.

Amundson, R. and 10 others. (2012). Geomorphic evidence for the late Pliocene onset of hyperaridity in the Atacama Desert. *Bulletin of the Geological Society of America*, doi:10.1130/B30445.1.

An, C.-B., Tang, L., Barton, L. and Chen, F.-H. (2005). Climate change and cultural response around 4000 cal yr B.P. in the western part of Chinese Loess Plateau. *Quaternary Research*, 63, 347–52.

An, C.-B, Zhao, J., Tao, S., Lv, Y., Dong, W. Li, H., Jin, M. and Wang, Z. (2011). Dust variation recorded by lacustrine sediments from arid central Asia since ~15 cal ka BP and its implication for atmospheric circulation. *Quaternary Research*, 75, 566–73.

Anderson, D., Goudie, A.S. and Parker, A.G. (2007). *Global Environments through the Quaternary*. Oxford: Oxford University Press.

Anderson, K., Wells, S. and Graham, R. (2002). Pedogenesis of vesicular horizons, Cima Volcanic Field, Mojave Desert, California. *Soil Science Society of America Journal*, 66, 878–87.

Anderson, R.F., Fleisher, M.Q. and Lao, Y. (2006). Glacial-interglacial variability in the delivery of dust to the central equatorial Pacific Ocean. *Earth and Planetary Science Letters*, 242, 406–14.

Andreae, M.O. (1995). Climatic effects of changing atmospheric aerosol levels. In *World Survey of Climatology vol. 16, Future Climates of the World*, ed. A. Henderson-Sellers. Amsterdam: Elsevier, pp. 341–92.

Andreotti, B., Claudin, P. and Douady, S. (2002). Selection of dune shapes and velocities part 1: Dynamics of sand, wind and barchans. *The European Physical Journal* B, 28, 321–39.

Andrews, J.E., Singhvi, A.K., Kailath, A.J., Kuhn, R., Dennis, P., Tandon, S.K. and Dhir, R.P. (1998). Do stable isotope data from calcrete record late Pleistocene monsoonal climate variation in the Thar Desert of India? *Quaternary Research*, 50, 240–51.

Angeli, M., Benavente, D., Bigas, J.-P., Menéndez, B., Hébert, R. and David, C. (2008). Modification of the porous network by salt crystallization in experimentally weathered sedimentary stones. *Materials and Structures*, 41, 1091–108.

Anton, D. and Vincent, P. (1986). Parabolic dunes of the Jafurah Desert, Eastern Province, Saudi Arabia. *Journal of Arid Environments*, 11, 187–98.

Applegarth, M.T. (2004). Assessing the influence of mountain slope morphology on pediment form, south-central Arizona. *Physical Geography*, 25, 225–36.

Arakel, A.V. and McConchie, D. (1982). Classification and genesis of calcrete and gypsite lithofacies in palaeodrainage systems of inland Australia and their relationship to carnotite mineralization. *Journal of Sedimentary Petrology*, 52, 1149–70.

Arbogast, A.F. (1996a). Late Quaternary evolution of a lunette in the central Great Plains: Wilson Ridge, Kansas. *Physical Geography*, 17, 354–70.

Arbogast, A.F. (1996b). Stratigraphic evidence for late-Holocene aeolian sand mobilisation and soil formation in south-central Kansas, U.S.A. *Journal of Arid Environments*, 34, 403–14.

Arboleya, M.-L., Babault, J., Owen, L.A., Teixell, A. and Finkel, R.C. (2008). Timing and nature of Quaternary fluvial incision in the Ouarzazate foreland basin, Morocco. *Journal of the Geological Society of London*, 165, 1059–73.

Ardon, K., Tsoar, H. and Blumberg, D.G. (2009). Dynamics of nebkhas superimposed on a parabolic dune and their effect on the dune dynamics. *Journal of Arid Environments*, 73, 1014–22.

Aref, M.M. (2003). Classification and depositional environments of Quaternary pedogenic gypsum crusts (gypcrete) from east of the Fayum Depression, Egypt. *Sedimentary Geology*, 155, 87–108.

Aref, M.M. and Refai, E. (1987). Paleokarst processes in the Eocene limestones of the Pyramids Plateau, Egypt. *Journal of African Earth Sciences*, 6, 367–77.

Aref, M.M., Awadalah, F. and Ahmed, S. (1986). Karst landform development in the Miocene rocks of the Red Sea coastal zone, Egypt. *Geologische Rundschau*, 75, 781–90.

Aref, M.M., El-Khoriby, E. and Hamdan, M.A. (2002). The role of salt weathering in the origin of the Qattara Depression, Western Desert, Egypt. *Geomorphology*, 45, 181–95.

Arias, J. (2003). On the origin of saltpetre, northern Chile coast. *Abstract, XVI INQUA Congress, Reno, Nevada*, paper 6–9, July 24, 2003.

Armbrust, D.V. and Dickerson, J.D. (1971). Temporary wind erosion control: Cost and effectiveness of 34 commercial materials. *Journal of Soil and Water Conservation*, 26, 154–7.

Armbrust, D.V. and Lyles, L. (1975). Soil stabilizers to control wind erosion. *Soil Science Society of America Special Publication*, 7, 77–82.

Armitage, S.J., Drake, N.A., Stokes, S., El-Hawat, A, Salem, M.J., White, K., Turner, P. and McLaren, S.J. (2007). Multiple phases of North African humidity recorded in lacustrine sediments from the Fazzan Basin, Libyan Sahara. *Quaternary Geochronology*, 2, 181–6.

Arnold, A. and Zehnder, K. (1988). Decay of stony materials by salts in humid atmosphere. *Proceedings 7th International Congress on the Deterioration and Conservation of Stone*, 1, 138–48.

Arnow, T. and Stephens, D. (1990). Hydrologic characteristics of the Great Salt Lake, Utah, 1847–1986. *U.S. Geological Survey Water-Supply Paper*, 2332.

Arp, G., Hofmann, J. and Reitner, J. (1998). Microbial fabric formation in spring mounds ('Microbialites') of alkaline salt lakes in the Badain Jaran Sand Sea, PR China. *Palaios*, 13, 581–92.

Arpe, K., Bengtsson, L., Golitsyn, G.S., Mokhov, I.I., Semenov, A. and Sporyshev, P.V. 2000. Connection between Caspian Sea level variability and ENSO. *Geophysical Research Letters*, 27, 2693–6.

Arz, H.W., Lamy, F. and Patzold, J. (2006). A pronounced dry event recorded around 4.2 ka in brine sediments from the northern Red Sea. *Quaternary Research*, 66, 432–41.

Ashbel, D. (1949). Frequency and distribution of dew in Palestine. *Geographical Review*, 39, 291–7.

Ashkenazy, Y., Yizhaq, H. and Tsoar, H. (2012). Sand dune mobility under climate change in the Kalahari and Australian deserts. *Climatic Change*, 112, 901–23.

Assallay, A.M., Rogers, C.D.F. and Smalley, I.J. (1996). Engineering properties of loess in Libya. *Journal of Arid Environments*, 32, 373–86.

Atapour, H. and Aftabi, A. (2002). Geomorphological, geochemical and geo-environmental aspects of karstification in the urban areas of Kerman city, southeastern Iran. *Environmental Geology*, 42, 783–92.

Atkinson, O.A.C., Thomas, D.S.G., Goudie, A.S. and Bailey, R.M. (2011). Late Quaternary chronology of major dune ridge development in the northeast Rub' al-Khali, United Arab Emirates. *Quaternary Research*, 76, 93–105.

Atkinson, O.A.C., Thomas, D.S.G., Goudie, A.S. and Parker, A.G. (2012). Holocene development of multiple dune generations in the northeast Rub' al-Khali, United Arab Emirates. *The Holocene*, 22, 179–81.

Auler, A. and Smart, P.L. (2003). The influence of bedrock-derived acidity in the development of surface and underground karst: Evidence from the Precambrian carbonates of semi-arid northeastern Brazil. *Earth Surface Processes and Landforms*, 28, 157–68.

Avni, Y. (2005). Gully incision as a key factor in desertification in an arid environment, the Negev Highlands, Israel. *Catena*, 63, 185–220.

Baas, A.C.W. and Nield, J.M. (2007). Modelling vegetated dune landscapes. *Geophysical Research Letters*, 34, L06405, doi:10.1029/2006GL029152.

Babikir, A.A.A. (1986). The vegetation of natural depressions in Qatar in relation to climate and soil. *Journal of Arid Environments*, 10, 165–73.

Baccari, N., Boussema, M.R., Lamachère, J-M. and Nasri, S. (2008). Efficiency of contour benches, filling-in and silting-up of a hillside reservoir in a semi-arid climate in Tunisia. *C.R. Geoscience*, 340, 38–48.

Bach, A.J., Brazel, A.J. and Lancaster, N. (1996). Temporal and spatial aspects of blowing dust in the Mojave and Colorado Deserts of southern California, (1973–1994). *Physical Geography*, 17, 329–53.

Bachhuber, F.W. (1971). Paleolimnology of Lake Estancia and the Quaternary history of the Estancia Valley, central New Mexico. PhD diss., University of New Mexico, Alburquerque.

Bacon, S.N., McDonald, E.V., Bakre, S.E., Caldwell, T.G. and Stullenbarger, G. (2008). Desert terrain characterization of landforms and surface materials within vehicle test courses at U.S. Army Yuma Proving Ground, USA. *Journal of Terramechanics*, 45, 167–83.

Bacon, S.N., McDonald, E.V., Caldwell, T.G. and Dalldorf, G.K. (2010). Timing and distribution of alluvial fan sedimentation in response to strengthening of late Holocene ENSO variability in the Sonoran Desert, southwestern Arizona, USA. *Quaternary Research*, 73, 425–38.

Badano, E.I., Cavieres, L.A., Molina-Montenegro, M.A. and Quiroz, C.L. (2005). Slope aspect influences plant association patterns in the Mediterranean matorral of central Chile. *Journal of Arid Environments*, 62, 93–108.

Badarinath, K.V.S., Kharol, S.K., Kaskaoutis, D.G., Sharma, A.R., Ramaswamy, V. and Kambezidis, H.D. (2010). Long-range transport of dust aerosols over the Arabian Sera and Indian region – A case study using satellite data and ground-based measurements. *Global and Planetary Change*, 72, 164–81.

Baddock, M.C., Bullard, J.E. and Bryant, R.G. (2009). Dust source identification using MODIS: A comparison of techniques applied to the Lake Eyre Basin, Australia. *Remote Sensing of Environment*, 113, 1511–28.

Baddock, M.C., Livingstone, I. and Wiggs, G.F.S. (2007). The geomorphological significance of airflow patterns in transverse dune interdunes. *Geomorphology*, 87, 322–36.

Baddock, M.C., Zobeck, T.M., van Pelt, R.S. and Fredrickson, E.L. (2011a). Dust emissions from undisturbed and disturbed, crusted playa surfaces: Cattle trampling effects. *Aeolian Research*, 3, 31–41.

Baddock, M.C., Gill, T.E., Bullard, J.E., Acosta M.D. and Rivera, N.I.R. (2011b). Geomorphology of the Chihuahuan Desert based on potential dust emissions. *Journal of Maps*, 2011, 249–59.

Bagine, R.K.N. (1984). Soil translocated by termites of the genus *Odontotermes* (Holmgren) (Isoptera: Macrotermitinae) in an arid area of northern Kenya. *Oecologia*, 64, 263–6.

Bagnold, R.A. (1935). The movement of desert sand. *Geographical Journal*, 85, 343–69.

Bagnold, R.A. (1936). The movement of desert sand. *Proceedings of the Royal Society of London*, Series A, 157, 594–620.

Bagnold, R.A. (1937). The size grading of sand by wind. *Proceedings of the Royal Society of London*, Series A, 163, 250–64.

Bagnold, R.A. (1941). *The Physics of Blown Sand and Desert Dunes*. London: Methuen.

Bagnold, R.A. (1953). The surface movement of blown sand in relation to meteorology. *Research Council of Israel, Special Publication*, 2, 89–93.

Bahafzullah, A., Fayed, L.A., Kazi, A. and Al Saify, M. (1993). Classification and distribution of the Red Sea coastal sabkhas near Jeddah – Saudi Arabia. Carbonates and Evaporites, 8, 23–38.
Bailey, S.D. and Bristow, C.S. (2004). Migration of parabolic dunes at Aberffraw, Anglesey, north Wales. *Geomorphology*, 59, 165–74.
Baillie, I.C., Faulkner, P.H., Espin, G.P., Levett, M.J. and Nicholson B. (1986). Problems of protection against piping and surface erosion in central Tunisia. *Environmental Conservation*, 13, 27–32.
Baker, C.L. (1915). Geology and underground waters of the northern Llano Estacado. *University of Texas Bulletin*, No. 57.
Baker, V.R., Pickup, G. and Polach, H.A. (1983). Desert palaeofloods in central Australia. *Nature*, 301, 502–4.
Balboni, E., Espinosa-Marzal, R.M., Doehne, E. and Scherer, G.W. (2011). Can drying and rewetting of magnesium sulfate salts lead to damage of stone? *Environmental Earth Science*, 63, 1463–73.
Balestrieri, M.L., Stuart, F.M., Persano, C., Abbate, E. and Bigazzi, G. (2005). Geomorphic development of the escarpment of the Eritrean margin, southern Red Sea from combined apatite fission-track and (U-Th)/He thermochronometry. *Earth and Planetary Science Letters*, 231, 97–110.
Ballenger, J.A.M., Holliday, V.T., Kowler, A.L., Reitze, W.T., Prasciunas, M.M., Miller, D.S. and Windingstad, J.D. (2011). Evidence for Younger Dryas global climate oscillation and human response in the American Southwest. *Quaternary International*, 242, 502–19.
Balling, R.C. and Wells, S.G. (1990). Historical rainfall patterns and arroyo activity within the Zuni River drainage basin, New Mexico. *Annals of the Association of American Geographers*, 80, 603–17.
Balme, M.R., Whelley, P.L. and Greeley, R. (2003). Mars: Dust devil track survey in Argyre Planitia and Hellas Basin. *Journal of Geophysical Research*, 108, E8, doi:10.1029/2003JE002096.
Balme, M., Berman, D.C., Bourke, M.C. and Zimbelman, J.R. (2008). Transverse aeolian ridges (TARs) on Mars. *Geomorphology*, 101, 703–20.
Bani-Hani, K. and Barakat, S. (2006). Analytical evaluation of repair and strengthening measures of Qasr al-Bint historical monument – Petra, Jordan. *Engineering Structures*, 28, 1355–66.
Bao, H., Michalski, G.M. and Thiemens, M.H. (2001a). Sulfate oxygen-17 anomalies in desert varnishes. *Geochimica et Cosmochimica Acta*, 65, 2029–36.
Bao, H., Thiemens, M.H. and Heine, K. (2001b). Oxygen-17 excesses of the central Namib gypcretes: Spatial distribution. *Earth and Planetary Science Letters*, 192, 125–35.
Bao, H. and 6 others. (2000). Anomalous ^{17}O compositions in massive sulphate deposits on the Earth. *Nature*, 406, 176–8.
Barchyn, T.E. and Hugenholtz, C.H. (2012). Aeolian dune field geomorphology modulates the stabilization rate imposed by climate. *Journal of Geophysical Research*, 117, F02035, doi:10/1029/2011JF002274.
Bari, M.A. and Schofield, N.J. (1992). Lowering of a shallow, saline water table by extensive eucalyptus reforestation. Journal of Hydrology, 133, 273–91.
Barkan, J. and Alpert, P. (2010). Synoptic analysis of a rare event of Saharan dust reaching the Arctic region. *Weather*, 65, 208–11.
Barlow, P.M., and Reichard, E.G. (2010). Saltwater intrusion in coastal regions of North America. *Hydrogeology Journal*, 18, 247–60.
Bar-Matthews, M., Ayalon, A. and Kaufman, A. (1997). Late Quaternary paleoclimate in the eastern Mediterranean region from stable isotope analysis of speleothems at Soreq Cave, Israel. *Quaternary Research*, 47, 155–68.

Barnes, F.A. (1987). *Canyon Country Arches and Bridges*. Moab, UT: Canyon Country Publishers.

Barnes, J. (2001). Barchan dunes on the Kuiseb River Delta, Namibia. *South African Geographical Journal*, 83, 283–92.

Bar-Or, R., Erlick, C. and Gildor, H. (2008). The role of dust in glacial-interglacial cycles. *Quaternary Science Reviews*, 27, 201–8.

Barth, H.-J. (1998). Sabkhas als Ausbruck von Landschaftsdegradation im Zentralen Küstentiefland der Ostprovinz Saudi-Arabiens. *Regensburger Geographische Schriften*, 29.

Bartley, R. and 7 others. (2006). Runoff and erosion from Australia's tropical semi-arid rangelands: Influence of ground cover for differing time and space scales. *Hydrological Processes*, 20, 3317–33.

Barton, D.C. (1916). Notes on the disintegration of granite in Egypt. *Journal of Geology*, 24, 382–93.

Bashan, Y., Salazar, B.G., Moreno, M., Lopez, B.R. and Linderman, R.G. (2012). Restoration of eroded soil in the Sonoran Desert with native leguminous trees using plant growth-promoting microorganisms and limited amounts of compost and water. *Journal of Environmental Management*, 102, 26–36.

Bateman, M.D., Boulter, C.H., Carr, A.S., Frederick, C.D., Peter, D. and Wilder, M. (2007). Preserving the palaeoenvironmental record in drylands: Bioturbation and its significance for luminescence-derived chronologies. *Sedimentary Geology*, 195, 5–19.

Bateman, M.D., Bryant, R.G., Foster, I.D.L., Livingstone, I. and Parsons, A.J. (2012). On the formation of sand ramps: A case study from the Mojave Desert. *Geomorphology*, 161–162, 93–109.

Baumhauer, R. (2010). Scarpfoot depressions in south-central Sahara – State of knowledge and critical review. *Zeitschrift für Geomorphologie*, 54, 377–92.

Bautista S., Bellot, J. and Ramón Vallejo, V. (1996). Mulching treatment for postfire soil conservation in a semi-arid ecosystem. *Arid Soil Research and Rehabilitation*, 10, 235–42.

Beadnell, H.J.L. (1910). The sand-dunes of the Libyan Desert. *Geographical Journal*, 35, 379–92.

Beaty, C.B. (1963). Origin of alluvial fans, White Mountains, California and Nevada. *American Association of Geographers, Annals*, 53, 516–35.

Beaumont, B. (1968). Salt weathering on the margins of the Great Kavir, Iran. *Bulletin of the Geological Society of America*, 79, 1683–94.

Becker, T. and Getzin, S. (2000). The fairy circles of Kaokoland (north-west Namibia) – Origin, distribution, and characterisics. *Basic and Applied Ecology*, 1, 149–59.

Beckinsale, R.P. (1972). The effect upon river channels of sudden changes in sediment load. *Acta Geographica Debrecina*, 10, 181–6.

Beckmann, G.G., Thompson, C.H. and Hubble, G.D. (1973). Australian landform example No. 22: Linear gilgai. *Australian Geographer*, 12, 363–6.

Bedford, D.R. and Small, E.E. (2008). Spatial patterns of ecohydrologic properties on a hillslope-alluvial fan transect, central New Mexico. *Catena*, 73, 34–48.

Belnap, J. and Gillette, D.A. (1997a). Vulnerability of desert biological soil crusts to wind erosion: The influence of crust development, soil texture and disturbance. *Journal of Arid Environments*, 39, 133–42.

Belnap, J. and Gillette, D.A. (1997b). Disturbance of biological soil crusts: Impacts on potential erodibility of sandy desert soils in southeastern Utah. *Land Degradation and Development*, 8, 355–62.

Belnap, J. and Warren, S.D. (2002). Patton's tracks in the Mojave Desert, USA: An ecological legacy. *Arid Lands Research and Management*, 16, 245–58.

Belski, A.J. (1996). Viewpoint: Western juniper expansion: Is it a threat to arid northwest ecosystems? *Journal of Range Management*, 49, 53–9.
Ben-Ami, Y., Koren, I., Rudich, Y., Artaxo, P., Martin, S.T. and Andreae, M.O. (2010). Transport of Saharan dust from the Bodélé Depression to the Amazon Basin: A case study. *Atmospheric Chemistry and Physics Discussions*, 10, 4345–72.
Bendix, J., Bendix, A. and Richter, M. (2000). El Niño 1997/1998 in Nordperu: Anzeichen eines Őkosystem-Wandels? *Petermanns Geographische Mitteilungen*, 144, 20–31.
Benhouhou, S.S., Dargie, T.C.D. and Gilbert, O.L. (2003). Vegetation associations in the Ougarta Mountains and dayas of the Guir Hamada, Algerian Sahara. *Journal of Arid Environments*, 54, 739–53.
Ben-Hur, M. and Lado, M. (2008). Effect of soil wetting conditions on seal formation, runoff and soil loss in arid and semiarid soils – A review. *Australian Journal of Soil Research*, 46, 191–202.
Benison, K.C. and Bowen, B.B. (2006). Acid saline lake systems give clues about past environments and the search for life on Mars. *Icarus*, 183, 225–9.
Bennett, C.M., McKendry, I.G., Kelly, S., Denike K. and Koch, T. (2006). Impact of the 1998 Gobi dust event on hospital admissions in the Lower Fraser Valley, British Columbia. *Science of the Total Environment*, 366, 918–25.
Bennett, H.H. (1938). *Soil Conservation*. New York: McGraw-Hill.
Benson, L. (1994). Carbonate deposition, Pyramid Lake subbasin, Nevada: 1. Sequence of formation and elevational distribution of carbonate deposits (tufas). *Palaeoclimatology, Palaeogeography, Palaeoecology*, 109, 55–87.
Benson, L., Lund, S.P., Burdett, J.W., Kashgarian, M., Rose, T.P., Smoot, J.P. and Schwartz, M. (1998). Correlation of late-Pleistocene lake-level oscillations in Mono Lake, California, with North Atlantic climate events. *Quaternary Research*, 49, 1–10.
Benson, L. and 9 others. (2011). The rise and fall of Lake Bonneville between 45 and 10.5 ka. *Quaternary International*, 235, 57–69.
Berkey, C.P. and Morris, F.K. (1927). *Geology of Mongolia*. New York: American Museum of Natural History.
Bernat Rebollal, M. and Pérez-González, A. (2008). Inland aeolian deposits of the Iberian Peninsula: Sand dunes and clay dunes of the Duero Basin and the Manchega Plain. Palaeoclimatic considerations. *Geomorphology*, 102, 207–20.
Besbes, M. (2006). Aquifer recharge by floods in ephemeral streams. *IASH Publication*, 305, 43–72.
Besler, H. (1979). Feldversuche zue aktuellen Granitverwitterung und Rindenbildung in der Namib (Südwestafrika/Namibia). *Stuttgarter Geographische Studien*, 93, 95–106.
Besler, H. (2008). *The Great Sand Sea in Egypt*. Amsterdam: Elsevier.
Besler, H. and Marker, M.E. (1979). Namib Sandstone: A distinct lithological unit. *Transactions Geological Society of South Africa*, 82, 155–60.
Bettis, E.A., Muhs, D.R., Roberts, H.M. and Wintle, A.G. (2003). Last Glacial loess in the conterminous USA. *Quaternary Science Reviews*, 22, 1907–46.
Beveridge, C., Kocurek, G., Ewing, R.C., Lancaster, N., Morthekai, P., Singhvi, A. and Mahan, S.A.(2006). Development of spatially diverse and complex dune-field patterns: Gran Desierto Dune Field, Sonora, Mexico. *Sedimentology*, 53, 1391–1409.
Beydoun, Z.R. (1980). Some Holocene geomorphological and sedimentological observations from Oman and their palaeogeological implications. *Journal of Petroleum Geology*, 2, 427–37.
Bhark, E.W. and Small, E.E. (2003). Association between plant canopies and the spatial patterns of infiltration in shrubland and grassland of the Chihuahuan Desert, New Mexico. *Ecosystems*, 6, 185–96.

Bhattachan, A., D'Odorico, P., Baddock, M.C., Okin, G.S. and Cassar, N. (2012). The southern Kalahari: A potential new dust source in the Southern Hemisphere? *Environmental Research Letters*, 024001, doi:10.1088/1748-9326/7/2/024001.

Bielders, C.L., Lamers, J.P.A. and Michels, K. (2001). Wind erosion control technologies in the West African Sahel: The effectiveness of windbreaks, mulching and soil tillage, and the perspective of farmers. *Annals of Arid Zone*, 40, 369–94.

Bielders, C.L., Michels, K. and Rajot, J.-R. (2000). On-farm evaluation of ridging and residue management practices to reduce wind erosion in Niger. *Journal of the Soil Science Society of America*, 64, 1776–85.

Bierman, P.R. and Caffee, M. (2001). Slow rates of rock surface erosion and sediment production across the Namib Desert and escarpment, southern Africa. *American Journal of Science*, 301, 326–58.

Bierman, P.R. and Caffee, M. (2002). Cosmogenic exposure and erosion history of Australian bedrock landforms. *Bulletin of the Geological Society of America*, 114, 787–803.

Bierman, P.R., Reuter, J.M., Pavich, M., Caffee, M.W. and Larsen, J. (2005). Using cosmogenic nuclides to contrast rates of erosion and sediment yield in a semi-arid dominated landscape, Rio Puerco Basin, New Mexico. *Earth Surface Processes and Landforms*, 30, 935–53.

Bijen, J.M. (1989). Maintenance and repair of concrete structures. *Heron (Delft)*, 34(2), 1–82.

Binda, P.L. (1983). On the skewness of some aeolian sands from Saudi Arabia. *Developments in Sedimentology*, 38, 27–39.

Binnie, S.A., Phillips, W.M., Summerfield, M.A., Fifield, L.K. and Spotila, J.A. (2010). Tectonic and climatic controls of denudation rates in active orogens: The San Bernadino Mountains, California. *Geomorphology*, 118, 249–61.

Birken, A.S. and Cooper, D.J. (2006). Processes of *Tamarix* invasion and floodplain development along the lower Green River, Utah. *Ecological Applications*, 16, 1103–20.

Birkett, C.M. (2000). Synergist remote sensing of Lake Chad: Variability of basin inundation. *Remote Sensing of Environment*, 72, 218–36.

Birkett. C.M., Murtugudde, R. and Allan, T. (1999). Indian Ocean climate event brings flood to East Africa's lakes and the Sudd Marsh. *Geophysical Research Letters*, 26, doi:10.1029/1999GL900165.

Birmili, W. and 17 others. (2007). An episode of extremely high PM concentrations over central Europe caused by dust emitted over the southern Ukraine. *Atmospheric Chemistry and Physics Discussions*, 7, 12231–88.

Bishop, M.A. (2010). Nearest neighbour analysis of mega-barchanoid dunes, Ar Rub' al Khali, sand sea: The application of geographical indices to the understanding of dune field self-organization, maturity and environmental change. *Geomorphology*, 120, 186–94.

Bishop, M.A. (2011). Aeolian scours as putative signatures of wind erosion and sediment transport direction on Mars. *Geomorphology*, 125, 558–68.

Biswas, S.K. (1971). The miliolite rocks of Kutch and Kathiawar (western India). *Sedimentary Geology*, 5, 147–64.

Black, D.M. (1955). Natural dams of Havasu Canyon, Supai, Arizona. *Science*, 121, 611–12.

Blackwelder, E. (1925). Exfoliation as a phase of rock weathering. *Journal of Geology*, 33, 793–806.

Blackwelder, E. (1926). Fire as an agent in rock weathering. *Journal of Geology*, 35, 134–40.

Blackwelder, E. (1929). Cavernous rock surfaces of the desert. *American Journal of Science*, 17, 393–9.

Blackwelder, E. (1931). Desert plains. *Journal of Geology*, 39, 133–40.

Blainey, J.B. and Pelletier, J.D. (2008). Infiltration on alluvial fans in arid environments: Influence of fan morphology. *Journal of Geophysical Research*, 113, F03008, doi:10.1029/2007JF000792.
Blair, R.W. (1987). Development of natural sandstone arches in southeastern Utah. In *World Geomorphology*, ed. V. Gardiner. Chichester: Wiley, pp. 597–604.
Blair, T.C. (1999). Case of dominance by sheetflood vs. debris-flow processes on two adjoining alluvial fans, Death Valley, California. *Sedimentology*, 46, 1015–28.
Blair, T.C. (2003). Features and origin of the giant Cucomonga Canyon alluvial fan, Eureka Valley, California. *Geological Society of America Special Paper*, 370, 105–26.
Blake, J.F. (1897). On some superficial deposits in Cutch. *Quarterly Journal of the Geological Society of London*, 53, 223–43.
Blake, W.P. (1855). On the grooving and polishing of land rocks and minerals by the sand. *American Journal of Science and Arts*, 20, 178–82.
Blake, W.P. (1904). Origin of pebble-covered plains in desert regions. *Transactions of the American Institute of Mining Engineers*, 34, 161–2.
Blanc, P.-L. (2000). Of sills and straits: A quantitative assessment of the Messinian Salinity Crisis. *Deep-Sea Research 1*, 47, 1429–60.
Blechschmidt, I., Matter, A., Preusser, F. and Rieke-Zapp, D. (2009). Monsoon triggered formation of Quaternary alluvial megafans in the interior of Oman. *Geomorphology*, 110, 128–39.
Blissenbach, E. (1954). Geology of alluvial fans in semiarid regions. *Bulletin of the Geological Society of America*, 66, 175–90.
Blume, H.-P. and Barth, H.K. (1979). Laterische Krustenstufen in Australien. *Zeitschrift für Geomorphologie*, Supplementband, 33, 46–56.
Blümel, W.D. (1982). Calcretes in Namibia and SE Spain: Relations to substratum, soil formation and geomorphic factors. *Catena Supplement*, 1, 67–82.
Boardman, J., Parsons, A.J., Holland, R., Holmes, P.J. and Washington, R. (2003). Development of badlands and gullies in the Sneeuberg, Great Karoo, South Africa. *Catena*, 50, 165–84.
Bochet, E., García-Fayos, P. and Poesen, J. (2009). Topographic thresholds for plant colonization on semi-arid eroded slopes. *Earth Surface Processes and Landforms*, 34, 1758–71.
Bochet, E., Rubio, J.L. and Poesen, J. (1998). Relative efficiency of three representative matorral species in reducing water erosion at the microscale in a semi-arid climate (Valencia, Spain). *Geomorphology*, 23, 139–50.
Boggs, D.A., Boggs, G.S., Eliot, I. and Knott, B. (2006). Regional patterns of salt lake morphology in the lower Yarra Yarra drainage system of Western Australia. *Journal of Arid Environments*, 64, 97–115.
Bohannon, J. (2010). The Nile Delta's sinking future. *Science*, 327, 1444–7.
Böhlke, J. and Michalski, G. (2002). Atmospheric and microbial components of desert nitrate deposits indicated by variations in δ^{18} O δ^{17} O δ^{16} O isotope ratios. *Eos Transactions of the AGU*, 83(47), Fall Meeting Supplement of Abstracts.
Böhlke, J.K., Ericksen, G.E. and Revesz, K. (1997). Stable isotope evidence for an atmospheric origin of desert nitrate deposits in northern Chile and southern California, U.S.A. *Chemical Geology*, 136, 135–52.
Böhner, J., Schafer, W., Conrad, O., Gross, J. and Ringeler, A. (2003). The WEELS model: Methods, results and limitations. *Catena*, 52, 289–308.
Bond, G., Showers, W., Cheseby, M. and others. (1997). A pervasive millennial-scale cycle in North Atlantic Holocene and glacial climates. *Science*, 278, 1257–65.
Bontognali, T.R.R and 7 others. (2010). Dolomite formation within microbial mats in the coastal sabkha of Abu Dhabi (United Arab Emirates). *Sedimentology*, 57, 824–44.

Boomer, I., Aladin, N., Plotnikov, I. and Whatley, R. (2000). The palaeolimnology of the Aral Sea: A review. *Quaternary Science Reviews*, 19, 1259–78.

Boroda, R., Amit, R., Matmon, A., ASTER Team, Finkle, R., Porat, N., Enzel, Y. and Eyal, Y. (2011). Quaternary-scale evolution of sequences of talus flatirons in the hyperarid Negev. *Geomorphology*, 127, 41–52.

Bory, A., Wolff, E., Mulvaney, R., Jagoutz, E., Wegner, A., Ruth, U. and Elderfield H. (2010). Multiple sources supply eolian mineral dust to the Atlantic sector of coastal Antarctica: Evidence from recent snow layers at the top of Berkner Island ice sheet. *Earth and Planetary Science Letters*, 291, 138–48.

Bosboom, R.E. and 10 others. (2011). Late Eocene sea retreat from the Tarim Basin (west China) and concomitant Asian paleoenvironmental change. *Palaeogeography, Palaeoclimatology, Palaeoecology*, 299, 385–98.

Bouchette, F. and 7 others. (2010). Hydrodynamics in Holocene Lake Mega-Chad. *Quaternary Research*, 73, 226–36.

Bourgeois, O. and 8 others. (2008). A surface dissolution/precipitation model for the development of lakes on Titan, based on an arid terrestrial analogue: The pans and calcretes of Etosha (Namibia). *Lunar and Planetary Science,* 39, 1733.

Bourke, M.C. (2010). Barchan dune asymmetry: Observations from Mars and Earth. *Icarus*, 205, 183–97.

Bourke, M.C. and Goudie, A.S. (2009). Varieties of barchan form in the Namib Desert and on Mars. *Aeolian Research*, 1, 45–54.

Bourke, M.C. and Pickup, G. (1999). Fluvial form variability in arid central Australia. In *Varieties of Fluvial Form*, ed. A.J. Miller and A. Gupta. Chichester: Wiley, pp. 249–71.

Bourke, M.C. and Viles, H.A. (2007). *A Photographic Atlas of Rock Breakdown Features in Geomorphic Environments*. Tucson, AZ: Planetary Science Institute.

Bourke, M.C., Lancaster, N., Fenton, L.K., Parteli, E.J.R., Zimbelman, J.R. and Radebaugh, J. (2010). Extraterrestrial dunes: An introduction to the special issue on planetary dune systems. *Geomorphology*, 121, 1–14.

Bourne, J.A. and Twidale, C.R. (2010). Playas of inland Australia. *Cadernos Lab. Xeolóxico de Laxe Coruña*, 35, 71–98.

Bouza, P.J., Simón, M., Aguilar, J., del Valle, H. and Rostagno, M. (2007). Fibrous-clay mineral formation and soil evolution in aridisols of northeastern Patagonia, Argentina. *Geoderma*, 139, 38–50.

Bowen, M.W. and Johnson, W.C. (2012). Late Quaternary environmental reconstructions of playa-lunette system evolution on the central High Plains of Kansas, United States. *Bulletin of the Geological Society of America*, 124, 146–61.

Bowen, M.W., Johnson, W.C., Egbert, S.L. and Klopfenstein, S.T. (2010). A GIS-based approach to identify and map playa wetlands on the High Plains, Kansas, USA. *Wetlands*, 30, 675–84.

Bowler, J.M. (1973). Clay dunes: Their occurrence, formation and environmental significance. *Earth-Science Reviews*, 9, 315–38.

Bowman, D. (2010). The Dead Sea Graben: Geomorphology of the lowest spot on Earth. In *Geomorphological Landscapes of the World*, ed. P. Migoń. Heidelberg: Springer, pp. 247–55.

Bowman, I. (1924). *Desert Trails of Atacama*. American Geographical Society.

Box, M.A., Radhi, M. and Box, G.P. (2010). The great Sydney dust event: Size-resolved chemical composition and comparison. *IOP Conference Series: Earth and Environmental Science*, 11, doi:10.1088/1755–1315/11/1/012015.

Boy, J. and Wilcke, W. (2008). Tropical Andean forest derives calcium and magnesium from Saharan dust. *Global Biogeochemical Cycles*, 22, GB1027, doi:10.1029/2007GB002960.

Bradbury, J.P. (1997). A diatom record of climate and hydrology for the past 200 ka from Owens Lake, California with comparison to other Great Basin records. *Quaternary Science Reviews*, 16, 203–19.
Bradley, W.C., Hutton, J.T. and Twidale, C.R. (1978). Role of salts in development of granitic tafoni, South Australia. *Journal of Geology*, 86, 647–54.
Bradstock, R.A., Auld, T.D., Ellis, M.E. and Cohn, J.S. (1992). Soil temperatures during bushfires in semi-arid, mallee shrublands. *Australian Journal of Ecology*, 17, 433–40.
Brain, C.K. and Koste, W. (1993). Rotifers of the genus *Proales* from saline springs in the Namib Desert, with the description of a new species. *Hydrobiologia*, 255–256, 449–54.
Branagan, D.F. (1983). The Sydney Basin and its vanished sequence. *Journal of the Geological Society of Australia*, 30, 75–84.
Brandmeier, M., Kuhlemann, J., Krumrei, I., Kappler, A. and Kubik, P.W. (2011). New challenges for tafoni research. A new approach to understand process and weathering rates. *Earth Surface Processes and Landforms*, 36, 839–52.
Brazel, A.I. (1991). Blowing dust and highways: The case of Arizona, USA. In *Highway Meteorology*, ed. A.H. Perry and J. Symons. London: E. and F.N. Spon, pp. 131–61.
Breed, C.S., Fryberger, S.G., Andrew, S., McCauley, C., Lennartz, F., Gebel, D. and Horsham, K. (1979). Regional studies of sand sea, using Landsat (ERTS) imagery. *U.S. Geological Survey Professional Paper*, 1059, 305–97.
Breslow, P.B. and Sailor, D.J. (2002). Vulnerability of wind power resources to climate change in the continental United States. *Renewable Energy*, 27, 585–98.
Breuning-Madsen, H. and Awadzi, T.W. (2005). Harmattan dust deposition and particle size in Ghana. *Catena*, 63, 23–38.
Bridges, N.T., Ayoub, F., Avouac, J.-P., Leprince, S., Lucas, A. and Mattson, S. (2012a). Earth-like sand fluxes on Mars. *Nature*, 485, 339–42.
Bridges, N.T., Razdan, A., Yin, X., Greeley, R., Ali, S. and Kushunapally, R. (2010). Quantification of shape and texture for wind abrasion studies: Proof of concept using analog targets. *Geomorphology*, 114, 213–26.
Bridges, N.T. and 12 others. (2012b). Planet-wide sand motion on Mars. *Geology*, 40, 31–4.
Brink, W. (1951). *Big Hugh: The Father of Soil Conservation*. New York: Macmillan.
Bristow, C.S. and Lancaster, N. (2004). Movement of a small slipfaceless dome dune in the Namib Sand Sea, Namibia. *Geomorphology*, 59, 189–96.
Bristow, C.S., Bailey, S.D. and Lancaster, N. (2000). The sedimentary structure of linear sand dunes. *Nature*, 406, 56–8.
Bristow, C.S., Drake, N. and Armitage, S. (2009). Deflation in the dustiest place on Earth: The Bodélé Depression Chad. *Geomorphology*, 105, 50–8.
Bristow, C.S., Duller, G.A.T. and Lancaster, N. (2007). Age and dynamics of linear dunes in the Namib Desert. *Geology*, 35, 555–8.
Bristow, C.S., Lancaster, N. and Duller, G.A.T. (2005). Combining ground penetrating radar surveys and optical dating to determine dune migration in Namibia. *Journal of the Geological Society*, 162, 315–21.
Brock, E.J. and Twidale, C.R. (2011). J.T. Jutson and the work of wind in shaping the landscape. *Geographical Research*, 49, 47–58.
Broecker, W.S. (2002). Dust: Climate's Rosetta Stone. *Proceedings American Philosophical Society*, 146, 77–80.
Broecker, W.S. and Liu, T. (2001). Rock varnish: Recorder of desert wetness? *GSA Today*, 11(8), 4–10.
Bromblet, P. (1993). Relations entre les variations des conditions environmentales et les processus de dégradation successive des temples de Karnak (Egypt). In *Conservation of Stone and Other Materials*, ed. M.J. Theil. London: E. and F.N. Spon, 1, pp. 91–8.

Bronger, A., Winter, R. and Heinkele, T. (1998). Pleistocene climatic history of East and Central Asia based on paleopedological indicators in loess-paleosol sequences. *Catena*, 34, 1–17.

Brook, G.A., Marais, E. and Cowart, J.B. (1999). Evidence of wetter and drier conditions in Namibia from tufas and submerged speleothems. *Cimbebasia*, 15, 29–39.

Brook, G.A., Ellwood, B.B., Railsback, L.B. and Cowart, J.B. (2006). A 164 ka record of environmental change in the American Southwest from a Carlsbad Cavern speleothem. *Palaeogeography, Palaeoclimatology, Palaeoecology*, 237, 483–507.

Brook, G.A., Marais, E., Srivastava, P. and Jordan, T. (2007). Timing of lake-level changes in Etosha Pan, Namibia, since the middle Holocene from OSL ages of relict shorelines in the Okondeka region. *Quaternary International*, 175, 29–40.

Brooke, B. (2001). The distribution of carbonate eolianite. *Earth-Science Reviews*, 55, 135–64.

Brookes, I.A. (1993). Geomorphology and Quaternary geology of the Dakhla Oasis Region, Egypt. *Quaternary Science Review*, 12, 529–52.

Brookes, I.A. (2001). Aeolian erosional lineations in the Libyan Desert, Dakhla Region, Egypt. *Geomorphology*, 39, 189–209.

Brooks, N. (2006). Cultural responses to aridity in the Middle Holocene and increased social complexity. *Quaternary International*, 151, 29–49.

Brooks, N., Chiapello, I., Di Lernia, S., Drake, N., Legrand, M., Moulin, C. and Prospero, J. (2005). The climate-environment-society nexus in the Sahara from prehistoric times to the present day. *Journal of North African Studies*, 10, 253–92.

Brown, B.B. and Benison, K.C. (2009). Geochemical characteristics of naturally acid and alkaline lakes in southern Western Australia. *Applied Geochemistry*, 24, 268–84.

Brown, C.N. (1956). The origin of caliche in the northeastern Llano Estacado. *Journal of Geology*, 64, 1–15.

Brown, P.W., Robbins, C.R. and Clifton, J.R. (1979). Adobe II: Factors affecting the durability of adobe structures. *Studies in Conservation*, 24, 23–39.

Brüchert V., Currie, B. and Peard, K.R. (2009). Hydrogen sulphide and methane emissions on the central Namibian shelf. *Progress in Oceanography*, 83, 169–79.

Bruhn, R.L., Brown, F.H., Gathogo, P.N. and Haileab, B. (2011). Pliocene volcano-tectonics and paleography of the Turkana Basin, Kenya and Ethiopia. *Journal of African Earth Sciences*, 59, 295–312.

Brunelle, A., Minckley, T.A., Blissett, S., Cobabe, S.K. and Guzman, B.L. (2010). A $\sim$8000 year fire history from an Arizona/Sonora borderland ciénega. *Journal of Arid Environments*, 74, 475–81.

Brunotte, E. and Sander, H. (2008). Loess accumulation and soil formation in Kaokoland (northern Namibia) as indicators of Quaternary climatic change. *Global and Planetary Change*, 26, 67–75.

Brunotte, E., Maurer, B., Fischer, P., Lomax, J. and Sander, H. (2009). A sequence of fluvial and aeolian deposits (desert loess) covering the last 60 ka in the Opuwo Basin (Kaokoland/Kunene Region, Namibia) based on luminescence dating. *Quaternary International*, 196, 71–85.

Bryan, C.R., Helean, K.B., Marshall, B.D. and Brady, P.V. (2009). Feldspar dissolution rates in the Topopah Spring Tuff, Yucca Mountain, Nevada. *Applied Geochemistry*, 24, 2133–43.

Bryan, K. (1923). Wind erosion near Lees Ferry, Arizona. *American Journal of Science*, 5th series, 6, 291–307.

Bryan, K. (1925). The Papago country. *United States Geological Survey Water Supply Paper*, 499.

Bryan, K. (1927). Persistence of features in an arid landscape: The Navajo Twins, Utah. *Geographical Review*, 17, 251–7.

Bryan, K. (1928a). Historical evidence of changes in the channel of Rio Puerco, a tributary of the Rio Grande in New Mexico. *Journal of Geology*, 36, 265–82.
Bryan, K. (1928b). Niches and other cavities in sandstones at Chaco Canyon, N. Mexico. *Zeitschrift für Geomorphologie*, 3, 125–40.
Bryan, K. (1929). Solution-facetted limestone pebbles. *American Journal of Science*, 5th series, 18, 193–208.
Bryan, K. (1941). Pre-Columbian agriculture in the Southwest as conditioned by periods of alluviation. *Annals of the Association of American Geographers*, 31, 219–42.
Bryan, R. and Yair, A. (1982). *Badlands Geomorphology and Piping*. Norwich: Geo Books.
Bryan, R.B. and Jones, J.A.A. (1997). The significance of soil piping processes: Inventory and prospect. *Geomorphology*, 20, 209–18.
Bryant, R.G. (2003). Monitoring hydrological controls on dust emissions: Preliminary observations from Etosha Pan, Namibia. *Geographical Journal*, 169, 131–41.
Bryant, R.G. and Rainey, M.P. (2002). Investigation of flood inundation on playas within the Zone of Chotts using a time-series of AVHRR. *Remote Sensing of Environment*, 82, 360–75.
Bryant, R.G., Bigg, G.R., Mahowald, N.M., Eckardt, F.D. and Ross, S.G. (2007). Dust emission response to climate in southern Africa. *Journal of Geophysical Research*, 112, D09207, doi:10.1029/2005JD007025.
Bubenzer, O. and Reimer, H. (2007). Holocene climate change and human settlement between the central Sahara and the Nile Valley: Archaeological and geomorphological results. *Geoarchaeology*, 22, 607–20.
Bubenzer, O., Bolten, A. and Darius, F. (2007). *Atlas of Cultural and Environmental Change in Arid Africa*. Köln: Heinrich-Barth-Institut.
Buchbinder, B., Margaritz, M. and Buchbinder, L.G. (1983). Turonian to Neogene palaeokarst in Israel. *Palaeogeography, Palaeoclimatology, Palaeoecology*, 43, 329–50.
Buck, B.J. and Van Hoesen, J.G. (2002). Snowball morphology and SEM analysis of pedogenic gypsum, southern New Mexico, U.S.A. *Journal of Arid Environments*, 51, 469–87.
Büdel, B., Weber, B., Kühl, M., Pfanz, H., Sültemeyer, D. and Wessels, D. (2004). Reshaping of sandstone surfaces by cryptoendolithic cyanobacteria: Bioalkalization causes chemical weathering in arid landscapes. *Geobiology*, 2, 261–8.
Büdel, J. (1982). *Climatic Geomorphology* (trans. by L. Fischer and D. Busche). Princeton, NJ: Princeton University Press.
Buis, E., Temme, A.J.A.M., Veldkamp, A., Boeken, B., Jongmans, A.G., van Breeren, N. and Schoorl, J.M. (2010). Shrub mound formation and stability on semi-arid slopes in the northern Negev Desert of Israel: A field study and simulation study. *Geoderma*, 156, 363–71.
Bull, L.J. and Kirkby, M.J. (eds.). (2002). *Dryland Rivers*. Wiley: Chichester.
Bull, W.B. (1963). Alluvial-fan deposits in western Fresno County, California. *Journal of Geology*, 71, 243–51.
Bull, W.B. (1964). Geomorphology of segmented alluvial fans in western Fresno County, California. *U.S. Geological Survey Professional Paper*, 532F, 89–128.
Bull, W.B. (1977). The alluvial-fan environment. *Progress in Physical Geography*, 1, 222–70.
Bull, W.B. (1997). Discontinuous ephemeral streams. *Geomorphology*, 19, 227–76.
Bullard, J.E. (1997). Vegetation and dryland geomorphology. In *Arid Zone Geomorphology: Process, Form and Change in Drylands* (2nd ed.), ed. D.S.G. Thomas. Chichester: Wiley, pp. 109–31.
Bullard, J.E. and Livingstone, I. (2002). Interactions between aeolian and fluvial systems in dryland environments. *Area*, 34, 8–16.

Bullard, J.E. and Livingstone, I. (2010). Classics in physical geography revisited. Wasson RJ and Hyde R (1983). Factors determing desert dune type. Nature 304: 337-9. *Progress in Physical Geography*, 34, 857–62.

Bullard, J.E. and McTainsh, G.H. (2003). Aeolian-fluvial interactions in dryland environments: Examples, concepts and Australia case study. *Progress in Physical Geography*, 27, 471–501.

Bullard, J.E. and Nash, D.J. (2000). Valley-marginal sand dunes in the south-west Kalahari: Their nature, classification and possible origins. *Journal of Arid Environments*, 45, 369–83.

Bullard J.E. and White, K. (2005). Dust production and the release of iron oxides resulting from the aeolian abrasion of natural dune sands. *Earth Surface Processes and Landforms*, 30, 95–106.

Bullard, J.E., McTainsh, G.H. and Pudmenzky, C. (2004). Aeolian abrasion and modes of fine particle production from natural red dune sands: An experimental study. *Sedimentology*, 51, 1103–25.

Bullard, J.E., White, K.W. and Livingstone, I. (2011). Morphometric analysis of aeolian bedforms in the Namib Sand Sea using ASTER data. *Earth Surface Processes and Landforms*, 36, 1534–49.

Bullard, J.E., Baddock, M., McTainsh, G. and Leys, J. (2008). Sub-basin scale dust source geomorphology detected using MODIS. *Geophysical Research Letters*, 35, L15404, doi:10.1029/2008GL033928.

Bullard, J.E., Thomas, D.S.G., Livingstone, I. and Wiggs, G.F.S. (1995). Analysis of linear sand dune morphological variability, southwestern Kalahari Desert. *Geomorphology*, 11, 189–203.

Bullard, J.E., Thomas, D.S.G., Livingstone, I. and Wiggs, G.F.S. (1996). Wind energy variations in the southwestern Kalahari Desert and implications for linear dunefield activity. *Earth Surface Processes and Landforms*, 21, 263–78.

Bullard, J.E., Thomas, D.S.G., Livingstone, I. and Wiggs, G.F.S. (1997). Dunefield activity and interactions with climatic variability in the southwest Kalahari Desert. *Earth Surface Processes and Landforms*, 22, 165–74.

Bulley, B.G. (1986). The engineering geology of Swakopmund. *Communications Geological Survey SW Africa/Namibia*, 2, 7–12.

Bullock, M.S., Larnet, F.J., Cesar Izaurralde, R. and Feng, Y. (2001). Overwinter changes in wind erodibility of clayloam soils in southern Alberta. *Journal of the Soil Science Society of America*, 65, 423–30.

Burckhardt. J.L. (1829). *Travels in Arabia*. London: African Association.

Burnett, B.N., Meyer, G.A. and McFadden, L.D. (2008). Aspect-related microclimatic influences on slope forms and processes, northeastern Arizona. *Journal of Geophysical Research*, 113, F03002, doi:10.1029/2007JF000789.

Burrough, S.L., Thomas, D.S.G. and Bailey, R.M. (2009). Mega-lake in the Kalahari: A late Pleistocene record of the Palaeolake Makgadikgadi system. *Quaternary Science Reviews*, 28, 1392–411.

Burrough, S.L., Thomas, D.S.G., Bailey, R.M. and Davies, L. (2012). From landform to process: Morphology and formation of lake-bed barchan dunes, Makgadikgadi, Botswana. *Geomorphology*, 161–162, 1–14.

Busche, D. (1983). Silcrete in der zentralen Sahara (Murzuk–Bechen, Djado–Plateau und Kaouar; Süd-Libyen and Nord-Niger). *Zeitschrift für Geomorphologie*, Supplementband, 48, 35–49.

Busche, D. (1998). *Die Zentrale Sahara*. Gotha: Justus Perthes Verlag.

Busche, D. (2001). Early Quaternary landslides of the Sahara and their significance for geomorphic and climatic history. *Journal of Arid Environments*, 49, 429–48.
Busche, D. and Erbe, W. (1987). Silicate karst landforms of the southern Sahara (northeastern Niger and southern Libya). *Zeitschrift für Geomorphologie*, Supplementband, 64, 57–72.
Busche, D. and Sponholz, B. (1992). Morphological and micromorphological aspects of the sandstone karsts of eastern Niger. *Zeitschrift für Geomorphologie*, Supplementband, 85, 1–18.
Butler, B.E. (1961). Ground surfaces and the history of the Riverine Plain. *Australian Journal of Science*, 24, 39–40.
Butler, B.E. (1974). A contribution towards the better specification of parna and some other aeolian clays in Australia. *Zeitschrift für Geomorphologie*, Supplementband, 20, 106–16.
Butler, P.R. and Mount, J.F. (1986). Corroded cobbles in southern Death Valley: Their relationship to honeycomb weathering and lake shorelines. *Earth Surface Processes and Landforms*, 11, 377–87.
Butt, C.R.M. (1985). Granite weathering and silcrete formation on the Yilgarn block, Western Australia. *Australian Journal of Earth Sciences*, 32, 415–32.
Butzer, K.W. and Hansen, C.L. (1968). *Desert and River in Nubia: Geomorphology and Prehistoric Environments at the Aswan Reservoir*. Madison: University of Wisconsin Press.
Bywater-Reyes, S., Carrapa, B., Clementz, M. and Schoenbohm, L. 2010. Effect of late Cenozoic aridification on sedimentation in the Eastern Cordillera of northwest Argentina (Angastaco Basin). *Geology*, 38, 235–8.
Cácares, L., Gómez-Silva, B., Garró, X., Rodriguez, V., Monardes, V. and McKay, C.P. (2007). Relative humidity patterns and fog water precipitation in the Atacama Desert and biological implications. *Journal of Geophysical Research*, 112, G04S14, doi:10.1029/2006JG000344.
Cadol, D., Rathburn, S.L. and Cooper, D.J. (2010). Aerial photographic analysis of channel narrowing and vegetation expansion in Canyon de Chelly National Monument, Arizona, USA, 1935–2004. *River Research and Applications*, 27, 841–56.
Cagle, F.R. (1975). Evaporite deposits of the central Namib Desert. MSc thesis, University of New Mexico, Albuquerque.
Cakmur, R.V., Miller, R.L., Perlwitz, J., Geogdzhayev, I.V., Ginoux, P., Koch, D., Kohfeld, K.E., Tegen, I. and Zender, C.S. (2006). Constraining the magnitude of the global dust cycle by minimizing the difference between a model and observations. *Journal of Geophysical Research*, 111, D06207, doi:10.1029/2005JD005791.
Callot, Y., Marticorena, B. and Bergametti, G. (2000). Geomorphologic approach for modelling the surface features of arid environments in a model of dust emissions: Application to the Sahara Desert. *Geodinamica Acta*, 13, 245–70.
Calov, R., Ganopolski, A., Claussen, M., Petoukhov, V. and Greve, R. (2005). Transient simulation of the Last Glacial inception. Part 1: Glacial inception as a bifurcation in the climate system. *Climate Dynamics*, 24, 545–61.
Cammeraat, L.H., Willott, S.J., Compton, S.G. and Incoll, L.D. (2002). The effects of ants' nests on the physical, chemical and hydrological properties of a rangeland soil in semi-arid Spain. *Geoderma*, 105, 1–20.
Campbell, E.M. (1968). Lunettes in South Australia. *Transactions Royal Society of South Australia*, 92, 83–109.
Campbell, I. (1997). Badlands and badland gullies. In *Arid Zone Geomorphology*, ed. D.S.G. Thomas. Chichester: Wiley, pp. 261–91.

Camuffo, D. (1993). Controlling the aeolian erosion of the Great Sphinx. *Studies in Conservation*, 38, 198–205.

Cánton, Y., Domingo, F., Sole Benet, A. and Puigdefabregas, J. (2001). Hydrological and erosion response of a badlands system in semi-arid SE Spain. *Journal of Hydrology*, 252, 65–84.

Cánton, Y., Solé-Benet, A., Queralt, I. and Pini, R. (2001). Weathering of a gypsum-calcareous mudstone under semi-arid environment at Tabernas, SE Spain: Laboratory and field-based experimental approaches. *Catena*, 44, 111–32.

Cao, S., Tian, T., Chen, L., Dong, X., Yu, X. and Wang, G. (2010). Damage caused to the environment by reforestation policies in arid and semi-arid areas of China. *Ambio*, 39, 279–83.

Capo, R.C. and Chadwick, O.A. (1999). Sources of strontium in desert soil and calcrete. *Earth and Planetary Science Letters*, 170, 61–72.

Capot-Rey, R. (1939). Pays du Mzab et régions des Dayas. Etude sur le relief de la dorsale Saharienne. *Annales de Géographie*, 48, 41–63.

Carlston, C.W. (1966). The effect of climate on drainage density and streamflow. *Hydrological Sciences Journal*, 11, 62–9.

Carter, H.J. (1849). On foraminifera, their organization and their existence in a fossilized state in Arabia, Sindh, Kutch and Khattyawar. *Journal of the Bombay Branch of the Royal Asiatic Society*, 3, 158.

Casanova, J. (1991). Biosedimentology of Quaternary stromatolites in intertropical Africa. *Journal of African Earth Sciences*, 12, 409–15.

Casanova, J. and Hillaire-Marcel, C. (1992). Chronology and paleohydrology of late Quaternary high lake levels in the Manyara Basin (Tanzania) from isotopic data (^{18}O, ^{13}C, ^{14}C, Th/U) on fossil stromatolites. *Quaternary Research*, 38, 205–26.

Castañeda, I.S., Mulitza, S., Schefuss, E., Lopes dos Santos, R.A.L., Damsté, J.S.S. and Schouten, S. (2009). Wet phases in the Sahara/Sahel region and human migration patterns in North Africa. *Proceedings of the National Academy of Sciences*, 106, 20159–63.

Castellani, V. and Dragoni, W. (1986). Evidence for karstic mechanisms involved in the evolution of Moroccan hamadas. *International Journal of Speleology*, 15, 57–71.

Castiglia, P.J. and Fawcett, P.J. (2006). Large Holocene lakes and climate change in the Chihuahuan Desert. *Geology*, 34, 113–16.

Castillo, V.M., Mosch, W.M., Conesa García, C., Barberá, G.G., Navarro Cano, J.A. and López-Bermúdez, F. (2007). Effectiveness and geomorphological impacts of check dams for soil erosion control in a semi-arid Mediterranean catchment: El Cárcavo (Murcia, Spain). *Catena*, 70, 416–27.

Caton-Thompson, G. and Gardner, E.W. (1932). The prehistoric geography of Kharga Oasis. *Geographical Journal*, 80, 369–406.

Cattle, S.R., McTainsh, G.H. and Elias, S. (2009). Aeolian dust deposition rates, particle sizes and contributions to soils along a transect in semi-arid New South Wales, Australia. *Sedimentology*, 56, 765–83.

Catto, N. and Bachhuber, F. (2003). Aeolian geomorphic response to climate change: An example from the Estanica Valley, central New Mexico, USA. *Advances in Global Change Research*, 6, 171–92.

Cayan, D.R., Redmond, K.T. and Riddle, L.G. (1999). ENSO and hydrological extremes in the western United States. *Journal of Climate*, 12, 2881–93.

Cereceda, P. and Schemenauer, R.S. (1991). The occurrence of fog in Chile. *Journal of Applied Meteorology*, 30, 1097–105.

Cereceda, P., Larrain, H., Osses, P. Farías, M. and Egaña, I. (2008). The spatial and temporal variability of fog and its relation to fog oases in the Atacama Desert, Chile. *Atmospheric Research*, 87, 312–23.
Cereceda, P., Osses, P., Larrain, H., Farías, M. Lagos, M., Pinto, R. and Schemenauer, R.S. (2002). Advective, orographic and radiation fog in the Tarapacá region, Chile. *Atmospheric Research*, 64, 261–71.
Cerling, T.E., Webb, R.H., Poreda, R.J., Rigby, A.D. and Melis, T.S. (1999). Cosmogenic ^{3}He ages and frequency of late Holocene debris flows from Prospect Canyon, Grand Canyon, USA. *Geomorphology*, 27, 93–111.
Chakraborty, R. (2009). Keeping water in the wadis of Arabia's perfume capital. *Proceedings of the Institute of Civil Engineers*, 162, 114–21.
Chan, M.A., Beitler Bowen, B., Parry, B.W.T, Ormö, J. and Komatsu, G. (2005). Red rock and red planet diagenesis: Comparisons of Earth and Mars concretions. *GSA Today*, 15, 4–10.
Chan, M.A., Yonkee, W.A., Netoff, D.I., Seiler, W.M. and Ford, R.L. (2008). Polygonal cracks in bedrock on Earth and Mars: Implications for weathering. *Icarus*, 194, 65–71.
Changnon, S.A. (ed.) (2000). *El Niño 1997–1998*. New York: Oxford University Press.
Chapman, R.W. (1980). Salt weathering by sodium chloride in the Saudi Arabian Desert. *American Journal of Science*, 280, 116–29.
Chappell, M.A. and Bartholomew, G.A. (1981). Standard operative temperatures and thermal energetics of the antelope ground squirrel, *Ammospermophitous leucurus*. *Physiological Zoology*, 54, 81–93.
Charles, J.A. (1994). Collapse compression fills on inundation. In *Geotechnical Engineering: Emerging trends in design and practice*, ed. K.R. Saxena. Rotterdam: Balkema, pp. 353–79.
Chawner, W.D. (1935). Alluvial fan flooding: The Montrose, California, flood of 1934. *Geographical Review*, 25, 255–63.
Chen, F.-H. and 8 others. (2010). Moisture changes over the last millennium in arid central Asia: A review, synthesis and comparison with monsoon region. *Quaternary Science Reviews*, 29, 1055–68.
Chen, H.-Y. and Chen, L.-D. (2008). Importance of anthropogenic inputs and continental derived dust for the distribution and flux of water-soluble nitrogen and phosphorus species in aerosol within the atmosphere over the East China Sea. *Journal of Geophysical Research*, 113, D11303, doi:10.1029/2007JD009491.
Chen, X.Y. (1997). Pedogenic gypcrete formation in arid central Australia. *Geoderma*, 77, 39–61.
Chepil, W.S. (1945). Dynamics of wind erosion. *Soil Science*, 60, 305–20; 397–411; 475–80.
Chepil, W.S. and Milne, R.A. (1939). Comparative study of soil drifting in the field and in a wind tunnel. *Scientific Agriculture*, 19, 249–57.
Chi, S.C. and Reilinger, R.E. (1984). Geodetic evidence for subsidence due to groundwater withdrawal in many parts of the United States of America. *Journal of Hydrology*, 67, 155–82.
Chico, R.J. (1963). Playa mud cracks: Regular and king size. *Geological Society of America, Special Paper*, 76.
Chin, A. and Gregory, K.J. (2001). Urbanization and adjustment of ephemeral stream channels. *Annals of the Association of American Geographers*, 91, 595–608.
Chin, M., Chu, A., Levy, R., Remer, L., Kaufman, Y., Holben, B., Eck, T., Ginoux P. and Gao, Q. (2004). Aerosol distribution in the Northern Hemisphere using ACE-Asia: Results from global model, satellite observations, and Sun photometer measurements. *Journal of Geophysical Research*, 109, D23S90, doi:10.1029/2004JD004829.

Chiquet, A., Colin, F., Hamelin, B., Michard, A. and Nahon, D. 2000. Chemical mass balance of calcrete genesis on the Toledo granite (Spain). *Chemical Geology*, 170, 19–35.
Chiquet, A., Michard, A., Nahon, D. and Hemelin, B. (1999). Atmospheric input vs in situ weathering in the genesis of calcretes: An Sr isotope study at Gálvez (central Spain). *Geochimica et Cosmochimica Acta*, 63, 311–23.
Chirino, E., Bonet, A., Bellot, J. and Sánchez, J. R. (2006). Effects of 30-year-old Aleppo pine plantations on runoff, soil erosion, and plant diversity in a semi-arid landscape in south eastern Spain. *Catena*, 65, 19–29.
Chojnacki, M., Moresch, J.E. and Burr, D.M. (2010). Climbimg and falling dunes in Valles Marineris, Mars. *Geophysical Research Letters*, 37, L08201, doi:10.1029/2009GL042263.
Chu, P.C., Chen, Y., Lu, S., Li, Z. and Lu, Y. (2008). Particulate air pollution in Lanzhou China. *Environment International*, 34, 698–713.
Chudeau, R. (1909). *Sahara Soudanais*. Paris: Armand Colin.
Churchill, R.R. (1981). Aspect-related differences in badlands slope morphology. *Annals of the Association of American Geographers*, 71, 374–88.
Clapp, E.M., Bierman, P.R. and Caffee, M. (2002). Using ^{10}Be and ^{26}Al to determine sediment generation rates and identify sediment source areas in an arid region drainage basin. *Geomorphology,* 45, 89–104.
Clapp, E.M., Bierman, P.R., Schick, A.P., Lekach, J., Enzel, Y. and Caffee, M. (2000). Sediment yield exceeds sediment production in arid region drainage basins. *Geology*, 28, 995–8.
Clark, I.D., and Fontes, J.-C. (1990). Palaeoclimatic reconstruction in northern Oman based on carbonates from hypersaline groundwaters. *Quaternary Research*, 33, 320–36.
Clarke, J. (2011). Extraterrestrial arid surface processes. In *Arid Zone Geomorphology*, ed. D.S.G. Thomas. Chichester: Wiley Blackwell, pp. 61–82.
Clarke, J. D. (2006). Antiquity of aridity in the Chilean Atacama Desert. *Geomorphology*, 73, 101-14.
Clarke, J.D.A. (1994). Geomorphology of the Kambalda region, Western Australia. *Australian Journal of Earth Sciences*, 41, 229–39.
Clarke, J.D.A. and Stoker, C. (2003). Mound spring complexes in central Australia: An analog for Martian groundwater fed outflow channels? *Lunar and Planetary Science*, 34, 1504.
Clarke, M.L. and Rendell, H.M. (2010). Climate-driven decrease in erosion in extant Mediterranean badlands. *Earth Surface Processes and Landforms*, 35, 1281–8.
Clarke, M.L., Wintle, A.G. and Lancaster, N. (1996). Infra-red stimulated luminescence dating of sands from the Cronese Basins, Mojave Desert. *Geomorphology*, 17, 199–205.
Clauzon, G., Suc, J.-P., Gautier, F. and Loutre, M.-F. (1996). Alternate interpretation of the Messinian salinity crisis: Controversy resolved? *Geology*, 24, 363–6.
Clement, A.C., Seager, R. and Cane, M.A. (2000). Suppression of El Niño during the mid-Holocene by changes in the Earth's orbit. *Paleoceanograhy*, 15, 731–7.
Clements, T., Stone, R.O., Mann, J.F. and Eymann, J.L. (1963). A study of windborne sand and dust in desert areas. *US Army Natik Laboratories, Earth Science Division, Technical Report*, **ES-8**.
Closson, D., Abou Karaki, N. and Hallot, F. (2010). Landslides along the Jordanian Dead Sea coast triggered by the lake level lowering. *Environmental Earth Science*, 59, 1417–30.
Closson, D., LaMoreaux, P.E., Karaki, N.A. and al-Fugha, H. (2007). Karst system developed in salt layers of the Lisan Peninsula, Dead Sea, Jordan. *Environmental Geology*, 52, 155–72.
Cloudsley-Thomson, J.L. and Chadwick, M.J. (1964). *Life in Deserts*. London: Foulis.

Cockburn, H., Seidl, H.A. and Summerfield, M.A. (1999). Quantifying denudation rates on inselbergs in the central Namib Desert using in-situ produced cosmogenic ^{10}Be and ^{26}Al. *Geology*, 27, 399–402.
Codilean, A.T., Fenton, C.R., Fabel, D., Bishop, P. and Xu, S. (In press). Discordance between cosmogenic nuclide concentrations in amalgamated sands and individual fluvial pebbles in an arid zone catchment. *Quaternary Geochronology*.
Coe, J.A., Glancy, P.A. and Whitney, J.W. (1997). Volumetric analysis and hydrologic characterization of a modern debris flow near Yucca Mountain, Nevada. *Geomorphology*, 20, 11–28.
Cohen, H. and Laronne, J.B. (2005). High rates of sediment transport by flashfloods in the southern Judean Desert, Israel. *Hydrological Processes*, 19, 1687–702.
Cohen, H., Laronne, J.B. and Reid, I. (2010). Simplicity and complexity of bed load response during flash floods in a gravel bed ephemeral river: A 10 year field study. *Water Resources Research*, 46, W11542, doi:10.1029/2010WR009160.
Cohen, T.J., Nanson, G.C., Larsen, J.R., Jones, B.G., Price, D.M., Coleman, M. and Pietsch, T.J. (2010). Late Quaternary aeolian and fluvial interactions on the Cooper Creek Fan and the association between linear and source-bordering dunes, Strzelecki Desert, Australia. *Quaternary Science Reviews*, 29, 455–71.
Cohen, T.J. and 10 others. (2011). Continental aridification and the vanishing of Australia's megalakes. *Geology*, 39, 167–70.
Colton, R.B. (1978). Geologic map of the Boulder-Fort Collins-Greeley area, Colorado. *U.S. Geological Survey Miscellaneous Investigations Series* Map I–855-G.
Conca, J.L. and Astor, A.M. (1987). Capillary moisture flow and the origin of cavernous weathering in the dolerites of Bull Pass, Antarctica. *Geology*, 15, 151–4.
Conca, J.L. and Rossman, G.R. (1982). Case hardening of sandstone. *Geology*, 10, 520–23.
Conca, J.L. and Rossman, G.R. (1985). Core softening in cavernous weathered tonalite. *Journal of Geology*, 93, 59–73.
Contreras, S., Lange, C.B., Pantoja, S., Lavik, G., Rincón-Martínez, D. and Kuypers, M.M. (2010). A rainy northern Atacama Desert during the last interglacial. *Geophysical Research Letters*, 37, L23612, doi:10.1029/2010GL045728.
Conway, D., Kroi, M., Alcamo, J. and Hulme, M. (1996). Future availability of water in Egypt: The interaction of global, regional and basin scale driving forces in the Nile Basin. *Ambio*, 25, 336–42.
Cook, B.I., Miller, R.L. and Seager, R. (2009). Amplification of the North American 'Dust Bowl' drought through human-induced land degradation. *Proceedings of the National Academy of Sciences*, 106, 4997–5001.
Cook, B.I., Seager, R. and Miller, R.L. (2011). The impact of devegetated dune fields on North American climate during the late Medieval Climate Anomaly. *Geophysical Research Letters*, 38, L14704, doi:10.1029/2011GL047566.
Cook, K.L., Whipple, K.X., Heimsath, A.M. and Hanks, T.C. (2009). Rapid incision of the Colorado River in Glen Canyon – Insights from channel profiles, local incision rates, and modelling of lithologic controls. *Earth Surface Processes and Landforms*, 34, 994–1010.
Cooke, R.U. (1970). Stone pavements in deserts. *Annals of the Association of American Geographers*, 60, 560–77.
Cooke, R. U. (1981). Salt weathering in deserts. *Proceedings, Geologists' Association of London*, 92, 1–16.
Cooke, R.U. (1984). *Geomorphological Hazards in Los Angeles*. London: Allen and Unwin.
Cooke, R.U. (1994). Salt weathering and the urban water table in deserts. In *Rock Weathering and Landform Evolution*, ed. D.A. Robinson and R.BG. Williams. Chichester: Wiley, pp. 193–205.

Cooke, R.U. and Smalley, I.J. (1968). Salt weathering in deserts. *Nature*, 220, 1226–7.
Cooke, R.U. and Reeves, R.W. (1976). *Arroyos and Environmental Change in the American South-West*. Oxford: Clarendon Press.
Cooke, R.U. and Warren, A. (1973). *Geomorphology in Deserts*. London: Batsford.
Cooke, R.U., Warren, A. and Goudie, A. (1993). *Desert Geomorphology*. London: UCL Press.
Cooke, R.U., Brunsden, D., Doornkamp, J.C. and Jones, D.K.C. (1982). *Urban Geomorphology in Drylands*. Oxford: Oxford University Press.
Cools. J. and 9 others. (2012). An early warning system for flash floods in hyper-arid Egypt. *Natural Hazards and Earth System Sciences*, 12, 443–57.
Coque, R. (1961). *La Tunisie Présaharienne, Etude Géomorphologique*. Paris: Colin.
Coque, R. (1979). Sur la place du vent dans l'érosion en milieu aride. L'example des lunettes (bourrelets éolien) de la Tunisie. *Méditerranée*, 15–21.
Coque-Delhuille, B.C. and Gentele, P.P. (1998). Aeolian dust and superficial formations in the arid part of Yemen. In *Quaternary Deserts and Climate Change*, ed. A. Alsharhan, K.W. Glennie, G.L. Whittle and C.G.St.C. Kendall. Rotterdam: Balkema, pp. 199–208.
Corbett, I. (1993). The modern and ancient pattern of sandflow through the southern Namib deflation basin. *International Association of Sedimentologists Special Publication*, 16, 45–60.
Cornelis, W.M. and Gabriels, D. (2005). Optimal windbreak design for wind-erosion control. *Journal of Arid Environments*, 61, 315–32.
Correns, C.W. (1949). Growth and dissolution of crystals under linear pressure. *Discussions of the Faraday Society*, 5, 267–71.
Corte, A.E. (1963). Particle sorting by repeated freezing and thawing. *Science*, 142, 499–501.
Corte, A.E. and Higashi, A. (1964). Experimental research on desiccation cracks in soil. *Cold Regions Research and Engineering Laboratory, U.S. Army Material Command, Research Report*, 66.
Corwin, E.J., Alhadeff, S.C., Oggel, S.P. and Shlemon, R.J. (1991). Earth fissures, urbanization and litigation: A case study from the Terrcula area, southwestern Riverside County, California. *IAHS Publication*, 200, 291–9.
Costa, J.E. (1983). Paleohydraulic reconstruction of flash-flood peaks from boulder deposits in the Colorado Front Range. *Bulletin of the Geological Society of America*, 94, 986–1004.
Cotton, C.A. (1942). *Climatic Accidents in Landscape Making*. Christchurch: Whitcombe and Tombs.
Coudé-Gaussen, G. (1987). The periSaharan loess: Sedimentological characterisation and palaeoclimatic significance. *GeoJournal*, 15, 177–83.
Coudé-Gaussen, G. (1990). The loess and loess-like deposits along the sides of the western Mediterranean Sea: Genetic and palaeoclimatic significance. *Quaternary International*, 5, 1–8.
Coudé-Gaussen, G. (1991). *Les Poussières Sahariennes*. Montrouge: John Libbey Eurotext.
Coudé-Gaussen, G., Rognon, P. and Federoff, N. (1984). Piegeage de poussières éoliennes dans des fissures de granitoides du Sinai oriental. *Académie des Sciences de Paris II, Comptes Rendus*, 298, 369–74.
Coudé-Gaussen, G., Mosser, C., Rognon, P. and Tourenq, J. (1982). Une accumulation de loess du Pléistocène superior dans la Sud-Tunisien: La coupe de Téchine. *Bulletin Société Géologique de France*, 24, 283–92.
Courtillot, V., Armijo, R. and Tapponnier, P. (1987). The Sinai triple junction revisited. *Tectonophysics*, 141, 181–90.
Cox, G.W. (1990). Soil mining by pocket gophers along topographic gradients in a mima moundfield. *Ecology*, 71, 837–43.

Cox, G.W. and Allen, D.W. (1987). Sorted stone nets and circles of the Columbia Plateau: A hypothesis. *Northwest Science*, 61, 179–85.

Cox, G.W. and Gakahu, C.G. (1986). A latitudinal test of the fossorial rodent hypothesis of Mima mound origin. *Zeitschrift für Geomorphologie* NF, 30, 485–501.

Craddock, R.A. (2011). Aeolian processes on the terrestrial planets: Recent observations and future focus. *Progress in Physical Geography*, 36, 110–24.

Crammond, N.J. (1985). Thausamite in failed cement mortars and renders from exposed brickwork. *Cement and Concrete Research*, 15, 1039–50.

Crandell, D.R. (1958). Geology of the Pierre area, South Dakota. *U.S. Geological Survey Professional Paper*, 307.

Cremaschi, M. (1996). The rock varnish in the Messak Settafet (Fezzan, Libyan Sahara), age, archaeological context and paleo-environmental implication. *Geoarchaeology*, 11, 393–421.

Cremaschi, M., Zerboni, A., Spötl, C. and Felletti, F. (2010). The calcareous tufa in the Tadrart Acacus Mt. (SW Fezzan, Libya). An early Holocene palaeoclimate archive in the central Sahara. *Palaeogeography, Palaeoclimatology, Palaeoecology*, 287, 81–94.

Croll, J.G.A. (2009). Possible role of thermal ratchetting in alligator cracking of asphalt pavements. *International Journal of Pavement Engineering*, 10, 447–53.

Crouvi, O., Amit, R., Enzel, Y., Porat, N. and Sandler, A. (2008). Sand dunes as a major proximal dust source for late Pleistocene loess in the Negev Desert, Israel. *Quaternary Research*, 70, 275–82.

Crouvi, O., Amit, R., Porat, N., Gillespie, A.R., McDonald, E.V. and Enzel, Y. (2009). Significance of primary hilltop loess in reconstructing dust chronology, accretion rates and sources: An example from the Negev Desert, Israel. *Journal of Geophysical Research*, 114, F02017, doi:10.1029/2008JF001083.

Cruikshank, K.M. and Aydin, A. (1994). Role of fracture localization in arch formation, Arches National Park, Utah. *Bulletin of the Geological Society of America*, 106, 879–91.

Cui, Z., Li, H., Nan, L., and Li, D. (1999). The discovery and environmental significance of Chifeng wind route and huge pots of Inner Mongolia and Hebei province. *Chinese Science Bulletin*, 44, 1531–6.

Cullen, H.M., de Menocal, P.B., Hemming, S., Hemming, G., Brown, F.H., Guilderson, T. and Sirocko, F. (2000). Climate change and the collapse of the Akkadian empire: Evidence from the deep sea. *Geology*, 28, 379–82.

Currey, D.R. and Sack, D. (2009a). Hemiarid lake basins: Hydrographic patterns. In *Geomorphology of Desert Environments* (2nd ed.), ed. A.J. Parsons and A.D. Abrahams. Heidelberg: Springer, pp. 471–87.

Currey, D.R. and Sack, D. (2009b). Hemiarid lake basins: Geomorphic patterns. In *Geomorphology of Desert Environments* (2nd ed.), ed. A.J. Parsons and A.D. Abrahams. Heidelberg: Springer, pp. 489–514.

Czerewko, M.A. and Cripps, J.C. (2001). Assessing the durability of mudrocks using the modified jar slake index test. *Quarterly Journal of Engineering Geology and Hydrology*, 34, 153–63.

Dahan, O., Tatarsky, B., Enzel, Y., Kulls, C., Seely, M. and Benito, G. (2008). Dynamics of flood water infiltration and groundwater recharge in hyperarid desert. *Ground Water*, 46, 450–61.

Dai, A. (2011). Drought under global warming: A review. *Interdisciplinary Reviews: Climate Change*, 2, 45–65.

D'Almeida, G.A. (1986). A model for Saharan dust transport. *Journal of Climatology and Applied Meteorology*, 25, 903–16.

Dangavs, N.V. (1979). Presence de dunas de argilla fosiles en la Pampa Deprimada. *Revista Asociacion Geologica Argentina*, 34, 31–5.

Danin, A., Gerson, R., Marton, K. and Garty, J. (1982). Patterns of limestone and dolomite weathering by lichens and blue-green algae and their palaeoclimatic significance. *Palaeogeography, Palaeoclimatology, Palaeoecology*, 37, 221–33.

Darborn, M. (2000). The observation and implications of rock surface temperatures across the central Namib and a method for their accurate simulation in the laboratory. BA diss., University of Oxford.

Dare-Edwards, A.J. (1984). Aeolian clay deposits of south-eastern Australia: Parna or loessic clay? *Transactions of the Institute of British Geographers*, NS9, 337–44.

Dasgupta, P. and Bandyopadhyay, S. (2008). Carbonate aeolianites of western Saurashtra, India: Experimental decipherment of the depositional mechanisms. *Sedimentology*, 55, 1361–74.

David-Novak, H.B., Morin, E. and Enzel, Y. (2004). Modern extreme storms and the rainfall thresholds for initiating debris flows on the hyperarid western escarpment of the Dead Sea, Israel. *Bulletin of the Geological Society of America*, 116, 718–28.

Davila, A.F., Gómez-Silva, B., de los Rios, A., Ascaso, C., Olivares, H., McKay, C.P. and Wierzchos, J. (2008). Facilitation of endolithic microbial survival in the hyperarid core of the Atacama Desert by mineral deliquescence. *Journal of Geophysical Research*, 113, G01028, doi:10.1029/2007JG000561.

Davis, W.M. (1905). The geographical cycle in an arid climate. *Journal of Geology*, 13, 381–407.

Davis, W.M. (1938). Sheetfloods and streamfloods. *Bulletin of the Geological Society of America*, 48, 1337–416.

Dawod, G.M., Mirza, M.N. and Al-Ghamdi, K.A. (2011a). GIS-based spatial mapping of flash flood hazard in Makkah City, Saudi Arabia. *Journal of Geographic Information Systems*, 3, 225–31.

Dawod, G.M., Mirza, M.N. and Al-Ghamdi, K.A. (2011b). Assessment of several flood estimation methodologies in Makkah metropolitan area, Saudi Arabia. *Arabian Journal of Geoscience*, doi:10.1007/s12517-011-0405-5.

Day, J.A. and Seely, M.K. (1988). Physical and chemical conditions in an hypersaline spring in the Namib Desert. *Hydrobiologia*, 160, 141–53.

Dean, D.J. and Schmidt, J.C. (2011). The role of feedback mechanisms in historic channel changes of the lower Rio Grande in the Big Bend region. *Geomorphology*, 126, 333–49.

Dearing, J.A., Livingstone, I.P. and Zhou, L.P. (1996). A late Quaternary magnetic record of Tunisian loess and its climatic significance. *Geophysical Research Letters*, 23, 189–92.

Dearing, J.A., Livingstone, I.P., Bateman, M.D. and White, K. (2001). Palaeoclimate records from OIS 8.0–5.4 recorded in loess in palaeosol sequences on the Matmata Plateau, southern Tunisia, based on mineral magnetism and new luminescence dating. *Quaternary International*, 76–77, 43–56.

DeBano, L.F. (2000). The role of fire and soil heating on water repellency in wildland environments: A review. *Journal of Hydrology*, 231–232, 195–206.

Deblauwe, V., Couteron, P., Bogaert, J. and Barbier, N. (2012). Determinants and dynamics of banded vegetation pattern migration in arid climates. *Ecological Monographs*, 82, 3–21.

De Deckker, P., Norman, M., Goodwin, I.D., Wain, A. and Gingele, F.X. (2010). Lead isotopic evidence for an Australian source of aeolian dust to Antarctica at times over the last 170,000 years. *Palaeogeography, Palaeoclimatology, Palaeoecology*, 285, 205–23.

De la Montagne, J. (1953). Geomorphology of the Centennial-Big Hollow area, southeastern Wyoming. *Wyoming Geologists' Association Guidebooks – 8th Annual Field Conference*, pp. 77–80.

DeLong, S., Pelletier, J.D. and Arnold, L.J. (2011). Late Holocene alluvial history of the Cuyama River, California, USA. *Bulletin of the Geological Society of America*, 123, 2160–76.

De Longueville, F., Hountondji, Y.-C., Henry, S. and Ozer, P. (2010). What do we know about effects of desert dust on air quality and human health in West Africa compared to other regions? *Science of the Total Environment*, 409, 1–8.
Del Pilar Alvarez, M., Weiler, N.E. and Hernández, M.A. (2010). Linking geomorphology and hydrodynamics: A case study from Península Valdés, Patagonia, Argentina. *Hydrogeology Journal*, 18, 473–86.
Del Valle, H.F., Elissalde, N.O., Gagliardini, D.A. and Milovich, J. (1998). Status of desertification in the Patagonian region: Assessment and mapping from satellite imagery. *Arid Soil Research and Rehabilitation*, 12, 95–121.
Del Valle, H.F., Rostagno, C.M., Coronato, F.R., Bouza, P.J. and Blanco, P.D. (2008). Sand dune activity in north-eastern Patagonia. *Journal of Arid Environments*, 72, 411–22.
DeMenocal, P.B. (1995). Plio-Pleistocene African climate. *Science*, 270, 53–9.
DeMenocal, P.B. (2001). Cultural responses to climate change during the late Holocene. *Science*, 292, 667–73.
DeMenocal, P.B. and Rind, D. (1993). Sensitivity of Asian and African climate to variations in seasonal insolation, glacial ice cover, sea surface temperature and Asian orography. *Journal of Geophysical Research*, 98 (D4), 7265–87.
DeMenocal, P., Ortiz, J., Guilderson, T. and Sarnthein, M. (2000). Coherent high- and low-latitude climate variability during the Holocene Warm Period. *Science*, 288, 2198–202.
Denny, C.S. (1965). Alluvial fans in the Death Valley region, California and Nevada. *U.S. Geological Survey Professional Paper*, 466.
Denny, C.S. (1967). Fans and pediments. *American Journal of Science*, 265, 81–105.
Densmore, J., Ellett, K., Howle, J., Carpenter, M. and Sneed, M. (2010). Monitoring land-surface deformation on Bicycle Lake Playa, Fort Irwin, California, USA. *IAHS Publication*, 339, 39–43.
De Oliveira, P.E., Barreto, A.M.F. and Suguio, K. (1999). Late Pleistocene/Holocene climatic and vegetational history of the Brazilian caatinga: The fossil dunes of the middle São Francisco River. *Palaeogeography, Palaeoclimatology, Palaeoecology*, 152, 319–37.
Derbyshire, E. (2001). Geological hazards in loess terrain, with particular reference to the loess regions of China. *Earth-Science Reviews*, 54, 231–60.
Derbyshire, E., Meng, X. and Kemp, R.A. (1998). Provenance, transport and characteristics of modern aeolian dust in western Gansu Province, China, and interpretation of the Quaternary loess record. *Journal of Arid Environments*, 39, 497–516.
Derickson, D., Kocurek, G., Ewing R.C. and Bristow, C. (2008). Origin of a complex and spatially diverse dune-field pattern, Algodones, southeastern California. *Geomorphology*, 99, 186–204.
De Rouw, A. and Rajot, J.-L. (2004). Soil organic matter, surface crusting and erosion in Sahelian farming systems based on manuring or fallowing. *Agriculture, Ecosystems & Environment*, 104, 263–76.
Descroix, L., Viramontes, D., Estrada, J., Gonzales Barrios, J.-L. and Asseline, J. (2007). Investigating the spatial and temporal boundaries of Hortonian and Hewlettian runoff in northern Mexico. *Journal of Hydrology*, 346, 144–58.
Desert Processes Working Group. (1991). *Remote Sensing Field Guide – Desert. U.S. Army Corps of Engineers Engineer Topographic Laboratories Report ETL-0588*, Fort Belvoir, Virginia.
De Silva, S.L., Bailey, J.E., Mandt, K.E. and Viramonte, J.M. (2010). Yardangs in terrestrial ignimbrites: Synergistic remote and field observations on Earth with applications to Mars. *Planetary and Space Science*, 58, 459–71.
de Villiers, M. and van Heerden, J. (2011). Nashi dust storm over the United Arab Emirates. *Weather*, 66, 79–81.

Dhar, O.N., Rakhecha, P.R., Mandal, B.N. and Sangam, R.B. (1982). Was the July 1981 rainstorm over Rajasthan unprecedented? *Pageoph*, 120, 483–94.

Dhir, R.P. and Singhvi, A.K. (2012). The Thar Desert and its antiquity. *Current Science*, 102, 1001–8.

Dhir, R.P., Singhvi, A.K., Andrew, J.E., Kar, A., Sareen, B.K., Tandon, S.K., Kailath, A. and Thomas, J.V. (2010). Multiple episodes of aggradation and calcrete formation in late Quaternary aeolian sands, central Thar Desert, Rajasthan, India. *Journal of Asian Earth Sciences*, 37, 10–6.

Dia, A., Chauvel, C., Bulourde, M. and Gerard, M. (2005). Eolian contribution to soils on Mount Cameroon: Isotopic and trace element records. *Chemical Geology*, 226, 232–52.

Diaz, G.C., Mendoza, M., Garcia-Veigac, J., Pueyo, J.J. and Turner, P. (1999). Evolution and geochemical signatures in a Neogene forearc evaporitic basin: The Salar Grande (Central Andes of Chile). *Palaeogeography, Palaeoclimatology, Palaeoecology*, 151, 39–54.

Dick, G.S., Anderson, R.S. and Sampson, D.E. (1997). Controls on flash flood magnitude and hydrograph slope, Upper Blue Hills badlands, Utah. *Geology*, 25, 45–8.

Dickerson, J.D., Woodruff, N.P. and Banbury, E.E. (1976). Techniques for improving survival and growth of trees in semiarid areas. *Journal of Soil and Water Conservation*, 31, 63–6.

Dickson, J.C. (2010). Origin of calcrete and dolocrete in the carbonate mantle of St Vincent Basin, southern South Australia. *Cadernos Lab. Xeolóxico de Laxe Coruña*, 35, 109–22.

Diester-Haass, L. and Schrader, H.J. (1979). Neogene coastal upwelling history off north-west and south-west Africa. *Marine Geology*, 29, 39–53.

Dietze, M. and Kleber, A. (2012). Control of lateral processes to stone pavement formation in deserts inferred from clast orientation patterns. *Geomorphology*, 139–140, 172–87.

Dietzel, M., Kolmer, P., Pölt, P. and Simic, S. (2008). Desert varnish and petroglyphs on sandstone – Geochemical composition and climate changes from Pleistocene to Holocene (Libya). *Chemie der Erde*, 68, 31–43.

Ding, Z., Rutter, N., Han, J. and Liu, T. (1992). A coupled environmental system formed at about 2.5 Ma in East Asia. *Palaeogeography, Palaeoclimatology, Palaeoecology*, 94, 223–42.

Ding, Z.L. and Yang, S.L. (2000). C3/C4 vegetation evolution over the last 7.0 Myr in the Chinese Loess Plateau: Evidence from pedogenic carbonate δ13C. *Palaeogeography, Palaeoclimatology, Palaeoecology*, 160, 292–9.

Doerr, S.H. and Davies, R. (2007). Combining remote sensing, geophysical and 'old-school' geomorphological tools to assess weathering rates of the world's largest exposed carbonate platform: The Nullarbor Plain, Australia. *GSA Abstracts with Programs*, 39(6), 248.

Doerr, S.H., Shakesby, R.A. and Walsh, R.P.D. (2000). Soil water repellancy: Its causes, characteristics and hydro-geomorphological significance. *Earth-Science Reviews*, 51, 33–65.

Doherty, O.M., Riemer, N. and Hameed, S. (2008). Saharan mineral dust transport into the Caribbean: Observed atmospheric controls and trends. *Journal of Geophysical Research*, 113, D07211, doi:10.1029/2007JD009171.

Dohrenwend, J.C. (1987). The Basin and Range. In *Geomorphic Systems of North America*, ed. W. Graf. Boulder, CO: Geological Society of America, Centennial Special Volume 2, 303–42.

Dohrenwend, J.C. (1994). Pediments in arid environments. In *Geomorphology of Desert Environments*, ed. A.J. Parsons and A.D. Abrahams. London: Chapman and Hall, pp. 321–53.

Domínguez-Villar, D., Arteaga, C., Garcia-Gimenez, R., Smith, E. and Pedraza, J. (2008). Diurnal and seasonal water variations of temperature, pH, redox potential and conductivity in gnammas (weathering pits): Implications for chemical weathering. *Catena*, 72, 37–48.

Dong, Z., Wang, T. and Wang, X. (2004a). Geomorphology of the megadunes in the Badain Jaran Desert. *Geomorphology*, 60, 191–203.

Dong, Z., Wang, X. and Chen, G., (2000). Monitoring sand dune advance in the Taklimakan Desert. Geomorphology, 35, 219–31.

Dong, Z., Chen, G., He, X., Han, Z. and Wang, X. (2004b). Controlling blown sand along the highway crossing the Taklimakan Desert. *Journal of Arid Environments*, 57, 329–44.

Dong, Z., Lv, P., Qian, G., Xia, X. and Mu, G. (2012). Research progress in China's Lop Nur. *Earth-Science Reviews*, 111, 142–63.

Dong, Z., Lv, P., Lu, J., Qian, G., Zhang, G. and Luo, W. (2012). Geomorphology and origin of yardangs in the Kumtagh Desert, northwest China. *Geomorphology*, 139–140, 145–54.

Dong, Z., Qian, G., Luo, W., Zhang, Z., Xiao, S. and Zhao, A. (2009). Geomorphological hierarchies for complex mega-dune evolution in the Badain Jaran Desert. *Geomorphology*, 106, 180–5.

Doolittle, W.E. (1985). The use of check dams for protecting downstream agricultural lands in the prehistoric Southwest: A contextual analysis. *Journal of Anthropological Research*, 41, 279–305.

Doornkamp, J.C., Brunsden, D. and Jones, D.K.C. (eds.). (1980). *Geology, Geomorphology and Pedology of Bahrain*. Norwich: Geo Abstracts.

Dorn, R.I. (1998). *Rock Coatings*. Amsterdam: Elsevier.

Dorn, R.I. (2003). Boulder weathering and erosion associated with a wildfire, Sierra Ancha Mountains, Arizona. *Geomorphology*, 55, 155–71.

Dorn, R.I. (2009). Desert rock coatings. In *Geomorphology of Desert Environments*, ed. A.J. Parsons and A.D. Abrahams. Heidelberg: Springer, pp. 153–86.

Dorn, R.I. (2011). Revising dirt cracking as a physical weathering process in warm deserts. *Geomorphology*, 135, 129–42.

Dorn, R.I. (2012). Do debris flows pose a hazard to mountain-front property in Metropolitan Phoeniz, Arizona? *Professional Geographer*, 64, 1–14.

Dougill, A.J. and Thomas, A.D. (2002). Nebkha dunes in the Molopo Basin, South Africa and Botswana: Formation controls and their validity as indicators of soil degradation. *Journal of Arid Environments*, 50, 413–28.

Downes, R.G. (1946). Tunnelling erosion in northeastern Victoria. *Journal of Council for Science and Industrial Research*, 19, 283–91.

Dragovich, D. (1969). The origin of cavernous surfaces (tafoni) in granitic rocks of southern South Australia. *Zeitschrift für Geomorphologie*, 13, 163–81.

Dragovich, D. (1993). Distribution and chemical composition of microcolonial fungi and rock coatings from arid Australia. *Physical Geography*, 14, 323–41.

Dragovich, D. (1994). Fire, climate and the persistence of desert varnish near Dampier, Western Australia. *Palaeogeography, Palaeoclimatology, Palaeoecology*, 111, 279–88.

Drake, N. and Bristow, C. (2006). Shorelines in the Sahara: Geomorphological evidence for an enhanced monsoon from palaeolake Megachad. *The Holocene*, 16, 901–11.

Drake, N.A., Eckardt, F.D. and White, K.H. (2004). Sources of sulphur in gypsiferous sediments and crusts and pathways of gypsum redistribution in southern Tunisia. *Earth Surface Processes and Landforms*, 29, 1459–71.

Drake, N.A., Blench, R.M., Armitage, S.J., Bristow, C.S. and White, K.H. (2011). Ancient watercourses and biogeography of the Sahara explain the peopling of the desert. *Proceedings of the National Academy of Sciences*, 108, 458–62.

Draut, A.E. (2012). Effects of river regulation on aeolian landscapes, Colorado River, southwestern USA. *Journal of Geophysical Research*, 117, F02022, doi:10.1029/2011JF002329.

Dresch, J. (1957). Pediments et glacis d'érosion, pédiplains et inselbergs. *Information Géographique*, 22, 183–96.

Duce, R.A. (1995). Sources, distributions, and fluxes of mineral aerosols and their relationship to climate. In *Aerosol Forcing of Climate*, ed. R.J. Charlson, and J. Heintzenberg. Chichester: Wiley, pp. 43–72.

Dühnforth, M., Densmore, A.L., Ivy-Ochs, S., Allen, P.A. and Kubik, P.W. (2007). Timing and patterns of debris flow deposition on Shepherd and Symmes creek fans, Owens Valley, California, deduced from cosmogenic ^{10}Be. *Journal of Geophysical Research*, 112, F03S15, doi:10.1029/2006JF000562.

Dunkerley, D.L. (1979). The morphology and development of rillenkarren. *Zeitschrift für Geomorphologie*, 23, 332–48.

Dunkerley, D.L. (1995). Surface stone cover on desert hillslopes; parameterizing characteristics relevant to infiltration and surface runoff. *Earth Surface Processes and Landforms*, 20, 207–18.

Dunkerley, D.L. (2008). Bank permeability in an Australian ephemeral dry-land stream: Variation with stage resulting from mud deposition and sediment clogging. *Earth Surface Processes and Landforms*, 33, 226–43.

Dunkerley, D.L. (2010). Ecogeomorphology in the Australian drylands and the role of biota in mediating the effects of climate change on landscape processes and evolution. *Geological Society of London Special Publication*, 346, 87–120.

Dunkerley, D.L. (2011). Desert soils. In *Arid Zone Geomorphology* (3rd ed.), ed. D.S.G. Thomas. Chichester: Wiley-Blackwell, pp. 101–29.

Dunkerley, D.L. and Brown, K.J. (1995). Runoff and runon areas in a patterned chenopod shrubland, arid western New South Wales, Australia: Characteristics and origin. *Journal of Arid Environments*, 30, 41–55.

Dunkerley, D.L. and Brown, K.J. (1999). Flow behaviour, suspended sediment transport and transmission losses in a small (sub-bank-full) flow event in an Australian desert stream. *Hydrological Processes*, 13, 1577–88.

Dunkerley, D. L. and Brown, K. J. (2002). Oblique vegetation banding in the Australian arid zone: implications for theories of pattern evolution and maintenance. *Journal of Arid Environments*, 51, 163–81.

Dunne, T. and Leopold, L.B. (1978). *Water in Environmental Planning*. San Francisco: Freeman.

Dupont-Nivet, G., Krijgsman, W., Langereis, C.G., Abels, H.A., Dai, S. and Fang, X. (2007). Tibetan plateau aridification linked to global cooling at the Eocene-Oligocene transition. *Nature*, 445, 635–8.

Durán, O., Claudin, B. and Andreotti, B. (2011). On aeolian transport: Grain-scale interactions, dynamical mechanisms and scaling laws. *Aeolian Research*, 3, 243–70.

Durán, O., Schatz, V., Herrmann, H.J. and Tsoar, H. (2008). Transformation of barchans into parabolic dunes under the influence of vegetation. Unpublished manuscript.

Durán, O., Schwämmle, V., Lind, P.G. and Herrmann, H.J. (2011). Size distribution and structure of barchan dune fields. *Nonlinear Processes in Geophysics*, 18, 455–67.

Durán Zuazo, V.H., Francia Martínez, J.R. and Martínez Raya, A. (2004). Impact of vegetative cover on runoff and soil erosion at hillslope scale in Lanjaron, Spain. *The Environmentalist*, 24, 39–48.

Durant, A.J., Harrison, S.P., Watson, I.M. and Balkanski, Y. (2009). Sensitivity of direct radiative forcing by mineral dust to particle characteristics. *Progress in Physical Geography*, 33, 80–102.

Du Toit, G. van N., Snyman, H.A. and Malan, P.J. (2009). Physical impact of grazing by sheep on soil parameters in the Nama Karoo subshrub/grass rangeland of South Africa. *Journal of Arid Environments*, 73, 804–10.

Duveyrier, H. (1864). *Exploration du Sahara: Les Touareg du Nord*. Paris: Challamel Ainé.

Dzulynski, S. and Kotarba, A. (1979). Solution pans and their bearing on the development of pediments and tors in granite. *Zeitschrift für Geomorphologie*, 23, 172–91.

Eastwood, E., Nield, J., Baas, A. and Kocurek, G. (2011). Modelling controls on aeolian dune-field pattern evolution. *Sedimentology*, 58, 1391–406.

Eckardt, F. (1996). The distribution and origin of gypsum in the central Namib Desert, Namibia. D.Phil. thesis, University of Oxford.

Eckardt, F. and Kuring, N. (2005). SeaWiFS identifies dust sources in the Namib Desert. *International Journal of Remote Sensing*, 26, 4159–67.

Eckardt, F. and Schemenauer, R.S. (1998). Fog water chemistry in the Namib Desert, Namibia. *Atmospheric Environment*, 32, 2595–9.

Eckardt, F., Washington, R. and Wilkinson, M.J. (2002). The origin of dust on the west coast of southern Africa. *Palaeoecology of Africa*, 27, 207–19.

Eckardt, F.D. and Spiro, B. (1999). The origin of sulphur in gypsum and dissolved sulphate in the central Namib Desert, Namibia. *Sedimentary Geology*, 123, 255–73.

Eckardt, F.D., Drake, N., Goudie, A.S., White, K. and Viles, H. (2001). The role of playas in pedogenic gypsum crust formation in the central Namib Desert: A theoretical model. *Earth Surface Processes and Landforms*, 26, 1177–93.

Edgell, H.S. (2006). *Arabian Deserts*. Dordrecht: Springer.

Edgett, K.S. and Lancaster, N. (1993). Volcaniclastic aeolian dunes: Terrestrial examples and application to Martian sands. *Journal of Arid Environments*, 25, 271–97.

Eghbal, M.K. and Southard, R.J. (1993). Mineralogy of aridisols on dissected alluvial fans, western Mojave Desert, California. *Soil Science Society of America Journal*, 57, 538–44.

Eitel, B., Blümel, W.D., Hüser, K. and Mauz, B. (2001). Dust and loessic alluvial deposits in northwestern Namibia (Damaraland, Kaokoveld): Sedimentology and palaeoclimatic evidence based on luminescence data. *Quaternary International*, 76/77, 57–65.

Ekdale, A.A., Brown, F.H. and Feibel, C.S. (1989). Nonmarine macroborings in early Pleistocene algal biolithites (stromatolites) of the Turkana Basin, northern Kenya. *Palaios*, 4, 389–96.

El-Bana, M., Nijs, I. and Kockelbergh, F. (2002). Microenvironmental and vegetational heterogeneity induced by phytogenic nebkhas in an arid coastal ecosystem. *Plant and Soil*, 247, 283–93.

El-Bastawey, M., White, K. and Nasr, A. (2009). Integration of remote sensing and GIS for modeling flash floods in Wadi Hudain catchment, Egypt. *Hydrological Processes*, 23, 1359–68.

El-Baz, F. and Maxwell, T.A. (eds.) (1982). *Desert landforms of southwest Egypt: A basis for comparison with Mars*. Washington DC: NASA.

El-Baz, F., Mainguet, M. and Robinson, C. (2000). Fluvio-aeolian dynamics in the north-eastern Sahara: The relationship between fluvial/aeolian systems and ground-water concentration. *Journal of Arid Environments*, 44, 173–83.

Elbelrhiti, H. (2012). Initiation and early development of barchans dunes: A case study of the Moroccan Sahara Desert. *Geomorphology*, 138, 181–8.

Elbelrhiti, H. and Douady, S. (2010). Equilibrium versus disequilibrium of barchans dunes. *Geomorphology*, 125, 558–68.

Elbelrhiti, H., Claudin, P. and Andreotti, B. (2005). Field evidence for surface-wave-induced instability of sand dunes. *Nature*, 437, 720–3.

Elbelrhiti, H., Andreotti, B. and Claudin, P. (2008). Barchan dune corridors: field characterization and investigation of control parameters. *Journal of Geophysical Research*, 113(F2), F02S15.

Eldridge, D.J. (1998). Trampling of microphytic crusts on calcareous soils, and its impact on erosion under rain-impacted flow. *Catena*, 33, 221–39.

Eldridge, D. J. and Greene, R. S. B. (1994). Microbiotic soil crusts-a review of their roles in soil and ecological processes in the rangelands of Australia. *Soil Research*, 32, 389–415.

Eldridge, D. J. and Leys, J. F. (2003). Exploring some relationships between biological soil crusts, soil aggregation and wind erosion. *Journal of Arid Environments*, 53, 457–66.

Eldridge, D.J. and Robson, A.D. (1997). Blade ploughing and exclosure influence soil properties in a semi-arid Australian woodland. *Journal of Range Management*, 50, 191–8.

Eldridge, D.J. and Rosentreter, R. (1999). Morphological groups: A framework for monitoring microphytic crusts in arid landscapes. *Journal of Arid Environments*, 41, 11–25.

Eldridge, D.J., Zaady, G. and Shachack, M. (2000). Infiltration through three contrasting biological soil crusts in patterned landscapes in the Negev, Israel. Catena, 40, 323–36.

Eldridge, D.J., Koen, T.B., Killgore, A., Huang, N. and Whitford, W.G. (2012). Animal foraging as a mechanism for sediment movement and soil nutrient development: Evidence from the semi-arid Austrralian woodlands and the Chihuahuan Desert. *Geomorphology*, 157–158, 131–41.

Elkins, N.Z., Sabol, G.V., Ward, T.J. and Whitford, W.G. (1986). The influence of subterranean termites on the hydrological characteristics of a Chihuahuan desert ecosystem. *Oecologia*, 68, 521–8.

Elliott, J.G., Gellis, A.C. and Aby, S.B. (1999). Evolution of arroyos: Incised channels of the southwestern United States. In *Incised River Channels*, ed. S.E. Darby and A. Simon. Chichester: Wiley, pp. 153–85.

Ellwein, A.L., Mahan, S.A. and McFadden, L.D. (2011). New optically stimulated luminescence ages provide evidence of MIS3 and MIS2 eolian activity on Black Mesa, northeastern Arizona, USA. *Quaternary Research*, 75, 395–8.

El Maarry, M.R., Markiewicz, W.J., Mellon, M.T., Goetz, W., Dohm, J.M. and Pack, A. (2010). Crater floor polygons: Desiccation patterns of ancient lakes on Mars? *Journal of Geophysical Research*, 115, E10006, doi:10.1029/2010JE003609.

Elmore, A.J., Kaste, J.M., Okin, G.S. and Fantle, M.S. (2008). Groundwater influences on atmospheric dust generation in deserts. *Journal of Arid Environments*, 72, 1753–65.

El-Sayed, M.I. (1991a). Petrology of calcrete in southern Kuwait. *Egyptian Journal of Geology*, 35, 51–74.

El-Sayed, M.I. (1991b). An overview of silicification in post-Pleistocene landscapes of southern Kuwait, Arabian Gulf. *Egyptian Journal of Geology*, 35, 303–24.

El-Sayed, M.I. (1993). Gypcrete of Kuwait – Field investigation, petrography and genesis. *Journal of Arid Environments*, 25, 199–209.

El-Sayed, M.I. (1999). Sedimentological characteristics and morphology of the aeolian sand dunes in the eastern part of the UAE: A case study from Ar Rub' Al Khali. *Sedimentary Geology*, 123, 219–38.

El-Sayed, M.I. (2000). The nature and possible origin of mega-dunes in Liwa, Ar Rub' Al Khali, UAE. *Sedimentary Geology*, 134, 305–30.

Elvidge, C. D. and Iverson, R. M. (1983). Regeneration of desert pavement and varnish, in Webb, R. H., and Wilshire, H. G., eds., *Environmental effects of off-road vehicles*. New York: Springer-Verlag, pp. 225–43.

Ely, L.L., Enzel, Y., Baker, V.R. and Cayan, D.R. (1993). A 5000-year record of extreme floods and climate change in the southwestern United States. *Science*, 262, 410–2.

Embabi, N.S. (1967). A morphological study of the Kharga oases depression, the Western Desert, Egypt. PhD diss., University of Bristol.

Embabi, N.S. (1978). Statistical relationships between barchan shape dimensions. *Egyptian Computer Journal*, 1, 1–15.

Embabi, N.S. (1986–87). Dune movement in the Kharga and Dakhla oases depressions, the Western Desert, Egypt. *Bulletin de la Société de Géographie d'Egypte*, 59–60, 35–70.

Embabi, N.S. (1999). Playas of the Western Desert, Egypt. *Annales Academiae Fennicae Geologica-Geographica*, 160, 5–47.

Embabi, N.S. (2004). *The Geomorphology of Egypt*, Volume 1. Cairo: The Egyptian Geographical Society.

Embabi, N.S. and Ashour, M.M. (1993). Barchan dunes in Qatar. *Journal of Arid Environments*, 25, 49–69.

Engel, C.E. and Sharp, R.P. (1958). Chemical data on desert varnish. *Bulletin of the Geological Society of America, Bulletin*, 69, 487–518.

Engelstaedter, S. (2001). *Dust storm frequencies and their relationship to land surface conditions. Diploma thesis*. Jena: Friedrich-Schiller-University, Institute for Geoscience.

Englestaedter, S. and Washington, R. (2007). Atmospheric controls on the annual cycle of north African dust. *Journal of Geophysical Research*, 112, D03103, doi:10.1029/2006JD007195.

Engelstaedter, S., Tegen, I. and Washington, R. (2006). North African dust emissions and transport. *Earth-Science Reviews*, 79, 73–100.

Enzel, Y. and 8 others. (2012). Late Quaternary weathering, erosion, and deposition in Nahal Yael, Israel: An 'impact of climatic change on an arid watershed'? *Bulletin of the Geological Society of America*, 124, 705–22.

Eppes, M.C. and Griffing, D. (2010). Granular disintegration of marble in nature: A thermal-mechanical origin for a grus and corestone landscape. *Geomorphology*, 117, 170–80.

Eppes, M.C. and McFadden, L. (2008). The influence of bedrock weathering on the response of drainage basins and associated alluvial fans to Holocene climates, San Bernadino Mountains, California, USA. *The Holocene*, 18, 895–905.

Eren, M. (2007). Genesis of tepees in the Quaternary hardpan calcretes, Mersin, S. Turkey. *Carbonates and Evaporites*, 22, 123–34.

Eren, M. and Hatipolgu-Bagci, Z. (2010). Karst surface features of the hard laminated crust (caliche hardpan) in the Mersin area, southern Turkey. *Acta Carsologica*, 39, 93–102.

Ericksen, G.E. (1981). Geology and origin of the Chilean nitrate deposits. *U.S. Geological Survey Professional Paper*, 1188.

Ericksen, G.E. (1994). Discussion of a petrographic study of the Chilean nitrates. *Geological Magazine*, 131, 849–52.

Ericksen, G.E. and Mrose, M.E. (1972). High purity veins of soda-niter, $NaNO_3$, and associated saline minerals in the Chilean nitrate deposits. *U.S. Geological Survey Professional Paper*, 800-B, 43–9.

Ericksen, G.E., Hosterman, J.W. and St. Amand, P. (1988). Chemistry, mineralogy, and origin of the clay-hill nitrate deposits, Amagosa River valley, Death Valley region, California, U.S.A. *Chemical Geology*, 67, 85–102.

Escalente, S.A. and Pimentel, A.S. (2008). Coastal dune stabilization using geotextile tubes at Las Colorados. *Geosynthetics*, 26, 16–24.

Evan, A.T., Dunion, J., Foley, J.A., Heidinger, A.K. and Velden, C.S. (2006). New evidence for a relationship between Atlantic tropical cyclone activity and African dust outbreaks. *Geophysical Research Letters*, 33, L19813, doi:10.1029/2006GL026408.

Evans, G. (1995). The Arabian Gulf: A modern carbonate-evaporite factory; a review. *Cuadernos de Geología Ibérica*, 19, 61–96.

Evans, G., Schmidt, V., Bush, P. and Nelson, H. (1969). Stratigraphy and geologic history of the Sabkha, Abu Dhabi, Persian Gulf. *Sedimentology*, 12, 145–59.

Evans, I.S. (1970). Salt crystallization and rock weathering, a review. *Revue de Géomorphologie Dynamique*, 19, 153–77.

Evans, J.P. (2010). Global warming impact on the dominant precipitation processes in the Middle East. *Theoretical and Applied Climatology*, 99, 389–402.

Evans, J.R. (1962). Falling and climbing sand dunes in the Cronese ('Cat') Mountain area, San Bernadino County, California. *Journal of Geology*, 70, 107–13.

Evans, J.W. (1911). Dreikanter. *Geological Magazine*, 8, 334–5.

Evenari, M., Shanan, L. and Tadmor, N.H. (1982). *The Negev: The Challenge of a Desert* (2nd ed.). Cambridge, MA: Harvard University Press.

Everard, C.E. (1963). Contrasts in the form and evolution of hill-side slopes in central Cyprus. *Institute of British Geographers, Transactions*, 32, 31–47.

Everitt, B. (1993). Channel responses to declining flow on the Rio Grande between Ft. Qutiman and Presidio, Texas. *Geomorphology*, 6, 225–42.

Ewing, R.C. and Kocurek, G. (2010). Aeolian dune-field pattern boundary conditions. *Geomorphology*, 114, 175–87.

Ewing, R.C., Kocurek, G. and Lake, L.W. (2006). Pattern analysis of dune-field parameters. *Earth Surface Processes and Landforms*, 31, 1176–91.

Ewing, S.A. and 9 others. (2006). A threshold in soil formation at Earth's arid-hyperarid transition. *Geochimica et Cosmochimica Acta*, 70, 5293–322.

Fabian, P., Rollenbeck, R., Spichtinger, N., Brothers, L., Dominguez, G. and Thiemens, M. (2009). Sahara dust, ocean spray, volcanoes, biomass burning: Pathways of nutrients into Andean rainforests. *Advances in Geosciences*, 22, 85–94.

Fahey, B.D. (1983). Frost action and hydration as rock weathering mechanisms on schist: A laboratory study. *Earth Surface Processes and Landforms*, 8, 535–45.

Fahey, B.D. (1986). A comparative study of salt crystallisation and salt hydration as potential weathering agents in deserts. *Geografiska Annaler*, 68A, 107–11.

Fairlie, T.D., Jacob, D.J. and Park, R.J. (2007). The impact of transpacific transport of mineral dust in the United States. *Atmospheric Environment*, 41, 1251–66.

Fan, Y.-X., Chen, F.-H., Wei, G.-X., Madsen, D.B., Oviatt, C.G., Zhao, H., Chun, X., Yang, L-P., Fan, T.-L. and Li, G.-Q. (2010). Potential water sources for late Quaternary Megalake Jilantai-Hetao, China, inferred from mollusk shell $^{87}Sr/^{86}Sr$ ratios. *Journal of Paleolimnology*, 43, 577–87.

Fanning, P.C., Holdaway, S.J., Rhodes, E.J. and Bryant, T.G. (2009). The surface archaeological record in arid Australia: Geomorphic controls on preservation, exposure and visibility. *Geoarchaeology*, 24, 121–46.

Fawcett, P.J. and 19 others. (2011). Extended megadroughts in the southwestern United States during Pleistocene interglacials. *Nature*, 470, 518–21.

Felton, A., Jewell, P.W., Chan, M. and Currey, D.T. (2006). Controls on tufa development in Pleistocene Lake Bonneville, Utah. *Journal of Geology*, 114, 377–89.

Fernández, D.S. (2005). The giant paleolandslide deposits of Tafi del Valle, Tucumán Province, Argentina. *Geomorphology*, 70, 97–111.

Field, J. (2001). Channel avulsion on alluvial fans in southern Arizona. *Geomorphology*, 37, 93–104.

Field, J.P. and 9 others. (2009). The ecology of dust. *Frontiers in Ecology and the Environment*, doi:10.1890/090050.

Filho, A.C., Schwartz, D., Tatumi, S.H. and Rosique, T. (2002). Amazonian paleodunes provide evidence for drier climate phases during the late Pleistocene-Holocene. *Quaternary Research*, 58, 205–9.

Finkel, H.J. (1959). The barchans of southern Peru. *Journal of Geology*, 67, 614–47.

Fisher, J.A., Krapf, C.B.E., Lang, S.C., Nichols, G.J. and Payenberg, T.H.D. (2008). Sedimentology and architecture of the Douglas Creek terminal splay, Lake Eyre, central Australia. *Sedimentology*, 55, 1915–30.

Fitzhugh, T.W. and Vogel, R.M. (2011). The impact of dams on flood flows in the United States. *River Research and Applications*, 27, 1192–215.

Fitzner, B., and Heinrichs, K. (1991). Weathering forms and rock characteristics of historical monuments carved from bedrock in Petra/Jordan. In *Science, Technology and European Cultural Heritage*, ed. N.S. Baer, C. Sabbioni and A.I. Sors. Oxford: Butterworth-Heinemann, pp. 908–11.

Fitzner, B. and Heinrichs, K. (1994). Damage diagnosis at monuments carved from bedrock in Petra, Jordan. In *Proceedings Third International Conference on the Conservation of Monuments in the Mediterranean Basin, Venice, 22–25 June, 1994*, ed. V. Fassina, H. Ott and F. Zezza, pp. 663–71.

Fitzner, B. and Heinrichs, K. (2002). Damage diagnosis at stone monuments – Weathering forms, damage categories and damage indices. In *Understanding and Managing Stone Decay*, ed. R. Prikryl and H.A. Viles. Prague: The Karolinum Press, pp. 11–56.

Fitzsimmons, K.E. (2007). Morphological variability in the linear dunefields of the Strzelecki and Tirari Deserts, Australia. *Geomorphology*, 91, 146–60.

Fitzsimmons, K.E., Magee, J.W. and Amos, K.J. (2009). Characterisation of aeolian sediments from the Strzelecki and Tirari Deserts, Australia: Implications for reconstructing palaeoenvironmental conditions. *Sedimentary Geology*, 218, 61–73.

Fitzsimmons, K.E., Miller, G.H., Spooner, N.A. and Magee, J.W. (2012). Aridity in the monsoon zone as indicated by desert dune formation in the Gregory Lakes Basin, northwestern Australia. Australian Journal of Earth Sciences, 59, 469–78.

Fitzsimmons, K.E., Rhodes, E.J., Magee. J.W. and Barrows, T.T. (2007). The timing of linear dune activity in the Strzelecki and Tirari Deserts, Australia. *Quaternary Science Reviews*, 26, 2598–616.

Flam, L. (1993). Fluvial geomorphology of the Lower Indus Basin (Sindh, Pakistan) and the Indus civilization. In *Himalaya to the Sea*, ed. J.F. Shroder. London: Routledge, pp. 265–87.

Fleisher, M., Liu, T., Broecker, W.S. and Moore, W. (1999). A clue regarding the origin of rock varnish. *Geophysical Research Letters*, 26, 103–6.

Fleitmann, D. and Matter, A. (2009). The speleothem record of climate variability in southern Arabia. *C.R. Geoscience*, 341, 633–42.

Fleitmann, D., Burns, S.J., Mudelsee, M., Neff, U., Kramers, J., Mangini, A. and Matter, A. (2003). Holocene forcing of the Indian monsoon recorded in a stalactite from southern Oman. *Science*, 300, 1737–9.

Fleitmann, D., Burns, S.J., Pekala, M., Mangini, A., Al-Subbary, A., Al-Aowah, M., Kramers, J. and Matter, A. (2011). Holocene and Pleistocene pluvial periods in Yemen, southern Arabia. *Quaternary Science Reviews*, 30, 783–7.

Fleitmann, D., Burns, S.J., Mangini, A., Mudelsee, M., Kramers, J., Viia, I., Neff, U., Al-Subbary, A.A., Buettner, A., Hippler, D. and Matter, A. (2007). Holocene ITCZ and Indian monsoon dynamics recorded in stalagmites from Oman and Yemen (Socotra). *Quaternary Science Reviews*, 26, 170–88.

Fleskens, L., Ataev, A., Mamedov, B. and Spaan, W.P. (2007). Desert water harvesting from Takyr surfaces: Assessing the potential of traditional and experimental technologies in the Karakum. *Land Degradation and Development*, 18, 17–39.

Flint, R.F. (1971). *Glacial and Quaternary Geology*. New York: Wiley.

Florea, N. and Al-Joumaa, Kh. (1998). Genesis and classification of gypsiferous soils of the Middle Euphrates floodplain, Syria. *Geoderma*, 87, 67–85.

Floyd, K.W. and Gill, T.E. (2011). The association of land cover with aeolian sediment production at Jornada Basin, New Mexico, USA. *Aeolian Research*, 3, 55–66.

Fluteau, F., Ramstein, G. and Besse, J. (1999). Simulating the evolution of the Asian and African monsoons during the past 30 Myr using an atmospheric general circulation model. *Journal of Geophysical Research*, 104 (D10), 11995–2018.

Fodde, E. (2007). Fired brick and sulphate attack. *Journal of Architectural Conservation*, 13, 69–80.

Fogg, T., Fogg, P., Waltham, A. (2002). Magharet Qasiur Hafit, a significant cave in the United Arab Emirates. *Tribulus*, 12, 5–14.

Folk, R.L. (1971). Longitudinal dunes of the northwestern edge of the Simpson Desert, Northern Territory, Australia I. Geomorphology and grain size relationships. *Sedimentology*, 16, 5–54.

Folk, R.L. (1978). Angularity and silica coatings on Simpson Desert sand grains, Northern Territory, Australia. *Journal of Sedimentary Petrology*, 48, 611–24.

Folk, R.L. and Ward, W.C. (1957). Brazos River bar – A study in the significance of grain size parameters. *Journal of Sedimentary Petrology*, 27, 3–27.

Foltz, G.R. and McPhaden, M.J. (2008). Trends in Saharan dust and tropical Atlantic climate during 1980–2006. *Geophysical Research Letters*, 35, L20706, doi:10.1029/2008GL035042.

Foody, G.M., Ghoneim, E.M. and Arnell, N.W. (2004). Predicting locations sensitive to flash flooding in an arid environment. *Journal of Hydrology*, 292, 48–58.

Fookes, P.G. and Collis, L. (1977). *Concrete in the Middle East*. London: Cement and Concrete Association.

Fookes, P.G., French, W.J. and Rice, S.M. (1985). The influence of ground and groundwater geochemistry on construction in the Middle East. *Quarterly Journal of Engineering Geology*, 18, 101–28.

Fookes, P.G., Lee, E.M. and Sweeney, M. (2001). Pipeline route selection and ground characterization, Algeria. *Geological Society, London, Engineering Geology Special Publications*, 18, 115–21.

Ford, T.D. and Pedley, H.M. (1996). A review of tufa and travertine deposits of the world. *Earth-Science Reviews*, 41, 117–75.

Forman, S.L. and Pierson, J. (2003). Formation of linear and parabolic dunes on the eastern Snake River Plain, Idaho in the nineteenth century. *Geomorphology*, 56, 189–200.

Forman, S.L., Oglesby, R. and Webb, R.S. (2001). Temporal and spatial patterns of Holocene dune activity on the Great Plains of North America: Megadroughts and climate links. *Global and Planetary Change*, 29, 1–29.

Forman, S.L., Nordt, L., Gomez, J. and Pierson, J. (2009). Late Holocene dune migration on the south Texas sand sheet. *Geomorphology*, 108, 159–70.

Forman, S.L., Sagintayev, Z., Sultan, M., Smith, S., Becker, R., Kendall, M. and Márín, L. (2008). The twentieth-century migration of parabolic dunes and wetland formation at Cape Cod National Sea Shore, Massachusetts, USA: Landscape response to a legacy of environmental disturbance. *The Holocene*, 18, 765–74.

Forman, S.L., Spaeth, M., Márin, L.O., Pierson, J., Gómez, J., Bunch, F. and Valdez, A. (2006). Episodic late Holocene dune movements on the sand-sheet area, Great Sand Dunes National Park and Preserve, San Luis Valley, Colorado, USA. *Quaternary Research*, 66, 97–108.

Forti, P. and Buzio, A. (1990). La nitratite della Grotta de las Manos (Argentina). *Revista Italiano Mineralogia Palaeontologia*., 63, 3–9.

Frank, A.B., Harris, D.G. and Willis, W.O. (1977). Growth and yields of spring wheat as influenced by shelter and soil water. *Agronomy Journal*, 69, 903–6.

Frechen, M. and Dodonov, A.E. (1998). Loess chronology of the middle and upper Pleistocene in Tadjikistan. *Geologische Rundschau*, 87, 2–20.
Frechen, M., Oches, E.A. and Kohfeld, K.E. (2003). Loess in Europe – Mass accumulation rates during the Last Glacial Period. *Quaternary Science Reviews*, 22, 1835–57.
Free, E.E. (1911). Desert pavements and analogous phenomena. *Science*, 33, 355.
French, C., Periman, R., Cummings, L.S., Hall, S., Goodman-Elgar, M. and Boreham, J. (2009). Holocene alluvial sequences, cumulic soils and fire signatures in the Middle Rio Puerco Basin at Guadelupe Ruin, New Mexico. *Geoarchaeology*, 24, 638–76.
French, R.H., Miller, J.J. and Dettling, C.R. (2005). Estimating playa lake flooding: Edwards Air Force Base, California, USA. *Journal of Hydrology*, 306, 146–60.
Freydier, R., Michard, A., De Lange, G. and Thomson, T. (2001). Nd isotopic compositions of eastern Mediterranean sediments: Traces of the Nile influence during sapropel SI formation? *Marine Geology*, 177, 45–62.
Freytet, P., Baltzer, F., Conchon, O., Plaziat, J.C. and Purser, B.H. (1994). Hydrologic significance of Quaternary continental carbonates from the Red Sea coast of the Egyptian Eastern Desert. *Bulletin de la Société Géologique de France*, 165, 593–601.
Friedmann, E.J. (1980). Endolithic microbial life in hot and cold deserts. *Origins of Life*, 10, 223–35.
Frumkin, A. and 6 others. (2009). Gravitational deformations and fillings of aging caves: The example of Qesem karst system, Israel. *Geomorphology*, 106, 154–64.
Fryberger, S.G. (1979). Dune forms and wind regime. *United States Geological Survey Professional Paper*, 1052, 137–69.
Fryberger, S.G. (1980). Dune forms and wind regime, Mauritania, West Africa: Implications for past climate. *Palaeoecology of Africa*, 12, 79–96.
Fryberger, S.G. and Dean, G. (1979). Dune forms and wind regime. *United States Geological Survey Professional Paper*, 1052, 305–97.
Fryberger, S.G. and Goudie, A.S. (1981). Arid geomorphology. *Progress in Physical Geography*, 5, 420–8.
Fryberger, S.G., Ahlbrandt, T.S. and Andrews, S. (1979). Origin, sedimentary features, and significance of low-angle eolian 'sand sheet' deposits, Great Sand Dunes National Monument and vicinity, Colorado. *Journal of Sedimentary Petrology*, 49, 733–46.
Fryberger, S.G., Hesp, P. and Hastings, K. (1992). Aeolian granule ripple deposits, Namibia. *Sedimentology*, 39, 319–31.
Frye, J.C. and Leonard, A.B. (1957). Ecological interpretation of Pliocene and Pleistocene stratigraphy of the Great Plains region. *American Journal of Science*, 255, 1–11.
Fryrear, D.W., Saleh, A. and Bilbro, J.D. (1998). A single event wind erosion model. *Transactions of the ASAE*, 41, 1369–74.
Fuchs, E.H., Wood, M.K., Jones, T.M. and Racher, B. (2003). Impacts of tracked vehicles on sediment from a desert soil. *Journal of Range Management*, 56, 342–52.
Fujioka, T. and Chappell, J. (2010). History of Australian aridity: Chronology in the evolution of arid landscapes. *Geological Society of London Special Publication*, 346, 121–39.
Fujioka, T. and Chappell, J. (2011). Desert landscape processes on a timescale of millions of years, probed by cosmogenic nuclides. *Aeolian Research*, 3, 157–64.
Fujioka, T., Chappell, J., Fifield, L.K. and Rhodes, E.J. (2009). Australian desert dune fields initiated with Pliocene-Pleistocene global climatic shift. *Geology*, 37, 51–4.
Fujioka, T., Chappell, J., Honda, M., Yatsevich, I., Fifield, K. and Fabel, D. (2005). Global cooling initiated stony deserts in central Australia 2–4 Ma, dated by cosmogenic ^{21}Ne-^{10}Be. *Geology*, 33, 993–6.
Fuller, B.M., Sklar, L.S., Compson, Z.G., Adams, K.J., Marks, J.C. and Wilcox, A.C. (2011). Ecogeomorphic feedbacks in regrowth of travertine step-pool morphology after dam decommissioning, Fossil Creek, Arizona. *Geomorphology*, 126, 314–32.

Furbish, D.J., Childs, E.M., Haff, P.K. and Schmeeckle, M.W. (2009). Rain splash of soil grains as a stochastic advection-dispersion process, with implications for desert plant-soil interactions and land-surface evolution. *Journal of Geophysical Research*, 114, F00A03, doi:10.1029/2009JF001265.

Gabriel, A. (1938). The southern Lut and Iranian Baluchistan. *Geographical Journal*, 92, 192–210.

Gale, S.J. (1992). Long-term landscape evolution in Australia. *Earth Surface Processes and Landforms*, 17, 323–43.

Gale, S.J., Drysdale, R.N., Scherrer, N.C. and Fischer, M.J. (1997). The lost city of north-west Queensland: A test of the model of giant Grikeland development in semi-arid karst. *Australian Geographer*, 28, 107–15.

Gallala, W., Gaied, M.H., Essefi, E. and Montacre, M. (2010). Pleistocene calcretes from eastern Tunisia: The stratigraphy, the microstructure and the environmental significance. *Journal of African Earth Sciences*, 58, 445–56.

Gallart, F., Llorens, P., Latron, J. and Regüés, D. (2002). Hydrological processes and their seasonal controls in a small Mediterranean mountain catchment in the Pyrenees. *Hydrology and Earth System Sciences*, 6, 527–37.

Ganor, E., Osetinsky, I., Stupp, A. and Alpert, P. (2010). Increasing trend of African dust, over 49 years, in the eastern Mediterranean. *Journal of Geophysical Research*, 115, D070201, doi:10.1029/2009JD012500.

Gao, H. and Washington, R. (2009). The spatial and temporal characteristics of TOMS AI over the Tarim Basin, China. *Atmospheric Environment*, 43, 1106–15.

Gao H. and Washington, R. (2010). Arctic oscillation and the interannual variability of dust emissions from the Tarim Basin: A TOMS AI based study. *Climate Dynamics*, 35, 511–22.

García, M., Riquelme, R., Farías, M., Hérail, G. and Charrier, R. (2011). Late-Miocene-Holocene canyon incision in the western Altiplano, northern Chile: Tectonic or climatic forcing? *Journal of the Geological Society, London*, 168, 1047–60.

García-Ruiz, J.M., Lasanta, T. and Alberto, F. (1997). Soil erosion by piping in irrigated fields. *Geomorphology*, 20, 269–78.

Garcin, Y., Junginger, A., Melnick, D., Olago, D.O. and Strecker, M.R. (2009). Late Pleistocene-Holocene rise and collapse of Lake Suguta. *Quaternary Science Reviews*, 28, 911–25.

Garcin, Y., Melnick, D., Strecker, M.R., Olago, D. and Tiercelin, J.-J. (2012). East African mid-Holocene wet-dry transition recorded in palaeo-shorelines of Lake Turkana, northern Kenya rift. *Earth and Planetary Science Letters*, 331/332, 322–34.

Gardner E.W. (1932). Some problems of the Pleistocene hydrography of Kharga Oasis, Egypt. *Geological Magazine*, 69, 386–421.

Gardner, R. (2009). Trees as technology: Planting shelterbelts on the Great Plains. *History and Technology*, 25, 325–41.

Gardner, R.A.M. (1988). Aeolianites and marine deposits of the Wahiba Sands: Character and paleoenvironments. *Journal of Oman Studies Special Report*, 3, 75–94.

Gardner, R.A.M. and Pye, K. (1981). Nature, origin and palaeoenvironmental significance of red coastal and desert dune sands. *Progress in Physical Geography*, 5, 514–34.

Garland, G.G. (2010). Rising sea level and long term sustainability of near-shore islands of the United Arab Emirates: An approach to establishing setback lines for Abu Dhabi. *WIT Transactions on Ecology and the Environment*, 130, 135–46.

Garreaud, R.D., Molina, A. and Farias, M. (2010). Andean uplift, ocean cooling and Atacama hyperaridity. *Earth and Planetary Science Letters*, 292, 39–50.

Gartner, J.E., Cannon, S.H., Santi, P.M. and Dewolfe, V.G. (2008). Empirical models to predict the volumes of debris flows generated by recently burned basins in the western U.S. *Geomorphology*, 96, 339–54.

Garvie, L.A.J., Burt, D.M. and Buseck, P.R. (2008). Nanometer-scale complexity, growth, and diagenesis in desert varnish. *Geology*, 36, 215–18.

Garzanti, E., Ando, S., Vezzoli, G., Lustrino, M., Boni, M. and Vermeesch, P. (2012). Petrology of the Namib Sand Sea: Long-distance transport and compositional variability in the wind-displaced Orange Delta. *Earth-Science Reviews*, 112, 173–89.

Gasse, F. (1978). Late Quaternary lake level fluctuations and environments of the northern Rift Valley and Afar Region (Ethiopia and Djibouti). *Palaeogeography, Palaeoclimatology, Palaeoecology*, 24, 279–326.

Gasse, F. and Van Campo, E. (1994). Abrupt post-glacial climate events in West Africa and North Africa monsoon domains. *Earth and Planetary Science Letters*, 126, 435–56.

Gassó, S. and Stein, A.F. (2007). Does dust from Patagonia reach the sub-Antarctic Atlantic Ocean? *Geophysical Research Letters*, 34, L01801, doi:10.1029/2006GL027693.

Gaupp, R., Kött, A. and Wörner, G. (1999). Palaeoclimatic implications of Mio-Pliocene sedimentation in the high-altitude intra-arc Lauca Basin of northern Chile. *Palaeogeography, Palaeoclimatology, Palaeoecology*, 151, 79–100.

Gauri, K.L., Sinai, J.J. and Bandyopadhyay, J.K. (1995). Geologic weathering and its implications on the age of the Sphinx. *Geoarchaeology*, 10, 119–33.

Gautier, E.F. (1908). *Sahara Algérien*. Paris: Armand Colin.

Gavish, E. (1980). Recent sabkhas marginal to the southern coasts of Sinai, Red Sea. *Developments in Sedimentology*, 28, 233–51.

Gaylord, D.R. (1990). Holocene palaeoclimatic fluctuations revealed from dune and interdune strata in Wyoming. *Journal of Arid Environments*, 18, 123–38.

Gayo, E.M., Latorre, C., Santoro, C.M., Maldonado, A. and Pol-Holz, R de. (2012). Hydroclimate variability in low-elevation Atacama Desert over the last 2500 yr. *Climate Past*, 8, 287–306.

Geddes, N. and Dunkerley, D. (1999). The influence of organic litter on the erosive elements of raindrops and gravity drops released from desert shrubs. *Catena*, 36, 303–13.

Geikie, A. (1882). *Geological Sketches at Home and Abroad*. London: Macmillan.

Gellis, A.C. and 6 others. (2012). Erosion, storage, and transport of sediment in two subbasins of the Rio Puerco, New Mexico. *Bulletin of the Geological Society of America*, 124, 817–41.

Gelt, J. (1992). Land subsidence, earth fissures change Arizona's landscape. *Arroyo* 6(2).

George, W. (1986). The thermal niche: Desert sand and desert rock. *Journal of Arid Environments*, 10, 213–24.

Ghayoumian, J., Saravi, M.M., Feiznia, S., Nouri, B. and Malekian, A. (2007). Application of GIS techniques to determine areas most suitable for artificial groundwater recharge in a coastal aquifer in southern Iran. *Journal of Asian Earth Sciences*, 30, 364–74.

Ghazifard, A. and Khorashadizadeh, Y. (2010). Effect of geological conditions on road construction in Jandagh Playa of Iran. *Environmental Earth Science*, doi:10.1007/s12665-010-0527–1.

Ghebreab, W. (1998). Tectonics of the Red Sea Region reassessed. *Earth-Science Reviews*, 45, 1–44.

Ghoneim, E., Benedetti, M. and El-Baz, F. (2012). An intergrated remote sensing and GIS analysis of the Kufrah Paleoriver, eastern Sahara. *Geomorphology*, 139–140, 242–57.

Gibling, M.R., Nanson, G.C. and Maroulis, J.C. (1998). Anastomosing river sedimentation in the Channel Country of central Australia. *Sedimentology*, 45, 595–619.

Gilbert, G.K. (1876). The Colorado Plateau province as a field for geological study. *American Journal of Sciences and Arts*, 12, 85–104.

Gilbert, G.K. (1877). *Report on the Geology of the Henry Mountains*. Washington, DC: Government Printing Office.
Gilbert, G.K. (1882). Contribution to the history of Lake Bonneville. *U.S. Geological Survey Second Annual Report*, 167–200.
Gilbert, G.K. (1890). Lake Bonneville. *U.S. Geological Survey Monograph* 1.
Gilbert, G.K. (1895). Lake basins created by wind erosion. *Journal of Geology*, 3, 47–9.
Gile, L.H. and Hawley, J.W. (1966). Periodic sedimentation and soil formation in an alluvial fan-piedmont in southern New Mexico. *Soil Science Society of America, Proceedings*, 30, 261–8.
Gile, L.H., Peterson, F.F. and Grossman, R.B. (1966). Morphological and genetic sequences of carbonate accumulation in desert soils. *Soil Science*, 101, 347–60.
Gilewitch, D.A. (2004). The effect of military operations on desert pavement. In *Studies in Military Geography and Geology*, ed. D.R. Caldwell, J. Ehlen and R.S. Harmon. Heidelberg: Springer, pp. 243–58.
Gill, T.E. (1996). Eolian sediments generated by anthropogenic disturbance of playas: Human impacts on the geomorphic system and geomorphic impacts on the human system. *Geomorphology*, 17, 207–28.
Gill, T.E. and Cahill, T.A. (1992). Playa-generated dust storms from Owens Lake. In *The History of Water: Eastern Sierra Nevada, Owens Valley, White-Inyo Mountains*, ed. C.A. Hall, V. Doyle-Jones and B. Widawski. University of California, Los Angeles, pp. 63–73.
Gillet, J.B. (1941). The plant formations of western British Somaliland and the Harar Province of Abyssinia. *Kew Bulletin*, 1941, 37–199.
Gillette, D.A. and Hanson, K.J. (1989). Spatial and temporal variability of dust production caused by wind erosion in the United States. *Journal of Geophysical Research*, 94 (D2), 2197–206.
Gillies, J.A., Etyemezian, V., Kuhns, H., McAlpine, J.D., King, J., Uppapalli, S., Nikolich, G. and Englebrecht, J. (2010). Dust emissions created by low-level rotary-winged aircraft flight over desert surfaces. *Atmospheric Environment*, 44, 1043–53.
Ginoux, P., Prospero, J.M., Torres, O. and Chin, M. (2004). Long-term simulation of global dust distribution with the GOCART model: Correlation with North Atlantic Oscillation. *Environmental Modelling and Software*, 19, 113–28.
Ginoux, P.M., Chin, I., Tegen, I., Prospero, J., Holben, M., Dubovik, O. and Lin, S.-J. (2001). Global simulation of dust in the troposphere: model description and assessment. *Journal of Geophysical Research*, 106, 20,255–73.
Giraudi, C. (2004). The Upper Pleistocene to Holocene sediments on the Mediterranean island of Lampedusa (Italy). *Journal of Quaternary Science*, 19, 537–45.
Giraudi, C. (2005). Eolian sand in peridesert northwestern Libya and implications for late Pleistocene and Holocene Sahara expansions. *Palaeogeography, Palaeoclimatology, Palaeoecology*, 218, 161–73.
Glenn, N.F. and Finley, C.D. (2010). Fire and vegetation type effects on soil hydrophobicity and infiltration in the sagebrush-steppe: 1. Field analysis. *Journal of Arid Environments*, 74, 653–9.
Godfrey, A.E., Everitt, B.L. and Duque, J.F.M. (2008). Episodic sediment delivery and landscape connectivity in the Mancos Shale badlands and Fremont River system, Utah, USA. *Geomorphology*, 102, 242–51.
Goldstein, P.S. and Magilligan, F.J. (2011). Hazard, risk and agrarian adaptations in a hyperarid watershed: El Niño floods, streambank erosion, and the cultural bounds of vulnerability in the Andean Middle Horizon. *Catena*, 85, 155–67.
Golubev, G.N. (1998). Environmental policy-making for sustainable development of the Caspian Sea area. In *Central Eurasian Water Crisis: Caspian, Aral and Dead Seas*, ed. I. Kobori and M.H. Glantz. Tokyo: United Nations University Press, pp. 91–104.

Gomez, F., Beauchamp, W. and Barazangi, M. (2000). Role of the Atlas Mountains (northwest Africa) within the African-Eurasian plate-boundary. *Geology*, 28, 775–8.

Gómez-Heras, M. and Fort, R. (2007). Patterns of halite NaCl crystallisation in building stone conditioned by laboratory heating regimes. *Environmental Geology*, 52, 259–67.

Gómez-Heras, M., Smith, B.J. and Fort, R. (2006). Surface temperature differences between minerals in crystalline rocks: Implications for granular disintegration of granites through thermal fatigue. *Geomorphology*, 78, 236–49.

Gonzalez, M.A. (2001). Recent formation of arroyos in the Little Missouri Badlands of southwestern Dakota. *Geomorphology*, 38, 63–84.

Gonzalez, M.A. (2010). Badlands of the northern Great Plains: Hell with the fires out. In *Geomorphological Landscapes of the World*, ed. P. Migoń Heidelberg: Springer, pp. 29–38.

Goode, J.R., Luce, C.H. and Buffington, J.M. (2012). Enhanced sediment delivery in a changing climate in semi-arid mountain basins: Implications for water resource management and aquatic habitat in the northern Rocky Mountains. *Geomorphology*, 139–140, 1–15.

Goosens, D. (2003). On-site and off-site effects of wind erosion. In *Wind erosion on Agricultural Land in Europe*, ed. A. Warren. Luxembourg: European Commission, pp. 29–38.

Goossens, D. (2004). Effect of soil crusting on the emission and transport of wind-eroded sediment: Field measurements on loamy sandy soil. *Geomorphology*, 58, 145–60.

Goossens, D. and Buck, B. (2009). Dust dynamics in off-road vehicle trails: Measurements on 16 arid soil types, Nevada, USA. *Journal of Environmental Management*, 90, 3458–69.

Goudie, A.S. (1969). Statistical laws and dune ridges in southern Africa. *Geographical Journal*, 135, 404–6.

Goudie, A.S. (1970). Notes on some major dune types in southern Africa. *South African Geographical Journal*, 52, 93–101.

Goudie, A.S. (1972a). Climate, weathering, crust formation, dunes and fluvial features of the central Namib Desert, near Gobabeb, south west Africa. *Madoqua* series II, 1, 15–31.

Goudie, A.S. (1972b). The concept of post-glacial progressive desiccation. *Research Paper of the School of Geography, University of Oxford*, No. 4.

Goudie, A.S. (1972c). On the definition of calcrete deposits. *Zeitschrift für Geomorphologie*, 16, 464–8.

Goudie, A.S. (1973). *Duricrusts in tropical and subtropical landscapes*. Oxford: Oxford University Press.

Goudie, A. S. (1974). Further experimental investigation of rock weathering by salt and other mechanical processes. *Zeitschrift für Geomorphologie*, Supplementband, 21, 1–12.

Goudie, A.S. (1977). Sodium sulphate weathering and the disintegration of Mohenjo-Daro, Pakistan. *Earth Surface Processes*, 2, 75–86.

Goudie, A.S. (1978). Dust storms and their geomorphological implications. *Journal of Arid Environments*, 1, 291–310.

Goudie, A.S. (1983a). Dust storms in space and time. *Progress in Physical Geography*, 7, 502–30.

Goudie, A.S. (1983b). Calcrete. In *Chemical Sediments and Geomorphology*, ed. A.S. Goudie and K. Pye. London: Academic Press, pp. 93–131.

Goudie, A.S. (1984a). *The Nature of the Environment*. Oxford: Blackwell.

Goudie, A.S. (1984b). Salt efflorescences and salt weathering in the Hunza Valley, Karakoram Mountains, Pakistan. In *Proceedings International Karakoram Project*, vol. 2, ed. K.J. Miller. Cambridge: Cambridge University Press, pp. 607–15.

Goudie, A.S. (1985). Duricrusts and landforms. In *Geomorphology and Soils*, ed. K.S. Richards, R.R. Arnett and S. Ellis. London: Allen and Unwin, pp. 37–57.

Goudie, A.S. (1986). Laboratory simulation of the 'wick effect' in salt weathering of rock. *Earth Surface Processes and Landforms*, 11, 275–85.

Goudie, A.S. (1992). *Environmental Change* (3rd ed.). Oxford: Oxford University Press.

Goudie, A.S. (1993a). *The Nature of the Environment* (3rd ed.). Oxford: Blackwell.

Goudie, A.S. (1993b). Salt weathering simulation using a single-immersion technique. *Earth Surface Processes and Landforms*, 18, 369–76.

Goudie, A.S. (1994). Deserts in a warmer world. In *Environmental Change in Drylands: Biogeographical and Geomorphological perspectives*, ed. A.C. Millington and K. Pye. Chichester: Wiley, pp. 1–24.

Goudie, A.S. (1995). *The Changing Earth: rates of geomorphological processes* . Oxford: Blackwell.

Goudie, A.S. (1996a). Organic agency in calcrete development. *Journal of Arid Environments*, 32, 103–10.

Goudie, A.S. (1996b). Climate: Past and present. In *The Physical Geography of Africa*, ed. W.M. Adams, A.S. Goudie and A.R. Orme. Oxford: Oxford University Press, pp. 34–59.

Goudie, A.S. (1999a). A comparison of the relative resistance of limestones to frost and salt weathering. *Permafrost and Periglacial Processes*, 10, 309–16.

Goudie, A.S. (1999b). Experimental salt weathering of limestones in relation to rock properties. *Earth Surface Processes and Landforms*, 24, 715–24.

Goudie, A.S. (1999c). The history of desert dune studies over the last 100 years. In *Aeolian Environments, Sediments and Landforms*, ed. A.S. Goudie, I. Livingstone and S. Stokes. Chichester: Wiley, pp. 1–13.

Goudie, A.S. (1999d). Wind erosional landforms: Yardangs and pans. In *Aeolian Environments, Sediments and Landforms*, ed. A.S. Goudie, I. Livingstone and S. Stokes. Chichester: Wiley, pp. 167–80.

Goudie, A.S. (1999e). The Ice Age in the tropics and its human implications. In *Environments and Historical Change*, ed. P. Slack. Oxford: Oxford University Press, pp. 10–22.

Goudie, A.S. (2002). *Great Warm Deserts of the World – Landscapes and Evolution.* Oxford: Oxford University Press.

Goudie, A.S. (ed.) (2004). *Arid Lands Geomorphology*. London: Routledge.

Goudie, A.S. (2005). The drainage of Africa since the Cretaceous. *Geomorphology*, 67, 437–56.

Goudie, A.S. (2006). *The human impact on the natural environment.* 6th edition. Oxford: Blackwell Publishing.

Goudie, A.S. (2007a). Mega-yardangs: A Global Analysis. *Geography Compass*, 1, 65–81.

Goudie, A.S. (2007b). Desert landforms in Namibia – A Landsat interpretation. In *Geomorphological Variations*, ed. A.S. Goudie and J. Kalvoda. Prague: P3K, pp. 19–36.

Goudie, A.S. (2008a). The history and nature of wind erosion in deserts. *Annual Review of Earth and Planetary Science*, 36, 97–119.

Goudie, A.S. (2008b). *Wheels across the Desert*. London: Silphium Press.

Goudie, A.S. (2010). Dayas: Distribution and morphology of dryland solutional depressions developed in limestones. *Zeitschrift für Geomorphologie*, 54, 145–59.

Goudie, A.S. (2012). Charles Rollin Keyes and extravagant eolation. *Aeolian Research*, 4, 51–3.

Goudie, A.S. and Cooke, R.U. (1984). Salt efflorescences and saline lakes: A distributional analysis. *Geoforum*, 15, 563–82.

Goudie, A.S. and Day M.J. (1980). Disintegration of fan sediments in Death Valley, California, by salt weathering. *Physical Geography*, 1, 126–37.

Goudie, A.S. and Eckardt, F. (1999). The evolution of the morphological framework of the central Namib Desert, Namibia, since the early Cretaceous. *Geografiska Annaler*, 81A, 443–58.

Goudie, A.S. and Heslop, E. (2007). Sodium nitrate deposits and efflorescences. In *Geochemical Sediments and Landscapes*, ed. D.J. Nash and S.J. McLaren. Oxford: Blackwell Publishing, pp. 391–408.
Goudie, A.S. and Middleton, N.J. (1992). The changing frequency of dust storms through time. *Climate Change*, 20, 197–225.
Goudie, A.S. and Middleton, N.J. (2001). Saharan dust storms: Nature and consequences. *Earth-Science Reviews*, 56, 179–204.
Goudie, A.S., and Middleton, N.J. (2006). *Desert Dust in the Global System*. Heidelberg: Springer Verlag.
Goudie, A.S. and Migón, P. (1997). Weathering pits in the Spitzkoppe area, central Namib Desert. *Zeitschrift für Geomorphologie*, 41, 417–44.
Goudie, A.S., and Parker, A.G. (1998). Experimental simulation of rapid rock block disintegration by sodium chloride in a foggy coastal desert. *Journal of Arid Environments*, 40, 347–55.
Goudie, A.S. and Seely, M. (2011). *World Heritage Desert Landscapes: Potential Priorities for the Recognition of Desert Landscapes and Geomorphological Sites on the World Heritage List*. Gland: IUCN.
Goudie, A.S. and Sperling, C.H.B. (1977). Long distance transport of foraminiferal tests by wind in the Thar Desert, northwest India. *Journal of Sedimentary Petrology*, 47, 630–33.
Goudie, A.S. and Thomas, D.S.G. (1985). Pans in southern Africa with particular reference to South Africa and Zimbabwe. *Zeitschrift fur Geomorphologie*, 29, 1–19.
Goudie, A.S. and Thomas, D.S.G. (1986). Lunette dunes in southern Africa. *Journal of Arid Environments*, 10, 1–12.
Goudie, A.S. and Viles, H.A. (1995). The nature and pattern of debris liberation by salt weathering: A laboratory study. *Earth Surface Processes and Landforms*, 20, 437–49.
Goudie, A.S. and Viles, H.A. (1997). *Salt Weathering Hazard*. Chichester: Wiley.
Goudie, A.S. and Viles, H.A. (2000). The thermal degradation of marble. *Acta Universitas Carolinae 2000 Geographica*, 35, Supplementum, 7–16.
Goudie, A. S. and Watson, A. (1981). The shape of desert sand dune grains. *Journal of Arid Environments*, 4, 185–90.
Goudie, A.S. and Watson, A. (1984). Rock block monitoring of rapid salt-weathering in southern Tunisia. *Earth Surface Processes and Landforms*, 9, 95–8.
Goudie, A.S. and Wells, G.L. (1995). The nature, distribution and formation of pans in arid zones. *Earth-Science Reviews*, 38, 1–69.
Goudie, A.S., Allchin, B. and Hegde, K.T.M. (1973). The former extensions of the Great Indian Sand Desert. *Geographical Journal*, 139, 243–57.
Goudie, A.S., Allison, R.J. and McLaren, S.J. (1992). The relations between modulus of elasticity and temperature in the context of the experimental simulation of rock weathering by fire. *Earth Surface Processes and Landforms*, 17, 605–15.
Goudie, A.S., Cooke, R.U. and Doornkamp, J.C. (1979). The formation of silt from quartz dune sand by salt weathering processes in deserts. *Journal of Arid Environments*, 22, 102–12.
Goudie, A.S., Cooke, R.U. and Evans, I. (1970). Experimental investigation of rock weathering by salts. *Area*, 1970, 42–8.
Goudie, A.S., Rendell, H.A. and Bull, P.A. (1984). The loess of Tajik SSR. In *Proceedings of the International Karakoram Project*, vol. 1, ed. K. Miller. Cambridge: Cambridge University Press, pp. 399–412.
Goudie, A.S., Sands, M.J.S. and Livingstone, I. (1992). Aligned linear gilgai in the west Kimberley District, western Australia. *Journal of Arid Environments*, 23, 157–67.
Goudie, A.S., Viles, H.A. and Parker, A.G. (1997). Monitoring of rapid salt weathering in the central Namib using limestone blocks. *Journal of Arid Environments*, 37, 581–98.

Goudie A.S., Wright E., and Viles, H.A. (2002). The roles of salt (sodium nitrate) and fog in weathering: A laboratory simulation of conditions in the northern Atacama Desert, Chile. *Catena*, 48, 255–66.

Goudie, A.S., Migoń, P., Allison, R.J. and Rosser, N.J. (2002). Sandstone geomorphology of the Al-Quwayra area of south Jordan. Zeitschrift für Geomorphologie, 46, 365–90.

Goudie, A.S., Warren, A., Jones, D.K.C. and Cooke, R.U. (1987). The character and possible origins of the aeolian sediments of the Wahiba Sand Sea. *Geographical Journal*, 153, 231–56.

Goudie, A.S., Parker, A.G., Bull, P.A., White, K. and Al-Farraj, A. (2000). Desert loess in Ras Al Khaimah, United Arab Emirates. *Journal of Arid Environments*, 46, 123–35.

Goudie, A.S., Stokes, S., Cook, J., Samieh, S. and El-Rashidi, O.A. (1999). Yardang landforms from Kharga Oasis, south-western Egypt. *Zeitschrift für Geomorphologie*, Supplementband, 116, 1–16.

Goudie, A.S., Stokes, S., Livingstone, I., Bailiff, I.K. and Allison, R.J. (1993). Post-depositional modification of the linear sand ridges of the west Kimberley area of north-west Australia. *Geographical Journal*, 159, 306–17.

Goudie, A.S., Colls, A., Stokes, S., Parker, A., White, K. and Al-Farraj, A. (2000). Latest Pleistocene and Holocene dune construction at the north-eastern edge of the Rub Al Khali, United Arab Emirates. Sedimentology, 47, 1011–21.

Goudie, A.S., Viles, H.A., Allison, R.J., Day, M., Livingstone, I. and Bull, P.A. (1990). The geomorphology of the Napier Range, Western Australia. *Transactions of the Institute of British Geographers*, NS 15, 308–22.

Graber, E.R., Fine, P. and Levy, G.J. (2006). Soil stabilization in semiarid and arid land agriculture. *Journal of Materials in Civil Engineering*, 18, 190–205.

Graf, J.B., Webb, R.H. and Hereford, R. (1991). Relation of sediment load and flood-plain formation to climatic variability, Paria River drainage basin, Utah and Arizona. *Bulletin of the Geological Society of America*, 103, 1405–15.

Graf, W.L. (1978). Fluvial adjustments to the spread of tamarisk in the Colorado Plateau region. *Bulletin of the Geological Society of America*, 86, 1491–1501.

Graf, W.L. (1981). Channel instability in a sand-river bed. *Water Resources Research*, 17, 1087–94.

Graf, W.L. (1988). *Fluvial Processes in Dryland Rivers*. Berlin: Springer Verlag.

Graham, R.C. and O'Geen, A.T. (2010). Soil mineralogy trends in California landscapes. *Geoderma*, 154, 418–37.

Graham, R.C., Hirmas, D.R., Wood, Y.A. and Armhein, C. (2008). Large near-surface nitrate pools in soils capped by desert pavement in the Mojave Desert, California. *Geology*, 36, 259–62.

Grams, P.E., Schmidt, J.C. and Topping, D.J. (2007). The rate and pattern of bed incision and bank adjustment on the Colorado River in Glen Canyon downstream from the Glen Canyon Dam, 1956–2000. *Bulletin of the Geological Society of America*, 119, 556–75.

Greeley, R. and Iversen, J.D. (1985). *Wind as a Geological Process on Earth, Mars, Venus and Titan*. Cambridge: Cambridge University Press.

Greeley, R., Christensen, P. and Carrasco, R. (1989). Shuttle radar images of wind streaks in the Altiplano, Bolivia. *Geology*, 17, 665–8.

Greeley, R. and 17 others. (2008). Columbia Hills, Mars: Aeolian features seen from the ground and orbit. *Journal of Geophysical Research*, 113, E06S06, doi:10.1029/2007JE002971.

Greenbaum, N., Schick, A. and Baker, V.R. (2000). The palaeoflood record of a hyperarid catchment, Nahal Zin, Negev Desert, Israel. *Earth Surface Processes and Landforms*, 25, 951–71.

Greenbaum, N., Ben-Zvi, A., Haviv, I. and Enzel, Y. (2006). The hydrology and paleohydrology of the Dead Sea tributaries. *Geological Society of America Special Paper*, 401, 63–93.

Greenbaum, N., Porat, N., Rhodes, E. and Enzel, Y. (2006). Large floods during Oxygen Isotope Stage 3, southern Negev Desert, Israel. *Quaternary Science Reviews*, 25, 704–19.

Greenville, A.C., Dickman, C.R. and Wardle, G.M. (2009). The fire history of an arid grassland: The influence of antecedent rainfall and ENSO. *International Journal of Wildland Fire*, 18, 631–9.

Gregory, H.E. (1917). Geology of the Navajo Country. *U.S. Geological Survey Professional Paper*, 93.

Gregory, H.E. (1938). The San Juan Country. *U.S. Geological Survey Professional Paper*, 188.

Gregory, J.W. (1906). *The Dead Heart of Australia*. London: John Murray.

Gregory, K.J. (1976). Drainage networks and climate. In *Geomorphology and Climate*, ed. E. Derbyshire. Chichester: Wiley, 289–315.

Gregory, K.J. and Gardiner, V. (1975). Drainage density and climate. *Zeitschrift für Geomorphologie* NF, 19, 287–98.

Griffin, D.L. (2006). The late Neogene Sahabi rivers of the Sahara and their climatic and environmental implications for the Chad Basin. *Journal of the Geological Society of London*, 163, 905–21.

Griffin, D.L. (2011). The late Neogene Sahabi rivers of the Sahara and the hamadas of the eastern Libya-Chad border area. *Palaeogeography, Palaeoclimatology, Palaeoecology*, 309, 176–85.

Griffiths, P.G., Webb R.H. and Melis, T.S. (1997). Initiation of debris flows in bedrock canyons of the Colorado River, USA. In *Debris-flow Hazards Mitigation: Mechanics, Prediction, and Assessment*, ed. C. Chen. New York: American Society of Civil Engineers, pp. 12–20.

Griffiths, P.G., Webb, R.H. and Melis, T.S. (2004). Frequency and initiation of debris flows in Grand Canyon, Arizona. *Journal of Geophysical Research*, 109, F04002, doi:1o/1029/2003JF000077.

Griggs, D.T. (1936). The factor of fatigue in rock exfoliation. *Journal of Geology*, 44, 781–96.

Grimm, N.B., Chacon, A., Dahm, C.N., Hostetler, S.W., Lind, O.T., Starkweather, P.L. and Wertsbaugh, W.W. (1997). Sensitivity of aquatic ecosystems to climatic and anthropogenic changes: The Basin and Range, American south-west and Mexico. In *Freshwater Ecosystems and Climate Change in North America: A Regional Assessment*, ed. C.E. Cushing. Chichester: Wiley, pp. 205–23.

Grosjean, M., Núñez, L., Castajena, I. and Messerli, B. (1997). Mid-Holocene climate and culture change in the Atacama Desert, northern Chile. *Quaternary Research*, 48, 239–46.

Grousset, F.E., Parra, M., Bory, A., Martinez, P., Bertrand, P., Shimmield, G. and Ellam, R.M. (1998). Saharan wind regimes traced by the Sr-Nd Isotopic composition of subtropical Atlantic sediments: Last Glacial Maximum vs today. *Quaternary Science Reviews*, 17, 395–409.

Grove, A.T. (1969). Landforms and climatic change in the Kalahari and Ngamiland. *Geographical Journal*, 135, 191–212.

Grove, A.T. and Goudie, A.S. (1971). Late Quaternary lake levels in the Rift Valley of southern Ethiopia and elsewhere in tropical Africa. *Nature*, 234, 403–5.

Grove, A.T. and Warren, A. (1968). Quaternary landforms and climate on the south side of the Sahara. *Geographical Journal*, 134, 194–208.

Grove, A. T., Street, F. A. and Goudie, A. S. (1975). Former lake levels and climatic change in the Rift Valley of southern Ethiopia. *Geographical Journal*, 141, 177–94.

Grove, J.M. (2004). *Little Ice Ages: Ancient and Modern*. London: Routledge.

Gruber, S. (2002). The fairy circles of Kaokoland (northwest Namibia) – Is the harvester termite *Hodotermes mossambicus* the prime causal factor in circle formation? *Basic and Applied Ecology*, 3, 367–70.

Guba, I. and Glennie, K.W. (1998). Geology and geomorphology. In *Vegetation of the Arabian Peninsula*, ed. S.A. Ghaznfar and M. Fisher . Dordrecht: Kluwer, pp. 39–62.

Gunal, H. and Ransom, M.D. (2006). Clay illuviation and calcium carbonate accumulation along a precipitation gradient in Kansas. *Catena*, 68, 59–69.

Gunatilaka, A. (2011). Holocene evolution of Arabian coastal sabkhas: A re-evaluation based on stable-isotope analysis, forty years after Shearman's first view of the sabkha. *International Association of Sedimentologists Special Publication*, 43, 113–32.

Gunatilaka, A. and Mwango, S. (1987). Continental sabkha pans and associated nebkhas in southern Kuwait, Arabian Gulf. *Geological Society of London Special Publication*, 35, 187–203.

Guo, G. and Xie, G. (2007). Dust storms in China decreased during the past 50 years. *International Journal of Remote Sensing*, 29, doi:10.1080/01431160500196919.

Guo, X. and Chafetz, H. (2011). Large tufa mounds, Searles Lake, California. *Sedimentology*, doi:10.111/j.1365–3091.2011.01315.x.

Guo, Z.T., Ruddiman, W.F., Hao, Q.Z., Wu, H.B., Qiao, Y.S., Zhu, R.X., Peng, S.Z., Wei, J.J., Yuan, B.Y. and Liu, T.S. (2002). Onset of Asian desertification by 22 Myr ago inferred from loess deposits in China. *Nature*, 416, 159–63.

Gupta, A., Anderson, D.M. and Overpeck, J.T. (2003). Abrupt changes in the Asian southwest monsoon during the Holocene and their links to the North Atlantic Ocean. *Nature*, 421, 354–6.

Gustavson, T.C., Holliday, V.T. and Hovorka, S.D. (1995). Origin and development of playa basins, sources of recharge to the Ogallala Aquifer, southern High Plains, Texas and New Mexico. *Bureau of Economic Geology of the University of Texas, Report of Investigations*, 229.

Gutiérrez, M. and Sesé Martínez, V.H. (2001). Multiple talus flatirons, variations of scarp retreat rates and the evolution of slopes in Almazán Basin (semi-arid central Spain). *Geomorphology*, 38, 19–29.

Gutiérrez, M., Desir, G. and Gutiérrez-Santolalla, F. (2002). Yardangs in the semiarid central sector of the Ebro Depression (NE Spain). *Geomorphology*, 44, 155–70.

Gutíerrez, M., Sancho, C. and Arauzo, T. (1998). Scarp retreat rates in semi-arid environments from talus flatirons (Ebro Basin, NE Spain). *Geomorphology*, 25, 111–21.

Gutiérrez, M., Sancho, C., Benito, G., Sirvent, J. and Desir, G. (1997). Quantitative study of piping processes in badland areas of the Ebro Basin, NE Spain. *Geomorphology*, 20, 237–53.

Gutiérrez-Elorza, M., Lucha, P., Gracia, F.-J., Marín, C. and Petit-Maire, N. (In press). Palaeoclimatic considerations of talus flatirons and aeolian deposits in northern Fuerteventura volcanic island (Canary Islands, Spain). *Geomorphology*.

Ha, Si. (2002). Network dunes in southeastern Tengger Desert: Morphology, sediments and dynamics. *Proceedings of ICAR5/GCTE-SEN Joint Conference, Lubbock*, 443–5.

Haberlah, D., Williams, M.A.J., Halverson, G., McTainsh, G.H., Hill, S.M., Hrstka, T., Jaime, P., Butcher, A.R. and Glasby, P. (2010). Loess and floods: High-resolution multi-proxy data of Last Glacial Maximum (LGM) slackwater deposition in the Flinders Ranges, semi-arid South Australia. *Quaternary Science Reviews*, 29, 2673–93.

Hack, J.T. (1941). Dunes of the western Navajo country. *Geographical Review*, 31, 240–63.

Haddon, I.G. and McCarthy, T.S. (2005). The Mesozoic-Cenozoic interior sag basins of central Africa: The late-Cretaceous-Cenozoic Kalahari and Okavango Basins. *Journal of African Earth Sciences*, 43, 316–33.
Hadley, R.F. (1961). Influence of riparian vegetation on channel shape, northeastern Arizona. *U.S. Geological Survey Professional Paper*, 424C, 30–31.
Haff, P.K. (2001). Desert pavement: An environmental canary? *Journal of Geology*, 109, 661–8.
Haff, P.K. and Werner, B.T. (1996). Dynamical processes on desert pavements and the healing of surficial disturbances. *Quaternary Research*, 45, 38–46.
Hagen, L.J. (1991). A wind erosion prediction system to meet user needs. *Journal of Soil and Water Conservation*, 46, 106–11.
Halfen, A.F., Fredlund, G.G. and Mahan, S.A. (2010). Holocene stratigraphy and chronology of the Casper Dune Field, Casper, Wyoming, USA. *The Holocene*, 20, 773–83.
Halfen, A.F., Johnson, W.C., Hanson, P.R., Woodburn, T.L., Young, A.R. and Ludvigson, G.A. (2012). Activation history of the Hutchinson dunes in east-central Kansas, USA during the past 2200 years. *Aeolian Research*, 5, 9–20.
Halimov, M. and Fezer, F. (1989). Eight yardang types in central Asia. *Zeitschrift fűr Geomorphologie*, 33, 205–17.
Hall, K. (1999). The role of thermal stress fatigue in the breakdown of rock in cold regions. *Geomorphology*, 31, 47–63.
Hall, K. (2011). Natural building stone composed of light-transmissive minerals: Impacts on thermal gradients, weathering and microbial colonization. A preliminary study, tentative interpretations, and future directions. *Environmental Earth Science*, 62, 289–97.
Hall, K. and André, M.-F. (2001). New insights into rock weathering from high density rock temperature data: An Antarctic study of weathering by thermal stress. *Geomorphology*, 41, 23–35.
Hall, S.R., Farber, D.L., Audin, L., Finkel, R.C. and Mériaux, A.-S. (2008). Geochronology of pediment surfaces in southern Peru: Implications for Quaternary deformation of the Andean forearc. *Tectonophysics*, 459, 186–205.
Hallbauer, D.K. and Jahns, H.M. (1977). Attack of lichen on quartzitic rock surfaces. *Lichenologist*, 9, 119–22.
Halliday, W.R. (2003). Caves and karsts of northeast Africa. *International Journal of Speleology*, 32, 19–32.
Hallsworth, E.G. and Beckmann, G.G. (1969). Gilgai in the Quaternary. *Soil Science*, 107, 409–20.
Hallsworth E.G., Robertson G.K. and Gibbons F.R. (1955). Studies in pedogenesis in New South Wales. VIII. The 'gilgai' soils. *Journal of Soil Science*, 6, 1–34.
Halsey, L.A., Catto, N.R. and Rutter, N.W. (1990). Sedimentology and development of parabolic dunes, Grande Prairie dune field, Alberta. *Canadian Journal of Earth Sciences*, 27, 1762–72.
Han, G., Zhang, G. and Dong, Y. (2007). A model of the active origin and development of source-bordering dunefields on a semi-arid fluvial plain: A case study from the Xiliaohe Plain, northeast China. *Geomorphology*, 86, 512–24.
Han, L., Tsunekawa, A. and Tsubo, M. (2011). Effect of frozen ground on dust outbreaks in spring on the eastern Mongolian Plateau. *Geomorphology*, 129, 412–16.
Han, Z., Wang, T., Dong, Z., Hu, Y. and Yao, Z. (2007). Chemical stabilization of mobile dunefields along a highway in the Taklimakan Desert of China. *Journal of Arid Environments*, 68, 260–70.
Han, Z., Wang,T., Sun, Q., Dong, Z. and Wang, X. (2003). Sand harm in Taklimakan Desert Highway and sand control. *Journal of Geographical Sciences*, 13, 45–53.

Hansen, C.J. and 11 others. (2011). Seasonal erosion and restoration of Mars' northern polar dunes. *Science*, 331, 575–8.

Hansen, N.C., Allen, B.L., Baumhardy, R.L. and Lyon, D.J. (2012). Research achievements and adoption of no-till, dryland cropping in the semi-arid U.S. Great Plains. *Field Crops Research*, 132, 196–203.

Hanson, P.R., Arbogast, A.F., Johnson, W.C., Joeckel, R.M. and Young, A.R. (2010). Megadroughts and late Holocene dune activation at the eastern margin of the Great Plains, north-central Kansas. *Aeolian Research*, 1, 101–10.

Hanson, P.R., Joeckle, R.M., Young, A.R. and Horn, J. (2009). Late Holocene dune activity in the eastern Platte River Valley, Nebraska. *Geomorphology*, 103, 555–61.

Happold, D.C.D. (1984). Small mammals. In *Sahara desert*. Ed. J.L.Cloudsley-Thompson, Oxford: Pergamon Press, pp. 251–75.

Harden, T., Macklin, M.G. and Baker, V.R. (2010). Holocene flood histories in south-western USA. *Earth Surface Processes and Landforms*, 35, 707–16.

Hardgrove, C., Moersch, J. and Whisner, S. (2009). Thermal imaging of alluvial fans: A new technique for remote classification of sedimentary features. *Earth and Planetary Science Letters*, 285, 124–30.

Haregeweyn, N., Poesen, J., Nyssen, J., de Wit, J., Haile, M., Govers, G. and Deckers, S. (2006). Reservoirs in Tigray (northern Ethiopia): Characteristics and sediment deposition problems. *Land Degradation and Development*, 17, 211–30.

Harper, R.J. and Gilkes, R.J. (2004). Aeolian influences on the soils and landforms of the southern Yilgarn Craton of semi-arid, southwestern Australia. *Geomorphology*, 59, 215–35.

Harris, R. and Allison, M.E. (2006). Hazardous cracks running through Arizona. *Geotimes,* August.

Harris, S.A. (1959). The classification of gilgaied soils: Some evidence from northern Iraq. *Journal of Soil Science*, 10, 27–33.

Harrison, S.P. and Dodson, J. (1993). Climates of Australia and New Guinea since 18,000 yr B.P. In *Global Climates since the Last Glacial Maximum*, ed. H.E. Wright, J.E. Kutzbach, T. Webb, W.F. Ruddiman, F.A. Street-Perrott and P.J. Bartlein. Minneapolis: University of Minnesota Press, pp. 265–93.

Hart, M.W. (2008). Structural and geomorphic characteristsics of landslides at Coyote Mountain, Anza-Borrego Desert State Park, California. *Environmental and Engineering Geoscience*, 14, 81–96.

Hartley, A.J. and Chong, G. (2002). Late Pliocene age for the Atacama Desert: Implications for the desertification of western South America. *Geology*, 31, 43–6.

Hartley, A.J. and May, G. (1998). Miocene gypcretes from the Calama Basin, northern Chile. *Sedimentology*, *45*, 351–64.

Harvey, A.M. (1982). The role of piping in the development of badlands and gulley systems in south-east Spain. In *Badland Geomorphology and Piping*, ed. R. Bryan and A. Yair. Norwich: Geo Books, pp. 317–35.

Harvey, A.M. (1997). The role of alluvial fans in arid zone fluvial systems. In *Arid Zone Geomorphology*, ed. D.S.G. Thomas. Wiley: Chichester pp. 231–59.

Harvey, A.M. (2002). The relationships between alluvial fans and fan channels within Mediterranean mountain fluvial systems. In *Dryland Rivers: Hydrology and Geomorphology of Semi-Arid Channels*, ed. L.J. Bull and M.J. Kirkby. Chichester: Wiley, pp. 205–26.

Harvey, A.M. and Wells, S.G. (1994). Late Pleistocene and Holocene changes in hillslope sediment supply to alluvial fan systems, Zzyxx, California. In *Environmental Change in Drylands*, ed. A.C. Millington and K. Pye. Chichester: Wiley, pp. 66–84.

Harvey, A.M., Mather, A.E. and Stokes, M. (eds.). (2005). Alluvial fans: Geomorphology, sedimentology, dynamics. *Geological Society of London Special Publication*, 251.

Harvey, A.M., Wigand, P.E. and Wells, S.G. (1999). Response of alluvial fan systems to the late Pleistocene to Holocene climatic transition: Contrast between the margins of pluvial lakes Lahontan and Mojave, Nevada and California, U.S.A. *Catena*, 36, 255–81.

Harvey, J.E. and Pederson, J.L. (2011). Reconciling arroyo cycle and paleoflood approaches to late Holocene alluvial records in dryland streams. *Quaternary Science Reviews*, 30, 855–66.

Harvey, J.E., Pederson, J.L. and Rittenour, T.M. (2011). Exploring relations between arroyo cycles and canyon paleoflood records in Buckskin Wash, Utah: Reconciling scientific paradigms. *Bulletin of the Geological Society of America*, 123, 2266–76.

Hastenrath, S. (1967). The barchans of the Arequipa Region, southern Peru. *Zeitschrift für Geomorphologie*, 11, 300–11.

Hawass, Z. (1993). The Egyptian monuments: Problems and solutions. In *Conservation of Stone and Other Materials*, ed. M.J. Theil. London: E. and F.N. Spon. 1, pp. 19–25.

Hayes, S.M., White, S.A., Thompson, T.L., Maier, R.M. and Chorover, J. (2009). Changes in lead and zinc lability during weathering-induced acidification of desert mine tailings: Coupling chemical and micro-scale analyses. *Applied Geochemistry*, 24, 2234–45.

Haynes, C.V. (1989). Bagnold's barchan: A 57-year record of dune movement in the eastern Sahara and its implications for dune origin and paleoclimate since Neolithic times. *Quaternary Research*, 32, 153–67.

Hbirkou, C., Martius, C., Khamzina, A., Lamers, J.P.A., Welp, G. and Amelung, W. (2011). Reducing topsoil salinity and raising carbon stocks through afforestation in Khorezm, Uzbekistan. *Journal of Arid Environments*, 75, 146–55.

Hearty, P.J. and O'Leary, M.J. (2008). Carbonate eolianites, quartz sands and Quaternary sea-level cycles, Western Australia: A chronostratigraphic approach. *Quaternary Geochronology*, 3, 26–55.

Heatwole, H. and Muir, M. (1979). Thermal microclimates in the pre-Saharan steppes of Tunisia. *Journal of Arid Environments*, 2, 119–36.

Hedin, S. (1903). *Central Asia and Tibet*. New York: Scribners.

Heidbüchel, I. (2007). Recharge processes in ephemeral streams derived from a coupled stream flow routing/groundwater model – Application to the Lower Kuiseb (Namibia). DoctoralThesis, Universität Freiburg.

Heil, C.W., King, J.W., Zárate, M.A. and Schultz, P.H. (2010). Climatic interpretation of a 1.9 Ma environmental magnetic record of loess deposition and soil formation in the central eastern Pampas of Buenos Aires, Argentina. *Quaternary Science Reviews*, 29, 2705–18.

Heine, K. and Walter, R. (1996). Gypcretes of the central Namib Desert, Namibia. *Palaeoecology of Africa*, 24, 173–99.

Heinrichs, K. (2008). Diagnosis of weathering damage on rock-cut monuments in Petra, Jordan. *Environmental Geology*, 56, 643–75.

Heldmann, J.L., Conley, C.A., Brown, A.J., Fletcher, L., Bishop, J.L. and McKay, C.P. (2010). Possible liquid water origin for Atacama Desert mudflow and recent gully deposits on Mars. *Icarus*, 206, 685–90.

Henschel, J.R. and Seely, M.K. (2008). Ecophysiology of atmospheric moisture in the Namib Desert. *Atmospheric Research*, 87, 362–68.

Hereford, R. (1984). Climate and ephemeral-stream processes: Twentieth-century geomorphology and alluvial stratigraphy of the Little Colorado River, Arizona. *Geological Society of America, Bulletin*, 95, 654–68.

Hereford, R. (1986). Modern alluvial history of the Paria River Drainage Basin, southern Utah. *Quaternary Research*, 25, 293–311.

Hereford, R., Jacoby, G.C. and McCord, V.A.S. (1995). Geomorphic history of the Virgin River in the Zion National Park area, southwest Utah. *U.S. Geological Survey Circular*, 95–515.

Hereher, M.E. (2010). Sand movement patterns in the Western Desert of Egypt: An environmental concern. *Environmental Earth Science*, 59, 1119–27.

Hermanns, R.L. and Strecker, M.R. (1999). Structural and lithological controls on large Quaternary rock avalanches (sturzstroms) in arid northwestern Argentina. *Bulletin of the Geological Society of America*, 111, 934–8.

Hermanns, R.L., Niedermann, S., Garcia, A.V., Gomez, J.S. and Strecker, M.R. (2001). Neotectonics and catastrophic failure of mountain fronts in the southern intra-Andean Puna Plateau, Argentina. *Geology*, 29, 619–23.

Hermanns, R.L., Trauth, M.H., Niedermann, S., McWilliams, M. and Strecker, M. (2000). Tephrochronologic constraints on temporal distribution of large landslides in northwest Argentina. *Journal of Geology*, 108, 35–52.

Hernandez-Marin, M. and Burbey, T.J. (2010). Controls on initiation and propagation of pumping-induced earth fissures: Insights from numerical simulations. *Hydrogeology Journal*, 18, 1773–85.

Herrero, J. and Porta, J. (2000). The terminology and the concepts of gypsum-rich soils. *Geoderma*, 96, 47–61.

Herrmann, H.J. (2002). Evolution and shape of dunes. *Comptes Rendus Physique*, 3, 197–206.

Herrmann, H.J., Sauermann, G., and Schwämmle, V. (2005). The morphology of dunes. *Physics A*, 358, 30–8.

Hersen, P., Andersen, K.H., Elbelrhiti, H., Andreotti, B. Claudin, P. and Douady, S. (2004). Corridors of barchan dunes: Stability and size selection. *Physical Review*, E 69, article number 011304.

Hesp, P.A. and Hastings, K. (1998). Width, height and slope relationships and aerodynamic maintenance of barchans. *Geomorphology*, 22, 193–204.

Hesp, P.A. and Thom, B.G. (1990). Geomorphology and evolution of active transgressive dunefields. In *Coastal Dunes: Form and Process*, ed. K.F. Nordstrom, N.P. Psuty and R.W.G. Carter. Chichester: John Wiley and Sons, pp. 253–88.

Hesp, P., Hyde, R., Hesp, V. and Zhengyu, Q. (1989). Longitudinal dunes can move sideways. *Earth Surface Processes and Landforms*, **14**, 447–51.

Hesse, P.P. (1994). The record of continental dust from Australia in Tasman Sea sediments. *Quaternary Science Review*, 13, 257–72.

Hesse, P.P. (2010). The Australian desert dunefields: Formation and evolution in an old, flat, dry continent. *Geological Society of London Special Publication*, 346, 141–64.

Hesse, P.P. (2011). Sticky dunes in a wet desert: Formation, stabilisation and modification of the Australian desert dunefields. *Geomorphology*, 134, 309–25.

Hesse, P.P. and McTainsh, G.H. (1999). Last Glacial Maximum to early Holocene wind strength in the mid-latitudes of the Southern Hemisphere from aeolian dust in the Tasman Sea. *Quaternary Research*, 52, 343–9.

Hesse, P.P. and McTainsh, G.H. (2003). Australian dust deposits: Modern processes and the Quaternary record. *Quaternary Science Reviews*, 22, 2007–35.

Hesse, P.P. and Simpson, R.L. (2006). Variable vegetation cover and episodic sand movement on longitudinal desert sand dunes. *Geomorphology*, 81, 276–91.

Hickey, B. and Goudie, A.S. (2007). The use of TOMS and MODIS to identify dust storm source areas: The Tokar Delta (Sudan) and the Seistan Basin (south west Asia). In *Geomorphological Variations*, ed. A.S. Goudie and J. Kalvoda. Prague: P3K, pp. 37–57.

Higgins, C.G. (1982). Drainage systems developed by sapping on Earth and Mars. *Geology*, 10, 147–52.

Hill, C.A. (1977). Niter and soda-niter in a lava tube, Socorro County, New Mexico. *Cave Research Fund Annual Report*, 19, 15.

Hillaire-Marcel, C. and Casanova, J. (1987). Isotopic hydrology and paleohydrology of the Magadi (Kenya) – Natron (Tanzania) Basin during the Quaternary. *Palaeoclimatology, Palaeogeography, Palaeoecology*, 58, 155–81.

Hills, E.S. (1939). The physiography of north-western Victoria. *Proceedings of the Royal Society of Victoria*, 51, 297–323.

Hills, E.S. (1940). The lunette: A new landform of aeolian origin. *Australian Geographer*, 3, 1–7.

Hipondoka, M.H.T., Busche, D. and Kempf, J. (2004). Source of lunette dune sediments: A geomorphic terrain analysis approach in Etosha National Park, Namibia. *Erdkunde*, 58, 212–20.

Hirmas, D.R. and Allen, B.L. (2007). Degradation of pedogenic calcretes in west Texas. *Soil Science Society of America Journal*, 71, 1878–88.

Hirmas, D.R., Graham, R.C. and Kendrick, K.J. (2011). Soil-geomorphic significance of land surface characteristics in an arid mountain rage, Mojave Desert, USA. *Catena*, 87, 408–20.

Hoelzmann, P., Keding, B., Berke, H., Kropelin, S. and Kruse, H.-J. (2001). Environmental change and archaeology: Lake evolution and human occupation in the eastern Sahara during the Holocene. *Palaeogeography, Palaeoclimatology, Palaeoecology*, 169, 193–217.

Hoetzl, H. (1995). Groundwater recharge in an arid karst area (Saudi Arabia). *IAHS Publication*, 232, 195–207.

Hoffmann, J.P., Pool, D.R., Konieczki, A.D. and Carpenter, M.C. (1998). Causes of sinks near Tucson, Arizona, USA. *Hydrogeology Journal*, 6, 349–64.

Hogg, S.E. (1982). Sheetfloods, sheetwash, sheetflow, or . . . ? *Earth-Science Reviews*, 18, 59–76.

Hoke, G.D., Isacks, B.L., Jordan, T.E. and Yu, S. (2004). Groundwater-sapping origin for the giant *quebradas* of northern Chile. *Geology*, 32, 605–8.

Hollands, C.B., Nanson, G.C., Jones, B.G., Bristow, C.S., Price, D.M. and Pietsch, T.J. (2006). Aeolian-fluvial interaction: Evidence for late Quaternary channel changes and wind-rift linear dune formation in the northwestern Simpson Desert, Australia. *Quaternary Science Reviews*, 25, 142–62.

Höllerman, P. (1975). Formen Kavernöser Verwitterung ('Tafoni') auf Teneriffa. *Catena*, 2, 385–410.

Holliday, V.T. (1997). Origin and evolution of lunettes on the High Plains of Texas and New Mexico. *Quaternary Research*, 47, 54–69.

Holliday, V.T. and Rawling, J.E. (2006). Soil-geomorphic relations of lamellae in eolian sand on the High Plains of Texas and New Mexico. *Geoderma*, 131, 154–80.

Holliday, V.T., Hovorka, S.D. and Gustavson, T.C. (1996). Lithostratigraphy and geochronology of fills in small playa basins on the southern High Plains, United States. *Bulletin of the Geological Society of America*, 108, 953–65.

Holm, D.A. (1960). Desert geomorphology in the Arabian Peninsula. *Science*, 132, 413–31.

Holm, E. and Edney, E.B. (1973). Daily activity of Namib Desert arthropods in relation to climate. *Ecology*, 54, 45–56.

Holmes, P.J., Bateman, M.D., Thomas, D.S.G., Telfer, M.W., Barker, C.H. and Lawson, M.P. (2008). A Holocene-late Pleistocene record from lunette dunes of the western Free State panfield, South Africa. *The Holocene*, 18, 1193–205.

Hong, N. (1994). Corrosiveness of the arid saline soil in China. In *Engineering Characteristics of Arid Soils*, ed. P. Fookes and R.H.G. Parry. Rotterdam: Balkema, pp. 29–34.

Hooghiemstra, H. (1989). Variations of the NW African trade wind regime during the last 140,000 years: Changes in pollen influx evidenced by marine sediment records. In *Paleoclimatology and Paleometeorology: Modern and Past Patterns of Global Atmospheric Transport*, ed. M. Leinen and M. Sarnthein. Dordrecht: Kluwer, pp. 733–70.
Hooke, R. le.B. (1967). Processes on arid-region alluvial fans. *Journal of Geology*, 75, 438–60.
Hooke, R. le.B. (1968). Steady-state relationships on arid-region alluvial fans in closed basins. *American Journal of Science*, 226, 609–29.
Hoorn, C. and Cremaschi, M. (2004). Late Holocene palaeoenvironmental history of Khawr Rawri and Khawr Al Balid (Dhofar, Sultanate of Oman). *Palaeoclimatology, Palaeogeography, Palaeoecology*, 213, 1–36.
Horta, J.C. de O.S. (1980). Calcrete, gypcrete and soil classification in Algeria. *Engineering Geology*, 15, 15–52.
Horta, J.C. de O.S. (1985). Salt heaving in the Sahara. *Géotechnique*, 35, 329–37.
Horton, R.E. (1945). Erosional development of streams and their drainage basins: Hydrophysical approach to quantitative morphology. *Geological Society of America, Bulletin*, 56, 275–370.
House, P.K. (2005). Using geology to improve flood hazard management on alluvial fans – An example from Laughlin, Nevada. *Journal of the American Water Resources Association*, 41, 1431–47.
Houston, J. (2006). The great Atacama flood of 2001 and its implications for Andean hydrology. *Hydrological Processes*, 20, 591–610.
Houston, S.L., Houston, W.N., Zapata, C.E. and Lawrence, C. (2001). Geotechnical engineering practice for collapsible soils. *Geotechnical and Geological Engineering*, 19, 333–55.
Howard, A.D. (1994). Badlands. In *Geomorphology of Desert Environments*, ed. A.D. Abrahams and A.J. Parsons. London: Chapman and Hall, pp. 213–42.
Howard, A.D. (2009). Badlands and gullying. In *Geomorphology of Desert Environments* (2nd ed.), ed. A.J. Parsons and A.D. Abrahams. Heidelberg: Springer, pp. 265–99.
Howard, A.D. and Kochel, R.C. (1988). Introduction to cuesta landforms and sapping processes on the Colorado Plateau. In *Sapping Features of the Colorado Plateau*, ed. A.D. Howard, R.C. Kochel and H.E. Holt. NASA Special Publication, SP 491, pp. 6–56.
Howard, A.D., Spiga, A. and Moore, J.M. (2012). The deepest basin on Mars is formed by aeolian erosion: Western Hellas Planitia. *43rd Lunar and Planetary Science Conference*, 1105.pdf.
Howard, A. D., Morton, J. B., Gad-el-hak, M. and Pierce, D. B. (1978). Sand transport model of barchan dune equilibrium. *Sedimentology*, 25, 307–38.
Huang, C.C., Pang, J. and Zhao, J. (2000). Chinese loess and the evolution of the East Asian monsoon. *Progress in Physical Geography*, 24, 75–96.
Huang, X., Oberhänsli, H., von Suchodoletz, H. and Sorrel, P. (2011). Dust deposition in the Aral Sea: Implications for changes in atmospheric circulation in central Asia during the past 2000 years. *Quaternary Science Reviews*, 30, 3661–74.
Hubble, G.D., Isbell, R.F. and Northcote, K.-H. (1983). Features of Australian Soils. In *Soils: an Australian Viewpoint*. Melbourne: Division of Soils, CSIRO, and Academic Press, pp. 17–47.
Huckleberry, G. and Duff, A.I. (2008). Alluvial cycles, climate and Pueblan settlement shifts near Zuni Salt Lake, New Mexico, USA. *Geoarchaeology*, 23, 107–30.
Hugenholtz, C.H. and Barchyn, T.E. (2010). Spatial analysis of sand dunes with a new global topographic dataset: New approaches and opportunities. *Earth Surface Processes and Landforms*, doi:10,1002/esp.2013.

Hugenholtz, C.H. and Barchyn, T.E. (2012). Real barchan dune collisions and ejections. *Geophysical Research Letters*, 39, L02306, doi:10.1029/2011GLO50299.
Hugenholtz, C.H. and Wolfe, S.A. (2005a). Biogeomorphic model of dunefield activation and stabilization on the northern Great Plains. *Geomorphology*, 70, 53–70.
Hugenholtz, C.H. and Wolfe, S.A. (2005b). Recent stabilization of active sand dunes on the Canadian prairies and relation to recent climate variations. *Geomorphology*, 68, 131–47.
Hugenholtz, C.H., Bender, D. and Wolfe, S.A. (2010). Declining sand dune activity in the southern Canadian prairies: Historical context, controls and ecosystem implications. *Aeolian Research*, 2, 71–82.
Hugenholtz, C.H., Moorman, B.J. and Wolfe, S.A. (2007). Ground penetrating radar (GPR) imaging of the internal structure of an active parabolic sand dune. *Geological Society of America Special Paper*, 432, 35–45.
Hugenholtz, C.H.,Wolfe, S.A. and Moorman, B.J. (2008). Effect of sand supply on the morphodynamics and stratigraphy of active parabolic dunes, Bigstick Sand Hills, southwestern Saskatchewan. *Canadian Journal of Earth Sciences*, 45, 321–35.
Hülle, D. and 7 others (2010). OSL dating of sediments from the Gobi Desert, southern Mongolia. *Quaternary Geochronology*, 5, 107–13.
Hume, W.F. (1925). *Geology of Egypt. Vol. 1*. Cairo: Government Press.
Hunt, C.B. (1956). Cenozoic geology of the Colorado Plateau. *U.S. Geological Survey Professional Paper*, 279.
Hunt, C.B. (1975). *Death Valley, Geology, Ecology, Archaeology*. Berkeley: University of California Press.
Hunt, C.B. and Washburn, A.L. (1966). Patterned ground. *U.S. Geological Survey Professional Paper*, 494B, 104–33.
Hunter, R.E. (1977). Terminology of cross-stratified sedimentary layers and climbing-ripple structures. *Journal of Sedimentary Petrology*, 47, 697–706.
Huntington, E. (1907). *The Pulse of Asia*. Boston and New York: Houghton Mifflin.
Huntington, E. (1914). The climate factor as illustrated in arid America. *Carnegie Institution of Washington,* Publication 192.
Hussain, M., Al-Khalifah, F. and Khandaker, N.I. (2006). The Jabal Al Qarah caves of the Hofuf area, northeastern Saudi Arabia: A geological investigation. *Journal of Cave and Karst Studies*, 68, 12–21.
Idso, S.B. and Brazel, A.J. (1984). Rising atmospheric carbon dioxide concentrations may increase streamflow. *Nature*, 312, 51–3.
Ikeda, N. and 14 others. (1998). The vesicular layer and carbonate collars of desert soils and pavements: Formation, age and relation to climate. *Geomorphology*, 24, 101–45.
Inbar, M. (1992). Rates of fluvial erosion in basins with a Mediterranean climate type. *Catena*, 19, 393–409.
Inbar, M. and Risso, C. (2001). Holocene yardangs in volcanic terrains in the southern Andes, Argentina. *Earth Surface Processes and Landforms*, 26, 657–66.
Indoitu, R., Orlovsky, L. and Orlovsky, N. (2012). Dust storms in central Asia: Spatial and temporal variations. *Journal of Arid Environments*, 85, 62–70.
IPCC. (2007). *Climate Change: The Physical Science Basis. Contribution of Working Group I to the Fourth Assessment Report of the Intergovernmental Panel on Climate Change*, ed. S. Solomon et al. Cambridge and New York: Cambridge University Press.
Isenberg, O., Yizhaq, H., Tsoar, H., Wenkart, R., Karnieli, A., Kok, J.F. and Katra, I. (2011). Megaripple flattening due to strong winds. *Geomorphology*, 131, 69–84.
Israelevich, P., Ganor, E., Alpert, P., Kishcha, P. and Stupp, A. (2012). Predominant transport paths of Saharan dust over the Mediterranean Sea to Europe. *Journal of Geophysical Research*, 117, D02205, doi:10.1029/2011JD016482.

Issa, O.M., Le Bissonnais, Y., Défrage, C. and Trichet, J. (2001). Role of cyanobacterial cover on structural stability of sandy soils in the Sahelian part of western Niger. *Geoderma*, 101, 15–30.
Issa, O.M., Défarge, C., Trichet, J., Valentin, C. and Rajot, J.L. (2009). Microbiotic soil crusts in the Sahel of western Niger and their influence on soil porosity and water dynamics. *Catena*, 77, 48–55.
Issar, A. and Eckstein, Y. (1969). The lacustrine beds of Wadi Feiran, Sinai: Their origin and significance. *Israel Journal of Earth Science*, 18, 21–7.
Issawi, B. and McCauley, J.F. (1992). The Cenozoic rivers of Egypt: The Nile problem. In *The Followers of Horus*, ed. B. Adams and R. Friedman. Oxford: Oxbow Press, pp. 1–18.
Istanbulluoglu, E., Yetemen, O., Vivoni, E.R., Gutiérrez-Jurado, H.A. and Bras, R.L. (2008). Eco-geomorphic implications of hillslope aspect: Inferences from analysis of landscape morphology in central New Mexico. *Geophysical Research Letters*, 35, L14403, doi:10.1029/2008GL034477.
Jäckel, D. and Dronica, H. (1976). Ergebrisse von boden- und Gesteintemperatur-messungen in der Sahara. *Berliner Geographische Abhandlungen*, 24, 55–64.
Jacobs, P.M. and Mason, J.A. (2005). Impact of Holocene dust aggradation on A horizon characteristics and carbon storage in loess-derived mollisols of the Great Plains, USA. *Geoderma*, 125, 95–106.
Jacoby, Y., Grodek, T., Enzel, Y., Porat, N., McDonald, E.V. and Dahan, O. (2008). Late Holocene upper bounds of flood magnitudes and twentieth century large floods in the ungauged, hyperarid alluvial Nahal Arava, Israel. *Geomorphology*, 95, 274–94.
Jaenicke, R. (1979). Monitoring and critical review of the estimated source strength of mineral dust from the Sahara. In *Saharan Dust*, ed. C. Morales. Chichester: Wiley, pp. 233–42.
Jagoutz, E. (2006). Salt-induced rock fragmentation on Mars: The role of salt in the weathering of Martian rocks. *Advances in Space Research*, 38, 696–700.
Jansen, M. (1996). Save Mohenjo-Daro? In *Forgotten Cities on the Indus*, ed. M. Jansen, M. Mulloy and G. Urban. Mainz: Verlag Philipp von Zabern, pp. 220–34.
Januszke, R.M. and Booth, E.H.S. (1984). Soluble salt damage to sprayed seals on the Stuart Highway. *Australian Road Research Board Proceedings*, **12**(3), 18–31.
Jayant, K., Tripathi, V. and Rajamani, V. (1999). Geochemistry of the loessic sediments on Delhi Ridge, eastern Thar Desert, Rajasthan: Implications for exogenic processes. *Chemical Geology*, 155, 265–78.
Jebar, L., Berndtsson, R., Bahri, A. and Boufarouna, M. (2010). Spatial soil loss risk and reservoir siltation in semi-arid Tunisia. *Hydrological Sciences Journal*, 55, 121–37.
Jennings, J.N. and Sweeting, M.M. (1961). Caliche pseudo-anticlines in the Fitzroy Basin, Western Australia. *American Journal of Science*, 259, 635–9.
Jennings, J.N. (1985). *Karst Geomorphology*. Oxford: Basil Blackwell.
Jensen, A.M., and Hajej, M.S. (2001). The Road of Hope: Control of moving sand dunes in Mauritania. *Unasylva*, 52, 31–6.
Jerolmack, D.J., Reitz, M.D. and Martin, R.L. (2011). Sorting out abrasion in a gypsum dune field. *Journal of Geophysical Research*, 116, F02003, doi:10.1029/2010JF001821.
Jerolmack, D.J. and 7 others. (2012). Internal boundary layer model for the evolution of desert dune fields. *Nature Geoscience*, 5, 206–9.
Jerram, D.A., Mountney, N.P., Howell, J.A., Long, D. and Stollhofen, H. (2000). Death of a sand sea: An active aeolian erg systematically buried by the Etendeka flood basalts of NW Namibia. *Journal of the Geological Society, London*, 157, 513–16.
Jerwood, L.C., Robinson, D.A. and Williams, R.B.G. (1987). Frost and salt weathering as periglacial processes: The results and implications of some laboratory experiments. In

Periglacial Processes and Landforms in Britain and Ireland, ed. J. Boardman. Cambridge: Cambridge University Press, pp. 135–43.

Jewell, P.W. and Nicoll, K. (2011). Wind regimes and aeolian transport in the Great Basin, USA. *Geomorphology*, 129, 1–13.

Jiang, Y., Luo, Y., Zhao, Z. and Tao, S. (2009). Changes in wind speed over China during 1956–2004. *Theoretical and Applied Climatology*, doi:10.1007/s00704–009–0152–7.

Jilbert, T., Reichard, G.-J., Aeschlimann, B., Günther, D., Boer, W. and de Lange, G. (2010). Climate-controlled multidecadal variability in North African dust transport to the Mediterranean. *Geology*, 38, 19–22.

Jin, L., Chen, F., Morrill, C., Otto-Bliesner, B.L. and Rosenbloom, N. (2012). Causes of early Holocene desertification in arid central Asia. *Climate Dynamics*, 38, 1577–91.

Johannessen, C.L., Feiereisen, J.J. and Wells, A.N. (1982). Weathering of ocean cliffs by salt expansion in a midlatitude coastal environment. *Shore and Beach*, 50, 26–34.

Johnson, C.R., and 7 others. (2009). Tufa as a record of perennial fresh water in a semi-arid rift basin, Kapthurin Formation, Kenya. *Sedimentology*, 56, 1115–37.

Johnson, D.W. (1932a). Rock fans of arid regions. *American Journal of Science*, 23, 389–416.

Johnson, D.W. (1932b). Rock planes in arid regions. *Geographical Review*, 22, 656–65.

Jones, C.E. (1991). Characteristics and origin of rock varnish from the hyperarid coastal deserts of northern Peru. *Quaternary Research*, 35, 116–29.

Jones, D., Wilson, M.J. and Tait, J.M. (1980). Weathering of a basalt by *Pertusaria corallina*. *Lichenologist*, 12, 277–89.

Jones, D.K.C. (1980). British applied geomorphology: An appraisal. *Zeitschrift für Geomorphologie*, Supplementband, 36, 48–73.

Jones, D.K.C., Brunsden, D. and Goudie, A.S. (1983). A preliminary geomorphological assessment of part of the Karakoram Highway. *Quarterly Journal of Engineering Geology*, 16, 331–55.

Jung, S.J.A., Davies, G.R., Ganssen, G.M. and Kroon, D. (2004). Stepwise Holocene aridification in NE Africa deduced from dust-borne radiogenic isotope records. *Earth and Planetary Science Letters*, 221, 27–37.

Jutson, J.T. (1917). Erosion and the resulting landforms in sub-arid Western Australia, including the origin and growth of the dry lakes. *Geographical Journal*, 50, 418–34.

Jutson, J.T. (1918). The influence of salts in rock weathering in sub-arid Western Australia. *Proceedings, Royal Society of Victoria*, 80, 165–72.

Jutson, J.T. (1919). Sheet-flows, or sheet-floods, and their associated phenomena in the Niagara District of sub-arid south-central Western Australia. *American Journal of Science*, 48, 435–39.

Jutson, J.T. (1934). *The Physiography of Western Australia* (2nd ed.). Perth: Government Printer.

Jutson J.T. (1950). *The Physiography (Geomorphology) of Western Australia* (3rd ed.). Perth: Government Printer.

Juyal, N., Chamyal, L.S., Bhandari, S., Bhushan, R. and Singhvi, A.K. (2006). Continental record of the southwest monsoon during the last 130 ka: Evidence from the southern margin of the Thar Desert, India. *Quaternary Science Reviews*, 25, 2632–50.

Kabakci, H., Chevalier, P.M. and Papendick, R.I. (1993). Impact of tillage and residue management on dryland spring wheat development. *Soil and Tillage Research*, 26, 127–37.

Kádár, L. (1934). A study of the sand sea in the Libyan Desert. *Geographical Journal*, 83, 470–8.

Kadir, S. and Eren, M. (2008). The occurrence and genesis of clay minerals associated with Quaternary caliches in the Mersin area, southern Turkey. *Clays and Clay Minerals*, 56, 244–58.

Kamh, G.M.E., Kallash, A. and Azzam, R. (2008). Factors controlling building susceptibility to earthquakes: 14-year recordings of Islamic archaeological sites in Old Cairo, Egypt: A case study. *Environmental Geology*, Special Issue, DOI 10.10007/s00254–007–1162–3.

Kampunzu, A.B., Ringrose, S., Huntsman-Mapila, P., Harris, C., Vink, B.W. and Matheson, W. (2007). Origins and palaeo-environments of Kalahari duricrusts in the Moshaweng dry valleys (Botswana) as detected by major and trace element composition. *Journal of African Earth Sciences*, 48, 199–221.

Kang, S., Dahe, Q., Mayewski, P.A., Wake, C.P. and Ren, J. (2001). Climatic and environmental records from the far east Rongbuk ice core, Mt. Qomolangma (Mt. Everest). *Episodes*, 24, 176–81.

Kaniewski, D., Paulissen, E., Van Campo, E., Al-Maqdissi, M., Bretschneider, J. and Van Leberge, K. (2008). Middle East coastal ecosystem response to middle-to-late Holocene abrupt climate changes. *Proceedings of the National Academy of Sciences*, 105, 13941–6.

Kapp, P., Pelletier, J.R., Rohrmann, A., Heermance, R., Russell, J. and Lin, D. (2011). Wind erosion in the Qaidam Basin, central Asia: Implications for tectonics, paleoclimates and the source of the Loess Plateau. *GSA Today*, 21, 4–10.

Kar, A. (1990). Megabarchanoids of the Thar: Their environment, morphology and relationship with longitudinal dunes. *Geographical Journal*, 156, 51–61.

Kar, A. (1993). Aeolian processes and bedforms in the Thar Desert. *Journal of Arid Environments*, 25, 83–96.

Kar, A. and Takeuchi, K. (2004). Yellow dust: An overview of research and felt needs. *Journal of Arid Environments*, 59, 167–87.

Karátson, D., Telbisz, T. and Wörner, G. (2012). Erosion rates and erosion pattern of Neogene to Quaternary stratovolcanoes in the western cordillera of the Central Andes: An SRTM DEM based analysis. *Geomorphology*, 139–140, 122–35.

Karimi, A., Khademi, H., Kehl, M. and Jalalian, A. (2009). Distribution, lithology and provenance of peridesert loess deposits in northeastern Iran. *Geoderma*, 148, 241–50.

Kasanin-Grubin, M. and Bryan, R. (2007). Lithological properties and weathering response on badland hillslopes. *Catena*, 70, 68–78.

Kaspari, S., Mayewski, P.A., Handley, M., Kang, S., Hou, S., Sneed, S., Maasch, K. and Qin, D. (2009). A high-resolution record of atmospheric dust composition and variability since A.D. 1650 from a Mount Everest ice core. *Journal of Climate*, 22, 3910–25.

Kastanja, M.M., Diekmann, B. and Henrich, R. (2006). Controls on carbonate and terrigenous deposition in the incipient Benguela upwelling system during the middle to late Miocene (ODP Sites 1085 and 1087). *Palaeogeography, Palaeoclimatology, Palaeoecology*, 241, 515–30.

Katsuki, A. and Kikuchi, M. (2011). Simulation of barchan dynamics with inter-dune sand streams. *New Journal of Physics*, 13, 063049.

Kazmierczak, J., Kempe, S., Kremer, B., López-Garciá, P., Moreira, D. and Tavera, R. (2011). Hydrochemistry and microbialites of the alkaline crater Lake Alchichica, Mexico. *Facies*, 57, 543–70.

Keatings, K., Tassie, G.J., Flower, R.J., Hassan, F.A., Hamdan, M.A.T., Hughes, M. and Arrowsmith, C. (2007). An examination of groundwater within the Hawara Pyramid, Egypt. *Geoarchaeology*, 22, 533–54.

Keefer, W.R., Whitney, J.W. and Taylor, E.M. (eds.) (2004). Quaternary paleoseismology and stratigraphy of the Yucca Mountain area, Nevada. *U.S. Geological Survey Professional Paper*, 1689.

Kehew, A.E., Milewski, A. and Soliman, F. (2010). Reconstructing an extreme flood from boulder transport and rainfall-runoff modelling: Wadi Isla, South Sinai, Egypt. *Global and Planetary Change*, 70, 64–75.

Kellogg, C.A. and Griffin, D.W. (2006). Aerobiology and the global transport of desert dust. *Trends in Ecological Evolution*, 21, 638–44.
Kemp, J. (2010). Downstream channel changes on a contracting, anabranching river: The Lachlan, southeastern Australia. *Geomorphology*, 121, 231–44.
Kemp, J. and Rhodes, E.J. (2010). Episodic fluvial activity of inland rivers in southeastern Australia: Palaeochannel systems and terraces of the Lachlan River. *Quaternary Science Reviews*, 29, 732–52.
Kempe, S. and Dirks, H. (2008). Layal Lakes, Saudi Arabia: The world-wide largest lacustrine gypsum tufas. *Acta Carsologica*, 37, 7–14.
Kempf, J. (2010). Morphotectonics and denudation in the landform history of the Namibian Great Escarpment landscapes. *Zeitschrift für Geomorphologie*, 54, 347–76.
Kenea, N.H., Ebinger, C.J. and Rex, D.C. (2001). Late Oligocene volcanism and extension in the Red Sea Hills, Sudan. *Journal of the Geological Society of London*, 158, 285–94.
Keppel, M.N., Clarke, J.D.A., Halhan, T., Love, A.J. and Werner, A.D. (2011). Mound springs in the arid Lake Eyre south region of South Australia: A new depositional tufa model and its controls. *Sedimentary Geology*, 240, 55–70.
Kes, A.S. and Fedorovich, B.A. (1976). Process of forming of aeolian dust in space and time. *23rd International Geographical Congress, Section*, 1, 174–7.
Kesel, R.H. (1977). Some aspects of the geomorphology of inselbergs in central Arizona, U.S.A. *Zeitschrift für Geomorphologie*, 21, 119–46.
Keyes, C.R. (1912). Deflative scheme of the geographic cycle in an arid climate. *Bulletin of the Geological Society of America*, 23, 537–62.
Keys, J.R. and Williams, K. (1981). Origin of crystalline, cold desert salts in the McMurdo Region, Antarctica. *Geochimica et Cosmochimica Acta*, 45, 2299–309.
Khadkikar, A.S., Chamyal, L.S. and Ramesh, R. (2000). The character and genesis of calcrete in late Quaternary alluvial deposits, Gujarat, western India, and its bearing on the interpretation of ancient climates. *Palaeogeography, Palaeoclimatology, Palaeoecology*, 162, 239–61.
Khalaf, F.I. and Al-Awadhi, J.M. (2012). Sedimentological and morphological characteristics of gypseous coastal nebkhas on Bubiyan Island, Kuwait, Arabian Gulf. *Journal of Arid Environments*, 82, 31–43.
Khalaf, F.I., Misak, R. and Al-Dousari, A. (1995). Sedimentological and morphological characteristics of some nebkha deposits in the northern coastal plain of Kuwait, Arabia. *Journal of Arid Environments*, 29, 267–92.
Khresat, S.A. and Taimeh, A.Y. (1998). Properties and characterization of vertisols developed on limestones in a semi-arid environment. *Journal of Arid Environments*, 40, 235–44.
Kidron, G.J. (1999). Altitude dependent dew and fog in the Negev Desert, Israel. *Agricultural and Forest Meteorology*, 96, 1–8.
Kidron, G.J. (2005). Angle and aspect dependent dew and fog precipitation in the Negev Desert. *Journal of Hydrology*, 301, 66–74.
Kidron, G.J. and Yair, A. (1997). Rainfall-runoff relationship over encrusted dune surfaces, Nizzama, Western Negev, Israel. *Earth Surface Processes and Landforms*, 22, 1169–84.
Kidron, G.J., Barzilay, E. and Sachs, E. (2000). Microclimate control upon sand microbiotic crusts, western Negev Desert, Israel. *Geomorphology*, 36, 1–18.
Kiepe, P. (1996). Cover and barrier effect of Cassia siamea hedgerows on soil conservation in semi-arid Kenya. *Soil Technology*, 9, 161–71.
Kim, C.S., Stack, D.H. and Rytuba, J.J. (2012). Fluvial transport and surface enrichment of arsenic in semi-arid mining regions: Examples from the Mojave Desert, California. *Journal of Environmental Monitoring*, 14, 1798–813.

Kim, J.H., Schneider, R.R., Mulitza, S. and Mueller, P.J. (2003). Reconstruction of SE trade-wind intensity based on sea-surface temperatures in the southeast Atlantic over the last 25 kyr. *Geophysical Research Letters*, 30, doi:10.1029/2003GL017557.

Kindermann, K., Bubenzer, O., Nussbaum, S., Riemer, H., Darius, F., Pöllath, N. and Smettan, U. (2006). Palaeoenvironment and Holocene land use of Djara, Western Desert of Egypt. *Quaternary Science Reviews*, 25, 1619–37.

Kindle, E.M. (1917). Some factors affecting the development of mud-cracks. *Journal of Geology*, 25, 135–44.

Kinlaw, A. (1999). A review of burrowing by semi-fossorial vertebrates in arid environments. *Journal of Arid Environments*, 41, 127–45.

Kirchner, G. (1996). Cavernous weathering in the Basin and Range area, southwestern USA and northern Mexico. *Zeitschrift für Geomorphologie*, Supplementband, 106, 73–97.

Kirkham, A. (1997). Shoreline evolution, aeolian deflation and anhydrite distribution of the Holocene, Abu Dhabi. *GeoArabia*, 2, 403–16.

Kirono, D.G.C., Kent, D.M., Hennessy, K.J. and Mpelasoka, F. (2011). Characteristics of Australian droughts under enhanced greenhouse conditions: Results from 14 global climate models. *Journal of Arid Environments*, 75, 566–75.

Klaer, W. (1993). Beobachtungen zu Formen der Verwitterung und Abtragung auf Sansteinfelsen und Graniten in der algerischen Sahara. *Würzburger Geographische Arbeiten*, 87, 121–32.

Klammer, C. (1982). Die Palaeowüste des Pantannal von Mato Grosso and die Pleistozane Klimageschichte der Brazilianischen Randtropen. *Zeitschrift für Geomorphologie*, 26, 393–416.

Kleeman, R. and Power, S.B. (2000). Modulation of ENSO variability on decadal and larger timescales. In *El Niño and the Southern Oscillation*, ed. H.F. Diaz and V. Markgraf. Cambridge: Cambridge University Press, pp. 413–41.

Klose, M., Shao, Y., Karremann, M.K. and Fink, A.H. (2010). Sahel dust zone and synoptic background. *Geophysical Research Letters*, 37, L09802, doi:10.1029?2010GL042816.

Knight, J. (2008). The environmental significance of ventifacts: A critical review. *Earth-Science Reviews*, 86, 89–105.

Knight, M., Thomas, D.S.G. and Wiggs, G.F.S. (2004). Challenges of calculating dunefield mobility over the 21st century. *Geomorphology*, 59, 197–213.

Knighton, D. and Nanson, G.C. (1994). Flow transmission along an arid zone anastomosing river, Cooper Creek, Australia. *Hydrological Processes*, 8, 137–54.

Knighton, D. and Nanson, G. (1997). Distinctiveness, diversity and uniqueness in arid zone river systems. In *Arid Zone Geomorphology*, ed. D.S.G. Thomas. Wiley: Chichester, pp. 185–203.

Knighton, D. and Nanson, G.C. (2000). Waterhole form and process in the anastomosing channel system of Cooper Creek, Australia. *Geomorphology*, 35, 101–17.

Knippertz, P. and Todd, M.C. (2010). The central west Saharan dust hot spot and its relation to African easterly waves and extratropical disturbances. *Journal of Geophysical Research*, 115, D12117, doi:10.1029/2009JD012819.

Knippertz, P. and Todd, M.C. (2012). Mineral dust aerosols over the Sahara: Meteorological controls on emission and transport and implications for modeling. *Reviews of Geophysics*, 50, RG1007/2012.

Kober, F., Ivy-Ochs, S., Schlunegger, F., Baur, H., Kubik, P.W. and Wieler, R. (2007). Denudation rates and a topography-driven rainfall threshold in northern Chile: Multiple cosmogenic nuclide data and sediment yield budgets. *Geomorphology*, 83, 97–120.

Koch, A. and Siegesmund, S. (2004). The combined effect of moisture and temperature on the anomalous expansion behaviour of marble. *Environmental Geology*, 46, 350–63.

Kocurek, G. (1991). Interpretation of ancient eolian sand dunes. *Annual Review of Earth and Planetary Science*, 19, 43–75.
Kocurek, G. and Ewing, R.C. (2005). Aeolian dune field self-organization – Implications for the formation of simple versus complex dune-field patterns. *Geomorphology*, 72, 94–105.
Kocurek, G. and Havholm, K. (1993). Eolian sequence stratigraphy – A conceptual framework. *Memoir of the American Association of Petroleum Geologists*, 58, 393–409.
Kocurek, G. and Lancaster, N. (1999). Aeolian system sediment state: Theory and Mojave Desert Kelso dune field example. *Sedimentology*, 46, 505–15.
Kocurek, G. and Nielson, J. (1986). Conditions favourable for the formation of warm-climate aeolian sand sheets. *Sedimentology*, 33, 795–816.
Kohfeld, K.E., and Harrison, S.P. (2003). Glacial-interglacial changes in dust deposition on the Chinese Loess Plateau. *Quaternary Science Reviews* 22, 1859–78.
Kok, J.F. and Renno, N.O. (2006). Enhancement of the emission of mineral dust aerosols by electric forces. *Geophysical Research Letters*, 33, L19S10, doi:10.1029/2006GL026284.
Kok, J.F. and Renno, N.O. (2008). Electrostatics in wind-blown sand. *Physical Review Letters*, 100, doi:10.1103/Phys.Rev.Lett.100.014501.
Kolla, V. and Biscaye, P.C. (1977). Distribution and origin of quartz in the sediments of the Indian Ocean. *Journal of Sedimentary Petrology*, 47, 642–9.
Kong, P., Na, C., Fink, D., Ding, L. and Huang, F. (2007). Erosion in northwest Tibet from *in-situ*-produced cosmogenic ^{10}Be and ^{26}Al in bedrock. *Earth Surface Processes and Landforms*, 32, 116–25.
Koren, I., Joseph, J.H. and Israelevich, P. (2003). Detection of dust plumes and their sources in northeastern Libya. *Canadian Journal of Remote Sensing*, 29, 792–6.
Koren, I., Kaufman, Y.J., Washington, R., Todd, M.C., Rudich, Y., Martins, J.V. and Rosenfeld, D. (2006). The Bodélé Depression: A single spot in the Sahara that provides most of the mineral dust to the Amazon forest. *Environmental Research Letters*, 1, doi:10.1088/1748–9326/1/1/014005.
Kosmas, C. and 22 others. (1997). The effect of land use on runoff and soil erosion rates under Mediterranean conditions. *Catena*, 29, 45–59.
Kotlyakov, V.M. (1991). The Aral Sea Basin: A critical environmental zone. *Moscow Environment*, 33, 4–9, 36–8.
Kotwicki, V. and Allan, R. (1998). La Nina de Australia – Contemporary and palaeo-hydrology of Lake Eyre. *Palaeogeography, Palaeoclimatology, Palaeoecology*, 144, 265–80.
Kotwicki, V. and Isdale, P. (1991). Hydrology of Lake Eyre, Australia: El Niño link. *Palaeogeography, Palaeoclimatology, Palaeoecology*, 84, 87–98.
Krapf, C.B.E., Stollhofen, H. and Stanistreet, I.G. (2003). Contrasting styles of ephemeral river systems and their interaction with dunes of the Skeleton Coast Erg (Namibia). *Quaternary International*, 104, 41–52.
Kravtsova, V.I. and Tarasenko, T.V. (2010). Space monitoring of Aral Sea degradation. *Water Resources*, 37, 285–96.
Krebs, U., Park, W. and Schneider, B. (2011). Pliocene aridification of Australia caused by tectonically induced weakening of the Indonesian throughflow. *Palaeogeography, Palaeoclimatology, Palaeoecology*, 309, 111–17.
Krenkel, E. (1928). *Geologie Afrikas*. Berlin: Reimer.
Krinner, G., Boucher, O. and Balkanski, Y. (2006). Ice-free glacial northern Asia due to dust deposition on snow. *Climate Dynamics*, 27, 613–25.
Krinner, G., Lezine, A.M., Braconnot, P., Sepulchre, P., Ramstein, G., Grenier, C. and Gouttevin, I. (2012). A reassessment of lake and wetland feedbacks on the north African Holocene climate. *Geophysical Research Letters*, 39, L07701, doi:10.1029/2012GL050992.

Krinsley, D. (1998). Models of rock varnish formation constrained by high resolution transmission electron microscopy. *Sedimentology*, 45, 711–25.

Krinsley, D., Dorn, R.I. and DiGregorio, B. (2009). Astrobiological implications of rock varnish in Tibet. *Astrobiology*, 9, 551–62.

Kronfeld, J., Vogel, J.C., Rosenthal, E. and Weinstein-Evron, M. (1998). Age and paleoclimatic implications of the Bet Shean travertines. *Quaternary Research*, 30, 298–303.

Kroonenberg, S.B., Rusakov, G.V. and Svitoch, A.A. (1997). The wandering of the Volga Delta: A response to rapid Caspian sea-level change. *Sedimentary Geology*, 107, 189–209.

Kröpelin, S. (1987). Palaeoclimatic evidence from early to mid-Holocene playas in the Gilf Kebir (southwest Egypt). *Palaeoecology of Africa*, 18, 189–208.

Kröpelin, S. (1993a). Zur Rekonstruktion der Spatquartaren Umwelt am Unteren Wadi Howar (Sudostliche Sahara/NW-Sudan). *Berliner Geographische Abhandlungen*, 54.

Kröpelin, S. (1993b). Geomorphology, landscape evolution and paleoclimates of southwest Egypt. *Catena*, 26, 31–65.

Kröpelin, S. (2002). Damage to natural and cultural heritage by petroleum exploration and desert tourism in the Messak Settafet (central Sahara, southwest Libya). In *Tides of the Desert*, ed. Jennerstrasse 8. Cologne: Heinrich-Barth-Institut, pp. 405–23.

Kroy, K., Fischer, S. and Obermayer, B. (2005). The shape of barchan dunes. *Journal of Physics: Condensed Matter*, 17, S1229–S1235.

Krumbein, W.E. (1969). Über den Einfluss der Mikroflora auf die exogene Dynamik (Verwitterung und Krustenbildung). *Geologische Rundschau*, 58, 333–65.

Kuhlman, K.R., Venkat, P., La Duc, M.T., Kuhlman, G.M. and McKay, C.P. (2008). Evidence of a microbial community associated with rock varnish at Yungay, Atacama Desert, Chile. *Journal of Geophysical Research*, 113, G04022, doi:10.1029/2007JG000677.

Kuhn, N.J., Yair, A. and Grubin, M.K. (2004). Spatial distribution of surface properties, runoff generation and landscape development in the Zin Valley badlands, northern Negev, Israel. *Earth Surface Processes and Landforms*, 29, 1417–30.

Kuper, R. and Kröpelin, S. (2006). Climate-controlled Holocene occupation in the Sahara: Motor of Africa's evolution. *Science*, 313, 803–7.

Kurosaki, Y. and Mikami, M. (2004). Effect of snow cover on threshold wind velocity of dust outbreak. *Geophysical Research Letters*, 31, article no. L03106.

Kurosaki, Y., Shinoda, M. and Mikami, M. (2011). What caused a recent increase in dust outbreaks over East Asia? *Geophysical Research Letters*, 38, L11702, doi:10.1029/2011GL047494.

Kurth, G., Phillips, F.M., Reheis, M.C., Redwine, J.L. and Paces, J.B. (2011). Cosmogenic nuclide and uranium-series dating of old, high shorelines in the western Great Basin, USA. *Bulletin of the Geological Society of America*, 123, 744–68.

Kurtz, H.D. and Netoff, D.I. (2001). Stabilization of friable sandstone surfaces in a desiccating, wind-abraded environment of south-central Utah by rock surface microorganisms. *Journal of Arid Environments*, 48, 89–100.

Kustu, M.D., Fan, Y., and Robock, A. (2010). Large-scale water cycle perturbation due to irrigation pumping in the US High Plains: A synthesis of observed streamflow changes. *Journal of Hydrology*, 390, 222–44.

Kutiel, P. and Lavee, H. (1999). Effect of slope aspect on soil and vegetation properties along an aridity transect. *Israel Journal of Plant Sciences*, 47, 169–78.

Kutzbach, J. and Street-Perrott, F.A. (1985). Milankovitch forcing of fluctuations in the level of tropical lakes from 18 to 0 kyr BP. *Nature*, 317, 130–4.

Kwaad, F. J. M. (1970). Experiments on the disintegration of granite by salt action. *University Amsterdam Fysisch Geografisch en Bodemkundig Laboratorium Publicatie*, 16, 67–80.

Labraga, J.C. and Villalba, R. (2008). Climate in the Monte Desert: Past trends, present conditions, and future projections. *Journal of Arid Environments*, 73, 154–63.
Lachenbruch, A.H. (1962). Mechanisms of thermal contraction cracks and ice-wedge polygons in permafrost. *Geological Society of America Special Paper* 70.
Lahlou, A. (1996). Environmental and socio-economic impacts of erosion and sedimentation in North Africa. *IAHS Publication*, 236, 491–500.
Laity, J.E. (1994). Landforms of aeolian erosion. In *Geomorphology of Desert Environments*, ed. A.D. Abrahams and A.J. Parsons. London: Chapman and Hall, pp. 506–35.
Laity, J.E. (2003). Aeolian destabilization along the Mojave River, Mojave Desert, California: Linkages among fluvial, groundwater, and aeolian systems. *Physical Geography*, 24, 196–221.
Laity, J.E. (2008). *Deserts and Desert Environments*. Chichester: Wiley-Blackwell.
Laity, J.E. and Bridges, N.T. (2009). Ventifacts on Earth and Mars: Analytical, field and laboratory studies supporting sand abrasion and windward feature development. *Geomorphology*, 105, 202–17.
Laity, J.E. and Malin, M.C. (1985). Sapping processes and the development of theatre-headed valley networks on the Colorado Plateau. *Bulletin of the Geological Society of America*, 96, 203–17.
Lakhdar, R., Soussi, M., Ismail, H.M.B. and M'Rabet, A. (2006). A Mediterranean Holocene restricted coastal lagoon under arid climate: Case of the sedimentary record of Sabkha Boujmel (SE Tunisia). *Palaeogeography, Palaeoclimatology, Palaeoecology*, 241, 177–91.
Lamb, M.P., Dietrich, W.E., Aciego, S.M., DePaolo, D.J. and Manga, M. (2008). Formation of Box Canyon, Idaho, by megaflood: Implications for seepage erosion on Earth and Mars. *Science*, 320, 1067–70.
Lamb, M.P., Howard, A.D., Johnson, J., Whipple, K.X., Dietrich, W.E. and Perron, J.T. (2006). Can springs cut canyons into rock? *Journal of Geophysical Research*, 111, E07002, doi:10.1029/2005JE002663.
Lamb, S., Hoke, L., Kenna, L. and Dewey, J. (1997). Cenozoic evolution of the Central Andes in Bolivia and northern Chile. *Geological Society of London Special Publication*, 121, 237–64.
Lambert, F. and 9 others (2008). Dust-climate couplings over the past 800,000 years from the EPICA Dome C ice core. *Nature*, 452, 616–19.
Lamotte, D.F. de, Leturmy, P., Missenard, Y., Khomis, S., Ruiz, G., Saddiqi, O., Guillocheau, F. and Michard, A. (2009). Mesozoic and Cenozoic vertical movements in the Atlas system (Algeria, Morocco, Tunisia): An overview. *Tectonophysics*, 475, 9–28.
Lamplugh, G.W. (1907). Geology of the Zambezi Basin around Batoka Gorge. *Quarterly Journal of the Geological Society of London*, 63, 162–216.
Lancaster, N. (1978). The pans of the southern Kalahari, Botswana. *Geographical Journal*, 144, 80–98.
Lancaster, N. (1988). Controls of eolian dune size and spacing. *Geology*, 16, 972–5.
Lancaster, N. (1989a). *The Namib Sand Sea*. Rotterdam: Balkema.
Lancaster, N. (1989b). Late Quaternary palaeoenvironments of the southern Kalahari. *Palaeogeography, Palaeoclimatology, Palaeoecology*, 70, 367–76.
Lancaster, N. (1995a). *Geomorphology of Desert Dunes*. London and New York: Routledge.
Lancaster, N. (1995b). Origin of the Gran Desierto Sand Sea: Sonora, Mexico: Evidence from dune morphology and sedimentology. In *Desert Aeolian Processes*, ed. V.P. Tchakerian. London: Chapman and Hall, pp. 11–35.
Lancaster, N. (1996). Desert environments. In *The Physical Geography of Africa*, ed. W.M. Adams, A.S. Goudie and A.R. Orme. Oxford: Oxford University Press, pp. 211–37.

Lancaster, N. (1997). Response of eolian geomorphic systems to minor climate change: Examples from the southern Californian deserts. *Geomorphology*, 19, 333–47.
Lancaster, N. (2006). Linear dunes on Titan. *Science*, 312, 702–3.
Lancaster, N. (2008). Desert dune dynamics and development: Insights from luminescence dating. *Boreas*, 37, 559–73.
Lancaster, N. (2009). Dune morphology and dynamics. In *Geomorphology of Desert Environments* (2nd ed.), ed. A.J. Parsons and A.D. Abrahams. Heidelberg: Springer, pp. 557–95.
Lancaster, N. and Mahan, S.A. (2012). Holocene dune formation at Ash Meadows National Wildlife Area, Nevada. *Quaternary Research*, 78, 266–74.
Lancaster, N. and Ollier, C.D. (1983). Sources of sand for the Namib Sand Sea. *Zeitschrift für Geomorphologie*, Supplementband, 45, 71–83.
Lancaster, N. and Tchakerian, V.P. (1996). Geomorphology and sediments of sand ramps in the Mojave Desert. *Geomorphology*, 17, 151–65.
Lancaster, J., Lancaster, N. and Seely, M.K. (1984). Climate of the central Namib Desert. *Madoqua*, 14, 5–61.
Lane, D.S. (2006). Thermal properties of aggregates. In *Significance of Tests and Properties of Concrete and Concrete-making Materials*, ed. J.F. Lamond and J.H. Pielert. West Conshohocken, PA: ASTM, pp. 425–31.
Langbein, W.B. and Schumm, S.A. (1958). Yield of sediment in relation to mean annual precipitation. *Transactions American Geophysical Union*, 39, 1076–84.
Langford, R.P. (2000). Nabkha (coppice dune) fields of south-central New Mexico, USA. *Journal of Arid Environments*, 46, 25–41.
Langford, R.P., Rose, J.M. and White, D.E. (2009). Groundwater salinity as a control on development of eolian landscape: An example from the White Sands of New Mexico. *Geomorphology*, 105, 39–49.
Langford-Smith, T. (1960). The dead river systems of the Murrumbidgee. *Geographical Review*, 50, 368–89.
Langford-Smith, T. (ed.). (1978). *Silcrete in Australia*. Armidale, New South Wales: Department of Geography, University of New England.
Langford-Smith, T. (1982). The geomorphic history of the Australian deserts. *Striae*, 17, 4–19.
Langston, G. and McKenna Neuman, C. (2005). An experimental study on the susceptibility of crusted surfaces to wind erosion: A comparison of the strength properties of biotic and salt crusts. *Geomorphology*, 72, 40–53.
Laronne, J.B. and Reid, I. (1993). Very high rates of bedload sediment transport by ephemeral desert rivers. *Nature*, 366, 148–50.
Lasanta, T., García-Ruez, J.M., Pérez-Rontomé, C. and Sancho-Marcén. (2000). Runoff and sediment yield in a semi-arid environment: The effect of land management after farmland abandonment. *Catena*, 38, 265–78.
Lateef, A.S.A. (1988). Distribution, provenance, age and palaeoclimatic record of the loess in central north Iran. In *Loess*, ed. D.N. Eden and R.J. Furkert. Rotterdam: Balkema, pp. 93–101.
Latif, M. and Keenlyside, N.S. (2009). El Nino/Southern Oscillation response to global warming. *Proceedings of the National Academy of Sciences*, 106, 20578–83.
Latrubesse, E.M. and Nelson, B.W. (2001). Evidence for late Quaternary aeolian activity in the Roraima-Guyana Region. *Catena*, 43, 63–80.
Lavenu, A., Fournier, M. and Sebrier, M. (1984). Existence de deux nouveaux episodes lacustraes quaternaires dans l'Altiplano peruvo-bolivien. *Cahiers ORSTOM, ser. Géologie*, 14, 103–14.

Lawrence, C.R. and Neff, J.C. (2009). The contemporary physical and chemical flux of aeolian dust: A synthesis of direct measurements of dust deposition. *Chemical Geology*, 267, 46–63.

Lawson, A.C. (1915). The epigene profiles of the desert. *University of California Department of Geology Bulletin*, 9, 23–48.

Lawson, M.P. and Thomas, D.S.G. (2002). Late Quaternary lunette dune sedimentation in the southwestern Kalahari Desert, South Africa: Luminescence based chronologies of aeolian activity. *Quaternary Science Reviews*, 21, 825–36.

Lazarenko, A.A. (1984). The loess of central Asia. In *Late Quaternary Environments of the Soviet Union*, ed. A. Velichko. Harlow: Longman, pp. 125–31.

Lázaro, R., Cantón, Y., Solé-Benet, A., Bevan, J., Alexander, R., Sancho, L.G. and Puigdefábregas, J. (2008). The influence of competition between lichen colonization and erosion on the evolution of soil surfaces in the Tabernas badlands (SE Spain) and its landscape effects. *Geomorphology*, 102, 252–66.

Lebedeva-Verba, M.P. and Gerasimova, M.I. (2010). Micromorphology of takyrs and the desert 'papyrus' of southwestern Turkmenia. *Eurasian Soil Science*, 43, 1220–9.

Lecce, S. (1991). Influence of lithographic erodibility on alluvial fan areas, western White Mountain, California and Nevada. *Earth Surface Processes and Landforms*, 16, 11–8.

Leduc, C., Favreau, G. and Schoeter, P. (2000). Long term rise in a Sahelian water-table: The Continental Terminal in south-west Niger. *Journal of Hydrology*, 243, 43–54.

Lee, C.F., Zhang, J.M. and Zhang, Y.X. (1996). Evolution and origin of the ground fissures in Xian, China. *Engineering Geology*, 43, 45–55.

Lee, J.A., Wigner, K.A. and Gregory, J.M. (1993). Drought, wind and blowing dust on the southern High Plains of the United States. *Physical Geography*, 14, 56–67.

Lee, J.A., Baddock, M.C., Mbuh, M.J. and Gill, T.E. (2012). Geomorphic and land cover characteristics of aeolian dust sources in west Texas and eastern New Mexico, USA. *Aeolian Research*, 3, 459–66.

Lee, Y.C., Yang, X. and Wenig, M. (2010). Transport of dusts from East Asian and non-East Asian sources to Hong Kong during dust storm related events, 1996–2007. *Atmospheric Environment*, 44, 3728–38.

Le Gall, A. and 7 others. (2012). Latitudinal and altitudinal controls of Titan's dune field morphometry. Icarus, 217, 231–42.

Le Heron, D.P., Armstrong, H.A., Wilson, C., Howard, J.P. and Gindre, L. (2010). Glaciation and deglaciation of the Libyan desert: The Late Ordovician record. *Sedimentary Geology*, 223, 100–25.

Lehmkuhl, F. (1997). The spatial distribution of loess and loess-like sediments in the mountain areas of central and high Asia. *Zeitschrift für Geomorphologie*, Supplementband, 111, 97–116.

Lehmkuhl, F., Klinge, M., Rees-Jones, J. and Rhodes, E.J. (2000). Late Quaternary aeolian sedimentation in central and south-east Tibet. *Quaternary International*, 68–71, 117–32.

Lei, J-Q. and 10 others. (2008). Classification and regionalization of the forming environment of windblown sand disasters along the Tarim Desert Highway. *Chinese Science Bulletin*, 53, Supplement 2, 1–7.

Leinen, M. and Heath, G.R. (1981). Sedimentary indicators of atmospheric activity in the northern hemisphere during the Cenozoic. *Palaeogeography, Palaeoclimatology, Palaeoecology*, 36, 1–21.

Lekouch, I., Muselli, M., Kabbachi, B., Ouazzani, J., Melnytchouk-Milimouk, I. and Beysens, D. (2010). Dew, fog and rain as supplementary sources of water in south-west Morocco. *Energy*, doi:10.1016/j.energy.2010.03.017.

Lele, V.S. (1974). The miliolite limestone at Saurashtra, western India. *Sedimentary Geology*, 10, 301–10.

Leopold, L.B. (1951). Rainfall frequency: An aspect of climatic variation. *Transactions of the American Geophysical Union*, 32, 347–57.

Leopold, M., Völkel, J. and Heine, K. (2006). A ground-penetrating radar survey of late Holocene fluvial sediments in NW Namibian river valleys: Characterization and comparison. *Journal of the Geological Society, London*, 163, 923–36.

Leprun, J.C. (1999). The influences of ecological factors on tiger bush and dotted bush patterns along a gradient from Mali to northern Burkina Faso. *Catena*, 37, 25–44.

Lesschen, J.P., Cammeraat, L.H. and Nieman, T. (2008). Erosion and terrace failure due to agricultural land abandonment in a semi-arid environment. *Earth Surface Processes and Landforms*, 33, 1574–84.

Le Tensorer, J.-M., Jagher, R., Rentzel, P., Hauck, T., Ismail-Meyer, K., Pümpin, C. and Wojtczak, D. (2007). Long-term site formation processes at the natural springs Nadaouiyeh and Hummal in the El Kown Oasis, central Syria. *Geoarchaeology*, 22, 621–39.

Letey, J. (2001). Causes and consequences of fire-induced soil water repellency. *Hydrological Processes*, 15, 2867–75.

Levy, J.S., Marchant, D.R. and Head, J.W. (2010). Thermal contraction crack polygons on Mars: A synthesis from HiRise, Phoenix, and terrestrial analog studies. *Icarus*, 206, 229–52.

Leys, J. (1999). Wind erosion on agricultural land. In *Aeolian Environments, Sediments and Landforms*, ed. A.S. Goudie, I. Livingstone and S. Stokes. Chichester: Wiley, pp. 143–66.

Leys, J., Heidenreich, S.K., Strong, C.L., McTainsh, G.H. and Quigley, S. (2011). PM_{10} concentrations and mass transport during 'Red Dawn' – Sydney 23 September 2009. *Aeolian Research*, 3, 327–42.

Lézine, A.-M. (2009). Timing of vegetation changes at the end of the Holocene Humid Period in desert areas at the northern edge of the Atlantic and Indian monsoon systems. *CR Geoscience*, 341, 750–9.

Lézine, A-M., Hély, C., Grenier, C., Braconnot, P. and Krinner, G. (2011). Sahara and Sahel vulnerability to climate changes, lessons from Holocene hydrological data. *Quaternary Science Reviews*, 30, 3001–12.

Lézine, A.-M. and 11 others. (2010). Climate change and human occupation in the southern Arabian lowlands during the last deglaciation and the Holocene. *Global and Planetary Change*, 72, 412–28.

Li, B.Y. and Zhu, L.P. (2001). 'Greatest lake period' and its palaeo-environment on the Tibetan Plateau. *Journal of Geographical Sciences (Acta Geographica Sinica)*, 11, 34–42.

Li, J., Okin, G.S. and Epstein, H.E. (2009). Effects of enhanced wind erosion on surface soil texture and characteristics of windblown sediments. *Journal of Geophysical Research*, 114, G02003, doi:10.1029/2008GJ000903.

Li, J., Xie, S.-P., Cook, E.R., Huang, G., D'Arrigo, R., Liu, F., Ma, J. and Zheng, X.-T. (2011). Interdeadal modulation of El Nino amplitude during the past millennium. *Nature Climate Change*, 1, 114–18.

Li, S.-H., Sun, J.-M. and Zhao, H. (2002). Optical dating of dune sands in the northeastern deserts of China. *Palaeogeography, Palaeoclimatology, Palaeoecology*, 181, 419–29.

Li, X., Ge, L., Dong, Y. and Chang, H.-C. (2010). Estimating the greatest dust storm in eastern Australia with MODIS. *Geoscience and Remote Sensing Symposium (IGARSS), 2010 IEEE International, Honolulu*, 1039–42.

Li, X.-Y. and Liu, L.-Y. (2003). Effect of gravel mulch on aeolian dust accumulation in the semiarid region of northwest China. *Soil and Tillage Research*, 70, 73–81.

Li, X.-Y., Liu, L.-Y. and Gong, J.-D. (2001). Influence of pebble mulch on soil erosion by wind and trapping capacity for windblown sediment. *Soil and Tillage Research*, 59, 137–42.

Li, Y., Cui, J., Zhangh, T., Okuro, T. and Drake, S. (2009). Effectiveness of sand-fixing measures on desert land restoration in Kerqin Sandy Land, northern China. *Ecological Engineering*, 35, 118–27.

Liao, H., Seinfeld, J.H., Adams, P.J. and Mickley, L.J. (2004). Global radiative forcing of coupled tropospheric ozone and aerosols in a unified general circulation model. *Journal of Geophysical Research*, 109, D16207, doi:10.1029/2003JD004456.

Lichvar, R.W., Ochs, W.R. and Gaines, S.M. (2008). Evaluation of surface features for delineating the ordinary high water boundary on playas in the arid western United States. *Wetlands*, 28, 68–80.

Lim, J., and Matsumoto, E. (2006). Bimodal grain-size distribution of aeolian quartz in a maar of Cheju Island, Korea, during the last 6500 years: Its flux variation and controlling factor. *Geophysical Research Letters*, 33, L21816, doi:10.1029/2006GL027432.

Lin, J.C., Broecker, W.S., Anderson, R.F., Hemming, S., Rubenstone, J.L. and Bonani, G. (1996). New 230^{Th}/U and 14C ages from Lake Lahontan carbonates, Nevada, USA, and a discussion of the origin of initial thorium. *Geochimica et Cosmochimica Acta*, 60, 2817–32.

Lin, J.C., Broecker, W.S., Hemming, S.R., Hajdas, I., Anderson, R.S., Smith, G.L., Kelley, M. and Bonani, G. (1998). A reassessment of U-Th and 14C ages for Late-Glacial high-frequency hydrological events at Searles Lake, California. *Quaternary Research*, 49, 11–23.

Linsenbarth, A. (1991). Sand desert forms in Libya – Remote sensing analysis. In *The Geology of Libya, Volume IV*, ed. M.J. Salem, A.M. Sbeta and M.R. Bakbak. Amsterdam: Elsevier, pp. 2235–55.

Lintern, M.J., Sheard, M. and Buller, N. (2011). The gold-in-calcrete anomaly at the ET gold prospect, Gawler Craton, South Australia. *Applied Geochemistry*, 26, 2027–43.

Lintern, M.J., Sheard, M.J. and Chivas, A.R. (2006). The source of pedogenic carbonate associated with gold-calcrete anomalies in the western Gawler Craton, South Australia. *Chemical Geology*, 235, 299–324.

Lisker, S., Vaks, A., Bar-Matthews, M., Porat, R. and Frumkin, A. (2009). Stromatolites in caves of the Dead Sea fault escarpment: Implication to latest Pleistocene lake levels and tectonic disturbances. *Quaternary Science Reviews*, 28, 80–92.

Littmann, T. (1991a.) Dust storm frequency in Asia: Climatic control and variability. *International Journal of Climatology*, 11, 393–412.

Littmann, T. (1991b). Rainfall, temperature and dust storm anomalies in the African Sahel. *Geographical Journal*, 157, 136–60.

Littmann, T. (1991c). Recent African dust deposition in West Germany – Sediment characteristics and climatological aspects. *Catena*, Supplement, 20, 57–73.

Liu, C. and Fu, G. (1996). The impact of climatic warming on hydrological regimes in China: An overview. In *Regional Hydrological Response to Climate Change*, ed. J.A.A. Jones, C. Liu, M.-K. Woo and H.-T. Kung. Dordrecht: Kluwer, pp. 133–51.

Liu, H., Han, X., Li, L., Huang, J., Liu, H. and Li, X. (2009). Grazing density effects on cover, species composition, and nitrogen fixation of biological soil crust in an Inner Mongolia steppe. *Rangeland Ecology and Management*, 62, 321–7.

Liu, T. and Broecker, W.S. (2008). Rock varnish microlamination dating of late Quaternary geomorphic features in the drylands of western USA. *Geomorphology*, 93, 501–23.

Liu, T. and Dorn, R.I. (1996). Understanding spatial variability in environmental changes in drylands with rock varnish microlaminations. *Annals of the Association of American Geographers*, 86, 187–212.

Liu, Z., Kutzbach, J. and Wu, L. (2000). Modeling climate shift of El Niño variability in the Holocene. *Geophysical Research Letters*, 27, 2265–8.

Livingston, R.A. (1989). The geological origins of the Great Sphinx and implications for its preservation. Abstract of paper presented to International Geological Congress, Washington, D.C., July 9–19, 1989.

Livingstone, D. (1857). *Missionary Travels and Researches in South Africa*. London: Murray.

Livingstone, I. (1986). Geomorphological significance of wind flow patterns over a Namib linear dune. In *Aeolian Geomorphology*, ed. W.G. Nickling. Boston: Allen and Unwin, pp. 87–112.

Livingstone, I. (1988). New models for the formation of linear sand dunes. *Geography*, 73, 105–15.

Livingstone, I. (1989). Monitoring change on a Namib linear dune. *Earth Surface Processes and Landforms*, 14, 317–32.

Livingstone, I. (1993). A decade of surface change on a Namib linear dune. *Earth Surface Processes and Landforms*, 18, 661–4.

Livingstone, I. (2003). A twenty-one-year record of surface change on a Namib linear dune. *Earth Surface Processes and Landforms*, 28, 1025–31.

Livingstone, I. and Warren, A. (1996). *Aeolian geomorphology: An Introduction*. Harlow: Longman.

Livingstone, I., Wiggs, G.F.S. and Weaver, C.M. (2007). Geomorphology of desert sand dunes: A review of recent progress. *Earth-Science Reviews*, 80, 239–57.

Livingstone, I., Bullard, J.E., Wiggs, G.F.S. and Thomas, D.S.G. (1999). Grain-size variation in the southwest Kalahari, southern Africa. *Journal of Sedimentary Research*, 69, 546–52.

Livingstone, I., Bristow, C., Bryant, R.G., Bullard, J., White, K., Wiggs, G.F.S., Baas, A.C.W., Bateman, M.D. and Thomas, D.S.G. (2010). The Namib Sand Sea digital database of aeolian dunes and key forcing variables. *Aeolian Research*, 2, 93–104.

Loáciga, H.A., Pedreros, D. and Roberts, D. (2001). Wildfire-streamflow interactions in a chaparral watershed. *Advances in Environmental Research*, 5, 295–305.

Logan, J. (2004). Laboratory and case studies of thermal cycling and stored strain on the stability of selected marbles. *Environmental Geology*, 46, 456–67.

Logan R.F. (1960). *The central Namib Desert, south west Africa*. Washington, DC: National Academy of Sciences/National Research Council Publication, 758.

Lohuizen-De Leeuw, J.E. Van. (1973). Moenjo Daro – A cause of common concern. *South Asian Archaeology 1973*. Leiden: Brill, pp. 1–11.

Lokier S. and Steuber, T. (2008). Quantification of carbonate-ramp sedimentation and progradation rates for the late Holocene Abu Dhabi shoreline. *Journal of Sedimentary Research*, 78, 423–31.

Lokier, S. and Steuber, T. (2009). Large-scale intertidal polygonal features of the Abu Dhabi coastline. *Sedimentology*, 56, 609–21.

Lomax, J., Hilgers, A. and Radtke, U. (2011). Palaeoenvironmental change recorded in the palaeodunefields of the western Murray Basin, South Australia – New data from single grain OSL-dating. *Quaternary Science Reviews*, 30, 723–6.

Londoño, A.C., Forman, S.L., Eichler, T. and Pierson, J. (2012). Episodic eolian deposition in the past ca. 50,000 years in the Alto Ilo dune field, southern Peru. *Palaeogeography, Palaeoclimatology, Palaeoecology*, 346/347, 12–14.

Long J.T. and Sharp R.P. (1964). Barchan dune movement in Imperial Valley, California. *Bulletin of the Geological Society of America*, 75, 149–56.

Lopez, B.R., Tinoco-Ojanguren, C., Bacilio, M., Mendoza, A. and Bashan, Y. (2012). Endophytic bacteria of the rock-dwelling cactus *Mammillaria fraileana* affect plant growth and mobilization of elements from rocks. *Environmental and Experimental Botany*, 81, 26–36.

Lopez, L.L. and Johnson, B.A. (1987). Morfologiá de cavidades en la costa de Chile Central. *Revista de Geografía Norte Grande*, 14, 21–34.

López, M.V., Sabre, M., Gracia, R., Arrúe, J.L. and Gomes, L. (1998). Tillage effects on soil surface conditions and dust emission by wind erosion in semiarid Aragón (NE Spain). *Soil and Tillage Research*, 45, 91–105.

Lorenz, R.D., Jackson, B. and Hayes, A. (2010). Racetrack and Bonnie Claire: Southwestern US playa lakes as analogs for Ontario Lacus, Titan. *Planetary and Space Science*, 58, 724–31.

Lorenz, W. (2002). The 'Calcreta Principal' at the border of the provinces of La Pampa and Buenos Aires. *Zeitschrift für Angewandte Geologie*, 48, 44–51.

Loubere, P. (2000). Marine control of biological production in the eastern equatorial Pacific Ocean. *Nature*, 406, 497–500.

Louderback, L.A., Grayson, D.K. and Llobera, M. (2010). Middle-Holocene climates and human populations in the Great Basin, western USA. *The Holocene*, 21, 366–73.

Lovegrove, B.G. and Siegfried, W.R. (1989). Spacing and origin(s) of mima-like earth mounds in the Cape Province of South Africa. *South African Journal of Science*, 85, 108–22.

Lowry, D.C. and Jennings, J.N. (1974). The Nullarbor karst, Australia. *Zeitschrift für Geomorphologie*, 18, 35–81.

Lu, H. and Shao, Y.P. (2001). Toward quantitative prediction of dust storms: An integrated wind erosion modeling system and its applications. *Environmental Modelling and Software*, 16, 233–49.

Lu, H., Burbank, D.W. and Li, Y. (2010). Alluvial sequence in the north piedmont of the Chinese Tian Shan over the past 550 kyr and its relationship to climate change. *Palaeogeography, Palaeoclimatology, Palaeoecology*, 285, 343–53.

Lu, H., Wang, X. and Li, L. (2010). Aeolian sediment evidence that global cooling has driven late Cenozoic stepwise aridification in central Asia. *Geological Society of London Special Publication*, 342, 29–44.

Lu, H., Mason, J.A., Stevens, T., Zhou, Y., Yi, S. and Miao, X. (2011). Response of surface processes to climatic change in the dunefields and Loess Plateau of North China during the late Quaternary. *Earth Surface Processes and Landforms*, 36, 1590–603.

Lukas, W. (1975). Betonzer störung durch SO_3 – Angriff uter bildung von Thaumasit und Woodfordit. *Cement and Concrete Research*, 5, 503–18.

Luna, M.C.M. de M., Parteli, E.J.R., Durán, O. and Herrmann, H.J. (2011). Model for the genesis of coastal dune fields with vegetation. *Geomorphology*, 129, 215–24.

Luo, C., Mahowald, N.M. and del Corral, J. (2003). Sensitivity study of meteorological parameters on mineral aerosol mobilization, transport and distribution. *Journal of Geophysical Research*, 108, doi:10.1029/2003JD003483.

Luo, W., Arvidson, R.E., Sultan, M., Becker, R., Crombie, M.K., Sturchio, N. and El Affy, Z. (1997). Groundwater-sapping processes, Western Desert, Egypt. *Bulletin of the Geological Society of America*, 109, 43–62.

Lupker, M., Aciego, S.M., Bourdon, B., Schwander, J. and Stocker, T.F. (2010). Isotopic tracing (Sd, Nd, U and Hf) of continental and marine aerosols in an 18th century section of the Dye-3 ice core (Greenland). *Earth and Planetary Science Letters*, 295, 277–86.

Luque, A., Ruiz-Agiudo, E., Cultrone, G., Sebastián, E. and Siegesmund, S. (2011). Direct observation of microcrack development in marble caused by thermal weathering. *Environmental Earth Science*, 62, 1375–86.

Lustig, L.K. (1969). Trend surface analysis of the Basin and Range Province. *U.S. Geological Survey Professional Paper*, 500-D.

Lyell, C. (1834). Observations on the loamy deposit called 'loess' of the basin of the Rhine. *Edinburgh New Philosophical Journal*, 17, 110–13, 118–20.

Lyles, L., Tatarko, J. and Dickerson, J.D. (1983). Windbreak effects on soil water and wheat yield. *American Society of Agricultural Engineers Paper*, 83, 69–72.

Mabbutt, J.A. (1955). Erosion surfaces in Namaqualand and the ages of surface deposits in the south-western Kalahari. *Transactions of the Geological Society of South Africa*, 58, 13–29.

Mabbutt, J.A. (1965). Stone distribution in a Stony Tableland soil. *Australian Journal of Soil Research*, 3, 131–42.

Mabbutt, J.A. (1966). Mantle-controlled planation of pediments. *American Journal of Science*, 264, 78–91.

Mabbutt, J.A. (1969). Landforms of arid Australia. In *Arid Lands of Australia*, ed. R.O. Slayter and R.A. Perry. Canberra: ANU Press, pp. 11–32.

Mabbutt, J.A. (1971). The Australian arid zone as a pre-historic environment. In *Aboriginal Man and Environment in Australia*, ed. D.J. Mulvaney and P.J. Golson. Canberra: ANU Press, pp. 66–79.

Mabbutt, J.A. (1977a). *Desert Landforms*. Cambridge, MA: MIT Press.

Mabbutt, J.A., (1977b). Desert lands. In *The Natural Environment*, ed. D.N. Jeans. Sydney: Sydney University Press, pp. 180–202.

Mabbutt, J.A. and Fanning, P.C. (1987). Vegetation bandings in arid western Australia. *Journal of Arid Environments*, 12, 41–59.

MacFadyen, W.A. (1950). Vegetation patterns in British Somaliland. *Nature*, 165, 121.

Machette, M.N. (1985). Calcic soils of the southwestern United States. *Geological Society of America Special Paper*, 203, 1–22.

Mackay, J.R. (1984). The frost heave of stones in the active layer above permafrost with downward and upward freezing. *Arctic and Alpine Research*, 16, 439–46.

Madella, M. and Fuller, D.Q. (2006). Palaeoecology and the Harappan Civilisation of South Asia: A reconsideration. *Quaternary Science Reviews*, 25, 1283–301.

Maestre, F.T. and 8 others. (2011). Ecology and functional roles of biological soil crusts in semi-arid ecosystems of Spain. *Journal of Arid Environments*, 75, 1282–91.

Magee, A.W., Bull, P.A. and Goudie, A.S. (1988). Chemical textures on quartz grains: An experimental approach using salts. *Earth Surface Processes and Landforms*, 13, 665–76.

Magilligan, F.J. and Goldstein, P.S. (2001). El Niño floods and culture change: A late Holocene flood history for the Rio Moquegua, southern Peru. *Geology*, 29, 431–4.

Magirl, C.S., Griffiths, P.G. and Webb, R.H. (2010). Analyzing debris flows with the statistically calibrated empirical model LAHARZ in southeastern Arizona, USA. *Geomorphology*, 119, 111–24.

Mahowald, N.M. and Luo, C. (2003). A less dusty future? *Geophysical Research Letters*, 30, doi:10.1029/2003GL017880.

Mahowald, N.M., Muhs, D.R., Levis, S., Rasch, P.J., Yoshioka, M., Zender, C.S. and Luo, C. (2006). Change in atmospheric mineral aerosols in response to climate: Last Glacial Period, preindustrial, modern and doubled carbon dioxide climates. *Journal of Geophysical Research*, 111 (D), doi:10.1029/2005JD006653.

Mahowald, N.M., Kohfeld, K., Hansson, M., Balkanski, Y., Harrison, S.P., Prentice, I.C., Schulz, M. and Rodhe, H. (1999). Dust sources and deposition during the Last Glacial Maximum and current climate: A comparison of model results with paleodata from ice cores and marine sediments. *Journal of Geophysical Research*, 104(D), 15895–916.

Mahowald, N.M., Baker, A.R., Bergametti, G., Brooks, N., Duce, R.A., Jickells, T.D., Kubilay, N., Prospero, J.M. and Tegen, I. (2005). Atmospheric global dust cycle and iron inputs to the ocean. *Global Biogeochemical Cycles*, 19, GB4025, doi:10.1029/2004GB002402.

Mahowald, N.M. and 17 others. (2008). Global distribution of atmospheric phosphorus sources, concentrations and deposition rates, and anthropogenic impacts. *Global Biogeochemical Cycles*, 22, GB4026, doi:10.1029/2008GB003240.

Mahowald, N.M. and 19 others. (2010). Observed 20th century desert dust variability: Impact on climate and biogeochemistry. *Atmospheric Chemistry and Physics*, 10, 10875–93.

Mainguet, M. (1972). *Le Modélé des Grès*. Paris: IGN.

Mainguet, M. (1975). Etude comparée des ergs à l'échelle continentale (Sahara et déserts d'Australie). *Bulletin de l'Association Géographie Français*, 25, 135–40.

Mainguet, M. (1977). Analyse quantitative de l'extrémité sahélien du courant éolien transporteur de sable au Sahara nigérien. *Comptes Rendus de l'Academie des Sciences, Paris*, 285, 1029–32.

Mainguet, M. (1983). Tentative mega-geomorphology study of the Sahara. In *Mega-geomorphology*. ed. R.A.M. Gardner and H. Scoging. Oxford: Oxford University Press, pp. 113–33.

Mainguet, M. (1984). A classification of dunes based on aeolian dynamics and the sand budget. In *Deserts and Arid Lands*, ed. F. El-Baz. The Hague: Martinus Nijhoff, pp. 31–58.

Mainguet, M. (1999). *Aridity, Droughts and Human Development*. Heidelberg: Springer.

Mainguet, M. and Chemin, M.-C. (1986). Wind system and sand dunes in the Taklamakan Desert (People's Republic of China). Paper presented at the Twentieth International Symposium on Remote Sensing of Environment, Nairobi, Kenya, 827–33.

Mainguet, M. and Chemin, M.-C. (1990). Le massif du Tibesti dans le système éolien du Sahara. *Berliner Geographische Studien*, 30, 261–76.

Mainguet, M. and Jacqueminent, C. (1984). Le Grand Erg Occidental et le Grand Erg Oriental. *Travaux de L'Institut de Géographie de Reims*, 59–60, 29–48.

Mainguet, M., Dumay, F., Elhacen, M.L.O. and Georges, J.-C. (2008). Changement de l'état de surface des ergs au nord de Nouakchott (1954–2000). Conséquences sur la désertification et l'ensablement de la capitale. *Géomorphologie*, 3, 144–52.

Maizels, J.K. (1987). Plio-Pleistocene raised channel systems of the western Sharquiya (Wahiba), Oman. *Geological Society of London Special Publication*, 35, 31–50.

Maizels, J.K. (1988). Palaeochannels: Plio-Pleistocene raised channel systems of the western Sharkiyah. *Journal of Oman Studies, Special Report*, 3, 95–112.

Malik, J.N., Khadikhar, A.S. and Merh, S.S. (1999). Allogenic control on late Quaternary continental sedimentation in the Mahi River Basin, western India, *Journal of the Geological Society of India*, 53, 299–314.

Malin, M.C. (1974). Salt weathering on Mars. *Journal of Geophysical Research*, 79, 3889–94.

Maman, S., Blumberg, D.G., Tsoar, H., Mamedov, B. and Porat, N. (2011). The central Asian ergs: A study by remote sensing and geographic information systems. *Aeolian Research*, 3, 353–66.

Mamedov, A.V. (1997). The late Pleistocene-Holocene history of the Caspian Sea. *Quaternary International*, 41–42, 161–6.

Manabe, S. and Wetherald, R.T. (1986). Reduction in summer soil wetness by an increase in atmospheric carbon dioxide. *Science*, 232, 626–8.

Mann, D.H. and Meltzer, D.J. (2007). Millennial-scale dynamics of valley fills over the past 12000 ^{14}C yr in northeastern New Mexico. *Bulletin of the Geological Society of America*, 119, 1433–48.

Mann, G.A., Clarke, J.D.A. and Gostin, V.A. (2004). Surveying for Mars analogue research sites in the central Australian deserts. *Australian Geographical Studies*, 42, 116–24.

Mao, R., Ho, C.-H., Shao, Y., Gong, D-Y. and Kim, J. (2011). Influence of Arctic Oscillation on dust activity over northeast Asia. *Atmospheric Environment*, 45, 326–37.
Maphangwa, K.W., Musil, C.F., Raitt, L. and Zedda, L. (2012). Differential interception and evaporation of fog, dew and water vapour and elemental accumulation by lichens explain their relative abundance in a coastal desert. *Journal of Arid Environments*, 82, 71–80.
Marcato, G., Bossi, G., Rivelli, F. and Borgatti, L. (2012). Debris flood hazard documentation and mitigation on the Tilcara alluvial fan (quebrada de Humahuaca, Jujuy Province, north-west Argentina). *Natural Hazards and Earth System Sciences*, 12, 1873–82.
Marchetti, D.W. and Cerling, T.E. (2005). Cosmogenic ^{3}He exposure ages of Pleistocene debris flows and desert pavements in Capitol Reef National Park, Utah. *Geomorphology*, 67, 423–35.
Márin, L., Forman, S.L., Valdez, A. and Bunch, F. (2005). Twentieth century dune migration at the Great Sand Dunes National Park and Preserve, Colorado, relation to drought variability. *Geomorphology*, 70, 163–83.
Marion, G.M., Schlesinger, W.H. and Fonteyn, P.J. (1985). CALDEP: A regional model for soil $CaCO_3$ (caliche) deposition in southwestern deserts. *Soil Science*, 139, 468–81.
Marker, M. (1982). Aspects of Namib geomorphology: A doline karst. *Palaeoecology of Africa*, 15, 187–99.
Marlow, J.R., Lange, C.B., Wefer, G. and Rosell-Melé, A. (2000). Upwelling intensification as part of the Pliocene-Pleistocene climate transition. *Science*, 290, 2288–91.
Marsh, G.P. (1864). *Man and Nature*. New York: Scribner.
Marston, R.A. and Dolan, L.S. (1999). Effectiveness of sediment control structures relative to spatial patterns of upland soil loss in an arid watershed, Wyoming. *Geomorphology*, 31, 313–23.
Marticorena, B. (2008). A tribute to Dale Gillette. *Journal of Geophysical Research*, 113, doi:10.1029/2007JF000785.
Marticorena, B. and Bergametti, G. (1995). Modeling the atmospheric dust cycle: 1. Design of a soil-derived dust emission scheme. *Journal of Geophysical Research*, 100 (D8), 16415–30.
Marticorena, B. and Bergametti, G. (1996). Two-years simulations of seasonal and interannual changes of the Saharan dust emissions. *Geophysical Research Letters*, 23, 1921–24.
Martin, H. (1963). A suggested theory for the origin and a brief description of some gypsum deposits of south west Africa. *Transactions Geological Society of South Africa*, 66, 345–50.
Martinéz Raya, A., Duran Zuazo, V.H. and Francia Martinez, J.R. (2006). Soil erosion and runoff response to plant-cover strips on semiarid slopes (SE Spain). *Land Degradation and Development*, 17, 1–11.
Martonne, E. de. (1927). Regions of interior drainage. *Geographical Review*, 17, 397–414.
Marx, S.K., Kamber, B.S. and McGowan, H.A. (2005a). Estimates of Australian dust flux into New Zealand: Quantifying the eastern Australian dust plume pathway using trace element calibrated ^{210}Pb as a monitor. *Earth and Planetary Science Letters*, 239, 336–51.
Marx, S.K., Kamber, B.S. and McGowan, H.A. (2005b). Provenance of long-travelled dust determined with ultra-trace-element composition: A pilot study with samples from New Zealand glaciers. *Earth Surface Processes and Landforms*, 30, 699–716.
Marx, S.K., Kamber, B.S., McGowan, H.A. and Denholm, J. (2011). Holocene dust deposition rates in Australia's Murray-Darling Basin record the interplay between aridity and the position of the mid-latitude westerlies. *Quaternary Science Reviews*, 30, 3290–305.

Mason, J.A. (2001). Transport direction of Peoria Loess in Nebraska and implications for loess sources on the central Great Plains. *Quaternary Research*, 56, 79–86.
Mason, J.A. and Kuzila, M.S. (2000). Episodic Holocene loess deposition in central Nebraska. *Quaternary International*, 67, 119–31.
Mason, J.A., Jacobs, P.M., Hanson, P.R., Miao, X. and Goble, R.J. (2003). Sources and paleoclimatic significance of Holocene Bignell loess, central Great Plains, USA. *Quaternary Research*, 60, 330–9.
Masoud, A.A. (2011). Runoff modeling of the wadi systems for estimating flash flood and groundwater recharge potential in southern Sinai, Egypt. *Arabian Journal of Geoscience*, 4, 785–801.
Masson, O., Piga, D., Gurriaran, R. and D'Amico, D. (2010). Impact of an exceptional Saharan dust outbreak in France: PM_{10} and artificial radionuclides in air and in dust deposit. *Atmospheric Environment*, 44, 2478–86.
Matmon, A. and 7 others. (2005). Landscape development in an hyperarid sandstone environment along the margins of the Dead Sea fault: Implications from dated rock falls. *Earth and Planetary Science Letters*, 240, 803–17.
Matmon, A. and 7 others. (2009). Desert pavement-coated surfaces in extreme deserts present the longest-lived landforms on earth. *Bulletin of the Geological Society of America*, 121, 688–97.
Matsuoka, N. (1995). Rock weathering processes and landform development in the Sør Rondane Mountains, Antarctica. *Geomorphology*, 12, 323–39.
Maurer, T., Herrmann, L. and Stahr, K. (2010). Wind erosion characteristics of Sahelian surface types. *Earth Surface Processes and Landforms*, 35, 1386–401.
Maxwell, T.A. (1982). Sand sheet and lag deposits in the southwestern desert. In *Desert Landforms of Southwest Egypt: A Basis for Comparison with Mars*, ed, F. El-Baz and T.A. Maxwell. Washington DC: NASA, pp. 157–73.
Maxwell, T.A. and Haynes, C.V. (2001). Sand sheet dynamics and Quaternary landscape evolution of the Selima Sand Sheet, southern Egypt. *Quaternary Science Reviews*, 20, 1623–47.
May, G., Hartley, A.J., Stuart F.M. and Chong, G. (1999). Tectonic signatures in arid continental basins: An example from the upper Miocene-Pleistocene Calama Basin, Andean forearc, northern Chile. *Palaeogeography, Palaeoclimatology, Palaeoecology*, 151, 55–77.
Mayer, M., McFadden, L.D. and Harden, J.W. (1988). Distribution of calcium carbonate in desert soils: A model. *Geology*, 16, 303–6.
Mazzoni, E. (2001). Space distribution and geomorphological characterization of small closed basins of the Extra-Andean Patagonia. *Anales del Instituto de la Patagonia, Serie Ciencias Naturales (Chile)*, 29, 5–24.
McBride, E.F. and Picard, M.D. (2000). Origin and development of tafoni in Tunnel Spring Tuff, Crystal Peak, Utah, USA. *Earth Surface Processes and Landforms*, 25, 869–79.
McCabe, S., Smith, B.J. and Warke, P.A. (2007). Sandstone response to salt weathering following simulated fire damage: A comparison of the effects of furnace heating and fire. *Earth Surface Processes and Landforms*, 32, 1874–83.
McCall, J. (2010). Lake Bogoria, Kenya: Hot and warm springs, geysers and Holocene stromatolites. *Earth-Science Reviews*, 103, 71–9.
McCarthy, T.S., Stannistreat, I.G., Cairncross, B., Ellery, W.N., Ellery, K., Oelofse, R. and Grobicki, T.S.A. (1988). Incremental aggradation on the Okavango Delta-fan, Botswana. *Geomorphology*, 1, 267–78.
McCauley, J.F., Breed, C.S. and Grolier, M.J. (1982a). The interplay of fluvial, mass-wasting and eolian processes in the Eastern Gilf Kebir region. In: *Desert Landforms of*

Southwest Egypt: A Basis for Comparison with Mars, ed. F. El-Bazs and T.A. Maxwell. Washington DC: NASA, pp. 207–39.

McCauley, J.F., Grolier, M.J. and Breed, C.S. (1977a). Yardangs. In *Geomorphology in Arid Regions*, ed. D.O. Doehring. *Proceedings 8th Annual Geomorphology Symposium*, pp. 233–69.

McCauley, J.F., Grolier, M.J. and Breed, C.S. (1977b). Yardangs of Peru and other desert regions. *USGS Interagency Report: Astrogeology*, 81.

McCauley, J.F., Schaber, G.G., Breed, C.S., Grollier, M.J., Haynes, C.V., Issawi, B., Elachi, C. and Blom, R. (1982b). Subsurface valleys and geoarchaeology of the eastern Sahara revealed by Shuttle radar. *Science*, 218, 1004–20.

McConnell, J.R., Aristarain, A.J., Banta, J.R., Edwards, P.R. and Simões, J.C. (2007). 20th-century doubling in dust archived in an Antarctic Peninsula ice core parallels climate change and desertification in South America. *Proceedings National Academy of Sciences*, 104, 5743–8.

McCoy, Z.L. (2011). The distribution and origin of silcrete in the Ogallala Formation, Garza County, Texas. PhD diss., Texas Tech University, Lubbock.

McDonald, A.K., Kinucan, R.J. and Loomis, L.E. (2009). Ecohydrological interactions within banded vegetation in the northeastern Chihuahuan Desert, USA. *Ecohydrology*, 2, 66–71.

McFadden, L.D. and McAuliffe, J.R. (1997). Lithologically influenced geomorphic responses to Holocene climatic changes in the southern Colorado Plateau, Arizona: A soil-geomorphic and ecologic perspective. *Geomorphology*, 19, 303–32.

McFadden, L.D. and Tinsley, J.C. (1985). Rate and depth of pedogenic carbonate accumulation in soils: Formulation and testing of a compartment model. *Geological Society of America Special Paper*, 203, 23–41.

McFadden, L.D., Wells, S.G. and Jercinovich, M.J. (1987). Influences of eolian and pedogenic processes on the origin and evolution of desert pavements. *Geology*, 15, 504–8.

McFadden, L.D., Eppes, M.C., Gillespie, A.R. and Hallet, B. (2005). Physical weathering in arid landscapes due to diurnal variation in the direction of solar heating. *Bulletin of the Geological Society of America*, 117, 161–73.

McFadden, L.D., McDonald, E.V., Wells, S.G., Anderson, K., Quake, J. and Forman, S.L. (1998). The vesicular layer and carbonate collars of desert soils and pavements: Formation, age and relation to climate change. *Geomorphology*, 24, 101–45.

McFarlane, M.J. and Twidale, C.R. (1987). Karstic features associated with tropical weathering profiles. *Zeitschrift für Geomorphologie*, Supplementband, 64, 73–95.

McGee, D., Broecker, W.S. and Winckler, G. (2010). Gustiness: The driver of glacial dustiness? *Quaternary Science Reviews*, 29, 2340–50.

McGee, D., Marcantonio, F. and Lynch-Stieglitz, J. (2007). Deglacial changes in dust flux into the eastern equatorial Pacific. *Earth and Planetary Science Letters*, 257, 215–30.

McGee, WJ. (1897). Sheetflood erosion. *Bulletin of the Geological Society of America*, 8, 87–112.

McGreevy, J.P. (1982). 'Frost and salt' weathering: Further experimental results. *Earth Surface Processes and Landforms*, 7, 475–88.

McInnis, C. and Whiting, J.D. (1979). The frost resistance of concrete subjected to a deicing agent. *Cement and Concrete Research*, 9, 325–36.

McIntosh, R.J. (1983). Floodplain geomorphology and human occupation of the upper Inland Delta of the Niger. *Geographical Journal*, 149, 182–201.

McKay, C.P., Molaro, J.L. and Marinova, M.M. (2009). High-frequency rock temperature data from hyper-arid and desert environments in the Atacama and the Antarctic Dry Valleys and implications for rock weathering. *Geomorphology*, 110, 182–7.

McKee, E.D. (ed.). (1979). A study of global sand seas. U.S. Geological Survey Professional Paper, **1052**.

McKee E.D. and Moiola R.J. (1975). Geometry and growth of the White Sands Dune Field. *Journal of Research of the United States Geological Survey*, 3, 59–66.

McKendrick, K.J. and McFadden, L.D. (1996). Comparison and contrast of processes of soil formation in the San Timoteo badlands with chronosequences in California. *Quaternary Research*, 46, 149–60.

McKendry, I.G., Strawbridge, K.B., O'Neill, N.T., Macdonald, A.M., Liu, P.S.K., Leaitch, W.R., Anlauf, K.G., Jaegle, L., Fairlie, T.D. and Westphal, D.L. (2007). Trans-Pacific transport of Saharan dust to western North America: A case study. *Journal of Geophysical Research*, 112, D01103, doi:10.1029/2009JD007129.

McKenna, W. (2007). An evolutionary model of parabolic dune development: From blowout to mature parabolic, Padre Island National Seashore, Texas. MSc thesis, Louisiana State University.

McKenna-Neuman, C., Lancaster, N. and Nickling, W.G. (1997). Relations between dune morphology, air flow, and sediment flux on reversing dunes, Silver Peak, Nevada. *Sedimentology*, 44, 1103–13.

McKenna-Neuman, C., Lancaster, N. and Nickling, W.G. (2000). The effect of unsteady winds on sediment transport on the stoss slope of a transverse dune, Silver Peak, NV, USA. *Sedimentology*, 47, 211–26.

McKenna-Neuman, C., Maxwell, C.D. and Bouton, J.W. (1996). Wind transport of sand surface crusted with photoautotrophic micro organisms. *Catena*, 27, 229–47.

McKirdy, D.M., Hayball, A.J., Warren, J.K., Edwards, D. and Borch, Ch.C. von der (2010). Organic facies of Holocene carbonates in North Stromatolite Lake, Coorong region, South Australia. *Cadernos Laboratorio Xeoloxico de Laxe, Coruna*, 35, 127–46.

McLaren, S.J. (2004). Characteristics, evolution and distribution of Quaternary channel calcretes, southern Jordan. *Earth Surface Processes and Landforms*, 29, 1487–507.

McLaren, S.J. (2007). Aeolianite. In *Geochemical Sediments and Landscapes*, ed. D.J. Nash and S.J. McLaren. Blackwell: Oxford, pp. 144–72.

McLaren, S. and Wallace, M.W. (2010). Plio-Pleistocene climate change and the onset of aridity in southeastern Australia. *Global and Planetary Change*, 71, 55–72.

McLaren, S., Wallace, M.W. and Reynolds, T. (2012). The Late Pleistocene evolution of palaeo megalake Bungunnia, southeastern Australia: A sedimentary record of fluctuating lake dynamics, climate change and the formation of the modern Murray River. *Palaeogeography, Palaeoclimatology, Palaeoecology*, 317–18, 114–27.

McLaren, S., Wallace, M.W., Gallagher, S.J., Miranda, J.A., Holdgate, G.R., Gow, L.J., Snowball, I. and Sandgren, P. (2011). Palaeogeographic, climatic and tectonic change in southeastern Australia: The late Neogene evolution of the Murray Basin. *Quaternary Science Reviews*, 30, 1086–111.

McLure, H.A. (1976). Radiocarbon chronology of late Quaternary lakes in the Arabian desert. *Nature*, 263, 755–6.

McMahon, H. (1906). Recent survey and exploration in Seistan. *Geographical Journal*, 28, 209–28.

McMahon, T.A., Murphy, R.E., Peel, M.C., Costelloe, J.F. and Chiew, F.H.S. (2008a). Understanding the surface hydrology of the Lake Eyre Basin: Part 1 – Rainfall. *Journal of Arid Environments*, 72, 1853–68.

McMahon, T.A., Murphy, R.E., Peel, M.C., Costelloe, J.F. and Chiew, F.H.S. (2008b). Understanding the surface hydrology of the Lake Eyre Basin: Part 2 – Streamflow. *Journal of Arid Environments*, 72, 1869–86.

McQueen, K.G., Hill, S.M. and Foster, K.A. (1999). The nature and distribution of regolith carbonate accumulations in southeastern Australia and their potential as a sampling medium in geochemical exploration. *Journal of Geochemical Exploration*, 67, 67–82.
McTainsh, G.H. (1985). Dust processes in Australia and west Africa: A comparison. *Search*, 16, 104–6.
McTainsh, G.H. (1987). Desert loess in northern Nigeria. *Zeitschrift für Geomorphologie*, NF 31, 145–65.
McTainsh, G.H. (1989). Quaternary aeolian dust processes and sediments in the Australian region. *Quaternary Science Reviews*, 8, 235–53.
McTainsh, G.H. (1999). Dust transport and deposition. In *Aeolian Environments, Sediments and Landforms*, ed. A.S. Goudie, I. Livingstone and S. Stokes. Chichester: Wiley, pp. 181–211.
McTainsh, G.H. and Pitblado, J.R. (1987). Dust storms and related phenomena measured from meteorological records in Australia. *Earth Surface Processes and Landforms*, 12, 415–24.
McTainsh, G.H., Leys, J., Baker, M., Tews, K. and Barton, K. (2010). *Wind Erosion Histories, Model Input Data and Community DustWatch*. Brisbane: Griffith University.
Meadows, A. and Meadows, P. (eds.). (1999). *The Indus River: Biodiversity, Resources, Humankind*. Karachi: Oxford University Press.
Meadows, D.G., Young, M.H. and MacDonald, E.V. (2008). Influence of relative surface age on hydraulic properties and infiltration on soils associated with desert pavements. *Catena*, 72, 169–78.
Meadows, P.S. and Meadows, A. (1999). The environmental impact of the River Indus on the coastal and offshore zones of the Arabian Sea and the north-west Indian Ocean. In *The Indus River: Biodiversity, resources, humankind*, ed. A. Meadows and P.S. Meadows. Karachi: Oxford University Press.
Meckelein, W. (1959). *Forschungen in der Zentralen Sahara*. Braunschweig: Westermann.
Mehta, P.K. (1983). Mechanism of sulphate attack on Portland cement concrete – Another look. *Cement and Concrete Research*, 13, 401–6.
Meigs, P. (1953). World distribution of arid and semi-arid homoclimates. *UNESCO Arid Zone Research Series*, 1, 203–9.
Mellor, A., Short, J. and Kirkby, S.J. (1997). Tafoni in the El Chorro area, Andalucia, southern Spain. *Earth Surfaces Processes and Landforms*, 22, 817–33.
Melton, F.A. (1934). Linear and dendritic sink-hole patterns in southeastern New Mexico. *Science*, 80, 123–4.
Melton, M.A. (1957). An analysis of the relations among elements of climate, surface properties and geomorphology. *Office of Naval Research, New York, Report* NR 389–042.
Mendez, M.O. and Maier, R.M. (2008). Phytostabilization of mine tailings in arid and semiarid environments – An emerging remediation technology. *Environmental Health Perspectives*, 116, 278–83.
Menéndez, I., Díaz-Hernández, J.L., Mangas, J., Alonso, I. and Sánchez-Soto, P.J. (2007). Airborne dust accumulation and soil development in the north-east sector of Gran Canaria (Canary Islands, Spain). *Journal of Arid Environments*, 71, 57–81.
Mercer, J.H. (1976). Glacial history of southernmost South America. *Quaternary Research*, 6, 125–66.
Merrill, G.P. (1897). *A Treatise on Rocks, Rock-Weathering and Soils*. New York: Macmillan.
Merrill, S.D., Black, A.L., Fryrear, D.W., Saleh, A., Zobeck, T.M., Halvorson, A.D. and Tanaka, D.L. (1999). Soil wind erosion hazard of spring wheat-fallow as affected by long-term climate and tillage. *Journal of the Soil Science Society of America*, 63, 1768–77.

Merrison, J.P. (2012). Sand transport, erosion and granular electrification. *Aeolian Research*, 4, 1–16.

Miao, X., Mason, J.A., Goble, R.J. and Hanson, P.R. (2005). Loess record of dry climate and aeolian activity in the early- to mid-Holocene central Great Plains, North America. *The Holocene*, 15, 339–46.

Miao, X., Mason, J.A., Swinehart, J.B., Loope, D.B., Hanson, P.R., Goble, R.J. and Liu, X. (2007). A 10,000 year record of dune activity, dust storms, and severe drought in the central Great Plains. *Geology*, 35, 119–22.

Miao, Y., Herrmann, M., Wu, F., Yan, X. and Yang, S. (2012). What controlled mid-late Miocene long-term aridification in central Asia? Global cooling or Tibetan Plateau uplift: A review. *Earth-Science Reviews*, 112, 155–72.

Michalski, G., Savarino, J., Böhlke, J.K. and Thiemens, M. (2002). Determination of the total oxygen isotopic composition of nitrate and the calibration of a Δ^{17} O nitrate reference material. *Analytical Chemistry*, 74, 4989–93.

Michel, P. (1973). *Les Bassins des Fleuves Sénégal et Gambie, Etude Géomorphologique.* Paris: Mémoire ORSTOM, 3 vols.

Michels, K., Sivakumar, M.V.K. and Allison, B.E. (1995). Wind erosion control using crop residue: I. Effects on soil flux and soil properties. *Field Crops Research*, 40, 101–10.

Micklin, P.P. (1972). Dimensions of the Caspian Sea problem. *Soviet Geography*, 13, 589–603.

Middleton, N.J. (1984). Dust storms in Australia: Frequency, distribution and seasonality. *Search*, 15, 46–7.

Middleton, N.J. (1985). Effect of drought on dust production in the Sahel. *Nature*, 316, 431–4.

Middleton, N.J. (1986a). A geography of dust storms in south-west Asia. *Journal of Climatology*, 6, 183–96.

Middleton N.J. (1986b). The geography of dust storms. D.Phil. thesis, University of Oxford.

Middleton, N.J. (1986c). Dust storms in the Middle East. *Journal of Arid Environments*, 10, 83–96.

Middleton, N.J. (1990). Wind erosion and dust storm prevention. In *Desert Reclamation*, ed. A.S. Goudie. Chichester: Wiley, pp. 87–108.

Middleton, N.J. (1991). Dust storms in the Mongolian People's Republic. *Journal of Arid Environments*, 20, 287–97.

Middleton, N.J. (2001). *Going to Extremes: Mud, Sweat and Frozen Tears.* London: Channel 4 Books.

Middleton, N.J. and Goudie, A.S. (2001). Saharan dust: Sources and trajectories. *Transactions of the Institute of British Geographers*, NS 26, 165–81.

Middleton, N.J. and Thomas, D.S.G. (1997). *World Atlas of Desertification* (2nd ed.). London: Arnold.

Migón, P. and Goudie, A. (2000). Granite landforms of the central Namib. *Acta Universitatis Carolinae 2000 Geographica*, 35, Supplementum, 17–38.

Milana, J.P. (2009). Largest wind ripples on Earth? *Geology*, 37, 343–6.

Mildrexler, D.J., Zhao, M. amd Running, S.W. (2011). Satellite finds highest land skin temperature on Earth. *Bulletin of the American Meteorological Society*, 92, 855–60.

Miller, C.R., James, N.P. and Bone, Y. (2012). Prolonged carbonate diagensis under an evolving late Cenozoic climate; Nullarbor Plain, southern Australia. *Sedimentary Geology*, 261–262, 33–49.

Miller, D.A., Schmidt, K.M., Mahan, S.A., McGeehin, J.P., Owen, L.A., Barron, J.A., Lehmkuhl, F. and Löhrer, R. (2009). Holocene landscape response to seasonality of storms in the Mojave Desert. *Quaternary International*, 215, 45–61.

Miller, J. (1937). Drainage lines in bas-relief. *Journal of Geology*, 45, 432–8.

Miller, K.J. (ed.) (1984). *The International Karakoram Project*. Cambridge: Cambridge University Press, 2 vols.
Miller, L. and Douglas, B.C. (2004). Mass and volume contribution to twentieth-century global sea level rise. *Nature*, 428, 406–9.
Miller, R.L. and Tegen, I. (1998). Climate response to soil dust aerosols. *Journal of Climate*, 11, 3247–67.
Miller, R.L., Tegen, I. and Perlwitz, J. (2004). Surface radiative forcing by soil dust aerosols and the hydrological cycle. *Journal of Geophysical Research*, 109, (D), doi:10.1029/2003JD004085.
Miller, R.McG., Pickford, M. and Senut, B. (2010). The geology, palaeontology and evolution of the Etosha Pan, Namibia: Implications for terminal Kalahari deposition. *South African Journal of Geology*, 113, 307–34.
Miller, S.D. and 7 others. (2008). Haboob dust storms of the southern Arabian Peninsula. *Journal of Geophysical Research*, 113, D01202, doi:10.1029/2007JD008550.
Mills, J. (1998). Relating micrometer-scale weathering to landform morphology at Papago Park, Phoenix, Arizona. *Journal of the Arizona-Nevada Academy of Science*, 31, 117–26.
Miner, R.E., Nelson, S.T., Tingey, D.G. and Murrell, M.T. (2007). Using fossil spring deposits in the Death Valley region, USA, to evaluate palaeoflowpaths. *Journal of Quaternary Science*, 22, 373–86.
Miri, A., Moghaddamnia, A., Pahlavanravi, A. and Panjehkek, N. (2010). Dust storm frequency after the 1999 drought in the Sistan region, Iran. *Climate Research*, 41, 83–90.
Mishra, D.C, Singh, B, Tiwari, V.M., Gupta, S.B. and Rao, M.B.S.V. (2000). Two cases of continental collisions and related tectonics during the Proterozoic period in India – Insights from gravity modelling constrained by seismic and magnetotelluric studies. *Precambrian Research*, 99, 149–69.
Misra, A., Pande, D., Kumar, K.R., Nanda, L.K., Maithani, P.S. and Chaki, A. (2011). Calcrete-hosted surficial uranium occurrences in playa-lake environment at Lachhri, Nagaur District, Rajasthan, India. *Current Science*, 101, 84–8.
Mitasova, H., Overton, M. and Harmon, R.S. (2005). Geospatial analysis of a coastal sand dune field evolution: Jockey's Ridge, North Carolina. *Geomorphology*, 72, 204–21.
Mitchell C.W. and Willimott, S.G. (1974). Dayas of the Moroccan Sahara and other arid regions. *Geographical Journal*, 140, 441–53.
Mitchell, J. and Westaway, R. (1999). Chronology of Neogene and Quaternary uplift and magmatism in the Caucasus: Constraints from K-Ar dating of volcanism in Armenia. *Tectonophysics*, 304, 157–86.
Mitchell, R.M., Campbell, S.K. and Qin, Y. (2010). Recent increase in aerosol loading over the Australian arid zone. *Atmospheric Chemistry and Physics*, 10, 1689–99.
Moglen, G.E., Eltahir, E.A.B. and Bras, R.L. (1998). On the sensitivity of drainage density to climate change. *Water Resources Research*, 34, 855–62.
Mohamedou, A.O., Aventurier, A., Barbiero, L., Caruba, R. and Valles, V. (1999). Geochemistry of clay dunes and associated pan in the Senegal delta, Mauritania. *Arid Soil Research and Rehabilitation*, 13, 265–80.
Molaro, J.L. and McKay, C.P. (2010). Processes controlling rapid temperature variations on rock surfaces. *Earth Surface Processes and Landforms*, 35, 501–7.
Molnar, P. (2001). Climate change, flooding in arid environments, and erosion rates. *Geology*, 29, 1071–4.
Molnar, P. and Rajagopalan, B. (2012). Late Miocene upward and outward growth of eastern Tibet and decreasing monsoon rainfall over the northwestern Indian subcontinent since ~10 Ma. *Geophysical Research Letters*, 39, L09702, doi:10.1029/2012GL051305.

Molnar, P., England, P. and Martinod, J. (1993). Mantle dynamics, uplift of the Tibetan Plateau and its margins. *Review of Geophysics*, 31, 357–96.

Mooney, H.A. and Parsons, D.J. (1973). Structure and function of the California Chaparral – An example from San Dimas. *Ecological Studies*, 7, 83–112.

Moore, A.E. (1999). A reappraisal of epeirogenic flexure axes in southern Africa. *South African Journal of Geology*, 102, 363–76.

Moores, J.E., Pelletier, J.D. and Smith, P.H. (2008). Crack propagation by differential insolation on desert surface clasts. *Geomorphology*, 102, 472–81.

Morales, M.S. and 8 others. (2012). Precipitation changes in the South American Altiplano since 1300 AD reconstructed by tree-rings. *Climate Past*, 8, 653–66.

Moreiras, S.M. (2005). Landslide susceptibility zonation in the Rio Mendoza Valley, Argentina. *Geomorphology*, 66, 345–57.

Moreno, A. and Canals, M. (2004). The role of dust in abrupt climate change: Insights from offshore northwest Africa and Alboran Sea sediment records. *Contributions to Science*, 2, 485–98.

Moreno, A., Targarona, J., Henderiks, J., Canals, M., Freudenthal, T. and Meggers, H. (2001). Orbital forcing of dust supply to the North Canary Basin over the last 250 kyr. *Quaternary Science Reviews*, 20, 1327–39.

Morgan, G.A., Head, J.W., Forget, F., Madeleine, J.-B. and Spiga, A. (2010). Gully formation on Mars: Two recent phases of formation suggested by links between morphology, slope orientation and insolation history. *Icarus*, 208, 658–66.

Morgan, R.P.C. (1995). *Soil erosion and Conservation* (2nd ed.). Harlow: Longman.

Morozova, G.S. (2005). A review of Holocene avulsions of the Tigris and Euphrates Rivers and possible effects on the evolutions of civilizations in Lower Mesopotamia. *Geoarchaeology*, 20, 401–23.

Morrison, R.B. (1991). *Quaternary geology of the southern Basin and Range Province. In Quaternary Nonglacial Geology: Coterminous U.S.*. Boulder: Geological Society of America. Volume K-2, pp. 353–371.

Mortensen, H. (1933). Die 'Salzprengung' and ihre bedeutung für die regionalklimatische Gliederung der Wüsten. *Petermanns Geographische Mitteilungen*, 79, 130–35.

Mountney, N.P. (2011). A stratigraphic model to account for complexity in aeolian dune and interdune successions. *Sedimentology*, doi:10.111/j.1365–3091.2011.01287x.

Moursy, F.I., Gaber, A.I. and Samak, M. (2002). Sand drift potential in El-Hanka area, Egypt. *Water, Air and Soil Pollution*, 136, 225–42.

Moussa, A.M.A., Kantiranis, N., Voudouris, K.S., Startis, J.A., Ali, M.F. and Christaras, V. (2009). The impact of soluble salts on the deterioration of Pharaonic and Coptic wall paintings at Al Qurna, Egypt: Mineralogy and chemistry. *Archaeometry*, 51, 292–308.

Mudd, G.M. (2000). Mound springs of the Great Artesian Basin in South Australia: A case study from Olympic Dam. *Environmental Geology*, 39, 463–76.

Mudd, S.M. (2006). Investigation of the hydrodynamics of flash floods in ephemeral channels: Scaling analysis and simulation using a shock-capturing flow model incorporating the effects of transmission losses. *Journal of Hydrology*, 324, 65–79.

Muhs, D.R. (2004). Mineralogical maturity in dunefields of North America, Africa and Australia. *Geomorphology*, 59, 247–69.

Muhs, D.R. and Budahn, J.R. (2009). Geochemical evidence for African dust and volcanic dust inputs to terra rossa soils on carbonate reef terraces, northern Jamaica, West Indies. *Quaternary International*, 196, 13–35.

Muhs, D.R. and Holliday V.T. (1995). Evidence of active dune sand on the Great Plains in the 19th century from accounts of early explorers. *Quaternary Research*, 43, 198–208.

Muhs, D.R. and Maat, P.B. (1993). The potential response of eolian sands to greenhouse warming and precipitation reduction on the Great Plains of the United States. *Journal of Arid Environments*, 25, 351–61.

Muhs, D.R. and Wolfe, S.A. (1999). Sand dunes of the northern Great Plains of Canada and the United States. *Geological Survey of Canada Bulletin*, 534, 183–97.

Muhs, D.R., Budahn, J.R., Prospero, J.M. and Carey, S.N. (2007b). Geochemical evidence for African dust inputs to soils of western Atlantic islands: Barbados, the Bahamas and Florida. *Journal of Geophysical Research*, 112, F02009, doi:10.1029/2005JF000445.

Muhs, D.R., Bush, C.A., Cowherd, S.D. and Mahan S. (1995). Geomorphic and geochemical evidence for the source of sand in the Algodones Dunes, Colorado Desert, southeastern California. In *Desert Aeolian Processes*, ed. V.P. Tchakerian. London: Chapman and Hall, pp. 37–74.

Muhs, D.R., Reynolds, R.L., Been, J. and Skipp, G. (2003). Eolian sand transport pathways in the southwestern United States: Importance of the Colorado River and local sources. *Quaternary International*, 104, 3–18.

Muhs, D.R., Budahn, J., Avila, A., Skipp, R., Freeman, J. and Patterson, D. (2010a). The role of African dust in the formation of Quaternary soils on Mallorca, Spain, and implications for the genesis of Red Mediterranean soils. *Quaternary Science Reviews*, 29, 2518–43.

Muhs, D.R., Budahn, J., Reheis, M., Beann, J., Skipp, G. and Fisher, E. (2007a). Airborne dust transport to the eastern Pacific Ocean off southern California: Evidence from San Clemente Island. *Journal of Geophysical Research*, 112, D13203, doi:10.1029/2006JD007577.

Muhs, D.R., Budahn, J., Skipp, G., Prospero, J.M., Paytterson, D. and Bettis, E.A. (2010b). Geochemical and mineralogical evidence for Sahara and Sahel dust additions to Quaternary soils on Lanzarote, eastern Canary Islands, Spain. *Terra Nova*, 22, 399–410.

Muhs, D.R., Swinehart, J.B., Loope, D.B., Been, J. Mahan, S.A. and Bush, C.A. (2000). Geochemical evidence for an eolian sand dam across the North and South Platte Rivers in Nebraska. *Quaternary Research*, 53, 214–22.

Muhs, D.R., Aleinikoff, J.N., Stafford, T.W., Kihl, R., Been, J., Mahan, S.A and Cowherd S. (1999). Late Quaternary loess in north-eastern Colorado: Part 1 – Age and paleoclimatic significance. *Bulletin of the Geological Society of America*, 111, 1861–75.

Muhs, D.R., Stafford T.W., Been J., Mahan S.A., Burdett, J., Skipp G. and Rowland Z.M. (1997). Holocene eolian activity in the Minot dune field, North Dakota. *Canadian Journal of Earth Science*, 34, 1442–59.

Muhs, D.R. and 7 others. (1996). Origin of the late Quaternary dune fields of northeastern Colorado. *Geomorphology*, 17, 129–49.

Muhs, D.R. and 7 others. (1997). Late Holocene eolian activity in the mineralogically mature Nebraska Sand Hills. *Quaternary Research*, 48, 162–76.

Mulitza, S. and 10 others. (2010). Increase in African dust flux at the onset of commercial agriculture in the Sahel region. *Nature*, 466, 226–8.

Muller, A., Reason, C.J.C. and Fauchereau, N. (2008). Extreme rainfall in the Namib Desert during late summer 2006 and influences of regional ocean variability. *International Journal of Climatology*, 28, 1061–70.

Munson, S., Belnap, J. and Okin, G.S. (2011). Responses of wind erosion to climate-induced vegetation changes on the Colorado Plateau. *Proceedings of the National Academy of Sciences*, 108, 3854–9.

Munthe, J. (2002). *Arches of the Escalante Canyons and Kaiporowits Plateau*. Thompson, UT: Arch Hunter Books.

Mustoe, G.E. (1982). The origin of honeycomb weathering. *Geological Society of America, Bulletin*, 93, 108–15.

Muturi, G.M., Mohren, G.M.J. and Kimani, J.N. (2009). Prediction of Prosopis species invasion in Kenya using geographical information system techniques. *African Journal of Ecology*, 48, 628–36.

Nagy, B., Nagy, L.A., Rigali, M.J., Jones, W.D., Krinsley, D.H. and Sinclair, N. (1991). Rock varnish in the Sonoran Desert: Microbiologically mediated accumulation of manganiferous sediments. *Sedimentology*, 38, 1153–71.

Nahon, D.B. (1986). Evolution of iron crusts in tropical landscapes. In *Rates of Chemical Weathering of Rocks and Minerals*, ed. S.M. Colman and D.P. Dethier. London: Academic Press, pp. 169–91.

Naiman, Z., Quade, J. and Patchett, J. (2000). Isotopic evidence for eolian recycling of pedogenic carbonate and variations in carbonate dust sources throughout the southwest United States. *Geochimica et Cosmochimica Acta*, 64, 3099–109.

Namsaraev, Z.B. and 7 others. (2010). Water regime and variation in hydrochemical characteristics of the Soda Salt Lake Khilganta (southeastern Transbaikalia). *Water Resources*, 37, 513–19.

Nanson, G.C. and Tooth, S. (1999). Arid-zone rivers as indicators of climate change. In *Paleoenvironmental Reconstruction in Arid Lands*, ed. A.K. Singhvi and E. Derbyshire. New Delhi and Calcutta: Oxford and IBH, pp. 75–216.

Nanson, G.C., Callen, R.A. and Price, D.M. (1998). Hydroclimatic interpretation of Quaternary shorelines on South Australian playas. *Palaeogeography, Palaeoclimatology, Palaeoecology*, 144, 281–305.

Nanson, G.C., Chen, X.Y. and Price, D.M. (1992). Lateral migration, thermoluminescence chronology and colour variation of longitudinal dunes near Birdsville in the Simpson Desert, central Australia. *Earth Surface Processes and Landforms*, 17, 807–19.

Nanson, G.C., Chen, X.Y. and Price, D.M. (1995). Aeolian and fluvial evidence of changing climate and wind patterns during the past 100 Ka in the western Simpson Desert, Australia. *Palaeogeography, Palaeoclimatology, Palaeoecology*, 113, 87–102.

Nanson, G.C., Price, D.M. and Short, S.A. (1992). Wetting and drying of Australia over the past 300 Ka. *Geology*, 20, 791–4.

Nanson, G. C., Tooth, S. and Knighton, A. D. (2002). *A global perspective on dryland rivers: perceptions, misconceptions and distinctions*. Chichester: John Wiley and Sons.

Nash, C.R. (1972). Primary anhydrite in Precambrian gneisses from the Swakopmund District, South West Africa. *Contributions to Mineralogy and Petrology*, 36, 27–32.

Nash, D.J. (1996). On the dry valleys of the Kalahari: Documentary evidence for environmental change in central southern Africa. *Geographical Journal*, 162, 154–68.

Nash, D.J. (1997). Groundwater as a geomorphological agent in drylands. In *Arid Zone Geomorphology*, ed. D.S.G. Thomas. Chichester: Wiley, pp. 319–40.

Nash, D.J. and McLaren, S.J. (2003). Kalahari valley calcretes: Their nature, origins and environmental significance. *Quaternary International*, 111, 3–22.

Nash, D.J. and Shaw, P.A. (1998). Silica and carbonate relationships in silcrete-calcrete intergrade duricrusts from the Kalahari of Botswana and Namibia. *Journal of African Earth Sciences*, 27, 11–25.

Nash, D.J. and Ullyott, S. (2007). Silcrete. In *Geochemical Sediments and Landscapes*, ed. D.J. Nash and S.J. McLaren. Oxford: Blackwell, pp. 95–143.

Nash, D.J., McLaren, S.J. and Webb, J.A. (2004). Petrology, geochemistry, and environmental significance of silcrete-calcrete intergrade duricrusts at Kang Pan and Tswaane, central Kalahari, Botswana. *Earth Surface Processes and Landforms*, 29, 1559–86.

Nash, D.J., Shaw, P.A. and Thomas, D.S.G. (1994). Duricrust development and valley evolution: Process landform links in the Kalahari. *Earth Surface Processes and Landforms*, 19, 299–317.

Nash, D.J., Thomas, D.S.G. and Shaw, P.A. (1994). Timescales, environmental change and dryland valley development. In *Environmental Change in Drylands*, ed. A.C. Millington and K. Pye. Chichester: Wiley, pp. 25–41.

Natsagdorj, D.L., Jugder, D. and Chung, Y.S. (2003). Analysis of dust storms observed in Mongolia during 1937–1999. *Atmospheric Environment*, 37, 1401–11.

Naudé, Y., van Rooyen, M.W. and Rohwer, E.R. (2011). Evidence for a geochemical origin of the mysterious circles in the Pro-Namib desert. *Journal of Arid Environments*, 75, 446–56.

Navarro, M.C., Pérez-Sirvent, C., Martinez-Sánchez, M.J., Vidal, J., Tovar, P.J. and Bech, J. (2008). Abandoned mine sites as a source of contamination by heavy metals: A case study in a semi-arid zone. *Journal of Geochemical Exploration*, 96, 183–93.

Neal, J.T. (1965). Geology, mineralogy and hydrology of US playas. Environmental Research Paper, 96, U.S. Air Force Cambridge Research Laboratories, Bedford, Massachusetts.

Neal, J.T. (1968). Playa surface changes at Harper Lake, California, 1962–67. Environmental Research Paper, 283, U.S. Air Force Cambridge Research Laboratories, Bedford, Massachusetts.

Neave, M. and Abrahams, A.D. (2001). Impact of small mammal disturbances on sediment yield from grassland and shrubland ecosystems in the Chihuahuan Desert. *Catena*, 44, 285–303.

Needham, C.E. (1937). Ventifacts from New Mexico. *Journal of Sedimentary Petrology*, 7, 31–3.

Neev, D. and Emery, K.O. (1995). *The Destruction of Sodom, Gomorrah and Jericho*. New York: Oxford University Press.

Neff, J.C. and 9 others. (2008). Increasing eolian dust deposition in the western United States linked to human activity. *Nature Geoscience*, 1, 189–95.

Neff, U., Burns, S.J., Mangini, A., Mudelsee, M. et al. (2001). Strong coherence between solar variability and the monsoon in Oman between 9 and 6 kyr ago. *Nature*, 411, 290–3.

Nelson, S.T., Wood, M.J., Mayo, A.L., Tingey, D.G. and Eggett, D. (2005). Shoreline tufa and tufaglomerate from Pleistocene Lake Bonneville, Utah, USA: Stable isotopic and mineralogical records of lake conditions, processes and climate. *Journal of Quaternary Science*, 20, 3–19.

Nelson, S.T., Karlsson, H.R., Paces, J.B., Tingey, D.G., Ward, S. and Peters, M.T. (2001). Paleohydrologic record of spring deposits in and around Pleistocene pluvial Lake Tecopa, southeastern California. *Bulletin of the Geological Society of America*, 113, 659–70.

Netoff, D.I. and Shroba, R.R. (2001). Conical sandstone landforms cored with clastic pipes in Glen Canyon National Recreation Area, southeastern Utah. *Geomorphology*, 39, 99–110.

Netoff, D.I., Cooper, B.J. and Shroba, R.R. (1995). Giant sandstone weathering pits near Cookie Jar Butte, southeastern Utah. *Transactions and Proceedings Series NPS/NRNAU/NRJP*, 95/11, 25–53.

Netterberg, F. (1967). Some roadmaking properties of South African calcretes. *Proceedings 4th Regional Conference on Soil Mechanisms and Foundation Engineering*, 77–81.

Netterberg, F. (1971). Calcrete in road construction. *National Institute of Road Research Bulletin*, 10, CSIR, Pretoria.

Nettleton, W.D. and Chadwick, O.A. (1996). Late Quaternary, redeposited loess-soil developmental sequences, South Yemen. *Geoderma*, 70, 21–36.

Nettleton, W.D., Nelson, R.E., Brasjer, B.R. and Derr, P.S. (1982). Gypsiferous soils of the western United States. *Soil Science Society of Arizona Special Publication*, 10, 147–68.

Newberry, J.S. (1861). *Colorado exploring expedition*. Washington D.C.: Secretary of War.

Newman, B.D., Vivoni, E.R. and Groffman, A.R. (2006). Surface water – Groundwater interactions in semiarid drainages of the American Southwest. *Hydrological Processes*, 20, 3371–94.

Newsome, D. and Ladd, P. (1999). The use of quartz grain microtextures in the study of the origin of sand terrains in Western Australia. *Catena*, 35, 1–17.

Newsome, D. and Walden, J. (2000). Mineral magnetic evidence for heterogeneous sandplain regolith in Western Australia. *Journal of Arid Environments*, 45, 139–50.

Newton, M.S. and Grossman, E.L. (1988). Late Quaternary chronology of tufa deposits, Walker Lake, Nevada. *Journal of Geology*, 96, 417–33.

Nichols, G. (2009). Fluvial systems in desiccating endorheic baains. In *Sedimentary Processes, Environments and Basins*, ed. G. Nichols, E. Williams and C. Paola. doi:10.1002/9781444304411.ch23.

Nichols, M.H. (2006). Measured sediment yield rates from semiarid rangeland watersheds. *Rangeland Ecology and Management*, 59, 55–62.

Nicholson, S.E. (1996). Environmental change within the historical period. In *The Physical Geography of Africa*, ed. W.M. Adams, A.S. Goudie and A.R. Orme. Oxford: Oxford University Press, pp. 60–87.

Nicholson, S.E. (1998). Historical fluctuations of Lake Victoria and other lakes in the northern rift valley of East Africa. In *Environmental Change and Response in East African Lakes*, ed. J.T. Lehman. Dordrecht: Kluwer, pp. 7–35.

Nicholson, S.E. (2011). *Dryland Climatology*. Cambridge: Cambridge University Press.

Nickling, W.G. (1994). Aeolian sediment transport and deposition. In *Sediment Transport and Depositional Processes*, ed. K. Pye. Oxford: Blackwell Scientific Publications, pp. 293–350.

Nickling, W.G. and Gillies, J.A. (1989). Emission of fine-grained particulates from desert soils. In *Paleoclimatology and Paleometeorology: Modern and Past Patterns of Global Atmospheric Transport*, ed. M. Leinen and M. Sarnthein, Dordrecht: Kluwer, pp. 123–65.

Nickling, W.G. and McKenna-Neuman, C. (1999). Recent investigations of airflow and sediment transport over desert dunes. In *Aeolian Environments, Sediments and Landforms*, ed. A.S. Goudie, I. Livingstone, and S. Stokes. Chichester: Wiley, pp. 15–47.

Nickling, W.G. and Wolfe, S.A. (1994). The morphology and origin of nabkhas, region of Mopti, Mali, West Africa. *Journal of Arid Environments*, 28, 13–30.

Nicoll, K. (2001). Radiocarbon chronologies for prehistoric human occupation and hydroclimatic change in Egypt and northern Sudan. *Geoarchaeology*, 16, 47–64.

Nicoll, K., Giegengack, R. and Kleindeienst, M. (1999). Petrogenesis of artefact-bearing fossil-spring tufa deposits from Kharga Oasis, Egypt. *Geoarchaeology*, 14, 849–63.

Nield, J.M. and Baas, A.C.W. (2008). The influence of different environmental and climatic conditions on vegetated aeolian dune landscape development and response. *Global and Planetary Change*, 64, 76–92.

Nielson, J. and Kocurek, G. (1986). Climbing zibars of the Algodones. *Sedimentary Geology*, 48, 1–15.

Nielson, J. and Kocurek, G. (1987). Surface processes, deposits, and development of star dunes: Dumont dune field, California. *Bulletin of the Geological Society of America*, 99, 177–86.

Nigam, S., Guan, B. and Ruiz-Barradas, A. (2011). Key role of the Atlantic Multidecadal Oscillation in 20th century drought and wet periods over the Great Plains. *Geophysical Research Letters*, 38, L16713, doi:10.1029/2011GL048650.

Nilson, E. and Lehmkuhl, F. (2001). Interpreting temporal patterns in the late Quaternary dust flux from Asia to the North Pacific. *Quaternary International*, 76–77, 67–76.

Nishiizumi, K., Caffee, M.W., Finkel, R.C., Brimhall, G. and Mote, T. (2005). Remnants of a fossil alluvial fan landscape in the Atacama Desert of northern Chile using cosmogenic nuclide exposure dating. *Earth and Planetary Science Letters*, 237, 499–507.

Nordstrom, K.F. and Hotta, S. (2004). Wind erosion from cropland in the USA: A review of problems, solutions and prospects. *Geoderma*, 121, 157–67.

Norwick, S.A. and Dexter, L.R. (2002). Rates of development of tafoni in the Moenkopi and Kaibab formations in Meteor Crater and the Colorado Plateau, northeast Arizona. *Earth Surface Processes and Landforms*, 27, 11–26.

Nouaouria, M.S., Guenfoud, M. and Lafifi, B. (2008). Engineering properties of loess in Algeria. *Engineering Geology*, 99, 85–90.

Nyamweru, C.K. and Bowman, D. (1989). Climatic changes in the Chalbi Desert, northern Kenya. *Journal of Quaternary Science*, 4, 131–9.

Oba, G., Post, E. and Stenseth, N.C. (2001). Sub-Saharan desertification and productivity are linked to hemispheric climate variability. *Global Change Biology*, 7, 241–6.

Oberlander, T.M. (1977). Origin of segmented cliffs in massive sandstones of southeastern Utah. In *Geomorphology in Arid Regions*, ed. D.O. Doering. London: Allen and Unwin, pp. 79–114.

Oberlander, T.M. (1994). Global deserts: A geomorphic comparison. In *Geomorphology of Desert Environments*, ed. A.D. Abrahams and A.J. Parsons. London: Chapman and Hall, pp. 13–35.

Oberlander, T. M. (1997). Slope and pediment systems. In *Arid Zone Geomorphology: Process, Form and Change in Drylands*. ed. D.S.G. Thomas. Chichester: John Wiley and Sons, pp. 135–63.

O'Brien, G.R., Kaufman, D.S., Sharp, W.D., Atudorei, V., Parnell, R.A. and Crossey, L.J. (2006). Oxygen isotope composition of annually banded modern and mid-Holocene travertine and evidence of paleomonsoon floods, Grand Canyon, USA. *Quaternary Research*, 65, 366–79.

O'Connor, J.E., Ely, L.L., Wohl, E.E., Stevens, L.E., Melis, T.S., Kale, V.S. and Baker, V.R. (1994). A 4500-year record of large floods on the Colorado River in the Grand Canyon, Arizona. *Journal of Geology*, 102, 1–9.

Ogden, R., Spooner, N., Reid, M. and Head, J. (2001). Sediment dates with implications for the age of the conversion from palaeochannel to modern fluvial activity on the Murray River and its tributaries. *Quaternary International*, 83–85, 195–209.

Oglesby, R., Feng, S., Hu, Q. and Rowe, C. (2012). The role of the Atlantic multidecadal oscillation on medieval drought in North America: Synthesizing results from proxy data and climate models. *Global and Planetary Change*, 84–85, 56–65.

Ogunjobi, K.O., Oluleye, A. and Ajayi, V.O. (2012). A long-term record of aerosol index from TOMS observations and horizontal visibility in sub-Saharan West Africa. *International Journal of Remote Sensing*, 33, 6076–93.

O'Hara, S. L., Clarke, M. L. and Elatrash, M. S. (2006). Field measurements of desert dust deposition in Libya. *Atmospheric Environment*, 40, 3881–97.

O'Hara, S.L., Wiggs, G., Mamedov, B., Davidson, G. and Hubbard, R.B. (2000). Exposure to airborne dust contaminated with pesticide in the Aral Sea region. The Lancet, 355, 627–8.

O'Hara-Dhand, K., Taylor, R.L.S., Smalley, I.J., Krinsley, D.H. and Vita-Finzi, C. (2010). Loess and dust on Earth and Mars: Particle generation by impact mechanisms. *Central European Journal of Geosciences*, 2, 45–51.

Ohde, T. (2009). Investigation of hydrogen sulphide eruptions along the Namibian coastline using different remote sensing systems. *Central European Journal of Geosciences*, 1, 340–6.

Ohde, T., Siegel, H., Reissmann, J. and Gerth, M. (2007). Identification and investigation of sulphur plumes along the Namibian coast using the MERIS sensor. *Continental Shelf Research*, 27, 744–56.

Oke, A.M.C., Dunkerley, D. and Tapper, N.J. (2007b). Willy-willies in the Australian landscape: Sediment transport characteristics. *Journal of Arid Environments*, 71, 216–28.

Oke, A.M.C., Tapper, N.J. and Dunkerley, D. (2007a). Willy-willies in the Australian landscape: The role of key meteorological variables and surface conditions in defining frequency and spatial characteristics. *Journal of Arid Environments*, 71, 201–15.

Olivier, J. (1995). Spatial distribution of fog in the Namib. *Journal of Arid Environments*, 29, 129–38.

Ollier, C.D. (1963). Insolation weathering: Examples from central Australia. *American Journal of Science*, 261, 376–81.

Ollier, C.D. (1965). Dirt cracking – A type of insolation weathering. *Australian Journal of Science*, 27, 236–7.

Ollier, C.D. (1966). Desert gilgai. *Nature*, 212, 581–3.

Ollier, C.D. (1988). The regolith in Australia. *Earth-Science Reviews*, 25, 355–62.

Ollier, C.D. (1991). Aspects of silcrete formation in Australia. *Zeitschrift für Geomorphologie*, 35, 151–63.

Ollier, C.D. and Ash, J.E. (1983). Fire and rock breakdown. *Zeitschrift für Geomorphologie*, 27, 363–74.

Orlovsky, L., Orlovsky, N. and Durdyev, A. (2005). Dust storms in Turkmenistan. *Journal of Arid Environments*, 60, 83–97.

Orme, A.R. (2008). Pleistocene pluvial lakes of the American West: A short history of research. In *History of Geomorphology and Quaternary Geology*, ed. R.H. Grapes, D. Oldroyd and A. Grigelis. *Geological Society of London, Special Publication*, 301, 51–78.

Orts, W.J., Roa-Espinosa, A., Sojka, R.E., Glenn, G.M., Imam, S.H., Erlacher, K. and Pedersen, J.S. (2007). Use of synthetic polymers and biopolymers for soil stabilization in agricultural, construction and military applications. *Journal of Materials in Civil Engineering*, 19, 58–66.

Osborne, A.H., Vance, D., Rohling, E.J., Barton, N., Rogerson, M. and Fello, N. (2008). A humid corridor across the Sahara for the migration of early modern humans out of Africa 120,000 years ago. *Proceedings of the National Academy of Sciences*, 105, 16444–7.

Osterkamp, W.R. and Costa, J.E. (1987). Changes accompanying an extraordinary flood on a sand-bed stream. In *Catastrophic Flooding*, ed. L. Mayer and D. Nash. Boston: Allen and Unwin, pp. 201–24.

Osterkamp, W.R. and Wood, W.W. (1987). Playa-lake basins of the southern High Plains of Texas and New Mexico: Part 1. Hydrologic, geomorphic and geologic evidence for their development. *Bulletin of the Geological Society of America*, 99, 215–23.

Oviatt, C.G., Currey, D.R. and Sack, D. (1992). Radiocarbon chronology of Lake Bonneville, eastern Great Basin, U.S.A. *Palaeogeography, Palaeoclimatology, Palaeoecology*, 99, 225–41.

Oviatt, C.G., Madsen, D.B. and Schmitt, D.N. (2003). Late Pleistocene and early Holocene rivers and wetlands in the Bonneville Basin of western North America. *Quaternary Research*, 60, 200–10.

Owen, J.J., Amundson, R., Dietrich, W.E., Nishiizumi, K., Sutter, B. and Chong, G. (2011). The sensitivity of hillslope bedrock erosion to precipitation. *Earth Surface Processes and Landforms*, 36, 117–35.

Owen, L.A., Davis, T., Caffee, M.W., Budinger, F. and Nash, D. (2011). Surface ages and rates of erosion at the Calico Archaeological Site in the Mojave Desert, southern California. *Geomorphology*, 125, 40–50.

Oyarzún, J., Maturana, H., Paulo, A. and Pasieczna, A. (2003). Heavy metals in stream sediments from the Coquimbo Region (Chile): Effects of sustained mining and natural processes in a semi-arid Andean basin. *Mine Water and the Environment*, 22, 155–61.

Ozer, P. (2003). Fifty years of African mineral dust production. *Bulletin Scientifique Academie Royale Sciences d'Outre-Mer*, 49, 371–96.

Ozer, P., Laghdaf, M.B.O., Lemine, S.O.M. and Gassani, J. (2006). Estimation of air quality degradation due to Saharan dust at Nouakchott, Mauritania, from horizontal visibility data. *Water, Air, and Soil Pollution.* doi:10.1007/s11270–006-9152–8.

Pachur, H.J. and Kropelin, S. (1987). Wadi Hower: Paleoclimatic evidence from an extinct river system in the southeastern Sahara. *Science*, 237, 298–300.

Pagani, M., Freeman, K.H. and Arthur, M.A. (1999). Late Miocene atmospheric CO_2 concentrations and the expansion of C4 grasses. *Science*, 285, 876–9.

Page, K., Nanson, G. and Price, D. (1996). Chronology of Murrumbidgee River palaeochannels on the Riverine Plain, south eastern Australia. *Journal of Quaternary Science*, 11, 311–26.

Page, K., Dare-Edwards, A., Nanson, G. and Price, D. (1994). Late Quaternary evolution of Lake Urana, New South Wales, Australia. *Journal of Quaternary Science*, 9, 47–57.

Paillou, P., Schuster, M., Tooth, S., Farr, T., Rosenqvist, A., Lopez, S. and Malezieux, J.-M. (2009). Mapping of a major paleodrainage system in eastern Libya using orbital imaging radar: The Kufrah River. *Earth and Planetary Science Letters*, 277, 327–33.

Paillou, P. and 10 others (2006). An extended field of crater-shaped structures in the Gilf Kebir region, Egypt: observations Observations and hypotheses of their origin. *Journal of African Earth Sciences*, 46, 281–99.

Pain, C.F., Clarke, J.D.A. and Thomas, M. (2007). Inversion of relief on Mars. *Icarus*, 190, 478–91.

Paine, J.G. (1994). Subsidence beneath a playa basin of the southern High Plains, U.S.A.: Evidence from shallow seismic data. *Bulletin of the Geological Society of America*, 106, 233–42.

Painter, T.H., Barrett, A.P., Landry, C.C., Neff, J.C., Cassidy, M.P., Lawrence, C.R., McBride, K.E. and Farmer, G.L. (2007). Impact of disturbed desert soils on duration of mountain snow cover. *Geophysical Research Letters*, 34, L12502, doi:10.1029/2007GL030284.

Paquet, H., Coudé-Gaussen, G. and Rognon, P. (1984). Etude minéralogique de poussières sahariennes le long d'un itinéraire entre 19° et 35° de latitude nord. *Revue de Géologie Dynamique et de Géographie Physique*, 25, 257–65.

Paradise, T. (2005). Sandstone weathering and aspect in Petra, Jordan. *Zeitschrift für Geomorphologie*, 46, 1–17.

Parker, A.G. and Goudie, A.S. (2007). Development of the Bronze Age landscape in the southeastern Arabian Gulf: New evidence from a buried shell midden in the eastern extremity of the Rub' Al Khali desert, Emirate of Ras al-Khaimah, UAE. *Arabian Archaeology and Epigraphy*, 18, 132–8.

Parker, A.G. and Goudie, A.S. (2008). Geomorphological and palaeoenvironmental investigations in the southeastern Arabian Gulf region and the implication for the archaeology of the region. *Geomorphology*, 101, 458–70.

Parker, A.G., Wilkinson, T.J. and Davies, C. (2006b). The early-mid Holocene period in Arabia: Some recent evidence from lacustrine sequences in eastern and southwestern Arabia. *Proceedings of the Seminar for Arabian Studies*, 36, 243–55.

Parker, A.G., Goudie, A.S., Stokes, S., White, K. et al (2006a). A record of Holocene climate change from lake geochemical analyses in southeastern Arabia. *Quaternary Research*, 66, 465–76

Parker, A.G., Eckersley, L., Smith, M.M., Goudie, A.S., Stokes, S., Ward, S., White, K. and Hodson, M.J. (2004). Holocene vegetation dynamics in the northeastern Rub' al-Khali desert, Arabian peninsula: A phytolith, pollen and carbon isotope study. *Journal of Quaternary Science*, 19, 665–76.

Parker, G.G. (1963). Piping, a geomorphic agent in landform development in drylands. *IASH Publication*, 65, 103–13.

Parker, G.G. and Jenne, E.A. (1967). The structural failure of US highways caused by piping. USGS Water Resources Division (Washington, D.C.) for 46th annual meeting of the Highways Research Board.

Parkin, D.W. (1974). Trade-winds during the glacial cycles. *Proceedings of the Royal Society of London*, 337 A, 73–100.

Paron, P. and Goudie, A.S. (2007). Preliminary results about mapping and geomorphological correlation of tiger bush (*brousse tigrée*) in Somalia, from a remote sensing and GIS analysis perspective. In *Geomorphological Variations*, ed. A.S. Goudie and J. Kalvoda. Prague: Nakladatelství P3K, pp. 37–86.

Parsons, A.J. and Abrahams, A.D. (1984). Mountain mass denudation and piedmont formation in the Mojave and Sonoran Deserts. *American Journal of Science*, 284, 255–71.

Parsons, A.J. and Abrahams, A.D. (eds.). (2009). *Geomorphology of Desert Environments* (2nd ed.). Heidelberg: Springer.

Parsons, A.J., Abrahams, A.D. and Simanton, J.R. (1992). Microtopography and soil-surface materials on semi-arid piedmont hill-slopes, southern Arizona. *Journal of Arid Environments*, 22, 107–115.

Parsons, A.J., Abrahams, A.D. and Wainwright, J. (1996). Responses of interrill runoff and erosion rates to vegetation change in southern Arizona. *Geomorphology*, 14, 311–17.

Parteli, E.J.R., Durán, O. and Herrmann, H.J. (2007). Minimal size of a barchan dune. *Physical Review*, E, 75, article no. 011301.

Parteli, E.J.R., Schatz, V. and Herrmann, H.J. (2005). Barchan dunes on Mars and on Earth. Proceedings on Powders and Grains, 2, 959–62.

Parteli, E.J.R., Durán, O., Tsoar, H., Schwämmle, V. and Herrmann, H.J. (2009). Dune formation under bimodal winds. *Proceedings of the National Academy of Sciences*, 106, 22085–9.

Partridge, J. and Baker, V.R. (1987). Paleoflood hydrology of the Salt River, Arizona. *Earth Surface Processes and Landforms*, 12, 109–25.

Passarge S. (1904). *Die Kalahari.* Berlin: D. Reimer.

Patton, P.C., Biggar, N., Condit, C.D., Gillam, M.L., Love, D.W., Machette, M.N., Mayer, L., Morrison, R.B. and Rosholt, J.N. (1991). Quaternary geology of the Colorado Plateau. In Vol. K-2, *Quaternary Nonglacial Geology: Conterminous U.S.* Boulder, CO: Geological Society of America, pp. 363–406.

Pease, P.P. and Tchakerian, V.P. (2002). Composition and sources of sand in the Wahiba Sand Sea, Sultanate of Oman. *Annals of the Association of American Geographers*, 92, 416–34.

Pease, P.P., Bierly, G.D., Tchakerian, V.P. and Tindale, N.W. (1999). Mineralogical characterization and transport pathways of dune sand using Landsat TM data, Wahiba Sand Sea, Sultanate of Oman. *Geomorphology*, 29, 235–49.

Pease, P.P., Tchakerian, V.P. and Tindale, N.W. (1998). Aerosols over the Arabian Sea: Geochemistry and source areas for aeolian desert dust. *Journal of Arid Environments*, 39, 477–96.

Peck, J.A., Green, R.R., Shanahan, T., King, J.W., Overpeck, J.T. and Scholz, C.A. (2004). A magnetic mineral record of late Quaternary tropical climate variability from Lake Bosumtwi, Ghana. *Palaeogeography, Palaeoclimatology, Palaeoecology*, 215, 37–57.

Pedro, G. (1957). Nouvelle récherches sur l'influence des sels dans la désagrégation des roches. *Comptes Rendus Academie des Sciences*, 244, 2822–4.

Peel, R.F. (1941). Denudational landforms of the central Libyan Desert. *Journal of Geomorphology*, 4, 3–23.

Peel, R.F. (1966). The landscape in aridity. *Transactions of the Institute of British Geographers*, 38, 1–23.

Peel, R.F. (1974). Insolation weathering: Some measurements of diurnal temperature changes in exposed rocks in the Tibesti region, central Sahara. *Zeitschrift für Geomorphologie*, Supplementband, 21, 19–28.

Pell, S.D., Chivas, A.R. and Williams, I.S. (1999). Great Victorian Desert: development and sand provenance. *Australian Journal of Earth Sciences*, 46, 289–99.

Pell, S.D., Chivas, A.R. and Williams, I.S. (2000). The Simpson, Strzelecki and Tirari Deserts: Development and sand provenance. *Sedimentary Geology*, 130, 107–30.

Pell, S.D., Chivas, A.R. and Williams, I.S. (2001). The Mallee Dunefield: Development and sand provenance. *Journal of Arid Environments*, 48, 149–70.

Pelletier, J.D. (2006). Sensitivity of playa wind-blown dust emissions to climatic and anthropogenic change. *Journal of Arid Environments*, 66, 62–75.

Pelletier, J.D. (2010a). Numerical modelling of the late Cenozoic geomorphic evolution of Grand Canyon, Arizona. *Bulletin of the Geological Society of America*, 122, 595–608.

Pelletier, J.D. (2010b). How do pediments form? A numerical modeling investigation with comparison to pediments in southern Arizona. *Bulletin of the Geological Society of America*, 122, 1815–29.

Pelletier, J.D., Cline, M. and Delong, S.B. (2007). Desert pavement dynamics: Numerical modelling and field-based calibration. *Earth Surface Processes and Landforms*, 32, 1913–27.

Pelletier, J.D., Mitasova, H., Harmon, R. and Overton, M. (2009). The effects of interdune vegetation changes on eolian dune field evolution: A numerical-modeling case study at Jockey's Ridge, North Carolina, USA. *Earth Surace Processes and Landforms*, 34, 1245–54.

Pelletier, J.D., Delong, S.B., Cline, M.L., Harrington, C.D. and Keating, G.N. (2008). Dispersal of channel-sediment contaminants in distributary fluvial systems: Application to fluvial tephra and radionuclide redistribution following a potential volcanic eruption at Yucca Mountain. *Geomorphology*, 94, 226–46.

Pelletier, J.D., Pearthree, P.A., Houise, P.K., Demsey, K.A., Klawon, J.E. and Vincent, K.R. (2005). An integrated approach to flood hazard assessment on alluvial fans using numerical modelling, field mapping, and remote sensing. *Bulletin of the Geological Society of America*, 117, 1167–80.

Penck, W. (1953). *Morphological Analysis of Landforms* (trans. by H. Czech and K.C. Boswell). London: Macmillan.

Permenter, J.L. and Oppenheimer, C. (2007). Volcanoes of the Tibesti massif (Chad, northern Africa). *Bulletin of Volcanology*, 69, 609–26.

Perthuisot, J.-P. and Jauzein, A. (1975). Sebkhas et dunes d'argile: L'enclave endoreique de pont du Fahs, Tunisie. *Revue de Géographie Physique et de Géologie Dynamique*, 17, 295–306.

Peterson, F.F. (1981). Landforms of the Basin and Range Province defined for soil survey. *Nevada Agricultural Experimental Station, Technical Bulletin, 28.*

Peterson, M.J., Hargrove, W.W. and Efroymson, R.A. (2008). The Apache Longbow-Hellfire Missile Test at Yuma Proving Ground: Ecological risk assessment for tracked vehicle movement across desert pavement. *Human and Ecological Risk Assessment*, 14, 919–46.

Petit-Maire, N., Benfort, L. and Paye, N. (1997). Holocene climate change and man in the present day Sahara desert. In *Third Millennium BC Climate Change and Old World Collapse.*, ed. H.N. Dalfes, G. Kukla and H. Weiss. Berlin: Springer Verlag, pp. 297–308.

Petit-Maire, N., Burollet, P.F., Ballais, J.-L., Fontugne, M., Rosso, J-C. and Lazaar, A. (1999). Paléoclimats Holocènes du Sahara septentionale. Dépôts lacustres et terrasses alluviales en bordure du Grand Erg Oriental à l'extrême – Sud de la Tunisie. *Comptes Rendus Académie des Sciences*, 312, 1661–6.

Petit-Maire, N., Carbonel, P., Reyss, J.L., Sanlaville, P., Abed, A., Bourrouilh, R., Fontugne, M. and Yasin, S. (2010). A vast Eemian paleolake in southern Jordan (29°). *Global and Planetary Change*, 72, 1368–73.

Petryshyn, V.A. and Corsetti, F.A. (2011). Analysis of growth directions of columnar stromatolites from Walker Lake, western Nevada. *Geobiology*, 9, 423–35.

Petryshyn, V.A., Corsetti, F.A., Berelson, W.M., Beaumont, W. and Lund, S.P. (2012). Stromatolite lamination frequency, Walker Lake, Nevada: Implications for stromatolites as biosignatures. *Geology*, 40, 499–502.

Pettke, T., Halliday, A.N., Hall, C.M. and Rea, D.K. (2000). Dust production and deposition in Asia and the North Pacific Ocean over the past 12 Myr. *Earth and Planetary Science Letters*, 178, 397–413.

Petts, G.E. (1985). *Impounded Rivers: Perspectives for Ecological Management.* Chichester: Wiley.

Pfeffer, W.T., Harper, J.T. and O'Neel, S. (2008). Kinematic constraints on glacier contributions to 21st-century sea-level rise. *Science*, 321, 1340–3.

Phillips, R.E., Blevins, R.L., Thomas, G.W., Frye, W.W. and Phillips, S.H. (1980). No-tillage agriculture. *Science*, 208, 1108–13.

Pias, J. (1971). Les loess en Afghanistan Orietntal et leurs pédogenèses successives au Quaternaire recent. *Comples Rendus Academie de Sciences*, 272 (D), 1602–5.

Picker, M.D., Hoffman, M.T. and Leverton, B. (2007). Density of Microhodotermes viator (*Hodotermitidae*) mounds in southern Africa in relation to rainfall and vegetative productivity gradients. *Journal of Zoology*, 271, 37–44.

Picker, M.D., Ross-Gillespie, V., Vlieghe, K. and Moll, E. (2012). Ants and the enigmatic Namibia fairy circles – Cause and effect? *Ecological Entomology*, 37, 33–42.

Pierre, C., Berhgametti, G., Marticorena, B., Mougin, E., Bouet, C. and Schmechtig, C. (2012). Impact of vegetation and soil moisture seasonal dynamics on dust emissions over the Sahel. *Journal of Geophysical Research*, 117, D06114, doi:10.1029/2011JD016950.

Pietrasiak, N., Johansen, J.R. and Drenovsky, R.E. (2011). Geologic composition influences distribution of microbiotic crusts in the Mojave and Colorado Deserts at the regional scale. *Soil Biology & Biochemistry*, 43, 967–74.

Pineda-Martinez, L.F., Carbajal, N., Campos-Ramos, A.A., Noyola-Medrano, C. and Aragón-Piña, A. (2011). Numerical research of extreme wind-induced dust transport in a semi-arid human-impacted region of Mexico. *Atmospheric Environment*, 45, 4652–60.

Pinker, R.T., Sun, D., Miller, M. and Robinson, G.J. (2007). Diurnal cycle of land surface temperature in a desert encroachment zone as observed from satellites. *Geophysical Research Letters*, 34, L11809, doi:10.1029/2007GL030186.

Pitlick, J. (1994). Relation between peak flows, precipitation, and physiography for five mountainous regions in the western USA. *Journal of Hydrology*, 158, 219–40.

Placzek, C., Quade, J. and Betancourt, J.L. (2001). Holocene lake-level fluctuations of Lake Aricota, southern Peru. *Quaternary Research*, 56, 181–90.

Placzek, C., Quade, J. and Patchett, J. (2006). Geochronology and stratigraphy of late Pleistocene lake cycles on the southern Bolivian Altiplano: Implications for causes of tropical climate change. *Bulletin of the Geological Society of America*, 118, 515–32.

Placzek, C. and 8 others. (2009). Climate in the dry Central Andes over geologic, millennial, and interannual timescales. *Annals of the Missouri Botanic Garden*, 96, 386–97.

Poesen, J.W., Torri, D. and Bunte, K. (1994). Effects of rock fragments on soil erosion by water and different spatial scales: A review. *Catena*, 23, 141–66.

Poesen, J.W., Vandekerckhove, L., Nachtergaele, J., Osstwoud, J., Wijdenes, D., Verstraeten, J. and Van Wesemael, B. (2002). Gully erosion in dryland environments. In *Dryland Rivers: Hydrology and Geomorphology of Semi-arid Channels*, ed. L.J. Bull and M.J. Kirkby. Chichester: Wiley, pp. 229–62.

Poesen, J.WE., Nachtergaele, J., Verstraeten, G. and Valentin, C. (2003). Gully erosion and environmental change: Importance and research needs. *Catena*, 50, 91–133.

Polyakov, V.O., Nearing, M.A., Nichols, M.H., Scott, R.L., Stone, J.J. and McClaran, M.P. (2010). Long-term runoff and sediment yields from small semiarid watersheds in southern Arizona. *Water Resources Research*, 46, W09512, doi:10.1029WR009001.

Potter, R.M. and Rossman, G.R. (1977). Desert varnish: The importance of clay minerals. *Science*, 196, 1446–8.

Powell, J.W. (1895). *Canyonlands of the Colorado*. New York: Flood and Vincent.

Powers, M.C. (1953). A new roundness scale for sedimentary particles. *Journal of Sedimentary Petrology*, 23, 117–19.

Preston, G.W., Parker, A.G., Walkington, H., Leng, M.J. and Hodson, M.J. (2012). From nomadic herder-hunters to sedentary farmers: The relationship between climate change and ancient subsistence strategies in south-eastern Arabia. *Journal of Arid Environments*, 86, 122–130.

Preusser, F. (2009). Chronology of the impact of Quaternary climate change on continental environments in the Arabian Peninsula. *CR Geoscience*, 341, 621–32.

Price, W.A. (1925). Caliche and pseudo-anticlines. *Bulletin of the American Association of Petroleum Geologists*, 9, 1009–17.

Prigent, C., Tegen, I., Aires, F., Marticorena, B. and Zribi, M. (2005). Estimation of the aerodynamic roughness length in arid and semi-arid regions over the globe with the ERS scatterometer. *Journal of Geophysical Research*, 110, D09205, doi:10.1029/2004JD005370.

Prospero, J.M. (1996a). Saharan dust transport over the North Atlantic Ocean and Mediterranean: An overview. In *The Impact of Desert Dust across the Mediterranean*, ed. S. Guerzoni and R. Chester. Dordrecht: Kluwer, pp. 131–51.

Prospero, J.M. (1996b). The atmospheric transport of particles to the ocean. In *Particle Flux in the Ocean*, ed. V. Ittekkot, P. Schafer, S. Honjo, and P.J. Depetris. Chichester: Wiley, pp. 19–52.

Prospero, J.M., Ginoux, P., Torres, O., Nicholson, S.E. and Gill, T.E. (2002). Environmental characterisation of global sources of atmospheric soil dust identified with the Nimbus 7 Total Ozone Mapping Spectrometer (TOMS) absorbing aerosol product. *Reviews of Geophysics*, 40, 2–31.

Puckridge, J.T., Walker, K.F. and Costelloe, J.F. (2000). Hydrological persistence and the ecology of dryland rivers. *Regulated Rivers Research and Management*, 16, 385–402.

Puente, M.E., Bashan, Y., Li, C.Y. and Lebsky, V.K. (2004). Microbial populations and activities in the rhizoplane of rock-weathering desert plants. 1. Root colonization and weathering of igneous rocks. *Plant Biology*, 6, 629–42.

Pueyo, J.J., Chong, G. and Jensen, A. (2001). Neogene evaporites in desert volcanic environments: Atacama Desert, northern Chile. *Sedimentology*, 48, 1411–31.

Pueyo, J.J., Chong, G. and Vega, M. (1998). Mineralogia y evolucion de las salmueras madres en el yacimento de nitratos Pedro de Valdivia, Antofagasta, Chile. *Revista Geologica de Chile*, 25, 3–15.

Pulido-Villena, E., Wagener, T. and Guieu, C. (2008). Bacterial response to dust pulses in the western Mediterranean: Implications for carbon cycling in the oligotrophic ocean. *Global Biogeochemical Cycles*, 22, GB 1020, doi:10.1029/2007GB003091.

Pullen, A. and 7 others. (2011). Qaidam Basin and northern Tibetan Plateau as dust sources for the Chinese Loess Plateau and paleoclimate implications. *Geology*, 39, 1031–4.

Purkis, S.J., Rowlands, G.P., Riegl, B.M. and Renaud, P.G. (2010). The paradox of tropical karst morphology in the coral reefs of the arid Middle East. *Geology*, 38, 227–30.

Pye, K. (1983a). Early post-depositional modification of aeolian dune sands. In *Eolian Sediments and Processes*, ed. M.E. Brookfield and T.S. Ahlbrandt. Elsevier: Amsterdam, pp. 197–221.

Pye, K. (1983b). Red beds. In *Chemical Sediments and Geomorphology*, ed. A. Goudie and K. Pye. London: Academic Press, pp. 227–63.

Pye, K. (1987). *Aeolian Dust and Dust Deposits*. London: Academic Press.

Pye, K. and Sherwin, D. (1999). Loess. In *Aeolian Environments, Sediments and Landforms*, ed. A.S. Goudie, I. Livingstone and S. Stokes. Chichester: Wiley, pp. 214–38.

Pye, K. and Tsoar, H. (1990). *Aeolian Sand and Sand Dunes*. London: Unwin Hyman.

Pye, K., Goudie, A.S. and Thomas, D.S.G. (1984). A test of petrological control in the development of bornhardts and koppies on the Matopos batholith, Zimbabwe. *Earth Surface Processes and Landforms*, 9, 455–67.

Pye, K., Goudie, A.S. and Watson, A. (1986). Petrological influence on differential weathering and inselberg development in the Kora area of central Kenya. *Earth Surface Processes and Landforms*, 11, 41–52.

Qadir, M. and Schubert, S. (2002). Degradation processes and nutrient constraints in sodic soils. *Land Degradation and Development*, 13, 275–94.

Qian, G., Dong, Z., Luo, W. and Lu. J. (2011). Mean airflow pattern upwind of topographic obstacles and their implications for the formation of echo dunes: A wind tunnel simulation of the effects of windward slope. *Journal of Geophysical Research*, 116, F04026, doi:10.1029/2011JF002020.

Qian, G., Dong, Z., Zhang, Z., Luo, W. and Lu, J. (2012). Granule ripples in the Kumtagh Desert, China: Morphology, grain size and influencing factors. *Sedimentology*, doi: 10.111/j.1365–3091.2012.01330.x.

Qiang, X.K., Li, Z.X., Powell, C.McA. and Zheng, H.B. (2001). Magnetostratigraphic record of the late Miocene onset of the East Asian monsoon, and Pliocene uplift of northern Tibet. *Earth and Planetary Science Letters*, 187, 83–93.

Qiang, X.K. and 15 others. (2011). New eolian red clay sequence on the western Chinese Loess Plateau linked to onset of Asian desertification about 25 Ma ago. *Science China Earth Sciences*, 54, 136–44.

Qin, Y., Li, Y., Bao, H., Liu, F., Hou, K., Wan, D. and Zhang, C. (2012). Massive atmospheric nitrate accumulation in a continental interior desert, northwestern China. *Geology*, 40, 623–6.

Qu, J.J., Dong, R.G., Niu, Q.H., Jing, Z.F. and Han, Q.J. (2011). Feathered sand ridges in the Kumtagh Desert and their position in the classification system. *Science China Earth Sciences*, doi:10.1007/s11430–011-4209-y.

Quade, J. and Broecker, W.S. (2009). Dryland hydrology in a warmer world: Lessons from the Last Glacial period. *European Physics Journal Special Topics*, 176, 21–36.

Quade, J., Cerling, T.E. and Bowman, J.R. (1989). Development of Asian monsoon revealed by marked ecological shift during the latest Miocene in northern Pakistan. *Nature*, 342, 163–6.

Quade, J., Mifflin, M.D., Pratt, W.L., McCoy, W. and Burckle, L. (1995). Fossil spring deposits in the southern Great Plains and their implications for changes in water-table levels near Yucca Mountain, Nevada, during Quaternary time. *Bulletin of the Geological Society of America*, 107, 213–30.

Quick, D.J. and Chadwick, O.A. (2011). Accumulation of salt-rich dust from Owens Lake playa in nearly alluvial soils. *Aeolian Research*, 3, 23–9.

Quigley, M.C., Horton, T., Hellstrom, J.C., Cupper, M.L. and Sandiford, M. (2010). Holocene climate change in arid Australia from speleothem and alluvial records. *The Holocene*, 20, 1093–104.

Quillin, J.P., Zartman, R.E. and Fish, E.B. (2005). Spatial distribution of playa basins on the Texas High Plains. *The Texas Journal of Agriculture and Natural Resources*, 18, 1–14.

Quinton, J.N., Edwards, G.M. and Morgan, R.P.C. (1997). The influence of vegetation species and plant properties on runoff and soil erosion: Results from a rainfall simulation study in south east Spain. *Soil Use and Management*, 13, 143–8.

Radebaugh, J., Lorenz, R., Farr, T., Paillou, P., Savage, C. and Spencer, C. (2010). Linear dunes on Titan and Earth: Initial remote sensing comparisons. *Geomorphology*, 121, 122–32.

Radies, D., Hasiotis, S.T., Preusser, F., Neubert, E. and Matter, A. (2005). Paleoclimatic significance of early Holocene faunal assemblages in wet interdune deposits of the Wahiba Sand Sea, Sultanate of Oman. *Journal of Arid Environments*, 62, 109–25.

Rahn, P.H. (1967). Sheetfloods, streamfloods, and the formation of pediments. *Annals of the Association of American Geographers*, 57, 593–604.

Raji, B.A., Utovbisere, E.O. and Momodu, A.B. (2004). Impact of sand dune stabilization structures on soil and yield of millet in the semi-arid region of NW Nigeria. *Environmental Monitoring and Assessment*, 99, 181–96.

Ramakrishnan, D.E. and Tiwari, K.C. (2006). Calcretized ferricretes around the Jaisalmer area, Thar Desert, India: Their chemistry, mineralogy, micromorphology and genesis. *Turkish Journal of Earth Sciences*, 15, 211–23.

Ramsey, M.WS., Christensen, P.R., Lancaster, N., and Howard, D.A. (1999). Identification of sand sources and transport pathways at the Kelso Dunes, California, using thermal infrared sensing. *Bulletin of the Geological Society of America*, 111, 646–62.

Rango, A., Chopping, M., Ritchie, J., Havstad, K., Kustas, W. and Schmugge, T. (2000). Morphological characteristics of shrub coppice dunes in desert grasslands of southern New Mexico derived from scanning LIDAR. *Remote Sensing of the Environment*, 74, 26–44.

Rao, K.K., Wasson, R.J. and Kutty, M.K. (1989). Foraminifera from late Quaternary dune sands of the Thar Desert, India. *Palaios*, 4, 168–80.

Rao, V.B., Arai, E., Franchito, S.H., Shimabukuro, Y.E., Ramakrishna, S.S.V.S. and Naidu, C.V. (2011). The Thar, Rajputana Desert unprecedented rainfall in 2006 and 2010: Effect of climate change? *Geofisíca Internacional*, 50, 363–70.

Rashki, A., Rautenbach, C.J. de, Eriksson, P.G., Kaskaoutis, D.G. and Gupta, P. (2011). Temporal changes of particulate concentration in the ambient air over the city of Zahedan, Iran. *Air Quality and Atmsopheric Health*, doi:10.1007/s11869–011-0152–5.

Rashki, A., Kaskaoutis, D.G., Rautenbach, C.J. de W., Eriksson, P.G., Qiang, M. and Gupta, P. (2012). Dust storms and their horizontal dust loading in the Sistan region, Iran. *Aeolian Research*, 5, 51–62.

Rauh, W. (1985). The Peruvian-Chilean deserts. In: *Hot Deserts and Arid Shrublands*, ed. M. Evenari, I. Noy-Meir and D.W. Goodall. Amsterdam: Elsevier, pp. 239–67.

Ravi, S., Breshears, D.D., Huxman, T.E. and D'Odorico, P. (2010). Land degradation in drylands: Interactions among hydrologic-aeolian erosion and vegetation dynamics. *Geomorphology*, 116, 236–45.

Ravi, S., D'Odorico, P., Over, T.M. and Zobeck, T.M. (2004). On the effect of air humidity on soil susceptibility to wind erosion: The case of air-dry soils. *Geophysical Research Letters*, 31, article L09501.

Ravi, S., D'Odorico, P., Wang, J., White, C.S., Okin, G.S., Macko, S.A. and Collins, S.L. (2009). Post-fire resource redistribution in desert grasslands: A possible negative feedback on land degradation. *Ecosystems*, 12, 434–44.

Ravi, S. and 13 others. (2011). Aeolian processes and the biosphere. *Review of Geophysics*, 49, RG3001, doi:10.1029/2010RG000328.

Ravina, I. and Zaslavsky, D. (1974). The electrical double layer as a possible factor in desert weathering. *Zeitschrift für Geomorphologie*, 21, 13–18.

Raymo, M.E. and Ruddiman, W.F. (1992). Tectonic forcing of late Cenozoic climate. *Nature*, 359, 117–22.

Razo. I., Carrizales, L., Castro, J., Díaz-Barriga, F. and Monroy, M. (2004). Arsenic and heavy metal pollution of soil, water and sediments in a semi-arid climate mining area in Mexico. *Water, Air, and Soil Pollution*, 152, 129–52.

Rea, D.K. (1994). The paleoclimatic record provided by eolian deposition in the deep sea: The geologic history of wind. *Review of Geophysics*, 32, 159–95.

Rech, J.A., Quade, J. and Betancourt, J.L. (2002). Late Quaternary paleohydrology of the central Atacama Desert (lat 22°–24°S), Chile. *Bulletin of the Geological Society of America*, 114, 334–48.

Rech, J.A. and 8 others. (2010). Evidence for the development of the Andean rain shadow from a Neogene isotopic record in the Atacama Desert, Chile. *Earth and Planetary Science Letters*, 292, 371–82.

Reeves, C.C. (1966). Pluvial lake basins of west Texas. *Journal of Geology*, 74, 269–91.

Reeves, C.C. (1971). Relations of caliche to small natural depressions, southern High Plains, Texas and New Mexico. *Bulletin of the Geological Society of America*, 82, 1983–8.

Reffet, E., du Pont, C.S., Hersen, P. and Douady, S. (2010). Formation and stability of transverse and longitudinal sand dunes. *Geology*, 38, 491–4.

Reheis, M.C. (1987). Gypsic soils on the Kane alluvial fans, Bighorn County, Wyoming. *United States Geological Survey Bulletin*, 1590C.

Reheis, M.C. (1990). Influence of climate and eolian dust on the major-element chemistry and clay mineralogy of soils in the western Bighorn Basin, USA. *Catena*, 17, 219–48.

Reheis, M. (1997). Dust deposition of Owens (dry) Lake, 1991–1994: Preliminary findings. *Journal of Geophysical Research*, 102, 25999–6008.

Reheis, M. (1999). Highest pluvial-lake shorelines and Pleistocene climate of the western Great Basin. *Quaternary Research*, 52, 196–205.

Reheis, M. (2006). A 16-year record of eolian dust in southern Nevada and California, USA: Controls on dust generation and accumulation. *Journal of Arid Environments*, 67, 487–520.

Reheis, M. and Kihl, R. (1995). Dust deposition in southern Nevada and California (1984–1989): Relations to climate, source area and source lithology. *Journal of Geographical Research*, 100 (D5), 8893–918.

Reheis M.C., Budahn, J.R., Lamothe, P.J. and Reynolds, R. (2009). Compositions of modern dust and surface sediments in the desert Southwest, United States. *Journal of Geophysical Research*, 114, F01028, doi:10.1029/2008JF001009.

Reheis, M.C., Goodmacher, J.C., Harden, J.W., McFadden, L.D., Rockwell, T.K., Shroba, R.R, Sowers, J.M. and Taylor, E.M. (1995). Quaternary soils and dust deposition in southern Nevada and California. *Bulletin of the Geological Society of America*, 107, 1003–22.

Reheis, M.C., Sowers, J.M., Taylor, E.M., McFadden, L.D. and Harden, J.W. (1992). Morphology and genesis of carbonate soils in the Kyle Canyon Fan, Nevada, USA. *Geoderma*, 52, 303–42.

Reid, I. and Frostick, L.E. (1997). Channel form, flows and sediments in deserts. In *Arid Zone Geomorphology*, ed. D.S.G. Thomas. Chichester: Wiley, pp. 205–29.

Reid, I. and Laronne J.B. (1995). Bedload sediment transport in an ephemeral stream and a comparison with seasonal and perennial counterparts. *Water Resources Research*, 31, 773–81.

Reid, I., Laronne, J.B. and Powell, D.M. (1998). Flash-flood and bedload dynamics of desert gravel-bed streams. *Hydrological Processes*, 12, 543–57.

Reider, R.G., Huss, J.M. and Miller, T.W. (1996). A groundwater vortex hypothesis for mima-like mounds, Laramie Basin, Wyoming. *Geomorphology*, 16, 295–317.

Reitz, M.D., Jerolmack, D.J., Ewing, R.C. and Martin, R.L. (2010). Barchan-parabolic dune pattern transition from vegetation stability threshold. *Geophysical Research Letters*, 37, L19402, doi:10,1029/2010GL044957.

Rendell, H.M. (1984). New perspectives on the Pleistocene and Holocene sequences of the Potwar Plateau and adjacent areas of northern Pakistan. In *The International Karakoram Project*, ed. K.J. Miller. Cambridge: Cambridge University Press. *Vol.* 1, pp. 389–98.

Rendell, H.M. (1989). Loess deposition during the late Pleistocene in northern Pakistan. *Zeitschrift für Geomorphologie*, Supplementband, 76, 247–55.

Rendell, H.M. (1993). The palaeoclimatic significance of the loess deposits in northern Pakistan. In *Himalaya to the Sea*, ed. J.F. Shroder. London: Routledge, pp. 227–35.

Rendell, H.M. (1997). Tectonic frameworks. In *Arid Zone Geomorphology: Processes, Form and Change in Drylands* (2nd ed.), ed. D.S.G. Thomas. Chichester: Wiley, pp. 13–22.

Rendell, H.M., Clarke, M.L., Warren, A. and Chappell, A. (2003). The timing of climbing dune formation in southwestern Niger: Fluvio-aeolian interactions and the rôle of sand supply. *Quaternary Science Reviews*, 22, 1059–65.

Reneau, S.L., Raymond, R. and Harrington, C. (1992). Elemental relationships in rock varnish stratigraphic layers, Cima volcanic field, California: Implications for varnish development and the interpretation of varnish chemistry. *American Journal of Science*, 292, 684–723.

Resane, T., Annegarn, H. and Freiman, T. (2004). The day of the white rain: Origin of unusual dust deposition in Johannesburg, South Africa. *South African Journal of Science*, 100, 483–7.

Resch, F., Sunnu, A. and Afeti, G. (2007). Saharan dust flux and deposition rate near the Gulf of Guinea. *Tellus*, doi:10.iii/j.1600–0889.2007.00286.x.

Retallack, G.J. (2000). Depth to pedogenic carbonate horizon as a palaeoprecipitation indicator. *Geology*, 28, 572–3.

Retallack, G.J. and Huang, C. (2010). Depth to gypsic horizon as a proxy for paleoprecipitation in paleosols of sedimentary environments. *Geology*, 38, 403–6.

Revel, M., Ducassou, E., Grousset, F.E., Bernasconi, S.M., Migeon, S., Revillon, S., Mascle, J., Murat, A., Zaragosi, S. and Bosch, D. (2010). 100,000 years of African monsoon variability recorded in sediments of the Nile margin. *Quaternary Science Reviews*, 29, 1342–62.

Revelle, R.R. and Waggoner, P.E. (1983). Effect of a carbon dioxide-induced climatic change on water supplies in the western United States. In *Changing Climate*, Carbon Dioxide Assessment Committee, Washington DC: National Academy Press, pp. 419–32.

Revel-Rolland, M., De Deckker, P., Delmonte, B., Hesse, P.P., Magee, J.W., Basile-Doelsch, I., Grousset, F. and Bosch, D. (2006). Eastern Australia: A possible source of dust in East Antarctica interglacial ice. *Earth and Planetary Science Letters*, 249, 1–13.

Reynolds, R.L., Yount, J.C., Reheis, M., Goldstein, H., Chavez, P., Fulton, R., Whitney, J., Fuller, C. and Forester, R.M. (2007). Dust emission from wet and dry playas in the Mojave Desert, USA. *Earth Surface Processes and Landforms*, 32, 1811–27.

Rhoades, J.D. (1990). Soil salinity – Causes and controls. In *Techniques for Desert Reclamation*, ed. A.S. Goudie. Chichester: Wiley, pp. 109–34.

Rice, A. (1976). Insolation warmed over. *Geology*, 4, 61–2.

Rich, J. and Stokes, S. (2011). A 200,000-year record of late Quaternary eolian sedimentation on the southern High Plains and nearby Pecos River Valley, USA. *Aeolian Research*, 2, 221–40.

Rich J., Stokes S. and Wood W.W. (1999). Holocene chronology for lunette dune deposition on the southern High Plains, U.S.A.. *Zeitschrift für Geomorphologie*, Supplementband, 116, 165–80.

Richards, K.S. and Reddy, K.R. (2007). Critical appraisal of piping phenomena in earth dams. *Bulletin of Engineering Geology and Environment*, 66, 381–402.

Richer, R., Anchassi, D., El-Assaad, I., El-Matbouly, M., Ali, F., Makki, I. and Metcalf, J.S. (2012). Variation in the coverage of biological soil crusts in the state of Qatar. *Journal of Arid Environments*, 78, 187–90.

Richthofen, F. von. (1882). On the mode of origin of the loess. *Geological Magazine*, 9, series 2, 293–305.

Rickson, R.J. (2006). Controlling sediment at source: An evaluation of erosion control geotextiles. *Earth Surface Processes and Landforms*, 31, 550–60.

Ridgwell, A.J. (2003). Implications of the glacial CO_2 'iron hypothesis' for Quaternary climate change. *Geochemistry, Geophysics, Geosystems*, 4, article 1076.

Ridley, D.A., Heald, C.L. and Ford, B. (2012). North African dust export and deposition: A satellite and model perspective. *Journal of Geophysical Research*, 117, D02202, doi:10.1029/2011JD016794.

Rienks, S.M., Botha, G.A. and Hughes, J.C. (2000). Some physical and chemical properties of sediments exposed in a gully (*donga*) in northern KwaZulu-Natal, South Africa and their relationship to the erodibility of the colluvial layers. *Catena*, 39, 11–31.

Riksen, M., Spaan, W., Arrué, J.L. and López, M.V. (2003). What to do about wind erosion. In *Wind Erosion on Agricultural Land in Europe*, ed. A. Warren. Luxembourg: European Commission, pp. 39–52.

Ringrose, S., Harris, C., Huntsman-Mapila, P., Vink, B.W., Diskins, S., Vanderpost, C. and Matheson, W. (2009). Origins of strandline duricrusts around the Makgadikgadi Pans (Botswana Kalahari) as deduced from their chemical and isotope composition. *Sedimentary Geology*, 219, 262–79.

Riquelme, R., Martinod, J., Herail, G., Darrozes, J. and Charrier, R. (2003). A geomorphological approach to determining the Neogene to recent tectonic deformation in the Coastal Cordillera of northern Chile (Atacama). *Tectonophysics*, 361, 255–75.

Riser, J. (1985). Le role du vent au cours des derniers millenaires dans le basin Saharien D'Araouane (Mali). *Bulletin Association Géographique Français*, 62, 311–17.

Ritley, M.K., Onontya, E., Carson, R.J. and Wegmann, K.W. (2004). Yardangs and dome dunes northeast of Tavan Har, Gobi, Mongolia. *GSA Abstracts with Programs*, 36(4), 33.

Ritter, J.B., Miller, J.R. and Husek-Wulforst, J. (2000). Environmental controls on the evolution of alluvial fans in Buena Vista Valley, north central Nevada, during late Quaternary time. *Geomorphology*, 36, 63–87.

Rivera, N.I.R., Gill, T.E., Gebhart, K.A., Hand, J.L., Bleiweiss, M.P. and Fitzgerald, R.M. (2009). Wind modeling of Chihuahuan Desert dust outbreaks. *Atmospheric Environment*, 43, 347–54.

Roberts, C.R. and Mitchell, C.W. (1987). Spring mounds in Tunisia. *Geological Society of London Special Publication*, 35, 321–34.

Roberts, G.G. and White, N. (2010). Estimating uplift histories from river profiles using African examples. *Journal of Geophysical Research*, 115, B02406, doi:10.1029/2009JB006692.

Roberts, H.M., Muhs, D.R., Wintle, A.G., Duller, G.A.T. and Bettis, E. A. (2003). Unprecedented last-glacial mass accumulation rates determined by luminescence dating of loess from western Nebraska. *Quaternary Research*, 59, 411–19.

Roberts, N. (1983). Age, paleoenvironments and climatic significance of late Pleistocene Konya Lake, Turkey. *Quaternary Research*, 19, 154–71.

Robins, C.R., Buck, B.J., Williams, A.J., Morton, J.L., House, P.K., Howell, M.S. and Yonovitz, M.L. (2009). Comparison of flood hazard assessments on desert piedmonts and playas: A case study in Ivanpah Valley, Nevada. *Geomorphology*, 103, 520–32.

Robinson, B.W. and Gunatilaka, A. (1991). Stable isotope studies and the hydrological regimes of sabkhas in southern Kuwait, Arabian Gulf. *Sedimentary Geology*, 73, 141–59.

Robinson, D.A. and Williams, R.B.G. (1982). Salt weathering of rock specimens of varying shape. *Area*, 14, 293–9.

Robinson, D.A. and Williams, R.B.G. (1989). Polygonal cracking of sandstone at Fontainbleu, France. *Zeitschrift für Geomorphologie*, 33, 59–72.

Robinson, D.A and Williams, R.B.G. (1992). Sandstone weathering in the High Atlas, Morocco. *Zeitschrift für Geomorphologie*, 36, 413–29.

Robinson, D.A. and Williams, R.B.G. (2000). Accelerated weathering of a sandstone in the High Atlas Mountains of Morocco by an epilithic lichen. *Zeitschrift für Geomorphologie*, 44, 513–28.

Robinson, D.M. (1995). Concrete corrosion and slab heaving in a Sabkha environment: Long Beach–Newport Beach, California. *Environmental and Engineering Geoscience*, 1, 35–40.

Rodbell, D.T., Seltzer, G.O., Anderson, D.M., Abbott, M.B., Enfield, D.B. and Newmann, J.H. (1998). An ~15000-year record of El Niño-driven alluviation in southwestern Ecuador. *Science*, 283, 516–20.

Rodriguez, J. A. P., Tanaka, K. L., Yamamoto, A., Berman, D. C., Zimbelman, J. R., Kargel, J. S. and Miyamoto, H. (2010). The sedimentology and dynamics of crater-affiliated wind streaks in western Arabia Terra, Mars and Patagonia, Argentina. *Geomorphology*, 121, 30–54.

Rodriguez-Navarro, C., Doehne, E. and Sebastian, E. (2000). How does sodium sulfate crystallize? Implications for the decay and testing of building materials. *Cement and Concrete Research*, 30, 1527–34.

Rogers, R.D. and Schumm, S.A. (1991). The effect of sparse vegetative cover on erosion and sediment yield. *Journal of Hydrology*, 123, 19–24.

Rögner, K. (1986). Temperature measurements of rock surfaces in hot deserts (Negev, Israel). In *International Geomorphology 1986*, vol. 2, ed. V. Gardiner. Chichester: Wiley, pp. 1271–86.

Rögner, K. and Symkatz-Kloss, W. (1991). The deposition of eolian sediments in lacustrine and fluvial environments of central Sinai (Egypt). *Catena*, Supplement, 20, 75–91.

Rohrmann, A., Heermance, R., Kapp, P.A. and McCallister, A. (2010). First quantification of severe wind erosion in yardang fields using cosmogenic ^{10}Be within the western Qaidam Basin, China. *American Geophysical Union, Fall Meeting 2010, abstract* #EP41C-0716.

Rolland, G. (1890). Les grandes dunes de sable du Sahara. *Compte Rendu des Séances de la Société de Géographie*, Année (1890), 158–65.

Romero-Diaz, A., Belmonte-Serrato, F. and Ruiz-Sinoga, J.D. (2010). The geomorphic impact of afforestations on soil erosion in southeast Spain. *Land Degradation & Development*, 21, 188–95.

Roof, S. and Callagan, C. (2003). The climate of Death Valley, California. *Bulletin of the American Meteorological Society*, 84, 1725–39.
Rooyen, M.W. van, Theron, G.K., Rooyen, N. van, Jankowitz, W.J. and Matthews, W.S. (2004). Mysterious circles in the Namib Desert: Review of hypotheses on their origin. *Journal of Arid Environments*, 57, 467–85.
Roqué, C., Zarroca, M. and Linares, R. (In press). Subsurface initiation of tafoni in granite terrains – Geophysical evidence from NE Spain: Geomorphological implications. *Geomorphology*.
Roseb, M.B., Arehart, G.B. and Lico, M.S. (2004). Exceptionally fast growth rate of <100-yr-old tufa, Big Soda Lake, Nevada: Implications for using tufa as a paleoclimate proxy. *Geology*, 32, 409–12.
Rosenberg, N.J., Epstein, D.J., Wang, D., Vail, L., Srinivasan, R. and Arnold, J.G. (1999). Possible impacts of global warming on the hydrology of the Ogallala Aquifer region. *Climatic Change*, 42, 677–92.
Rosenberg, T.M., Preusser, F., Blechschmidt, I., Fleitmann, D., Jagher, R. and Matter, A. (2011). Late Pleistocene palaeolake in the interior of Oman: A potential key area for the dispersal of anatomically modern humans out-of-Africa. *Journal of Quaternary Science*, doi:10.1002/jqs.1560.
Rosenzweig, C. and Hillel, D. (1993). The Dust Bowl of the 1930s: Analogy of greenhouse effect in the Great Plains? *Journal of Environmental Quality*, 22, 9–22.
Roskin, J., Blumberg, D.G., Porat, N., Tsoar, H. and Rozenstein, O. (2012). Do dune sands redden with age? The case of the northwestern Negev dunefield, Israel. *Aeolian Research*, 5, 63–75.
Roskin, J., Porat, N., Tsoar, H., Blumberg, D.G. and Zander, A.M. (2011). Age, origin and climatic controls on vegetated linear dunes in the northwestern Negev Desert (Israel). *Quaternary Science Reviews*, 30, 1649–74.
Rösner, U. (1989). "Löss" am Rande der Wustensteppe? *Erdkunde*, 43, 233–42.
Rösner, U. (1998). Aolische sedimente und formen als indikatoren der spatpleistozanen bis aktuellen landschaftentwicklung beispiele aus den trockengebiete syriens. *Zeitschrift für Geomorphologie*, Supplementband, 111, 33–50.
Rostagno, C.M. and del Valle, H.F. (1988). Mounds associated with shrubs in aridic areas of northeastern Patagonia: Characteristics and possible origins. *Catena*, 15, 347–359.
Rothrock, E.P. (1925). On the force of crystallisation of calcite. *Journal of Geology*, 33, 80–2.
Rouchy, J.M., Servant, M., Fournier, M. and Causse, C. (1996). Extensive carbonate algal bioherms in upper Pleistocene saline lakes of the central Altiplano of Bolivia. *Sedimentology*, 43, 973–93.
Rousseau, D.D. and 11 others. (2002). Abrupt millennial climatic changes from Nussloch (Germany) upper Weichselian eolian records during the last glaciation. *Quaternary Science Reviews*, 21, 1577–82.
Royal Geographical Society. (1986). Oman Wahiba Sands Project, Rapid Assessment Document. Royal Geographical Society, London, unpublished.
Royer-Carfagni, G.F. (1999). On the thermal degradation of granite. *International Journal of Rock Mechanics and Mining Science*, 36, 119–26.
Rubin, D.M. (1990). Lateral migration of linear dunes in the Strzelecki Desert, Australia. *Earth Surface Processes and Landforms*, 15, 1–14.
Rubin, D.M. and Hesp, P.A. (2009). Multiple origins of linear dunes on Earth and Titan. *Nature Geoscience*, 2, 653–8.
Rubin, D. M. and Hunter, R. E. (1985). Why deposits of longitudinal dunes are rarely recognized in the geologic record. Sedimentology, 32, 147–57.
Rubin, D.M., Tsoar, H. and Blumberg, D.G. (2008). A second look at western Sinai seif dunes and their lateral migration. *Geomorphology*, 93, 335–42.

Ruddiman, W.F. (1997). Tropical Atlantic terrigenous fluxes since 25,000 yrs BP. *Marine Geology*, 136, 189–207.

Ruddiman, W.F. (2001). *Earth's Climate: Past and Future*. New York: W.H. Freeman and Company.

Ruiz-Agudo, E., Mees, F., Jacobs, P. and Rodriguez-Navarro, C. 2007. The role of saline solution properties on porous limestone salt weathering by magnesium and sodium sulphates. *Environmental Geology*, 52, 269–81.

Russell, I.C. (1885). Geological history of Lake Lahontan. *U.S. Geological Survey Monograph* 11.

Russell, W.L. (1929). Drainage alignment in the western Great Plains. *Journal of Geology*, 37, 240–55.

Rychagov, G.I. (1997). Holocene oscillations of the Caspian Sea, and forecasts based on palaeogeographical reconstructions. *Quaternary International*, 41–42, 167–72.

Sabin, Ty.J. and Holliday, V.T. (1995). Playas and lunettes of the southern High Plains: Morphometric and spatial relationships. *Annals of the Association of American Geographers*, 85, 286–305.

Sabtan, A.A. (2005). Geotechnical properties of expansive clay shale in Tabuk, Saudi Arabia. *Journal of Asian Earth Sciences*, 25, 747–57.

Saco, P.M., Willgoose, G.R. and Hancock, G.R. (2007). Eco-geomorphology of banded vegetation patterns in arid and semi-arid regions. *Hydrology and Earth System Sciences*, 11, 1717–30.

Sadiq, A.M. and Nasir, S.J. (2002). Middle Pleistocene karst evolution in the state of Qatar, Arabian Gulf. *Journal of Cave and Karst Studies*, 64, 132–9.

Sadooni, F.N., Howari, F., Edwards, H.G.M. and El-Saiy, A. (2010). Lithology, mineral assemblages and microbial fingerprints of the evaporate-carbonate sediments of Abu Dhabi and their extraterrestrial implications. *International Journal of Astrobiology*, 9, 147–56.

Saez, A., Cabrera, L., Jensen, A. and Chong, G. (1999). Late Neogene lacustrine record and palaeogeography in the Quillagua-Llamara Basin, Central Andean fore-arc (northern Chile). *Palaeogeography, Palaeoclimatology, Palaeoecology*, 151, 5–37.

Safar, M.I. (1985). *Dust and Duststorms in Kuwait*. Kuwait: Meteorological Department.

Saiko, T.A. and Zonn, I.S. (2000). Irrigation expansion and dynamics of desertification in the Circum-Aral region of central Asia. *Applied Geography*, 20, 349–67.

Saini, H.S. and Mutjaba, S.A.I. (2012). Depositional history and palaeoclimatic variations at the northeastern fringe of Thar Desert, Haryana Plains, India. *Quaternary International*, 250, 37–48.

Salman, A.B., Howari, F.M., El-Sankary, M.M., Wali, A.M. and Saleh, M.M. (2010). Environmental impact and natural hazards on Kharga Oasis monumental sites, Western Desert of Egypt. *Journal of African Earth Sciences*, 58, 341–53.

Sancho, C., Peña, J.L., Rivelli, F., Rhodes, E. and Muñoz, A. (2008). Geomorphological evolution of the Tilcara alluvial fan (Jujuy Province, NW Argentina): Tectonic implications and palaeoenvironmental considerations. *Journal of South American Earth Sciences*, 26, 68–77.

Sandercock, P.J., and Hooke, J.M. (2011). Vegetation effects on sediment connectivity and processes in an ephemeral channel in SE Spain. *Journal of Arid Environments*, 75, 239–54.

Sandweiss, D.H., Maasch, K.A., Burger, R.L., Richardson, J.B., Rollins, H.B. and Clement, A. (2001). Variation in Holocene El Niño frequencies: Climate records and cultural consequences in ancient Peru. *Geology*, 29, 603–6.

Sankey, J.B., Germino, M.J. and Glenn, N.F. (2009). Relationships of post-fire aeolian transport to soil and atmospheric conditions. *Aeolian Research*, 1, 75–85.

Santese, M., De Tomasi, F. and Perrone, M.E. (2007). Moderate Resolution Imaging Spectroradiometer (MODIS) and Aerosol Robotic Network (AERONET) retrievals during dust outbreaks over the Mediterranean. *Journal of Geophysical Research*, 112, D18201, doi:10.1029/2007JD008482.

Sapkota, A., Cheburkin, A.K., Bonani, G. and Shotyk, W. (2007). Six millennia of atmospheric dust deposition in southern South America (Isla Navarino, Chile). *The Holocene*, 17, 561–72.

Saqqa, W., and Atallah, M. (2004). Characterization of the aeolian terrain facies in Wadi Araba Desert, southwestern Jordan. *Geomorphology*, 62, 63–87.

Sarnthein, M. and Koopmann, P. (1980). Late Quaternary deep-sea core record of north-west African dust supply and wind circulation. *Palaeoecology of Africa*, 12, 239–53.

Sauermann, G., Rognon, P., Poliakov, A. and Herrmann, H.J. (2000). The shape of barchan dunes of southern Morocco. *Geomorphology*, 36, 47–62.

Sayward, J.M. (1984). Salt action on concrete. *U.S. Army Corps of Engineers, Cold Regions Research and Engineering Laboratory Special Report*, 84–25.

Schaefer, D.A. and Whitford, W.G. (1981). Nutrient cycling by the subterranean termite *Gnathamitermes tubiformans* in a Chihuahuan desert ecosystem. *Ecologia*, 48, 277–83.

Schattner, I. (1961). Weathering phenomena in the crystalline of the Sinai in the light of current notions. *Bulletin of the Research Council of Israel*, 10, 247–66.

Scheidt, S., Lancaster, N. and Ramsey, M. (2011). Eolian dynamics and sediment mixing in the Gran Desierto, Mexico, determined from thermal infrared spectroscopy and remote-sensing data. *Bulletin of the Geological Society of America*, 123, 1628–44.

Schepanski, K., Tegen, I. and Macke, A. (2012). Comparison of satellite based observations of Saharan dust source areas. *Remote Sensing of Environment*, 123, 90–7.

Schepanski, K., Tegen, I., Todd, M.C., Heinold, B., Bönisch, G., Laurent, B. and Macke, A. (2009). Meteorological processes forcing Saharan dust emission inferred from MSG-SEVIRI observations of subdaily dust source activation and numerical models. *Journal of Geophysical Research*, 114, D10201, doi:10.1029/2008JD01325.

Scherer, G.W. (2004). Stress from crystallization of salt. *Cement and Concrete Research*, 34, 1613–24.

Schiavon, N., Chiavari, G., Schiavon, G. and Fabbri, D. (1995). Nature and decay effects on urban soiling and granitic building stones. *The Science of the Total Environment*, 167, 87–101.

Schick, A. (1971). A desert flood: Physical characteristics; effects on man, geomorphic significance, human adaptation – A case study of the Southern Arava watershed. *Jerusalem Studies in Geography*, 2, 91–155.

Schick, A. (1988). Hydrologic aspects of floods in extreme arid environments. In *Flood Geomorphology*, ed. V.R. Baker, R.C. Kochel and P.C. Patton. New York: Wiley, pp. 189–203.

Schick, A. and Lekach, J. (1993). An evaluation of two ten-year sediment budgets, Nahal Yael, Israel. *Physical Geography*, 14, 225–38.

Schick, A., Grodek, T. and Wolman, M.G. (1999). Hydrologic processes and geomorphic constraints on urbanization of alluvial fan slopes. *Geomorphology*, 31, 325–35.

Schipull, K. Von. (1978). Waterpockets (Opferkessel) in Sandsteinen des zentralen Colorado-Plateaus. *Zeitschrift für Geomorphologie*, 22, 426–38.

Schlesinger, W.H. (1985). The formation of caliche in soils of the Mojave Desert, California. *Geochimica et Cosmochimica Acta*, 49, 57–66.

Schmidt, A., Quigley, M., Fattahi, M., Azizi, G., Maghsoudi, M. and Fazali, H. (2011). Holocene settlement shifts and palaeoenvironments on the central Iranian Plateau: Investigating linked systems. *The Holocene*, 21, 583–95.

Schmidt, K.-H. (1989). The significance of scarp retreat for Cenozoic landform evolution on the Colorado Plateau, U.S.A. *Earth Surface Processes and Landforms*, 14, 93–105.
Scholl, D.W. and Taft, W.H. (1964). Algae, contributors to the formation of calcareous tufa, Mono Lake, California. *Journal of Sedimentary Petrology*, 34, 309–19.
Scholz, C.A., Cohen, A.S., Johnson, T.C., King, J., Talbot, M.R. and Brown, E.T. (2010). Scientific drilling in the Great Rift Valley: The 2005 Lake Malawi Scientific Drilling Project – An overview of the past 145,000 years of climate variability in Southern Hemisphere East Africa. *Palaeogeography, Palaeoclimatology, Palaeoecology*, 303, 3–19.
Schooley, R.L. and Wiens, J.A. (2001). Dispersion of kangaroo rat mounds at multiple scales in New Mexico, USA. *Landscape Ecology*, 16, 267–77.
Schoewe, W.H. (1932). Experiments on the formation of wind-faceted pebbles. *American Journal of Science*, 24, 111–34.
Schumm, S.A. (1956). The role of creep and rainwash in the retreat of badland slopes. *American Journal of Science*, 254, 693–706.
Schumm, S.A. (1964). Seasonal variations of erosion rates and process on hillslopes in western Colorado. *Zeitschrift für Geomorphologie*, Supplementband, 5, 215–38.
Schumm, S.A. and Chorley, R.J. (1966). Talus weathering and scarp recession in the Colorado Plateau. *Zeitschrift für Geomorphologie*, 10, 11–36.
Schumm, S.A., Harvey, M.D. and Watson, C.C. (1984). *Incised Channels: Morphology, Dynamics and Control*. Littleton, CO: Water Resources Publications.
Schuster, M. and 10 others. (2009). Chad Basin: Paleoenvironments of the Sahara since the Late Miocene. *CR Geoscience*, 341, 603–11.
Schütz, L., Jaenicke, R. and Pietrek, H. (1981). Saharan dust transport over the North Atlantic Ocean. *Geological Society of America Special Paper*, 186, 87–100.
Schwämmle, V. and Herrmann, H.J. (2005). A model of barchan dunes including lateral shear stress. *The European Physical Journal* E, 16, 57–65.
Schwarz, H.E., Emel, J., Dickens, W.J., Rogers, P. and Thompson, J. (1991). Water quality and flows. In *The Earth as Transformed by Human Action*, ed. B.L. Turner. Cambridge:Cambridge University Press, pp. 253–70.
Scrivner, AE., Vance, D. and Rohling, E.J. (2004). New neodymium isotope data quantify Nile involvement in Mediterranean anoxic episodes. *Geology*, 32, 565–8.
Scuderi, L.A., Laudadio, C.K. and Fawcett, P.J. (2010). Monitoring playa lake inundation in the western United States: Modern analogues to late-Holocene lake level change. *Quaternary Research*, 73, 48–58.
Seager, R. and Vecchi, G.A. (2010). Greenhouse warming and the 21st century hydroclimate of southwestern North America. *Proceedings of the National Academy of Sciences*, 107, 21277–82.
Seager, R. and 11 others. (2007). Model projections of an imminent transition to a more arid climate in southwestern North America. *Science*, doi:10.1126/science.1139601.
Searl, A. and Rankin, S. (1993). A preliminary petrographic study of the Chilean nitrates. *Geological Magazine*, 130, 319–33.
Segerstrom, K. and Henriquez, H. (1964). Cavities or 'tafoni' in rock faces of the Atacama Desert, Chile. *United States Geological Survey Professional Paper*, 501C, 121–5.
Seifan, N. (2009). Long-term effects of anthropogenic activities on semi-arid sand dunes. *Journal of Arid Environments*, 73, 332–7.
Seifert, C.L., Xox, R.T., Forman, S.L., Foti, T.L., Wasklewicz, T.A. and McColgan, A.T. (2009). Relict nebkhas (pimple mounds) record prolonged late Holocene drought in the forested region of south-central United States. *Quaternary Research*, 71, 329–39.
Selwitz, C. (1990). Deterioration of the Great Sphinx: An assessment of the literature. *Antiquity*, 64, 853–9.

Senut B., Pickford M. and Ségalen, L. (2009). Neogene desertification of Africa. *C.R.Geoscience*, 341, 591–602.

Senut, B., Pickford, M. and Ward, J. (1994). Biostratigraphie de éolianites néogènes du Sud de la Sperrgebeit (Désert de Namib, Namibie). *Comptes Rendus de l'Academie des Sciences*, 318, 1001–7.

Sepúlveda, S.A., Rebolledo, S. and Vargas, G. (2006). Recent catastrophic debris flows in Chile: Geological hazard, climatic relationships and human response. *Quaternary International*, 158, 83–95.

Sereda, J., Bogard, M., Hudson, J., Helps, D. and Dessouki, T. (2010). Climate warming and the onset of salinization: Rapid changes in the limnology of two northern plains lakes. *Limnologica*, 41, 1–9.

Shahid, S.A. and Abdelfattah, M.A. (2009). Gypsum polymorphism in the desert environment of Abu Dhabi Emirate. *European Journal of Scientific Research*, 29, 237–48.

Shakesby, R.A., Doerr, S.H. and Walsh, R.P.D. (2000). The erosional impact of soil hydrophobicity: current problems and future research directions. Journal of Hydrology, 231–232, 178–91.

Shanyengana, E.S., Henschel, J.R., Seely, M.K. and Sanderson, R.D. (2002). Exploring fog as a supplementary water source in Namibia. *Atmospheric Research*, 64, 251–9.

Shao, M., Li, L., Wang, S., Wang, E. and Li, Z. (In press). Deterioration mechanisms of building materials of Jiaohe ruins in China. *Journal of Cultural Heritage*.

Shao, Y. (2000). *Physics and Modelling of Wind Erosion*. Dordrecht: Kluwer.

Shao, Y. (2008). *Physics and Modelling of Wind Erosion* (2nd ed.). Berlin and Heidelberg: Springer.

Shao, Y. and Wang, J.J. (2003). A climatology of northeast Asian dust events. *Meteorologische Zeitschrift*, 12, 175–83.

Shao, Y., Raupach, M. and Short, D. (1994). Preliminary assessment of wind erosion patterns in the Murray-Darling Basin. *Australian Journal of Soil and Water Conservation*, 7, 46–51.

Shao, Y. and 9 others. (2011). Dust cycle: An emerging core theme in Earth system science. *Aeolian Research*, 2, 181–204.

Shaqour, F. (1994). Hydrogeologic role in sinkhole development in the desert of Kuwait. *Environmental Geology*, 23, 201–8.

Sharon, D. (1962). On the nature of hamadas in Israel. *Zeitschrift für Geomorphologie*, 6, 129–47.

Sharon, D. (1972). The spotttiness of rainfall in a desert area. *Journal of Hydrology*, 17, 161–75.

Sharp, R.P. (1949). Pleistocene ventifacts east of the Bighorn Mountains, Wyoming. *Journal of Geology*, 57, 175–95.

Sharp, R.P. (1964). Wind-driven sand in the Coachella valley, California. *Bulletin of the Geological Society of America*, 75, 785–804.

Sharp R.P. (1966). Kelso Dunes, Mojave Desert, California. *Bulletin of the Geological Society of America*, 77, 1045–74.

Sharp, R.P. and Nobles, L.H. (1953). Mudflow of 1941 at Wrightwood, southern California. *Bulletin of the Geological Society of America*, 64, 547–60.

Sharpe, S.E. (2006). Using modern through mid-Pleistocene climate proxy data to bound future variations in infiltration at Yucca Mountain, Nevada. *Memoir of the Geological Society of America*, 99, 155–205.

Sharratt, B., Wendling, L. and Feng, G. (2010). Windblown dust affected by tillage intensity during summer fallow. *Aeolian Research*, 2, 129–34.

Shaw, A.I. and Goudie, A.S. (2002). Geomorphological evidence for the extension of the Mega-Kalahari into south-central Angola. *South African Geographical Journal*, 84, 182–94.

Shaw, A.I. and Goudie, A.S. (2004). Chemical sediments of the palaeo and active Okavango alluvial fan, Botswana: Evidence for variations in the processes of formation. *Zeitschrift für Geomorphologie*, Supplementband, 136, 89–112.

Shaw, P.A. (1985). The desiccation of Lake Ngami: An historical perspective. *Geographical Journal*, 151, 318–26.

Shaw, P.A. and De Vries, J.J. (1988). Duricrusts, groundwater and valley development in the Kalahari of southwest Botswana. *Journal of Arid Environments*, 14, 245–54.

Shaw, P.A. and Nash, D.J. (1998). Dual mechanisms for the formation of fluvial silcretes in the distal reaches of the Okavango Delta Fan, Botswana. *Earth Surface Processes and Landforms*, 23, 705–14.

Shaw, P.A. and Thomas, D.S.G. (1997). Pans, playas and salt lakes. In *Arid Zone Geomorphology: Process, Form and Change in Drylands* (2nd ed.), ed. D.S.G. Thomas. Chichester: Wiley, pp. 293–317.

Shaw, P.A., Thomas, D.S.G. and Nash, D.J. (1992). Late Quaternary fluvial activity in the dry valleys (megacha) of the middle and southern Kalahari, southern Africa. *Journal of Quaternary Science*, 7, 273–81.

Shehata, W. and Lotfi, H. (1993). Preconstruction solution for groundwater rise in Sabkha. *Bulletin of the International Association of Engineering Geology*, 47, 145–50.

Shehata, W., Bader, T., Irtem, O., Ali, A., Abdallah, M. and Aftab, S. (1992). Rate and mode of barchan dunes advance in the central part of the Jafurah sand sea. *Journal of Arid Environments*, 23, 1–17.

Sherif, M.M. and Singh, V.P. (1999). Effect of climate change on sea water intrusion in coastal aquifers. *Hydrological Processes*, 13, 1277–87.

Sherman, D.J. and Li, B. (2012). Predicting aeolian sand transport rates: A reevaluation of models. *Aeolian Research*, 3, 371–8.

Sherratt, A. (1997). Climatic cycles and behavioural revolutions: The emergence of modern humans and the beginning of farming. *Antiquity*, 71, 271–87.

Shi, N., Dupont, L.M., Beug, H.-J. and Schneider, R. (2000). Correlation between vegetation in southwestern Africa and oceanic upwelling in the past 21,000 years. *Quaternary Research*, 54, 72–80.

Shi, P., Yan, P., Yuan, Y. and Nearing, M.A. (2004). Wind erosion research in China: Past, present and future. *Progress in Physical Geography*, 28, 366–86.

Shiga, Y., Greene, R.S.B., Scott, K.M. and Stelcer, E. (2011). Recognising terrestrially-derived salt (NaCl) in ESE Australian dust. *Aeolian Research*, 2, 215–20.

Shiklomanov, I.A. (1999). Climate change, hydrology and water resources: The work of the IPCC (1988–1994). In *Impacts of Climate Change and Climate Variability on Hydrological Regimes*, ed. J.C. van Dam. Cambridge: Cambridge University Press, pp. 8–20.

Shroder, J.F. (1993). *Himalaya to the Sea*. London: Routledge.

Shroder, J.F., Schettler, M.J. and Weihs, B.J. (2011). Loess failure in northeast Afghanisatn. *Physics and Chemistry of the Earth*, 36, 1287–93.

Siesser, W.G. (1980). Late Miocene origin of the Benguela upwelling system off northern Namibia. *Science*, 208, 283–5.

Simon-Coinçon, R., Milnes, A.R., Thiry, M. and Wright, M.J. (1996). Evolution of landscapes in northern south Australia in relation to the distribution and formation of silcretes. *Journal of the Geological Society of London*, 153, 467–80.

Sims, D.B. and Francis, A.W. (2008). Mercury and cyanide used as indicators of sedimentary transport in ephemeral washes at the Techatticup mine and mill site, Nelson, Nevada. *International Journal of Soil, Sediment and Water*, 1, article 4.

Sims, D.B., Hooda, P.S. and Gillmore, G.K. (2012). Potential effects of environmental change on mining wastes in a hyperarid climate. *Environment and Pollution*, 1, doi:10.5539/ep.v1n2p61.

Sinclair, J.G. (1922). Temperatures of soil and air in a desert. *Monthly Weather Review*, 50, 142–4.

Singer, A. (1979). Palygorskite in sediments: Detrital diagenetic or neoformed: A critical review. *Geologische Rundschau*, 68, 996–1008.

Singer, A. (1980). The paleoclimatic interpretation of clay minerals in soils and weathering profiles. *Earth-Science Reviews*, 15, 303–26.

Singer, A. (1988). Illite in aridic soils, desert dusts and desert loess. *Sedimentary Geology*, 59, 251–9.

Singer, A., Zobeck, T., Poberezsky, L. and Argaman, E. (2003). The PM_{10} and $PM_{2.5}$ dust generation potential of soils/sediments in the southern Aral Sea Basin, Uzbekistan. *Journal of Arid Environments*, 54, 705–28.

Singer, M.J. and Shainberg, I. (2004). Mineral soil surface crusts and wind and water erosion. *Earth Surface Processes and Landforms*, 29, 1065–75.

Singh, G., Wasson, R.J. and Agrawal, D.P. (1990). Vegetational and seasonal climatic changes since the last full glacial in the Thar Desert, northwest India. *Review of Palaeobotany and Palynology*, 64, 351–8.

Singhvi, A.K. and Kar, A. (2004). The aeolian sedimentation record of the Thar. *Proceedings of the Indian Academy of Sciences (Earth and Planetary Science)*, 113, 371–401.

Singhvi, A.K. and Porat, N. (2008). Impact of luminescence dating on geomorphological and palaeoclimate research in drylands. *Boreas*, 37, 536–58.

Singhvi, A.K., and 9 others (2010). A $\sim$200ka record of climatic change and dune activity in the Thar Desert, India. *Quaternary Science Reviews*, 29, 3095–105.

Sirocko, F., Sarnthein, M., Lange, H. and Erlenkeuser, H. (1991). Atmospheric summer circulation and coastal upwelling in the Arabian Sea during the Holocene and the last glaciation. *Quaternary Research*, 36, 72–93.

Sivan, D. and Porat, N. (2004). Evidence from luminescence for late Pleistocene formation of calcareous aeolianite (kurkar) and paleosol (hamra) in the Carmel Coast, Israel. *Palaeogeography, Palaeoclimatology, Palaeoecology*, 211, 95–106.

Skidmore, E.L. (1986). Wind erosion climatic erosivity. *Climatic Change*, 9, 195–208.

Skinner, B.J. (1966). Thermal expansion. *Geological Society of America Memoir*, 97, 75–96.

Slattery, M.J. (1990). Barchan migration on the Kuiseb River Delta, Namibia. *South African Geographical Journal*, 72, 5–10.

Smale, D. (1973). Silcretes and associated silica diagenesis in southern Africa and Australia. *Journal of Sedimentary Petrology*, 43, 1077–89.

Small, I., van der Meer, J. and Upshur, R.E.G. (2001). Acting on an environmental health disaster: The case of the Aral Sea. *Environmental Health Perspectives*, 109, 547–9.

Smalley, I. J. (ed.). (1975). *Loess: lithology and genesis*. Stroudsburg: Dowden, Hutchinson and Ross.

Smalley, I.J. and Krinsley, D.H. (1978). Loess deposits associated with deserts. *Catena*, 5, 53–66.

Smalley, I.J. and Vita-Finzi, C. (1968). The formation of fine particles in sandy deserts and the nature of 'desert' loess. *Journal of Sedimentary Petrology*, 38, 766–74.

Smith, A.M. and Mason, T.R. (1991). Pleistocene, multiple-growth, lacustrine oncoids from the Poachers' Point Formation, Etosha Pan, northern Namibia. *Sedimentology*, 38, 591–9.

Smith, B.J. (1978). The origin and geomorphic implications of cliff foot recesses and tafoni on limestone hamadas in the northwest Sahara. *Zeitschrift für Geomorphologie*, 22, 21–43.

Smith, B.J. (1987). An integrated approach to the weathering of limestone in an arid area and its role in landscape evolution: A case study from southeast Morocco. In *International Geomorphology 1986*, ed. V. Gardiner. Chichester: Wiley, pp. 637–57.
Smith, B.J. (1988). Weathering of superficial limestone debris in a hot desert environment. *Geomorphology*, 1, 355–67.
Smith, B.J. (1994). Weathering processes and forms. In *Geomorphology of Desert Environments*, ed. A.D. Abrahams and A.J. Parsons. London: Chapman & Hall, pp. 39–63.
Smith, B.J. (2009). Weathering processes and forms. In *Geomorphology of Desert Environments* (2nd ed.), ed. A.J. Parsons and A.D. Abrahams. Heidelberg: Springer, pp. 69–100.
Smith, B.J., and McAlister, J.J. (1986). Observations on the occurrence and origins of salt weathering phenomena near Lake Magadi, southern Kenya. *Zeitschrift für Geomorphologie*, 30, 445–60.
Smith, B.J. and McGeevy, J.P. (1983). A simulation study of salt weathering in hot deserts. *Geografiska Annaler*, 65A, 127–33.
Smith, B.J., McGreevy, J.P. and Whalley, W.B. (1987). Silt production by weathering of a sandstone under hot, arid conditions: An experimental study. *Journal of Arid Environments*, 12, 199–214.
Smith, B.J., Warke, P.A. and Moses, C.A. (2000). Limestone weathering in contemporary arid environments: A case study from southern Tunisia. *Earth Surface Processes and Landforms*, 25, 1343–54.
Smith, B.J., Wright, J.S. and Whalley, W.B. (2002). Sources of non-glacial, loess-size quartz silt and the origins of 'desert loess'. *Earth-Science Reviews*, 59, 1–26.
Smith, G.D., Coughland, K.J., Yule, D.F., Laryea, K.B., Srivastava, K.L., Thomas, N.P. and Cogle, A.L. (1992). Soil management options to reduce runoff and erosion on a hardsetting Alfisol in the semi-arid tropics. *Soil and Tillage Research*, 25, 195–215.
Smith, G.I. and Street-Perrott, F.A. (1983). Pluvial lakes of the western United States. In *Late Quaternary Environments of the United States, 1. The Late Pleistocene*, ed. S. C. Porter. London: Longman, pp. 190–212.
Smith, G.I. (2009). Late Cenozoic geology and lacustrine history of Searles Valley, Inyo and San Bernadino Counties, California. *U.S. Geological Survey Professional Paper*, 1727.
Smith, J.J., Platt, B.F., Ludvigson, G.A. and Thomasson, J.R. (2011). Ant-nest ichnofossils in honeycomb calcretes, Neogene Ogallala Formation, High Plains region of western Kansas, U.S.A. *Palaeogeography, Palaeoclimatology, Palaeoecology*, 308, 383–94.
Smith, J.R., Giegengack, R., Schwarcz, H.P., McDonald, M.M.A., Kleindienst, M.R., Hawkins, A.L. and Churcher, C.S. (2004). A reconstruction of Quaternary pluvial environments and human occupations using stratigraphy and geochronology of fossil-spring tufas, Kharga Oasis, Egypt. *Geoarchaeology*, 19, 407–39.
Smith, L.L. (1941). Weather pits in granite of the southern Piedmont. *Journal of Geomorphology*, 4, 117–27.
Smith, R.M.H., Mason, T.R. and Ward, J.D. (1993). Flash-flood sediments and ichnofacies of the late Pleistocene Homeb Silts, Kuiseb River, Namibia. *Sedimentary Geology*, 85, 579–99.
Smith, S.E. (1986). An assessment of structural deterioration on ancient Egyptian monuments and tombs in Thebes. *Journal of Field Archaeology*, 13, 503–10.
Smolikowski, B., Puig, H. and Roose, E. (2001). Influence of soil protection techniques on runoff, erosion and plant production on semi-arid hillslopes of Cabo Verde. *Agriculture, Ecosystems and Environment*, 87, 67–80.
Smykatz-Kloss, W., Roscher, B. and Rögner, K. (2003). Pleistocene lakes in central Sinai, Egypt. *In Desertification in the Third Millennium*, ed. A.S. Alsharhan, W.W. Wood, A.S. Goudie, A. Fowler and E.M. Abdellatif. Lisse: Swets and Zeitlinger, pp. 111–16.

Snelder, D.J. and Bryan, R.B. (1995). The use of rainfall simulation tests to assess the influence of vegetation density on soil loss on degraded rangelands in the Baringo District, Kenya. *Catena*, 25, 105–16.

Soderblom, L.A. and 18 others. (2007). Topography and geomorphology of the Huygens landing site on Titan. *Planetary and Space Science*, 55, 2015–24.

Soleilhavoup, F. (2011). Microformes d'accumulation et d'ablation sur les surfaces désertiques du Sahara. *Géomorpholpogie*, 2, 173–86.

Soroka, I. (1993). *Concrete in hot environments*. London: Spon.

Sougnez, N., van Wesemeal, B. and Vanacker, V. (2011). Low erosion rates measured for steep, sparsely vegetated catchments in southeast Spain. *Catena*, 84, 1–11.

Soukup, D., Buck, B., Goossens, D., Ulery, B.T., McLaurin, B.T., Baron, D. and Teng, Y. (2012). Arsenic concentrations in dust emissions from wind erosion and off-road vehicles in the Nellis Dunes Recreational Area, Nevada, USA. *Aeolian Research*, 5, 77–89.

Soulard, C.E., Esque, T.C., Bedford, D.R. and Bond, S. (2012). The role of fire on soil mounds and surface roughness in the Mojave Desert. *Earth Surface Processes and Landforms*, doi:10.1002/esp/3264.

Soussa, H., El Feel, A.A., Alfy, S.Z. and Yousif, M.S.M. (2012). Flood hazard in Wadi Rahbaa area, Egypt. *Arabian Journal of Geoscience*, 5, 45–52.

Spate, O.H.K. (1957). *India and Pakistan: A General and Regional Geography* (2nd ed.). London: Methuen.

Spath, H. (1977). Rezente Verwitterung and Abtragung and Schicht- and Rumpfstufen im semi-ariden Westaustralien. *Zeitschrift für Geomorphologie*, Supplementband, 28, 81–100.

Spaulding, W.G. (1991). Pluvial climatic episodes in North America and North Africa: Types and correlation with global climate. *Palaeogeography, Palaeoclimatology, Palaeoecology*, 84, 217–29.

Spelz, R.M., Fletcher, J.M., Owen, L.A. and Caffee, M.W. (2008). Quaternary alluvial-fan development, climate and morphologic dating of fault scarps in Laguna Salada, Baja California, Mexico. *Geomorphology*, 102, 578–94.

Sperling, C.H.B. and Cooke, R.U. (1985). Laboratory simulation of rock weathering by salt crystallization and hydration processes in hot, arid environments. *Earth Surface Processes and Landforms*, 10, 541–55.

Sperling, C.H.B. and Goudie, A.S. (1975). The miliolite of western India: A discussion of the aeolian and marine hypotheses. *Sedimentary Geology*, 13, 71–5.

Springer, M.E. (1958). Desert pavement and vesicular layer of some desert soils in the desert of the Lahontan Basin. *Soil Science Society of America Proceedings*, 22, 63–6.

Sridhar, V., Loope, D.B., Swinehart, J.B., Mason, J.A., Oglesby, R.J. and Rowe, C.M. (2006). Large wind shift on the Great Plains during the Medieval Warm Period. *Science*, 313, 345–7.

Srivastava, P., Brook, G.A., Marais, E., Morthekai, P. and Singhvi, A.K. (2006). Depositional environment and OSL chronology of the Homeb silt deposits, Kuiseb River, Namibia. *Quaternary Research*, 65, 478–91.

Stager, J.C., Ryves, D.B., Chase, B.M. and Pausata, F.S.R. (2011). Catastrophic drought in the Afro-Asian monsoon region during Heinrich Event 1. *Science*, 331, 1299–302.

Stanley, D.J. (1996). Nile Delta: Extreme case of sediment entrapment on a delta plain and consequent coastal land loss. *Marine Geology*, 129, 189–95.

Staubwasser, M. and Weiss, H. (2006). Holocene climate and cultural evolution in late prehistoric–early historic West Asia. *Quaternary Research*, 66, 372–87.

Staubwasser, M., Sirocko, F., Grootes, P.M. and Segl, M. (2003). Climate change at the 4.2 ka BP termination of the Indus Valley Civilization and Holocene South Asian monsoon variability. *Geophysical Research Letters*, 30, 1425, doi:10.1029/2002GL016822.

Steffen, D., Schlunegger, F. and Preusser, F. (2009). Drainage basin response to climate change in the Pisco Valley, Peru. *Geology*, 37, 491–4.

Steffen, D., Schlunegger, F. and Preusser, F. (2010). Late Pleistocene fans and terraces in the Majes Valley, southern Peru, and their relation to climatic variations. *International Journal of Earth Sciences*, 99, 1975–89.

Steiger, M. and Asmussen, S. (2008). Crystallization of sodium sulphate phases in porous materials: The phase diagram Na_2SO_4-H_20 and the generation of stress. *Geochimica et Cosmochimica Acta*, 72, 4291–306.

Stein, A. (1912). *Ruins of Desert Cathay*. London: Macmillan.

Stein, A. (1942). A survey of ancient sites along the 'lost' Sarasvati River. *Geographical Journal*, 99, 173–82.

Stein, M., Torfstein, A., Gavriela, I. and Yechieli, Y. (2010). Abrupt aridities and salt deposition in the post-glacial Dead Sea and their North Atlantic connection. *Quaternary Science Reviews*, 29, 567–75.

Stein, R. (1985). The post-Eocene sediment record of DSDP Site 366: Implications for African climate and plate tectonic drift. *Geological Society of America Memoir*, 163, 305–15.

Stephen, L., Bellis, E. and Muir, A. (1956). Gilgai phenomena in tropical black clays of Kenya. *Journal of Soil Science*, 7, 1–9.

Stephens, C.G. (1971). Laterite and silcrete in Australia: A study of the genetic relationships of laterite and silcrete and their companion materials, and their collective significance in the formation of the weathered mantle, soils, relief and drainage of the Australian continent. *Geoderma*, 5, 5–52.

Stephens C.G. and Crocker R.L. (1946). Composition and genesis of lunettes. *Transactions of the Royal Society of South Australia*, 70, 302–12.

Sterk, G. (2003). Causes, consequences and control of wind erosion in Sahelian Africa: A review. *Land Degradation and Development*, 14, 95–108.

Sternberg, M. and Shoshany, M. (2001). Influence of slope aspect on Mediterranean woody formations: Comparison of a semiarid and an arid site in Israel. *Ecological Research*, 16, 335–45.

Stetler, L.D. and Gaylord, D.R. (1996). Evaluating eolian-climatic interactions using a regional climate model from Hanford, Washington (USA). *Geomorphology*, 17, 99–113.

Stevens, T., Palk, C., Carter, A., Lu, H. and Clift, P.D. (2010). Assessing the provenance of loess and desert sediments in northern China using U-Pb dating and morphology of detrital zircons. *Bulletin of the Geological Society of America*, 122, 1331–44.

Stirling, M., Ledgerwood, J., Liu, T. and Apted, M. (2010). Age of unstable bedrock landforms southwest of Yucca Mountain, Nevada, and implications for ground motions. *Bulletin of the Seismological Society of America*, 100, 74–86.

Stock, J.D., Schmidt, K.M. and Miller, D.M. (2007). Controls on alluvial fan long-profiles. *Bulletin of the Geological Society of America*, 120, 619–40.

Stokes, M., Nash, D.J. and Harvey, A.M. (2007). Calcrete 'fossilisation' of alluvial fans in SE Spain: The roles of groundwater, pedogenic processes and fan dynamics in calcrete development. *Geomorphology*, 85, 63–84.

Stokes, S. and Bray, H. (2004). Reconciling lateral and vertical dune migration rates and drift potential estimates: Some examples from the Rub Al Khali. *Geophysical Research Abstracts*, 6, 04012.

Stokes, S. and Gaylord, D.R. (1997). Optical dating of Holocene dune sands in the Ferris Dune Field, Wyoming. *Quaternary Research*, 39, 274–81.

Stokes, S. and Swinehart, J.B. (1997). Middle- and late-Holocene dune reactivation in the Nebraska Sand Hills, USA. *The Holocene*, 7, 263–72.

Stokes, S., Thomas, D.S.G. and Shaw, P.A. (1997b). New chronological evidence for the nature and timing of liner dune development in the southwest Kalahari Desert. *Geomorphology*, 20, 81–93.
Stokes, S., Thomas, D.S.G. and Washington, R. (1997a). Multiple episodes of aridity in southern Africa since the last Interglacial period. *Nature*, 388, 154–8.
Stokes, S., Haynes, G., Thomas, D.S.G., Horrocks, J.L., Higginson, M. and Malifa, M. (1998). Punctuated aridity in southern Africa during the Last Glacial cycle: The chronology of linear dune construction in the northeastern Kalahari. *Palaeogeography, Palaeoclimatology, Palaeoecology*, 137, 305–22.
Stokes, S., Goudie, A.S., Ballard, J., Gifford, C., Samieh, S., Embabi, N. and El-Rashidi, O.A. (1999). Accurate dune displacement and morphometric data using kinematic GPS. *Zeitschrift für Geomorphologie*, Supplementband, 116, 195–214.
Stone, A.E.C. and Thomas, D.S.G. (2008). Linear dune accumulation chronologies from the southwest Kalahari, Namibia: Challenges of reconstructing late Quaternary palaeoenvironments from aeolian landforms. *Quaternary Science Reviews*, 27, 1667–81.
Stone, A.E.C. and Thomas, D.S.G. (In press). Casting new light on late Quaternary environmental and paleohydrological change in the Namib Desert: A review of the application of optically stimulated luminescence in the region. *Journal of Arid Environments*.
Stone, A.E.C., Viles, H.A., Thomas, L. and van Calsteren, P. (2010). Can ^{234}U-^{230}Th dating be used to date large semi-arid tufas? Challenges from a study in the Naukluft Mountains, Namibia. *Journal of Quaternary Science*, 25, 1360–72.
Stout, J.E. (2010). Diurnal patterns of blowing sand. *Earth Surface Processes and Landforms*, 35, 314–8.
Stout, J.E. (2012). A field study of wind erosion following a grass fire on the Llano Estacado of North America. *Journal of Arid Environments*, 182, 165–74.
Strasser, M. and Schlunegger, F. (2005). Erosional processes, topographic length-scales and geomorphic evolution in arid climatic environments: The 'Lluta collapse', northern Chile. *International Journal of Earth Science*, 94, 433–46.
Street, F.A. (1980). The relative importance of climate and local hydrogeological factors in influencing lake-level fluctuations. *Palaeoecology of Africa*, 12, 137–58.
Street, F.A. and Grove, A.T. (1979). Global maps of lake-level fluctuations since 30,000 yr BP. *Quaternary Research*, 12, 83–118.
Street-Perrott, F.A. and Gasse, F. (1981). Recent development in research into the Quaternary climatic history of the Sahara. In *The Sahara: Geological Change and Early Economic History*, ed. J.A. Allen. London: SOAS, pp. 7–28.
Strong, C.L., Parsons, K., McTainsh, G.H. and Sheehan, A. (2011). Dust transporting wind systems in the lower Lake Eyre Basin, Australia: A preliminary study. *Aeolian Research*, 2, 205–14.
Strong, C.L., Bullard, J.E., Dubois, C., McTainsh, G.H. and Baddock, M.C. (2010). Impact of wildfire on interdune ecology and sediments: An example from the Simpson Desert, Australia. *Journal of Arid Environments*, 74, 1577–81.
Strudley, M.W. and Murray, A.B. (2006). Sensitivity analysis of pediment development through numerical simulation and selected geospatial query. *Geomorphology*, 88, 329–51.
Strudley, M.W., Murray, A.B. and Haff, P.K. (2006). Emergence of pediments, tors, and piedmont junctions from a bedrock weathering-regolith thickness feedback. *Geology*, 34, 805–8.
Stuut, J.-B.W., Prins, M.A., Schneider, R.R., Weltje, G.J., Jansen, J.H.F. and Postma, G. (2002). A 300-kyr record of aridity and wind strength in southwestern Africa:

Inferences from grain-size distributions of sediments on Walvis Ridge, SE Atlantic. *Marine Geology*, 180, 221–33.

Stüwe, K., Robl, J. and Matthai, S. (2009). Erosional decay of the Yucca Mountain Crest, Nevada. *Geomorphology*, 108, 200–8.

Subyani, A.M. (2011). Hydrologic behavior and flood probability for selected arid basins in Makkah area, western Saudi Arabia. *Arabian Journal of Geoscience*, 4, 817–24.

Subyani, A.M., Qari, M.H. and Matsah, M.I. (2012). Digital elevation model and multivariate statistical analysis of morphometric parameters of some wadis, western Saudi Arabia. *Arabian Journal of Geoscience*, 5, 147–57.

Suchodoletz, H. von, Oberhänsli, H., Faust, D., Fuchs, M., Blanchet, C., Goldhammer, T. and Zöller, L. (2010). The evolution of Saharan dust input on Lanzarote (Canary Islands) – Influenced by human activity during the early Holocene? *The Holocene*, 20, 169–79.

Suchodoletz, H. von and 8 others. (2011). The influence of Saharan dust deposits on La Palma soil properties (Canary Islands, Spain). *Catena*, 20, 169–79.

Sultan, B., Labadi, K., Guégan, J.-F. and Janicot, S. (2005). Climate drives the meningitis epidemics onset in West Africa. *PLoS Medicine*, 2, 44–9.

Summerfield, M.A. (1982). Distribution, nature and probable genesis of silcrete in arid and semi-arid southern Africa. *Catena*, Supplement, 1, 37–66.

Summerfield, M.A. (1983a). Petrology and diagenesis of silcrete from the Kalahari Basin and Cape Coastal Zone, southern Africa. *Journal of Sedimentary Petrology*, 53, 895–909.

Summerfield, M.A. (1983b). Silcrete as a palaeoclimatic indicator: Evidence from southern Africa. *Palaeogeography, Palaeoclimatology, Paleoecology*, 41, 65–79.

Sumner, P.D., Hedding, D.W. and Meiklejohn, K.I. (2007). Rock surface temperatures in southern Namib and implications for thermally-driven physical weathering. *Zeitschrift für Geomorphologie*, Supplementband, 51, 133–47.

Sumner, P.D., Meiklejohn, K.I. and Hedding, D.W. (2004). Thermal attributes of rock weathering: Zonal or azonal? A comparison of rock temperatures in different environments. *Polar Geography*, 28, 79–92.

Sun, J., Ye, J., Wu, W., Ni, X., Shundong, B., Zhang, Z., Liu, W. and Meng, J. (2010). Late Oligocene-Miocene mid-latitude aridification and wind patterns in the Asian interior. *Geology*, 38, 515–18.

Sun, J., Zhang, Z. and Zhang, L. (2009). New evidence on the age of the Taklimakan Desert. *Geology*, 37, 159–62.

Sunamura, T. and Aoki, H. (2011). Application of an S-shaped curve model to the temporal development of a tafoni of salt-weathering origin. *Earth Surface Processes and Landforms*, 36, 1624–31.

Svendsen, J., Stollhofen, H., Krapf, C.B.E. and Stannistreet, I.G. (2003). Mass and hyperconcentrated flow deposits record dune damming and catastrophic breakthrough of ephemeral rivers, Skeleton Coast Erg, Namibia. *Sedimentary Geology*, 160, 7–31.

Svendsen, J.B. (2002). Parabolic halite dunes on the Salar de Uyuni, Bolivia. *Sedimentary Geology*, 155, 147–56.

Swami, S.G. (2011). Temperature, strain and acoustic emission monitoring of a natural boulder exposed to the sun: A test of the efficacy of insolation in physical weathering. MSc thesis, University of North Carolina, Charlotte.

Swap, R., Ulanski, S., Cobbett, M. and Garstang, M. (1996). Temporal and spatial aspects of Saharan dust outbreaks. *Journal of Geophysical Research* D, 102, 4205–20.

Sweeney, M.R., McDonald, E.V. and Etyemezian, V. (2011). Quantifying dust emissions from desert landforms, eastern Mojave Desert, USA. *Geomorphology*, 135, 21–34.

Sweeting, M.M. and Lancaster, N. (1982). Solutional and wind erosion forms on limestone in the central Namib Desert. *Zeitschrift für Geomorphologie*, 26, 197–207.

Swezey, C. (1996). Structural controls on Quaternary depocentres within the Chotts trough region of southern Tunisia. *Journal of African Earth Sciences*, 22, 335–47.
Swezey, C. (1997). Climatic and tectonic controls on Quaternary eolian sedimentary sequences of the Chott Rharsa Basin, southern Tunisia. PhD diss., University of Texas at Austin.
Swezey, C. (2001). Eolian sediment responses to late Quaternary climate changes: Temporal and spatial patterns in the Sahara. *Palaeogeography, Palaeoclimatology, Palaeoecology*, 167, 119–55.
Swezey, C. (2009). Cenozoic stratigraphy of the Sahara, North Africa. *Journal of African Earth Sciences*, 53, 89–121.
Swezey, C., Lancaster, N., Koaurek, G., Deynoux, M., Blum, M., Price, D. and Pion, J.-C. (1999). Response of aeolian systems to Holocene climatic and hydrologic changes on the northern margin of the Sahara: A high-resolution record from the Chott Rharsa Basin, Tunisia. *The Holocene*, 9, 141–7.
Sylla, M.B., Gaye, A.T., Jenkins, G.S., Pal, J.S. and Giorgi, F. (2010). Consistency of projected drought over the Sahel with changes in the monsoon circulation and extremes in a regional climate model projections. *Journal of Geophysical Research*, 115, D16108, doi:10.1029/2009JD012983.
Symmons, P.M. and Hemming, C.F. (1968). A note on wind-stable stone-mantles in the southern Sahara. *Geographical Journal*, 134, 60–4.
Szabo, B.J. (1990). Ages of travertine deposits in eastern Grand Canyon National Park, Arizona. *Quaternary Research*, 34, 24–32.
Szynkiewicz, A., Ewing, R.C., Moore, C.H., Glamoclija, M., Bustos, D. and Pratt, L.M. (2009). Origin of terrestrial gypsum dunes – Implications for Martian gypsum-rich dunes of Olympia Undae. *Geomorphology*, 121, 69–83.
Ta, W., Xiao, H. and Dong, Z. (2008). Long-term morphodynamic changes of a desert reach of the Yellow River following upstream large reservoirs' operation. *Geomorphology*, 97, 249–59.
Taibi, A.N., Kemmouche, A. and Parrot, J.-F. (1999). Détermination des dynamismes d'évolution morphologique et végétale combinées des dayas du piedmont sud de L'Atlas saharien (Algérie) par télédétection. *Secheresse*, 10, 63–7.
Tanaka, T.Y. and Chiba, M. (2006). A numerical study of the contributions of dust source regions to the global dust budget. *Global and Planetary Change*, 52, 88–104.
Tate, G.P. (1910). *Seistan*. Calcutta: Government Printer.
Tatum, D.I. and Francke, J. (2012). Constructing hydrocarbon reservoir analogues of aeolian systems using ground penetrating radar. *Journal of Applied Geophysics*, 81, 21–8.
Taylor, M.P. and Hudson-Edwards, K.A. (2008). The dispersal and storage of sediment-associated metals in an arid river system: The Leichardt River, Mount Isa, Queensland, Australia. *Environmental Pollution*, 152, 193–204.
Taylor, M.P. and Kesterton, R.G.H. (2002). Heavy metal contamination of an arid river environment: Gruben River, Namibia. *Geomorphology*, 42, 311–27.
Tchakerian, V.P. (1997). North America. In *Arid Zone Geomorphology: Process Form and Change in Drylands* (2nd edition). Ed. D.S.G. Thomas. Chichester: Wiley, pp. 523–41.
Tchakerian, V.P. and Lancaster, N. (2001). Late Quaternary arid/humid cycles in the Mojave Desert and western Great Basin of North America. *Quaternary Science Reviews*, 21, 799–810.
Tegen, I. and Fung, I. (1994). Modelling of mineral dust in the atmosphere: Sources, transport and optical thickness. *Journal of Geophysical Research*, 99 (D11), 22879–914.
Tegen, I. and Schepanski, K. (2009). The global distribution of mineral dust. *IOP Conference Series: Earth and Environmental Science*, 7, doi:10.1088/1755–1307/7/7/1/012001.

Telfer, M.W. (2011). Growth by extension, and reworking, of a south-western Kalahari linear dune. *Earth Surface Processes and Landforms*, doi:10.1002/esp.2140.

Telfer, M.W. and Thomas, D.S.G. (2006). Complex Holocene lunette dune development, South Africa: Implications for paleoclimate and models of pan development in arid regions. *Geology*, 34, 853–6.

Telfer, M.W. and Thomas, D.S.G. (2007). Late Quaternary linear dune accumulation and chronostratigraphy of the southwestern Kalahari: Implications for aeolian palaeoclimatic reconstructions and predictions of future dynamics. *Quaternary Science Reviews*, 26, 2617–30.

Teller, J.T., Glennie, K.W., Lancaster, N. and Singhvi, A.K. (2000). Calcareous dunes of the United Arab Emirates and Noah's Flood: The postglacial reflooding of the Persian (Arabian) Gulf. *Quaternary International*, 68–71, 297–308.

Tengberg, A. and Chen, D. (1998). A comparative analysis of nebkhas in central Tunisia and northern Burkina Faso. *Geomorphology*, 22, 181–92.

Termina, M. and Kidron, G.J. (2011). Lichens as biomarkers for dew amount and duration in the Negev Desert. *Flora*, 206, 646–52.

Ternan, J.L., Williams, A.G., Elmes, A. and Fitzjohn, C. (1996). The effectiveness of bench-terracing and afforestation for erosion control on Rana sediments in central Spain. *Land Degradation and Development*, 7, 337–51.

Tetzlaff, G., Peters, M., Janssen, W. and Adams, L.J. (1989). Aeolian dust transport in West Africa. In *Palaeoclimatology and Palaeometeorology: Modern and Past Patterns of Global Atmospheric Transport*, ed. M. Leinen and M. Sarnthein. Dordrecht: Kluwer, pp. 1985–2003.

Thevenon, F., Chiaradia, M., Adatte, T., Hueglin, C. and Poté, J. (2011). Ancient versus modern mineral dust transported to high-altitude alpine glaciers evidences Saharan sources and atmospheric circulation changes. *Atmospheric Chemistry and Physics Discussions*, 11, 859–84.

Thiagarajan, N. and Lee, C.-T.Y. (2004). Trace-element evidence for the origin of desert varnish by direct aqueous atmospheric deposition. *Earth and Planetary Science Letters*, 224, 131–41.

Thiry, M. and Milnes, A.R. (1991). Pedogenic and groundwater silcretes at Stuart Creek Opal Field, South Australia. *Journal of Sedimentary Petrology*, 61, 111–27.

Thiry, M., Milnes, A.R., Rayot, V. and Simon-Coinçon, R. (2006). Interpretation of palaeoweathering features and successive silicifications in the Tertiary regolith of inland Australia. *Journal of the Geological Society of London*, 163, 723–36.

Thomas, D.S.G. (1984). Ancient ergs of the former arid zones of Zimbabwe, Zambia and Angola. *Transactions of the Institute of British Geographers*, NS 9, 75–88.

Thomas, D.S.G. (1987). Discrimination of depositional environments using sedimentary characteristics in the Mega Kalahari, central southern Africa. *Geological Society Special Publication*, 35, 293–306.

Thomas, D.S.G. (1992). Desert dune activity: Concepts and significance. *Journal of Arid Environments*, 22, 31–8.

Thomas, D.S.G. (ed.). (2011a). *Arid Zone Geomorphology* (3rd ed.). Chichester: Wiley-Blackwell.

Thomas, D.S.G. (2011b). Reconstructing paleoenvironments and palaeoclimates in drylands: What can landform analysis contribute? *Earth Surface Processes and Landforms*, doi:10.1002/esp.3190.

Thomas, D.S.G. and Dougill, A.J. (2007). Spatial and temporal distribution of cyanobacterial soil crusts in the Kalahari: Implications for soil surface properties. *Geomorphology*, 85, 17–29.

Thomas, D.S.G. and Shaw, P. (1991). *The Kalahari Environment*. Cambridge: Cambridge University Press.

Thomas, D.S.G. and Tsoar, H. (1990). The geomorphological role of vegetation in desert dune systems. In *Vegetation and Erosion*, ed. J.B. Thornes. Wiley: Chichester, pp. 471–89.

Thomas, D.S.G. and Wiggs, G.F.S. (2008). Aeolian system responses to global change: Challenges of scale, process and temporal integration. *Earth Surface Processes and Landforms*, 33, 1396–418.

Thomas, D.S.G., Burrough, S.L. and Parker, A.G. (2011). Extreme events as drivers of early human behaviour in Africa? The case for variability, not catastrophic drought. *Journal of Quaternary Science*, doi:10.1002/jqs.1557.

Thomas, D.S.G., Knight, M. and Wiggs, G.F.S. (2005). Remobilization of southern African desert dune systems by twenty-first century global warming. *Nature*, 435, 1218–21.

Thomas, D.S.G., Stokes, S. and Shaw, P.A. (1997). Holocene aeolian activity in the southwestern Kalahari Desert, southern Africa: Significance and relationship to late Pleistocene dune-building events. *The Holocene*, 7, 271–9.

Thomas, D.S.G., Bateman, M.D., Mehrshahi, D. and O'Hara, S.L. (1997). Development and environmental significance of an eolian sand ramp of Last-Glacial age, central Iran. *Quaternary Research*, 48, 155–61.

Thomas, M., Clarke, J.D.A. and Pain, C.F. (2005). Weathering, erosion and landscape processes on Mars identified from recent rover imagery, and possible Earth analogues. *Australian Journal of Earth Sciences*, 52, 365–78.

Thompson, L.G., Mosley-Thompson, E., Davis, J.F., Bolzan, J.F., Dai, J. and Klei, L. (1990). Glacial stage ice core records from the subtropical Dunde Ice Cap, China. *Annals of Glaciology*, 14, 288–97.

Thomson, A.M., Brown, R.A., Rosenberg, N.J., Srinivasan, R. and Izaurralde, R.C. (2005). Climate change impacts for the conterminous USA: An integrated assessment, Part 4: Water resources. *Climatic Change*, 69, 67–88.

Thornthwaite, C.W. (1948). An approach toward a rational classification of climate. *Geographical Review*, 38, 55–94.

Tickner, D.P., Angold, P.G., Gurnell, A.M. and Mountford, J.O. (2001). Riparian plant invasions: Hydrogeomorphological control and ecological impacts. *Progress in Physical Geography*, 25, 22–52.

Todd, M.C., Washington, R., Martins, J.V., Dubovik, O., Lizcano, G., M'Bainayel, S. and Engelstaedter, S. (2007). Mineral dust emission from the Bodélé Depression, northern Chad, during BoDEx 2005. *Journal of Geophysical Research*, 112, D06207, doi:10.1029/2006JD007170.

Todhunter, P.E. and Cihacek, L.J. (1999). Historical reduction of airborne dust in the Red River Valley of the North. *Journal of Soil and Water Conservation*, 54, 543–51.

Tolman, C.R. (1909). Erosion and deposition in the southern Arizona Bolson Region. *Journal of Geology*, 17, 136–63.

Tomasko, M.G. and 39 others. (2005). Rain, winds and haze during the Huygens probe's descent to Titan's surface. *Nature*, 438, 765–78.

Tongway, D.J. and Ludwig, J.A. (2001). Theories on the origins, maintenance, dynamics, and functioning of banded landscapes. *Ecological Studies*, 149, 20–31.

Tongway, D.J., Ludwig, J.A. and Whitford, W.G. (1989). Mulga log mounds: Fertile patches in the semi-arid woodlands of eastern Australia. *Australian Journal of Ecology*, 14, 263–8.

Toon, O.B. (2003). African dust in Florida clouds. *Nature*, 424, 623–4.

Tooth, S. (1999). Floodouts in central Australia. In *Varieties of Fluvial Form*, ed. A.J. Miller and A. Gupta. Chichester: Wiley, pp. 220–47.

Tooth, S. (2000). Process, form and change in dryland rivers: A review of recent research. *Earth-Science Reviews*, 51, 67–107.
Tooth, S. (2009). Arid geomorphology: Emerging research themes and new frontiers. *Progress in Physical Geography*, 33, 251–87.
Tooth, S. and Nanson, G.C. (1999). Anabranching rivers on the Northern Plains of arid central Australia. *Geomorphology*, 29, 211–33.
Toure, A.A., Rajot, J.L., Garba, Z., Marticorena, B., Petit, C. and Sebag, D. (2011). Impact of very low crop residues cover on wind erosion in the Sahel. *Catena*, 85, 205–14.
Trauth, M.H., Alonso, R.A., Haselton, K.R., Hermanns, R.A. and Strecker, M.R. (2000). Climate change and mass movements in the NW Argentina Andes. *Earth and Planetary Science Letters*, 179, 243–56.
Tricart, J. (1953). Influence des sols salées sur la deflation éolienne en basse Mauritanie et dans le delta du Sénégal. *Revue de Géomorphologie Dynamique*, 4, 124–32.
Tricart, J. (1960). Expériences de désagrégation de roches grantiques par la cristallisation du sel marin. *Zeitschrift für Geomorphologie*, Supplementband, 1, 239–40.
Tricart, J. (1969). Actions éoliennes dans la Pampa Deprimida (Rep. Argentina). *Revue de Géomorphologie Dynamique*, 19, 178–89.
Tricart, J. and Cailleux, A. (1969). *Le Modelé des Regions sèches*. Paris: Sedes.
Trigo, I.F., Monteiro, I.T., Olesen, F. and Kabsch, E. (2008). An assessment of remotely sensed land surface temperature. *Journal of Geophysical Research*, 113, D17108, doi:10.1029/2008JD010035.
Tripaldi, A. and Forman, S.L. (2007). Geomorphology and chronology of late Quaternary dune fields of western Argentina. *Palaeogeography, Palaeoclimatology, Palaeoecology*, 251, 300–20.
Tripaldi, A., Ciccioli, P.L., Alonso, M.S. and Forman, S.L. (2010). Petrography and geochemistry of late Quaternary dune fields of western Argentina: Provenance of aeolian materials in southern South America. *Aeolian Research*, 2, 33–48.
Trowbridge, A.C. (1911). The terrestrial deposits of Owens Valley, California. *Journal of Geology*, 19, 706–47.
Tsoar, H. (1978). The dynamics of longitudinal dunes. *Final Technical Report DA-ERO* 76-G-072, European Research Office, U.S. Army, London.
Tsoar, H. (1983). Wind tunnel modelling of echo and climbing dunes. *Developments in Sedimentology*, 38, 247–59.
Tsoar, H. and Blumberg, D.G. (2002). Formation of parabolic dunes from barchans and transverse dunes along Israel's Mediterranean coast. *Earth Surface Processes and Landforms*, 27, 1147–61.
Tsoar, H., and Pye, K. (1987). Dust transport and the question of desert loess formation. *Sedimentology*, 34, 139–53.
Tsoar, H., Blumberg, D.G. and Stoler, Y. (2004). Elongation and migration of sand dunes. *Geomorphology*, 57, 293–302.
Tsui, N., Flatt, R.J. and Scherer, G.W. (2003). Crystallization damage by sodium sulfate. *Journal of Cultural Heritage*, 4, 109–15.
Tucker, C.J., Dregne, H.E. and Newcomb, W.W. (1991). Expansion and contraction of the Sahara Desert from 1980 to 1990. *Science*, 253, 299–301.
Tucker, M.E. (1978). Gypsum crusts (gypcrete) and patterned ground from northern Iraq. *Zeitschrift für Geomorphologie*, 22, 89–100.
Tuenter, E., Weber, S.L., Hilgen, F.J. and Lourens, L.J. (2003). The response of the African summer monsoon to remote and local forcing due to precession and obliquity. *Global and Planetary Change*, 36, 219–35.
Turnbull, L., Wainwright, J. and Brazier, R.E. (2010). Changes in hydrology and erosion over a transition from grassland to shrubland. *Hydrological Processes*, 24, 393–414.

Turner, D., Ostendorf, B. and Lewis, M. (2008). An introduction to patterns of fire in arid and semi-arid Australia. *The Rangeland Journal*, 30, 95–107.
Twidale, C.R. (1972). Evolution of sand dunes in the Simpson Desert, central Australia. *Transactions of the Institute of British Geographers*, 56, 77–109.
Twidale, C.R. (1978). On the origin of Ayers Rock, central Australia. *Zeitschrift für Geomorphologie*, Supplementband, 31, 177–206.
Twidale, C.R. (1982). *Granite Landforms*. Amsterdam: Elsevier.
Twidale, C.R. (2000). Early Mesozoic (?Triassic) landscapes in Australia: Evidence, argument and implications. *Journal of Geology*, 108, 537–52.
Twidale, C.R. and Campbell, E.M. (1993). *Australian Landforms: Structure, Process and Time*. Adelaide: Gleneagles Publishing.
Twidale, C.R. and Corbin, E.M. (1963). Gnammas. *Revue de Géomorphologie Dynamique*, 14, 1–20.
Twidale, C.R. and Hutton, J.T. (1986). Silcrete as a climatic indicator: Discussion. *Palaeogeography, Palaeoclimatology, Palaeoecology*, 52, 351–60.
Twidale, C.R. and Vidal Romani, J.R. (2005). *Landforms and Geology of Granite Terrains*. Balkema: Leiden.
Tyler, S.W., Kranz, S., Parlange, M.B., Albertson, J., Katul, G.G. Cochram, G.F., Lyles, B.A. and Holder, G. (1997). Estimation of groundwater evaporation and salt flux from Owens Lake, California, USA. *Journal of Hydrology*, 200, 110–35.
Tyson, P.D. and Preston-Whyte, R.A. (2000). *The Weather and Climate of Southern Africa*. Cape Town: Oxford University Press.
Uck, B.J. and Van Hoesen, J.G. (2002). Snowball morphology and SEM analysis of pedogenic gypsum, southern New Mexico, U.S.A. *Journal of Arid Environments*, 51, 469–87.
Udden, J.A. (1894). Erosion, transportation and sedimentation performed by the atmosphere. *Journal of Geology*, 2, 318–31.
Uña Alvarez, E. de, and Vidal Romani, J.R. (2008). Some minor features (tafoni, cavernous forms) of Los Rojanos (Pampa de Achala, Sierra Grande de Cordoba, Republica Argentina). Geometric and morphologic properties. *Cadernos Laboratoria Xeologico de Laxe Coruna*, 33, 83–99.
UNESCO. (2008). *Operational Guidelines for the Implementation of the World Heritage Convention*. Paris: UNESCO World Heritage Centre.
Unger, P.W., Stewart, B.A., Parr, J.F. and Singh, R.P. (1991). Crop residue management and tillage methods for conserving soil and water in semi-arid regions. *Soil and Tillage Research*, 20, 219–40.
Uno, I. and 9 others. (2009). Asian dust transported one full circuit around the globe. *Nature Geosciences*, 2, 557–60.
Valentin, C. and Poesen, J. (eds.). (1999). The significance of soil, water and landscape processes in banded vegetation patterning. *Catena*, 37, 1–273.
Valentin, C., d'Herbès, J.M. and Poesen, J. (1999). Soil and water components of banded vegetation patterns. *Catena*, 37, 1–24.
Valentin, C., Rajot, J.-L. and Mitja, D. (2004). Response of soil crusting, runoff and erosion to fallowing in the sub-humid and semi-arid regions of West Africa. *Agriculture, Ecosystems and Environment*, 104, 287–302.
Valentine, G.A. and Harrington, C.D. (2006). Clast size controls and longevity of Pleistocene desert pavements at Lathrop Wells and Red Cone volcanoes, southern Nevada. *Geology*, 34, 533–6.
Valentine, G.A., Krier, D.J., Perry, F.V. and Heiken, G. (2007). Eruptive and geomorphic processes at the Lathrop Wells scoria cone volcano. *Journal of Volcanology and Geothermal Research*, 161, 57–80.

Valero-Garces, B.L., Arenas, C. and Delgado-Hoertas, A. (2001). Depositional environments of Quaternary lacustrine travertines, and stromatolites from high-altitude Andean lakes, northwestern Argentina. *Canadian Journal of Earth Sciences*, 38, 1263–83.

Van der Hove, S.J. and Quade, J. (2002). Tracing spatial and temporal variations in the sources of calcium in pedogenic carbonates in a semiarid environment. *Geoderma*, 108, 259–76.

Van der Wateren, F.M. and Dunai, T.J. (2001). Lake Neogene passive margin denudation history – Cosmogenic isotope measurements from the Central Namib Desert. *Global and Planetary Change*, 30, 271–307.

Van Etten, E.J.B. (2009). Inter-annual rainfall variability of arid Australia: Greater than elsewhere? *Australian Geographer*, 40, 109–20.

Vanmaercke, M., Zenebe, A., Poesen, J., Nyssen, J., Verstraeten, G. and Deckers, J. (2010). Sediment dynamics and the role of flash floods in sediment export from medium-sized catchments: A case study from the semi-arid tropical highlands in northern Ethiopia. *Journal of Soils and Sediments*, 10, 611–27.

Van Pelt, R.S., Zobeck, T.M., Potter, K.N., Stout, J.E. and Popham, T.W. (2004). Validation of the wind erosion stochastic simulator (WESS) and the revised wind erosion equation (RWEQ) for single events. *Environmental Modelling and Software*, 19, 191–8.

Vargas, E., Velloso, R.Q., Chávez, L.E., Gusmão, L. and Amaral, G.P. (2012). On the effect of thermally induced stresses in failures of some rock slopes in Rio de Janeiro, Brazil. *Rock Mechanics and Rock Engineering*, doi:10.1007/s00603–012-0247–9.

Vargas, G., Ortleib, L. and Rutland, J. (2000). Alvuniones históricas en Antofagasta y su relación con eventos El Niño/oscilación de Sur. *Revista Geológica de Chile*, 27, 157–76.

Várkonyi, P.L. and Laity, J.E. (2012). Formation of surface features on ventifacts: Modelling the role of sand grains rebounding within cavities. *Geomorphology*, 139–140, 220–9.

Vautard, R., Cattiaux, J., Yiou, P., Thépaut, J.-N. and Ciais, P. (2010). Northern Hemisphere atmospheric stilling partly attributed to an increase in surface roughness. *Nature Geoscience*, 3, 756–61.

Velázquez, V.F., Giannini, P.C.F., Riccomini, C., Sallun, A.E.M., Hachiro, J. and Gomes, C. de B. (2008). Columnar joints in the Patiño formation sandstones, eastern Paraguay: A dynamic interaction between dyke intrusion, quartz dissolution and cooling-induced fractures. *Episodes*, 31, 302–8.

Verbist, K., Baetens, J., Cornelis, W.M., Gabriels, D. and Torres, C. (2009). Hydraulic conductivity as influenced by stoniness in degraded drylands of Chile. *Soil Science Society of America Journal*, 73, 471–84.

Verdin, J.P. (1996). Remote sensing of ephemeral water bodies in western Niger. *International Journal of Remote Sensing*, 17, 733–48.

Verger, F. (1964). Mottureaux et gilgais. *Annales de Géographie*, 73, 413–30.

Vermeesch, P. and Drake, N. (2008). Remotely sensed dune celerity and sand flux measurements of the world's fastest barchans (Bodélé, Chad). *Geophysical Research Letters*, 35, L24404, doi:10.1029/2008GL035921.

Vermeesch, P., Fenton, C.R., Kober, F., Wiggs, G.F.S., Bristow, C.S. and Xu, S. (2010). Sand residence times of one million years in the Namib Sand Sea from cosmogenic nuclides. *Nature Geoscience*, 3, 862–5.

Verrecchia, E., Yair, A., Kidron, G.J. and Verrecchia, K. (1995). Physical properties of the psammophile cryptogamic crust and their consequences to the water regime of sandy soils, north-western Negev Desert, Israel. *Journal of Arid Environments*, 29, 427–37.

Verstappen, H.Th. (1968). On the origin of longitudinal (seif) dunes. *Zeitschrift für Geomorphologie*, 12, 200–20.

Verstappen, H.Th. (1970). Aeolian geomorphology of the Thar Desert and palaeoclimates. *Zeitschrift für Geomorphologie*, Supplementband, 10, 104–20.

Viles, H.A. (1995). Ecological perspectives on rock surface weathering: Towards a conceptual model. *Geomorphology*, 13, 21–35.
Viles, H.A. (2005). Microclimate and weathering in the central Namib Desert, Namibia. *Geomorphology*, 67, 189–209.
Viles, H.A. (2008). Understanding dryland landscape dynamics: Do biological crusts hold the key? *Geography Compass*, 2–3, 899–919.
Viles, H.A. (2011). Weathering systems. In *Arid Zone Geomorphology* (3rd ed.), ed. D.S.G. Thomas. Chichester: Wiley-Blackwell, pp. 85–100.
Viles, H.A. and Goudie, A.S. (1990). Reconnaissance studies of the tufa deposits of the Napier Range, N.W. Australia. *Earth Surface Processes and Landforms*, 15, 425–43.
Viles, H.A. and Goudie, A, S. (2000). Weathering, geomorphology and climatic variability in the central Namib Desert. In *Linking Climate Change to Land Surface Change*, ed. S.J. McLaren and D.R. Kniveton. Dordrecht: Kluwer, pp. 65–82.
Viles, H.A. and Goudie, A.S. (2007). Rapid salt weathering in the coastal Namib Desert: Implications for landscape development. *Geomorphology*, 85, 49–62.
Viles, H.A. and Goudie, A.S. (In press). Weathering in the central Namib Desert, Namibia: Controls, processes and implications. *Journal of Arid Environments*.
Viles, H.A., Taylor, M.P., Nicholl, K. and Neumann, S. (2007). Facies evidence of hydroclimatic regime shifts in tufa deposition sequences from the arid Naukluft Mountains, Namibia. *Sedimentary Geology*, 195, 39–53.
Viles, H., Ehlmann, B., Wilson, C.F., Cebula, T., Page, M. and Bourke, M. (2010). Simulating weathering of basalt on Mars and Earth by thermal cycling. *Geophysical Research Letters*, 37, L18201, doi:10/10292010GL043522.
Vincent, P. and Kattan, F. (2006). Yardangs on the Cambro-Ordovician Saq sandstones, north-west Saudi Arabia. *Zeitschrift für Geomorphologie*, 50, 305–20.
Vishkaee, F.A., Flamant, C., Cuesta, J., Oolman, L., Flamant, P. and Khalesifard, H.R. (2012). Dust transport over Iraq and northwest Iran associated with winter Shamal: A case study. *Journal of Geophysical Research*, 117, D03201, doi:10.1029/2011JD016339.
Vogel, M.B., Des Marais, D.J., Parenteau, M.N., Jahnke, L.L., Turk, K.A. and Kubo, M.D.Y. (2010). Biological influences on modern sulphates: Textures and composition of gypsum deposits from Guerro Negro, Baja California Sur, Mexico. *Sedimentary Geology*, 223, 265–80.
Vriend, N.M., Hunt, M.L. and Clayton, R.W. (2012). Sedimentary structures of large sand dunes: Examples from Dumont and Eureka Dunes, California. *Geophysical Journal International*, doi:10.111/j.1365–246X.2012.05514.x.
Wadge, G., Francis, P.W. and Ramirez, C.F. (1995). The Socompa collapse and avalanche event. *Journal of Volcanology and Geothermal Research*, 66, 309–36.
Wainwright, J., Parsons, A.J. and Abrahams, A.D. (1995). A simulation study of the role of raindrop erosion in the formation of desert pavement. *Earth Surface Processes and Landforms*, 20, 277–91.
Wainwright, J., Parsons, A.J. and Abrahams, A.D. (1999). Field and computer simulation experiments on the formation of desert pavement. *Earth Surface Processes and Landforms*, 24, 1025–37.
Wakindiki, I.I.C. and Ben-Hur, M. (2002). Indigenous soil and water conservation techniques: Effects on runoff, erosion, and crop yields under semi-arid conditions. *Australian Journal of Soil Research*, 40, 367–79.
Walden, J. and White, K. (1997). Investigation of the controls on dune colour in the Namib Sand Sea using mineral magnetic analyses. *Earth and Planetary Science Letters*, 152, 187–201.
Walden, J., White, K. and Drake, N.A. (1996). Controls on dune colour in the Namib Sand Sea: Preliminary results. *Journal of African Earth Sciences*, 22, 349–53.

Waldman, N., Torfstein, A. and Stein, M. (2010). Northwards intrusions of low- and mid-latitude storms across the Saharo-Arabian belt during past interglacials. *Geology*, 38, 567–70.

Wali, A.M.A. (1991). Sabkhas of the Bitter-Lakes, Egypt: Composition and origin. *Carbonates and Evaporites*, 6, 225–38.

Walker, M.J. (ed.). (2012). Hot deserts: Engineering, geology and geomorphology. *Geological Society of London Engineering Geology Special Publication*, 25.

Walker, R.T. and Fattahi, M. (2011). A framework of Holocene and late Pleistocene environmental change in eastern Iran inferred from dating of periods of alluvial fan abandonment, river terracing, and lake deposition. *Quaternary Science Reviews*, 30, 1256–71.

Walmsley, J.L. and Howard, A.D. (1985). Application of a boundary-layer model to flow over an aeolian dune. *Journal of Geophysical Research*, 90 (D6), 10631–40.

Walther, J. (1900). *Das Gesetz der Wüstenbildung in Gegenwart und Vorzeit*. Berlin: Quelle and Meyer.

Walther, J. (1997). *The Law of Desert Formation – Present and Past* (trans. by G. Meyer and ed. by E. Gischler and K.W. Glennie). Miami, FL: University of Miami, Geological Milestones, Vol. IV.

Wan, Z. (2008). New refinements and validation of the MODIS Land-Surface Temperature/Emissivity products. *Remote Sensing of the Environment*, 112, 59–74.

Wanas, H.A., Pickford, M., Mein, P., Soliman, H. and Segalen, L. (2009). Late Miocene karst system at Sheik Abdallah, between Bahariya and Farafra, Western Desert, Egypt: Implications for palaeoclimate and geomorphology. *Geologica Acta*, 7, 475–87.

Wang, H., Yang, Z., Saito, Y., Liu, J.P., Sun, X. and Wang, Y. (2007). Stepwise decreases of the Huanghe (Yellow River) sediment load (1950–2005): Impacts of climate change and human activities. *Global and Planetary Change*, 57, 331–54.

Wang, J., Wang, Y.J., Liu, Z.C., Li, J.Q. and Xi, P. (1999). Cenozoic environmental evolution of the Qaidam Basin and its implications for the uplift of the Tibetan Plateau and the drying of central Asia. *Palaeogeography, Palaeoclimatology, Palaeoecology*, 152, 37–47.

Wang, N. (2005). Decrease trend of dust event frequency over the past 200 years recorded in the Malan ice core from the northern Tibetan Plateau. *Chinese Science Bulletin*, 50, 2866–71.

Wang, T., Zhang, W., Dong, Z., Qu, J., Jing, Z., Wang, W. and Feng, J. (2005). The dynamic characteristics and migration of a pyramid dune. *Sedimentology*, 52, 429–40.

Wang, X., Zhou, Z. and Dong, Z. (2006a). Control of dust emissions by geomorphic conditions, wind environments and land use in northern China: An examination based on dust storm frequency from 1960–2003. *Geomorphology*, 81, 292–308.

Wang, X., Dong, Z., Zhang, J. and Chen, G. (2002). Geomorphology of sand dunes in the northeast Taklimakan Desert. *Geomorphology*, 42, 183–95.

Wang, X., Dong, Z., Zhang, J. and Liu, L. (2004). Modern dust storms in China: An overview. *Journal of Arid Environments*, 58, 559–74.

Wang, X., Eerdun, H., Zhou, Z. and Liu, X. (2007). Significance of variations in the wind energy environment over the past 50 years with respect to dune activity and desertification in arid and semiarid northern China. *Geomorphology*, 86, 252–66.

Wang, X., Dong, Z., Zhang, J., Qu, J. and Zhao, A. (2003). Grain size characteristics of dune sands in the central Taklimakan Sand Sea. *Sedimentary Geology*, 161, 1–14.

Wang, X., Wang, T., Dong, Z., Liu, X. and Qian, G. (2006b). Nebkha development and its significance to wind erosion and land degradation in semi-arid northern China. *Journal of Arid Environments*, 65, 129–41.

Wang, X., Zeng, L., Wiens, M., Schlossmacher, U., Jochum, P.K., Schröder, H.C. and Müller, W.E.G. (2011). Evidence for a biogenic, microorganismal origin of rock varnish from the Gangese Belt of Tibet. *Micron*, 42, 401–11.

Wang, X.M., Zhang, C.X., Hasi, E. and Dong, Z.B. (2010). Has the Three Norths Shelterbelt programme solved the desertification and dust storm problems in arid and semi-arid China? *Journal of Arid Environments*, 74, 13–22.

Wang, Y., Stein, A.F., Draxler, R.R., de la Rosa, J.D. and Zhang, X. (2011). Global sand and dust storms in 2008: Observation and HYSPLIT model verification. *Atmospheric Environment*, 45, 6368–81.

Wang, Z., Dong, Z., Yan, P., Zhang, J. and Qian, G. (2005). Wind energy environments and dunefield activity in the Chinese deserts. *Geomorphology*, 65, 33–48.

Wang, Z.-T., Sun, Q.-F., Sen, X.-R., Wang, T. and Chen, F.-F. (2009). Pseudo-feathery dunes in the Kumtagh Desert reclassified as linear dunes and zibars. *Aeolian Research*, 1, 87–9.

Wang, Z.-T., Wang, H.-T., Niu, Q.-H., Dong, Z.-B. and Wang, T. (2011). Abrasion of yardangs. *Physical Review*, E, 84, 031304.

Ward, A.W. and Greeley, R. (1984). Evolution of the yardangs at Rogers Lake, California. *Bulletin of the Geological Society of America*, 95, 829–37.

Ward, J.D. (1984). A reappraisal of the Cenozoic stratigraphy in the Kuiseb Valley of the central Namib Desert. In *Late Cainozoic Palaeoclimates of the Southern Hemisphere*, ed. J.C. Vogel. Rotterdam: Balkema, pp. 455–63.

Ward, J.D. (1988). Eolian, fluvial and pan (playa) facies of the Tertiary Tsondab Sandstone Formation in the central Namib Desert, Namibia. *Sedimentary Geology*, 55, 143–62.

Ward, J.D., Seely, M.K. and Lancaster, N. (1983). On the antiquity of the Namib. *South African Journal of Science*, 79, 175–83.

Warke, P.A. (2007). Complex weathering in drylands: Implications of 'stress' history for rock debris breakdown and sediment release. *Geomorphology*, 85, 30–48.

Warner, T.F. (2004). *Desert Meteorology*. Cambridge: Cambridge University Press.

Warren, A. (1971). Dunes in the Ténéré Desert. *Geographical Journal*, 137, 458–61.

Warren, A. (1976). Morphology and sediments of the Nebraska Sand Hills in relation to Pleistocene winds and the development of aeolian bedforms. *Journal of Geology*, 84, 685–700.

Warren, A. (1988). The dunes of the Wahiba Sands. *Journal of Oman Studies Special Report*, 3, 131–60.

Warren, A. and Allison, D. (1998). The palaeoenvironmental significance of dune size hierarchies. *Palaeogeography, Palaeoclimatology, Palaeoecology*, 137, 289–303.

Warren, A., Chappell, A., Todd, M.C., Bristow, C., Drake, N., Engelstaedter, S., Martins, V., M'bainayel, S. and Washington, R. (2007). Dust-raising in the dustiest place on Earth. *Geomorphology*, 92, 25–37.

Warren, J.K. (1989). *Evaporite Sedimentology*. Englewood Cliffs, NJ: Prentice Hall.

Warrick, J.A., Hatten, J.A., Pasternack, G.B., Gray, A.B., Goni, M.A. and Wheatcroft, R.A. (2012). The effects of wildfire on the sediment yield of a coastal California watershed. *Bulletin of the Geological Society of America*, doi:10.1130/B30451.1.

Washington, R. and Todd, M.C. (2005). Atmospheric controls on mineral dust emission from the Bodélé Depression, Chad: The role of the low level jet. *Geophysical Research Letters*, 32, L17701, doi:10.1029/2005GL023597.

Washington, R., Todd, M., Middleton, N. and Goudie, A.S. (2003). Global dust storm source areas determined by the Total Ozone Monitoring Spectrometer and ground observations. *Annals of the Association of American Geographers*, 93, 297–313.

Washington, R., Todd, M.C., Engelstaedter, S., Mbainayel, S. and Mitchell, F. (2006a). Dust and the low-level circulation over the Bodélé Depression, Chad: Observations from BoDEx 2005. *Journal of Geophysical Research*, 111, D03201, doi:10.1029/2005JD006502.

Washington, R., Todd, M.C., Lizcano, G., Tegen, I., Flamant, C., Koren, I., Ginoux, P., Engelstaedter, S., Bristow, C.S., Zender, C.S., Goudie, A.S., Warren, A. and Prospero, J.M. (2006b). Links between topography, wind, deflation, lakes and dust: The case of the Bodélé Depression, Chad. *Geophysical Research Letters*, 33, L09401, doi:10.1029/2006GL025827.

Wasklewicz, T., Mihir, M. and Whitworth, J. (2008). Surface variability of alluvial fans generated by disparate processes, eastern Death Valley, CA. *Professional Geographer*, 60, 207–23.

Wasson, R.J. (1983). Dune sediment type, sand colour, sediment provenance and hydrology in the Strzelecki-Simpson Dunefield, Australia. In *Eolian Sediments and Processes*, ed. M.E. Brookfield and T.S. Ahlbrandt. Elsevier: Amsterdam, pp. 165–95.

Wasson, R.J. (1984). Late Quaternary palaeoenvironments in the desert dunefields of Australia. In *Late Cainozoic Palaeoclimates of the Southern Hemisphere*, ed. J.C. Vogel. Rotterdam: Balkema, pp. 419–32.

Wasson, R.J. and Hyde, R. (1983). Factors determining desert dune type. *Nature*, 304, 337–9.

Wasson, R.J., Smith, G.I. and Agrawal, D.P. (1984). Late Quaternary sediments, minerals and inferred geochemical history of Didwana Lake, Thar Desert, India. *Palaeogeography, Palaeoclimatology, Palaeoecology*, 46, 345–72.

Wasson, R.J., Fitchett, K., Mackey, B. and Hyde, R. (1988). Large-scale patterns of dune type, spacing and orientation in the Australian continental dunefield. *Australian Geographer*, 19, 89–104.

Wasson, R.J., Rajaguru, S.N., Misra, V.N., Agrawal, D.P., Dhir, R.P., Singhvi, A.K. and Kameswara, R.K. (1983). Geomorphology, late Quaternary stratigraphy and palaeoclimatology of the Thar Dune Field. *Zeitschrift für Geomorphologie*, Supplementband, 45, 117–51.

Watchman, A. (2000). A review of the history of dating rock varnishes. *Earth-Science Reviews*, 49, 261–77.

Waters, M.R. and Haynes, C.V. (2001). Late Quaternary arroyo formation and climate change in the American Southwest. *Geology*, 29, 399–402.

Watson, A. (1979). Gypsum crusts in deserts. *Journal of Arid Environments*, 2, 3–20.

Watson, A. (1983). Gypsum crusts. In *Chemical Sediments and Geomorphology*, ed. A.S. Goudie and K. Pye. London: Academic Press, pp. 133–61.

Watson, A. (1985). Structure, chemistry and origin of gypsum crusts in southern Tunisia and the central Namib Desert. *Sedimentology*, 32, 855–75.

Watson, A. (1986). Grain-size variations on a longitudinal dune and a barchan dune. *Sedimentary Geology*, 46, 49–66.

Watson, A. (1988). Desert gypsum crusts as palaeoenvironmental indicators: A micropetrographic study of crusts from southern Tunisia and the central Namib Desert. *Journal of Arid Environments*, 15, 19–42.

Watson, A. (1990). The control of blowing sand and mobile desert dunes. In *Techniques for Desert Reclamation*, ed. A.S. Goudie. Chichester: Wiley, pp. 35–85.

Watson, A. and Pye, K. (1985). Pseudo-karstic micro-relief and other weathering features on the Mswati Granite (Swaziland). *Zeitschrift für Geomorphologie*, 29, 285–300.

Watts, N.L. (1977). Pseudo-anticlines and other structures in some calcretes of Botswana and South Africa. *Earth Surface Processes*, 2, 63–74.

Watts, N.L. (1980). Quaternary pedogenic calcretes from the Kalahari (southern Africa): Mineralogy, genesis and diagenesis. *Sedimentology*, 27, 661–86.

Watts, S.H. (1981). Near coastal and incipient weathering features in the Cape Herschel-Alexander Fjord Area, District of Franklin. *Geological Survey of Canada, Paper*, 81–1A, 389–94.

Webb, J.A. and Golding, S.D. (1998). Geochemical mass-balance and oxygen-isotope constraints on silcrete formation and its palaeoclimatic implications in southern Australia. *Journal of Sedimentary Research*, 68, 987–93.

Webb, J.A. and James, J.M. (2006). Karst evolution of the Nullarbor Plain, Australia. *Geological Society of America Special Paper*, 404, 65–78.

Webb, N.P. and McGowan, H.A. (2009). Approaches to modeling land erodibility by wind. *Progress in Physical Geography*, 33, 587–613.

Webb, N.P., McGowan, H.A., Phinn, S.R. and McTainsh, G.H. (2006). AUSLEM (AUStralian Land Erodibility Model): A tool for identifying wind erosion hazard in Australia. *Geomorphology*, 78, 179–200.

Webb, R.H. (1982). Off-road motorcycle effects on a desert soil. *Environmental Conservation*, 9, 197–208.

Webb, S. (2009). Palaeotrophic reconstruction and climatic forcing of mega-Lake Eyre in late Quaternary central Australia: A review. *Boreas*, 39, 312–24.

Wedekind, W. and Ruedrich, J. (2006). Salt-weathering, conservation techniques and strategies to protect the rock cut façades in Petra/Jordan. In *Heritage, Weathering, and Conservation*, ed. R. Fort, M.A. de Buergo, M. Gomez-Heras and C. Vasquez-Calvo. London: Taylor and Francis, pp. 261–8.

Wehmeier, E. (1998). Giant cracks and their formation on Willcox Playa in Sulphur Springs Valley, Arizona. *Petermanns Geographische Mitteilungen*, 142, 19–34.

Weinstein-Evron, M. (1987). Palynology of Pleistocene travertines from the Arava Valley, Israel. *Quaternary Research*, 27, 82–8.

Weisrock, A. (1986). Variations climatiques et periodes de sedimentation carbonatée a l'Holocene – L'age des depôts. *Mediterranée*, 10, 165–7.

Weiss H., Courty, M.A., Wetterstrom, W., Guichard, F., Senior R., Meadow, R. and Curnow, A. (1993). The genesis and collapse of third millennium Mesopotamian civilisation. *Science*, 261, 995–1004.

Weiss, T., Siegesmund, S., Kirchner, D. and Sippel, J. (2004). Insolation weathering and hygric dilatation: Two competitive factors in stone degradation. *Environmental Geology*, 46, 402–13.

Wellman, H.W. and Wilson, A.T. (1965). Salt weathering: A neglected geological erosive agent in coastal and arid environments. *Nature*, 205, 1097–8.

Wells, G.L. (1983). Late glacial circulation over North America revealed by aeolian features. In *Variation in the Global Water Budget*, ed. F.A. Street-Perrott, M. Beran and R. Ratcliffe. Dordrecht: Kluwer, pp. 317–30.

Wells, G.L. (1992). The aeolian landscape of central North America from the late Pleistocene. D.Phil. thesis. 3 vols. University of Oxford.

Wells, G.L. and Zimbelman, J.R. (1997). Extraterrestrial and surface processes. In *Arid Zone Geomorphology: Process, Form and Change in Drylands* (2nd ed.), ed. D.S.G. Thomas. Chichester: Wiley, pp. 659–90.

Wells, L.E. (1990). Holocene history of the El Niño phenomenon as recorded in flood sediments of northern coastal Peru. *Geology*, 18, 1134–7.

Wells, L.E. and Noller, J.S. (1999). Holocene coevolution of the physical landscape and human settlement in northern coastal Peru. *Geoarchaeology*, 14, 755–89.

Wells, S.G. and Gutierrez, A.A. (1982). Quaternary evolution of badlands in the southeastern Colorado Plateau, USA. In *Badland Geomorphology and Piping*, ed. R.B. Bryan and A. Yair. Norwich: Geo Books, pp. 238–58.

Wells, S.G., McFadden, L.D., Poths, J. and Olinger, C.T. (1995). Cosmogenic ^{3}He surface-exposure dating of stone pavements: Implications for landscape evolution in deserts. *Geology*, 23, 613–16.

Wells, S.G., McFadden, L.D., Renault, C.E. and Crowe, B.M. (1990). Geomorphic assessment of late Quaternary volcanism in the Yucca Mountain area, southern Nevada: Implications for the proposed high-level radioactive waste repository. *Geology*, 18, 549–53.

Werf, G.R. van der, Randerson, J.T., Giglio, L., Gobron, N. and Dolman, A.J. (2008). Climate controls on the variability of fires in the tropics and subtropics. *Global Biogeochemical Cycles*, 22, GB3028, doi:10.1029/2007GB003122.

Werner, B.T. (1995). Eolian dunes: Computer simulations and attractor interpretation. *Geology*, 23, 1107–10.

Werner, C.M., Mason, J.A. and Hanspon, P.R. (2010). Non-linear connections between dune activity and climate in the High Plains, Kansas and Oklahoma, USA. *Quaternary Research*, 75, 267–77.

Werner, M., Tegen, I., Harrison, S.P., Kohfeld, K.E., Prenctice, I.C., Balkanski, Y., Rodhe, H. and Roelandt, C. (2002). Seasonal and interannual variability of the mineral dust cycle under present and glacial climate conditions. *Journal of Geophysical Research*, 107, D24, doi: 10.1029/2002JD002365.

West, M.D., Clarke, J.D.A., Thomas, M., Pain, C.F. and Walter, M.R. (2010). The geology of Australian Mars analogue sites. *Planetary and Space Science*, 58, 447–58.

Westbeld, A., Klemm, O., Griessbaum, F., Sträter, E., Larrain, H., Osses, P. and Cereceda, P. (2009). Fog deposition to a Tillandsia carpet in the Atacama Desert. *Annales Geophysicae*, 27, 3571–6.

Wetherald, R.T. and Manabe, S. (2002). Simulation of hydrologic changes associated with global warming. *Journal of Geophysical Research*, 107, 4379, doi:10.1029/2001JD001195.

Whalley, W.B. (1983). Desert varnish. In *Chemical Sediments and Geomorphology*, ed. A.S. Goudie and K. Pye. London: Academic Press, pp. 179–226.

Whalley, W.B., McGreevy, J.P. and Ferguson, R.I. (1984). Rock temperature observations and chemical weathering in the Hunza region, Karakorum: Preliminary data. In *Proceedings of the International Karakorum Project*, vol. 2, ed. K.J. Miller. Cambridge: Cambridge University Press, pp. 616–33.

Wheaton, E.E. (1990). Frequency and severity of drought and dust storms. *Canadian Journal of Agricultural Economics*, 38, 695–700.

Whipple, K.X. and Dunne, T. (1992). The influence of debris-flow rheology on fan morphology, Owens Valley, California. *Bulletin of the Geological Society of America*, 104, 887–900.

White, E.M. (1961). Drainage alignment in western South Dakota. *American Journal of Science*, 259, 207–10.

White, E.M. and Bonestall, R.G. (1960). Some gilgaied soils in South Dakota. *Soil Science Society of America, Proceedings*, 24, 305–9.

White, K. (1991). Geomorphological analysis of piedmont landforms in the Tunisian Southern Atlas using ground data and satellite imagery. *Geographical Journal*, 157, 279–94.

White, K. (2004). Glacis d'erosion. In *Encyclopedia of Geomorphology*, ed. A.S. Goudie. London: Routledge, pp. 469–70.

White, K. and Bullard, J.E. (2009). Abrasion control on dune colour: Muleshoe Dunes, SW USA. *Geomorphology*, 105, 59–66.

White, K., Goudie, A.S., Parker, A. and Al-Farraj, A. (2001). Mapping the geochemistry of the northern Rub' Al Khali using multispectral remote sensing techniques. *Earth Surface Processes and Landforms*, 26, 735–48.

White, L.P. and Law, R. (1969). Channelling of alluvial depression soils in Iraq and Sudan. *Journal of Soil Science*, 20, 84–90.

Whitford, W.G. and Kay, F.R. (1999). Biopedturbation by mammals in deserts: A review. *Journal of Arid Environments*, 41, 203–30.
Whitney, J.W. (1983). Erosional history and surficial geology of western Saudi Arabia. *USGS Technical Record*, TR-04–1.
Whitney, J.W. (2006). Geology, water and wind in the lower Helmand Basin, southern Afghanistan. *U.S. Geological Survey Scientific Investigations Report*, 2006–5182.
Whitney, J.W. and Harrington, C.D. (1993). Relict colluvial boulder deposits as paleoclimatic indicators in the Yucca Mountain region, southern Nevada. *Bulletin of the Geological Society of America*, 105, 1008–18.
Wiggs, G.F.S., Baird, A.J. and Atherton, R.J. (2004). The dynamic effects of moisture on the entrainment and transport of sand by wind. *Geomorphology*, 59, 13–30.
Wiggs, G.F.S., Livingstone, I., Thomas, D.S.G. and Bullard, J.E. (1994). Effect of vegetation removal on airflow patterns and dune dynamics in the southwest Kalahari Desert. *Land Degradation and Rehabilitation*, 5, 13–24.
Wiggs, G.F.S., Livingstone, I., Thomas, D.S.G. and Bullard, J.E. (1996). Airflow and roughness characteristics over partially vegetated linear dunes in the southwest Kalahari Desert. *Earth Surface Processes and Landforms*, 21, 19–34.
Wiggs, G.F.S., Thomas, D.S.G., Bullard, J.E. and Livingstone, I. (1995). Dune mobility and vegetation cover in the southwest Kalahari Desert. *Earth Surface Processes and Landforms*, 20, 515–29.
Wiles, G.C., McAllister, R.P., Davi, N.K. and Jacoby, G.C. (2003). Eolian response to Little Ice Age climate change, Tana Dunes, Chugach Mountains, Alaska, U.S.A. *Arctic, Antarctic and Alpine Research*, 35, 67–73.
Wilhelmy, H. (1969). Das Urstromtal am Ostrand der Indusbene und der Sarasvati Problem. *Zeitschrift für Geomorphologie*, Supplementband, 8, 76–93.
Wilkinson, M.J. (1990). Palaeoenvironments in the Namib Desert. The lower Tumas Basin in the late Cainozoic. *University of Chicago Research Papers*, 231.
Wilkinson, M.J., Hems, D.R. and Whitehead, V.S. (1992). Albedo patterns and gypsum generation in the central Namib Desert: Land, sea and air interactions on an arid west coast. *Geoscience and Remote Sensing Symposium, 1992*, 2, 1565–7.
Williams, G.J. (1982). A preliminary Landsat interpretation of the relict landforms of western Zambia. *Paper Southern African Conference Commonwealth Geographical Bureau, Lusaka, 9th-15th June (1982).*
Williams, M.A.J. (1984). Geology. In *Sahara Desert*, ed. J.L. Cloudsley-Thompson. Oxford: Pergamon Press, pp. 31–9.
Williams, M.A.J. and Faure, H. (eds.). (1980). *The Sahara and the Nile*. Rotterdam: Balkema.
Williams, M.A.J., Dunkerley, D., De Deckker, P., Kershaw, P. and Chappell, J. (1998). *Quaternary Environments*. London: Arnold.
Williams, R.M.E., Irwin, R.P. and Zimbelman, J.R. (2009). Evaluation of paleohydrologic models for terrestrial inverted channels: Implications for application to Martian sinuous ridges. *Geomorphology*, 107, 300–15.
Williams, R.M.E., Zimbelman, J.R. and Johnston, A.K. (2006). Aspects of alluvial fan shape indicative of formation process: A case study in southwestern California with application to Mojave Crater fans on Mars. *Geophysical Research Letters*, 33, L10201, doi:10.1029/2005GL025618.
Williams, S.H. (1981). Calculated and inferred aeolian abrasion rates: Earth and Mars. MS thesis, Arizona State University.
Williams, S.H. and Zimbelman, J.R. (1994). Desert pavement evolution: An example of the role of sheetflood. *Journal of Geology*, 102, 243–8.

Wilshire, H.G. (1980). Human causes of accelerated wind erosion in California deserts. In *Geomorphic Thresholds*, ed. D.R. Coates and J.D. Vitek. Stroudsburg: Dowden, Hutchinson and Ross, pp. 415–33.

Wilshire, H.G., Nakata, J.K. and Hallet, B. (1981). Field observations of the December (1977) wind storm, San Joaquin Valley, California. In *Desert Dust: Origins, Characteristics and Effects on Man*, ed. T.L. Péwé. Boulder, CO: Geological Society of America, pp. 233–51.

Wilson, I.G. (1972). Aeolian bedforms – Their development and origins. *Sedimentology*, 19, 173–210.

Wilson, I.G. (1973). Ergs. *Sedimentary Geology*, 10, 77–106.

Wilson, J.S. and Pitts, J.P. (2010). Illuminating the lack of consensus among descriptions of Earth history data in the North American deserts: A resource for biologists. *Progress in Physical Geography*, 34, 419–41.

Winkler, E.M. and Singer, P.C. (1972). Crystallization pressure of salts in stone and concrete. *Bulletin of the Geological Society of America*, 83, 3509–14.

Winkler, E.M. and Wilhelm, E.J. (1970). Salt burst by hydration pressures in architectural stone in urban atmosphere. *Bulletin of the Geological Society of America*, 81, 567–72.

Winsborough, B.M., Seeler, J.-S., Golubic, S., Folk, R.L. and Maguire, B. (1994). Recent fresh-water lacustrine stromatolites, stromatolitic mats and oncoids from northeastern Mexico. In *Phanerozoic Stromatolites II*, ed. J. Bertrand-Safarti and C. Monti. Kluwer: Dordrecht, pp. 71–100.

Wohl, E.E., Fuertsch, S.J. and Baker, V.R. (1994). Sedimentary records of late Holocene floods along the Fitzroy and Margaret Rivers, Western Australia. *Australian Journal of Earth Sciences*, 41, 273–80.

Wolfe, S.A. and David, P.P. (1997). Parabolic dunes: Examples from the Great Sand Hills, southwestern Saskatchewan. *The Canadian Geographer*, 41, 207–13.

Wolfe, S.A. and Hugenholtz, C.H. (2009). Barchan dunes stabilized under recent climate warming on the northern Great Plains. *Geology*, 37, 1039–42.

Wolfenden, E., Ebinger, C., Yirgu, G., Renne, P.R. and Kelley, S.P. (2005). Evolution of a volcanic rifted margin: Southern Red Sea, Ethiopia. *Bulletin of the Geological Society of America*, 117, 846–64.

Wong, S., Dessler, A.E., Mahowald, N., Colarco, P.R. and da Silva, A. (2008). Long-term variability in Saharan dust transport and its link to North Atlantic sea surface temperature. *Geophysical Research Letters*, 35, L07812, doi:10.1029/2007GL032297.

Wood, A.P., Burke, K.M., Knott, B. and Kelly, B.P. (1991). Chemolithropic sulfur bacteria in sediments, mats, and stromatolites of Western Australian saline lakes. *Geomicrobiology Journal*, 9, 41–49.

Wood, W.W. and Imes, J.L. (1995). How wet is wet? Precipitation constraints on late Quaternary climate in the southern Arabian Peninsula. *Journal of Hydrology*, 164, 263–8.

Wood, W.W., Sanford, W.E. and Al Habshi, A.R.S. (2002). Sources of solutes to the coastal sabkha of Abu Dhabi. *Bulletin of the Geological Society of America*, 114, 259–68.

Wood, Y.A., Graham, R.C. and Wells, S.G. (2005). Surface control of desert pavement pedologic process and landscape function, Cima Volcanic field, Mojave Desert, California. *Catena*, 59, 205–30.

Woodruff, N.P. and Siddoway, F.H. (1965). A wind erosion equation. *Proceedings Soil Science Society of America*, 29, 602–8.

Woodruff, N.P., Chepil, W.S. and Lynch, R.D. (1957). Emergency chiselling to control wind erosion. *Kansas Agricultural Experimental Station Technical Bulletin*, 90.

Woodward, N.P. (1897). The dry lakes of Western Australia. *Geological Magazine* IV(4), 363–6.

Woolley, R.F. (1946). Cloudburst floods in Utah, 1850–1938. *United States Geological Survey Water Supply Paper*, 994.
Woolnough, W.G. (1927). The duricrust of Australia. *Journal and Proceedings Royal Society of New South Wales*, 61, 24–53.
Woolnough, W.G. (1930). The influence of climate and topography in the formation and distribution of products of weathering. *Geological Magazine*, 67, 123–32.
Wright, D.K., Forman, S.L., Waters, M.R. and Ravesloot, J.C. (2011). Holocene eolian activation as a proxy for broad-scale landscape change on the Gila River Indian Community, Arizona. *Quaternary Research*, 76, 10–21.
Wright, H.T. (2001). Cultural action in the Uruk world. In *Uruk Mesopotamia and Its Neighbours: Cross-Cultural Interactions in the Era of State Formation*, ed. M.S. Rothman. Santa Fe: School of American Research Advanced Seminar Series, pp. 123–47.
Wright, J.P., Jones, C.G., Boeken, B. and Schachak, M. (2006). Predictability of ecosystem engineering effects on species richness across environmental variability and spatial scales. *Journal of Ecology*, 94, 815–24.
Wright, J.S. (2000). Tufa accumulations in ephemeral streams: Observations from the Kimberley, north-west Australia. *Australian Geographer*, 31, 333–47.
Wright, J.S. (2001). 'Desert' loess versus 'glacial' loess: Quartz silt formation, source areas and sediment pathways in the formation of loess deposits. *Geomorphology*, 26, 231–56.
Wright, V.P. (2009). Calcrete. In *Geochemical Sediments and Landscapes*, ed. D.J. Nash and S.J. McLaren. Blackwell: Oxford, pp. 10–45.
Wright, V.P and Tucker, M.E. (1991). Calcretes: An introduction. In *Calcretes*, ed. V.P. Wright and M.E. Tucker. Oxford: Blackwell Scientific, pp. 1–22.
Wu, F., Tian, M. and Zhang, J. (eds.) (2005). *Magical Desert – Alxa.*
Wüst, R.A.J. and McLane, J. (2000). Rock deterioration in the Royal Tomb of Seti I, Valley of the Kings, Luxor, Egypt. *Engineering Geology*, 58, 163–90.
Wüst, R.A.J. and Schlüchter, C. (2000). The origin of soluble salts in rocks of the Thebes Mountains, Egypt: The damage potential to ancient Egyptian wall art. *Journal of Archaeological Science*, 27, 1161–72.
Xie, S. and 7 others. (2012). Microbial lipid records of highly alkaline deposits and enhanced aridity associated with significant uplift of the Tibetan Plateau in the Late Miocene. *Geology*, 40, 291–4.
Xiu, Z., Yu, X. and Lu, X.X. (2011). Factors controlling sediment yield in China's Loess Plateau. *Earth Surface Processes and Landforms*, 36, 816–26.
Xu, J. (2006). Sand-dust storms in and around the Ordos Plateau of China as influenced by land use change and desertification. *Catena*, 65, 279–84.
Xu, J., Hou, S., Dahe, Q., Shichang, K., Ren, J. and Ming, J. (2007). Dust storm activity over the Tibetan Plateau recorded by a shallow ice core from the north slope of Mt Qomolangma (Everest), Tibet-Himal region. *Geophysical Research Letters*, 34, L17504, doi:10.1029/2007GL030853.
Xu, X.-Z, Zhang, H.-W. and Zhang, O. (2004). Development of check-dam systems in gullies on the Loess Plateau, China. *Environmental Science and Policy*, 7, 79–86.
Xu, Y., Wang, H., Hong, L. and Jiang, D. (2011). Simulation of the direct radiative effect of mineral aerosol on the climate at the Last Glacial Maximum. *Journal of Climate*, 24, 843–8.
Xuan, J. (1999). Dust emission factors for environment of northern China. *Atmospheric Environment*, 33, 1767–76.
Yaalon, D. H. (1987). Saharan dust and desert loess: effect on surrounding soils. *Journal of African Earth Sciences*, 6, 569–71.

Yaalon, D. and Dan, J. (1974). Accumulation and distribution of loess-derived deposits in the semi-desert and desert fringe areas of Israel. *Zeitschrift für Geomorphologie*, Supplementband, 20, 91–105.
Yaalon, D.H. (1981). Pedogenic carbonate in aridic soils – Magnitude of the pool and annual fluxes (abstract). International Conference on Aridic Soils, Jerusalem.
Yair, A. (1974). Sources of runoff and sediment supplied by the slopes of a first order drainage basin in an arid environment (northern Negev, Israel). *Abhandlungen Der Akademie der Wissenschaften in Göttingen, Mathematisch-Physikalische Klasse* III, 29, 403–17.
Yair, A. (1990). The role of topography and surface cover upon soil formation along hill slopes in arid climates. *Geomorphology*, 3, 287–99.
Yair, A. (1995). Short and long-term effects of bioturbation on soil erosion, water resources and soil development in an arid environment. *Geomorphology*, 13, 87–99.
Yair, A. and Enzel, Y. (1987). The relationship between annual rainfall and sediment yield in arid and semi-arid areas. The case of the Negev. *Catena*, Supplement, 10, 121–35.
Yair, A. and Lavee, H. (1985). Runoff generation in arid and semi-arid zones. In *Hydrological Forecasting*, ed. M.G. Anderson and T.P. Burt. Chichester: Wiley, pp. 183–220.
Yair, A., Almog, R. and Veste, M. (2011). Differential hydrological response of biological topsoil crusts along a rainfall gradient in a sandy arid area: Northern Negev Desert, Israel. *Catena*, 87, 326–33.
Yair, A., Bryan, R.B., Lavee, H., Schwanghart, W. and Kuhn, N. (In press). The resilience of a badland area to climate change in an arid environment. *Catena*.
Yang, B., Bräuning, A., Zhang, Z., Dong, Z. and Epser, J. (2007). Dust storm frequency and its relation to climate changes in northern China during the past 1000 years. *Atmospheric Environment*, 41, 9288–99.
Yang, C.-Y. (2006). Effects of Asian dust storm events on daily clinical visits for conjunctivitis in Taipei, Taiwan. *Journal of Toxicology and Environmental Health A*, 69, 1673–80.
Yang, M., Yao, T. and Wang, H. (2006). Microparticle content records of the Dunde ice core and dust storms in northwestern China. *Journal of Asian Earth Sciences*, 27, 223–9.
Yang, X. (2001). Late Quaternary evolution and paleoclimates, western Alashan Plateau, Inner Mongolia, China. *Zeitschrift fűr Geomorphologie*, 45, 1–16.
Yang, X. and Scuderi, L.A. (2010). Hydrological and climatic change in deserts of China since the late Pleistocene. *Quaternary Research*, 73, 1–9.
Yang, X., Li, H. and Conacher, A. (2011). Large-scale controls on the development of sand seas in northern China. *Quaternary International*, 250, 74–83.
Yang, X., Scuderi, L., Paillou, P., Liu, Z., Li, H. and Ren, X. (2011). Quaternary environmental changes in the drylands of China – A critical review. *Quaternary Science Reviews*, 30, 3219–33.
Yang, X. and 9 others. (2011). Formation of the highest sand dunes on Earth. *Geomorphology*, 135, 108–16.
Yang, Y.Q., Hou, Q., Zhou, C.H., Liu, H.L., Wang, Y.Q. and Niu, T. (2008). Sand/dust storm processes in northeast Asia and associated large-scale circulations. *Atmospheric Chemistry and Physics*, 8, 25–33.
Yao, Z., Xiao, J., Li, C. and Zhu, K. (2010). Regional characteristics of dust storms observed in the Alxa Plateau of China from 1961 to 2005. *Environmental Earth Sciences*, doi:10.1007/s12665–010-0845–3.
Yao, Z.Y., Wang, T., Han, Z.W., Zhang, W.M. and Zhao, A.G. (2007). Migration of sand dunes on the northern Alxa Plateau, Inner Mongolia, China. *Journal of Arid Environments*, 70, 80–93.

Yechieli, Y., Abelson, M., Bein, A., Crouvi, O. and Shtivelman, V. (2006). Sinkhole 'swarms' along the Dead Sea coast: Reflection of disturbance of lake and adjacent groundwater systems. *Bulletin of the Geological Society of America*, 118, 1075–87.

Yilmaz, I. (2001). Gypsum/anhydrite: Some engineering problems. *Bulletin of Engineering Geology and Environment*, 59, 227–30.

Yizhaq, H. (2008). Aeolian megaripples: Mathematical model and numerical simulations. *Journal of Coastal Research*, 24, 1369–78.

Yizhaq, H., Ashkenazy, Y. and Tsoar, H. (2007). Why do active and stabilized dunes coexist under the same climatic conditions? *Physical Review Letters*, 98, doi:10.1103?PhysRevLett.98.188001.

Yizhaq, H., Ashkenazy, Y. and Tsoar, H. (2009). Sand dune dynamics and climate change: A modelling approach. *Journal of Geophysical Research*, 114, F01023, doi:10.1029/2008JF001138.

Yizhaq, H., Katra, I., Kok, J.F. and Isenberg, O. (2012). Transverse instability of megaripples. *Geology*, 40, 459–62.

Young, A.R.M. (1987). Salt as an agent in the development of cavernous weathering. *Geology*, 15, 962–6.

Young, M.H., McDonald, E.V., Caldwell, T.G., Benner, S.G. and Meadows, D.G. (2004). Hydraulic properties of a desert soil chronosequence in the Mojave Desert, USA. *Vadose Zone Journal*, 3, 956–63.

Young, R.A. (1985). Geomorphic evolution of the Colorado Plateau margin in west-central Arizona: A tectonic model to distinguish between the causes of rapid, symmetrical scarp retreat and scarp dissection. In *Tectonic Geomorphology*, ed. M. Morisawa and J.T. Hack. Boston: Allen and Unwin, pp. 261–78.

Young, R.G. (1964). Fracturing of sandstone cobbles in caliche cemented terrace gravels. *Journal of Sedimentary Petrology*, 34, 886–9.

Young, R.W. (1985). Silcrete distribution in eastern Australia. *Zeitschrift für Geomorphologie*, 29, 21–36.

Young, R.W. and Young, A. (1992). *Sandstone Landforms*. Berlin: Springer Verlag.

Young, R.W., Wray, R.A.L. and Young, A.R.M. (2009). *Sandstone landforms*. Cambridge: Cambridge University Press.

Youssef, A.M., Pradhan, B. and Hassan, A.M. (2010). Flash flood risk estimation along the St. Katherine road, southern Sinai, Egypt using GIS based morphometry and satellite imagery. *Environmental Earth Science*, 62, 611–23.

Yu, H., Dickinson, R.E., Chin, M., Kaufman, Y.J., Holben, B.N., Geogdzhayev, I.V. and Mischensko, M.I. (2003). Annual cycle of global distributions of aerosol optical depth from integration of MODIS retrievals and GOCART model simulations. *Journal of Geophysical Research*, 108 D, doi:10.1029/2002JD002717.

Yu, S. and Oguchi, C.T. (2009). Complex relationships between salt type and rock properties in a durability experiment of multiple-rock treatments. *Earth Surface Processes and Landforms*, 34, 2096–110.

Zaady, E and Offer, Z.Y. (2010). Biogenic soil crusts and soil depth: A long-term case study from the central Negev Desert highland. *Sedimentology*, 57, 351–8.

Zaady, E., Offer, Z.Y. and Shachak, M. (2001). The content and contribution of deposited aeolian organic matter in a dry land ecosystem of the Negev Desert, Israel. *Atmospheric Environment*, 35, 769–76.

Zahar, Y., Ghorbel, A. and Albergel, J. (2008). Impacts of large dams on downstream flow conditions of rivers: Aggradation and reduction of the Medjerda channel capacity downstream of the Sidi Salem Dam (Tunisia). *Journal of Hydrology*, 351, 318–30.

Zangvil, A. (1996). Six years of dew observations in the Negev Desert, Israel. *Journal of Arid Environments*, 32, 361–71.

Zaprowski, B.J., Pazzaglia, F.J. and Evenson, E.B. (2005). Climatic influences on profile concavity and river incision. *Journal of Geophysical Research*, 110, F03004, doi:10.1029/2004JF000138.

Zarasvandi, A., Carranza, E.J.M., Moore, F. and Rastmanesh, F. (2011). Spatio-temporal occurrences and mineralogical-geochemical characteristics of airborne dusts in Khuzestan Province (southwestern Iran). *Journal of Geochemical Exploration*, 111, 138–51.

Zárate, M.A. (2003). Loess of southern South America. *Quaternary Science Reviews*, 22, 1987–2006.

Zárate, M.A. and Tripaldi, A. (2012). The aeolian system of central Argentina. *Aeolian Research*, 3, 401–17.

Zarei, M., Raeisi, E. and Talbot, C.J. (2012). Karst development on a mobile substrate: Konarsiah salt extrusion, Iran. *Geological Magazine*, 149, 412–22.

Zdanowicz, C., Hall, G., Vaive, J., Amelin, Y., Percival, J., Girard, I., Biscaye, P. and Bory, A. (2006). Asian dustfall in the St. Elias Mountains, Yukon, Canada. *Geochimica et Cosmochimica Acta*, 70, 3493–507.

Zender, C.S. and Talamantes, J. (2006). Climate controls on valley fever incidence in Kern County, California. *International Journal of Biometeorolology*, 50, 174–82.

Zender, C.S., Bian, H. and Newman, D. (2003a). Mineral Dust Entrainment And Deposition (DEAD) model: Description and 1990s dust climatology. *Journal of Geophysical Research*, 108 (D14), 4416, doi:10.1029/2002JD002775.

Zender, C.S., Newman, D. and Torres, O. (2003b). Spatial heterogeneity in eolian erodibility: Uniform, topographic, geomorphic and hydrologic hypotheses. *Journal of Geophysical Research*, 108 (D), article 4543.

Zeng, N. and Yoon, J. (2009). Expansion of the world's deserts due to vegetation-albedo feedback under global warming. *Geophysical Research Letters*, 36, L17401, doi:10.1029/2009GL039699.

Zerboni, A. (2008). Holocene rock varnish on the Messak plateau (Libyan Sahara): Chronology of weathering processes. *Geomorphology*, 102, 640–51.

Zhang, D., Narteau, C., Rozier, O. and du Pont, S.C. (2012). Morphology and dynamics of star dunes from numerical modeling. *Nature Geoscience*, 5, 463–7.

Zhang, T.-H., Zhao, H.-L., Li, S.-G., Li, F.-R., Shirato, Y., Ohkuro, T. and Taniyama, I. (2004). A comparison of different measures for stabilizing moving sand dunes in the Horqin Sandy Land of Inner Mongolia, China. *Journal of Arid Environments*, 58, 203–14.

Zhang, X.Y., Arimoto, R. and An, Z.S. (1997). Dust emission from Chinese desert sources linked to variations in atmospheric circulation. *Journal of Geophysical Research*, 102 (D23), 28041–7.

Zhang, X.Y., Arimoto, R., Zhu, G.H. and Zhang, G.Y. (1998). Concentration, size-distribution and deposition of mineral aerosol over Chinese desert regions. *Tellus*, 50B, 317–30.

Zhang, Y.-G., Nearing, M.A., Liu, B.Y., Van Pelt, R.S., Stone, J.J., Wei, H. and Scott, R.L. (2011). Comparative rates of wind versus water erosion from a small semiarid watershed in southern Arizona, USA. *Aeolian Research*, 3, 197–204.

Zhang, Z. and Sun, J. (2011). Palynological evidence for Neogene environmental change in the foreland basin of the southern Tianshan range, northwestern China. *Global and Planetary Change*, 75, 56–66.

Zhao, Q., He, K., Rahn, K.A., Ma, Y., Yang, F., Duna, F., Lei, Y., Chen, Y., Cheng, Y., Liu, H. and Wang, S. (2010). Dust storms come to central and southwestern China, too: Implications from a major dust event in Chongqing. *Atmospheric Chemistry and Physics*, 10, 2615–30.

Zheng, H., Tada, R., Jia, J., Lawrence, C. and Wang, K. (2010). Cenozoic sediments in the southern Tarim Basin: Implications for the uplift of northern Tibet and evolution of the Taklimakan Desert. *Geological Society of London Special Publication*, 342, 67–78.

Zhou, J. and Chafetz, H.S. (2009.) Biogenic caliches in Texas: The role of organisms and effect of climate. *Sedimentary Geology*, 222, 207–25.

Zhou, J., Zhu, Y. and Yuan, C. (2012). Origin and lateral migration of linear dunes in the Qaidam Basin of MW China revealed by dune sediments, internal structures and optically stimulated luminescence ages, with implications for linear dunes on Titan. *Bulletin of the Geological Society of America*, doi:10.1130/B30550.1.

Zhou, W., Head, M.J., An, Z., De Deckker, P., Liu, Z., Liu, X., Li, X., Donahue, D., Jull, A.J.T. and Beck, W.W. (2001). Terrestrial evidence for a spatial structure of tropical-polar interconnections during the Younger Dryas episode. *Earth and Planetary Science Letters*, 191, 231–9.

Zhu, A., Ramanathan, V., Li, F. and Kim, D. (2007). Dust plumes over the Pacific, Indian and Atlantic Oceans: Climatology and radiative impact. *Journal of Geophysical Research*, 112, D16208, doi:10.1029/2007JD008427.

Zhu, C., Wang, B. and Qian, W. (2008). Why do dust storms decrease in northern China concurrently with the recent global warming? *Geophysical Research Letters*, 35, L18702, doi:10.1029/2008GL034886.

Zhu, T.X. (1997). Deep-seated, complex tunnel systems – A hydrological study in a semi-arid catchment, Loess Plateau, China. *Geomorphology*, 20, 255–67.

Zhu, T.X. (2012). Gully and tunnel erosion in the hilly Loess Plateau region, China. *Geomorphology*, 153–154, 144–55.

Zhu, Y., Lu, H., Zhang, J., Mason, J.A. and Zhou, L. (2009). Luminescence dating of sand-loess sequences and response of Mu Us and Otindag sand fields (north China) to climatic changes. *Journal of Quaternary Science*, 24, 336–44.

Zhu Z. (1984). Aeolian landforms in the Taklimakan Desert. In *Deserts and Arid Lands*, ed. F. El-Baz. The Hague: Nijhoff, pp. 133–44.

Zhuang, G., Hourigan, J.K., Koch, P.L., Ritts, B.D. and Kent-Corson, M.L. (2011). Isotopic constraints on intensified aridity in central Asia around 12 Ma. *Earth and Planetary Science Letters*, 312, 152–63.

Zimbelman, J.R. (2010). Transverse aeolian ridges on Mars: First results from HiRise images. *Geomorphology*, 121, 22–9.

Zimbelman, J.R. and Griffin, L.J. (2010). HiRise images of yardangs and sinuous ridges in the lower member of the Medusae Fossae Formation, Mars. *Icarus*, 205, 198–210.

Zimbelman, J.R. and Williams, S.H. (1996). Wind streaks: Geological and botanical effects on surface albedo contrast. *Geomorphology*, 17, 167–85.

Zimbelman, J.R., Irwin, R.P., Williams, S.H., Bunch, F., Valdez, A. and Stevens, S. (2009). The rate of granule ripple movement on Earth and Mars. *Icarus*, 203, 71–6.

Zonneveld, K.A.F., Ganssen, G., Troelrma, S., Versteegh, G.J.M. and Visser, H. (1997). Mechanisms forcing abrupt fluctuations of the Indian Ocean summer monsoon during the last deglaciation. *Quaternary Science Reviews*, 16, 187–201.

Index